Stallcup's Electrical Calculations Simplified

based on the 1999
national electrical code®

by James G. Stallcup

Grayboy & Associates

4308 Broadway • Fort Worth • Texas • 76117 • 817-831-6599
• Fax 817-831-6579 • http://www.grayboyassociates
• e-mail - grayboy 02@aol.com

Published by Grayboy Publishing

1 2 3 4 5 6 7 8 9

Library of Congress Catalog Card Number: 99-95716

ISBN: 1-885341-39-3

Written by: James G. Stallcup
Edited by: James W. Stallcup
Design, layout, and graphics by: Billy G. Stallcup

Grayboy & Associates
4308 Broadway
Fort Worth, Texas 76117
(817) 831-6599 FAX: (817) 831-6579

Introduction

Electrical Calculations Simplified is not intended to be used as a substitute for the NEC®. It is intended for use in conjunction with the NEC®, as an explanatory guide to expand upon and clarify NEC® rules through discussion, examples, and illustrations. Considerable effort has been made to condense the more complicated rules pertaining to calculating loads into a compact listing, which provide easier understanding of how to perform calculations according to the provisions of the NEC®.

A broad assortment of basic code calculations have been selected to represent the main principles of electric circuits. Determination of the most efficient and economical design and installation of all phases of electrical work hinges on sound, accurate, realistic and practical calculations, with the object always being to arrive at firm, accurate, numerical data.

Additional Help methods are included to illustrate different procedures on how to apply other formulas to calculate pertinent values which can be used in conjunction with a particular calculation in an Example Problem. Note that **Additional Help** methods are identified by boxed-in calculations which identify and separate them from the illustrations and Example Problems. **Additional Help** methods are listed in their own index for quick reference on how to apply the formulas. The author believes that such information will be very beneficial and add an extra feature to each calculation procedure of a particular Section.

JAMES G. STALLCUP

Table of Contents

Calculating Branch-Circuits

Branch-circuit conductors extend between the final overcurrent protection device protecting the circuit conductors and supplying power to the outlets. The equipment supplied is either hard-wired (permanent) or cord-and-plug connected.

Branch-circuits are computed at either continuous or noncontinuous operation. Overcurrent protection devices, conductors and other elements are sized and selected based upon the operation and use of the equipment.

Properly designed branch-circuits will supply the needed power and prevent the overheating of the elements in the equipment served. Loads are required to be calculated and elements selected accordingly.

Quick Reference

Voltage Drop - Branch-circuits 210-19(a), FPN 4

Conductors are sometimes increased in size to prevent excessive voltage drop due to long runs between the OCPD's and the load served. Voltage drop on branch-circuits is usually limited to 2 or 3 percent to the farthest outlet serving the load.

Example 1-1(a). What is the voltage drop for the branch-circuit in Illustration 1-1(a)?

Finding Voltage Drop for a single-phase circuit

Step 1: Selecting percentage
210-19(a), FPN 4
Branch-circuit = 3%

Step 2: Calculating VD
210-19(a), FPN 4; Table 8, Ch. 9
VD = 2 x R x L x I / 1,000
VD = 2 x 0.245 x 200' x 80 A / 1,000
VD = 7.84

Step 3: Calculating allowable VD
VD = supply V x 3%
VD = 240 V x 3%
VD = 7.2 V

Step 4: Checking percentage
210-19(a), FPN 4
% = VD / V
% = VD / V
% = 7.84 V / 240 V
% = .0327 or 3.27%

Solution: The voltage drop rating of 7.84 volts is greater than 7.2 volts and larger conductor must be used to reduce the 3.27 percent to 3 percent or less.

Using larger conductor to compensate for VD

Step 1: Selecting percentage
210-19(a), FPN 4
Branch-circuit = 3%

Step 2: Calculating VD
210-19(a), FPN 4; Table 8, Ch. 9
VD = 2 x R x L x I / 1,000
VD = 2 x 0.194 x 200' x 80 A / 1,000
VD = 6.2 V

Step 3: Checking percentage
210-19(a), FPN 4
% = 6.2 V / 240 V
% = .0258 or 2.58

Solution: The voltage drop rating of 6.2 volts is less than 7.2 volts, which is well below the 3 percent limit. The No. 2 conductor is large enough to reduce the voltage drop to 3 percent or less.

VOLTAGE DROP FORMULAS

To find PERCENT VOLTAGE DROP, divide the voltage drop by the volts in the circuit. Use the formula.

$$\%VD = \frac{VD}{V} \times 100$$

To find CONDUCTOR SIZE when the voltage drop, length of circuit, and current are known.

$$CM = \frac{2 \times K \times L \times I}{VD} \quad \text{(single-phase, 2-wire circuit)}$$

$$CM = \frac{2 \times K \times L \times I \times .866}{VD} \quad (\sqrt{3} = 1.732 \quad \frac{1.732}{2} = .866)$$

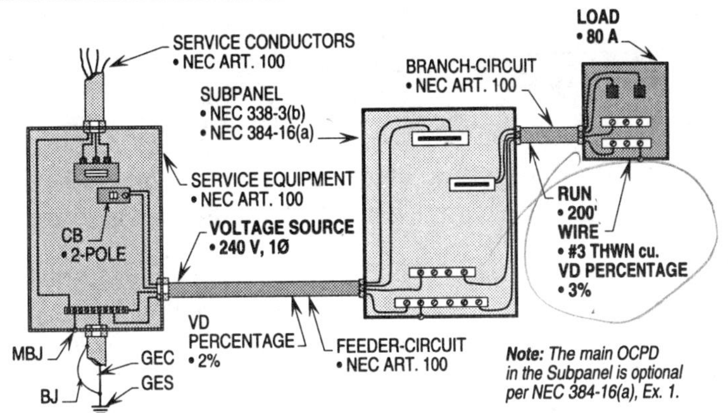

NEC 210-19(a), FPN 4

Branch-Circuits
- **NEC 210-19(a), FPN 4**

Feeder-Circuits
- **NEC 215-2(d), FPN 2**

Services
- **NEC 230-31(c), FPN**

EGC
- **NEC 250-122**

Grounded Conductor
- **NEC 240-24**

Note: The VD for feeder-circuits are calculated using the same procedure per NEC 215-2(d), FPN 2.

The values that may be used when calculating voltage drop are:

- VD = Voltage drop
- R = Resistivity for conductor material: (Use chapter 9, Table 8, Column 6 or 8 (uncoated ohm/MFT)]
- L = One-way length of circuit conductor in feet
- I = Current in conductor in amperage
- CM = Conductor area in circular mills (See Table 8 to Chapter 9.)
- 1,000 = Length of conductors based on Table 8 to Chapter 9

VD FORMULAS

To find LENGTH OF CIRCUIT when the voltage drop, conductor size and current are known:

$$L = \frac{VD \times CM}{2 \times K \times I} \quad \text{(single-phase)}$$

$$L = \frac{VD \times CM}{K \times I \times 1.73} \quad \text{(three-phase)}$$

$$\text{or } L = \frac{VD \times CM}{2 \times K \times I \times .866}$$

To find the RESISTANCE of a conductor:

L = length of conductor

R = resistance

K = constant (12) for copper, (18) for aluminum

$$R = \frac{KL}{CM} \qquad L = \frac{R \times CM}{K} \qquad CM = \frac{K \times L}{R}$$

VD FORMULAS

To find CURRENT when the voltage drop, length of circuit and conductor are known:

$$I = \frac{VD \times CM}{2 \times K \times L} \quad \text{(single-phase, 2-wire circuit)}$$

$$I = \frac{VD \times CM}{K \times L \times 1.73} \quad \text{(three-phase circuit)}$$

$$\text{or } I = \frac{VD \times CM}{2 \times K \times L \times .866}$$

Ill. 1-1(a) (Note: Ill. 1-1(a) is also used for Example 1-1(b))

Voltage Drop - Branch-circuits
210-19(a), FPN 4

Due to long runs of feeder-circuit conductors, the conductors are increased in size to compensate for poor voltage drop. The voltage drop on the feeder conductors should not exceed 3 percent at the farthest outlet supplying power to loads. The voltage drop on the feeder and branch-circuit conductors should not exceed 5 percent overall.

Example1-1(b). What is the voltage drop for the branch-circuit in Illustration 1-1(a) if the supply voltage was three-phase.

Finding Voltage Drop for a Three-Phase Circuit

Step 1: Selecting percentage
210-19(a), FPN 4
Branch-circuit = 3%

Step 2: Calculating VD
210-19(a), FPN 4; Table 8, Ch. 9
VD= 2x R x L x I / 1,000
VD= 2 x 0.245 x 200' x 80 A /1,000
VD= 7.84 x .866 = 6.79

Step 3: Calculating allowable VD
VD= supply V x 3%
VD= 240 V x 3%
VD= 7.2 V

Step 4: Checking percentage
210-19(a), FPN 4
% = VD / V
% = 6.79 V / 240 V
% = .0283 or 2.83%

Solution: The voltage drop rating of 6.79 volts is less than 7.2 volts and a larger conductor is not required.

Example 1-1(c). What is the CM rating for the branch-circuit in Illustration 1-1(b)?

Finding Voltage Drop

Step 1: Selecting percentage
210-19(a), FPN 4
Branch-circuit = 3%

Step 2: Calculating CM
210-19(a), FPN 4; Table 8, Ch. 9
CM = 1.732 x K x L x I / V
CM = 1.732 x 12 x 125' x 82.5 A / 208 x 3% (6.24)
CM =34,349

Solution: The CM rating of 34,349 requires No. 4 THWN copper conductors.

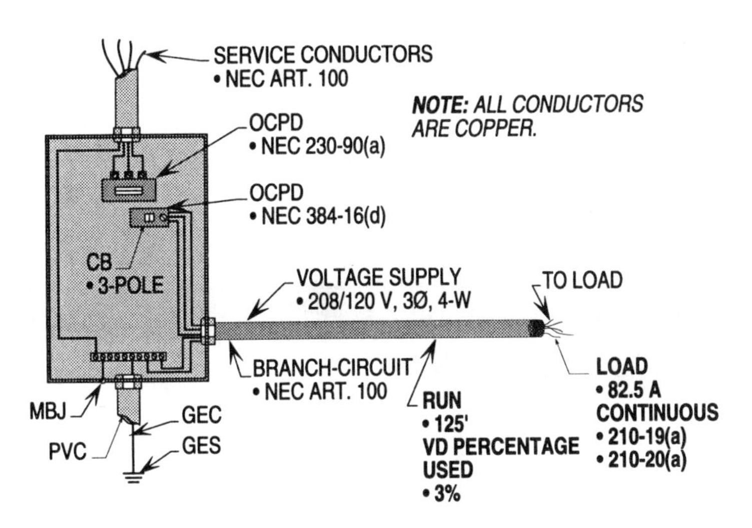

NEC 210-19(a), FPN 4

NEC 210-19(a), FPN 4
NEC 215-2(b), FPN 2
NEC 230-31(c), FPN
NEC 250-122
NEC 240-24

Note: *The CM for feeder-circuits are calculated using the same procedure per NEC 215-2(d), FPN 2.*

The values that may be used when calculating voltage drop are:

- VD = Voltage drop
- R = Ohms or resistance
- K = Constant 12 for cu. or 18 for alu.
- I = Amps
- CM = Circular mils
- L = One-way length of circuit conductor in feet

ADDITIONAL HELP 1:

Calculating VD in Ill. 1-1(b)

- $VD = \frac{1.732 \times K \times L \times I}{CM}$
- $VD = \frac{1.322 \times 12 \times 125' \times 82.5\ A}{34{,}349}$
- VD = 6.239

ADDITIONAL HELP 2:

- Calculating Resistance of #4 cu. in circuit of Ill. 1-1(b)

  cu. = 12; length = 250'
  CM = 34,349; 3Ø = .866

- Applying formulas

  $\frac{12 \times 250' \times .866}{34{,}349} = .0756R$

ADDITIONAL HELP 3:
Note: Not as correct as **Additional Help** 2.

- Table 8, Ch. 9
  The resistance based on 1,000'
  #4 cu. = .321R
  Length of conductors = 250'
  3Ø = .866

- Applying formulas

  $\frac{250' \times .866}{1{,}000} \times .321 = .0695R$

Ill. 1-1(b)

Voltage Drop - Feeder-Circuit 215-2(d), FPN 2

Conductors are sometimes increased in size to prevent excessive voltage drop due to long runs between the OCPD's and the load served. Voltage drop on feeder-circuits is usually limited to 2 or 3 percent to the farthest outlet serving the load.

Example 1-1(c). What is the voltage drop for the feeder-circuit in Illustration 1-1(d)?

Finding Voltage Drop for a single-phase circuit

Step 1: Selecting percentage
215-2(d), FPN 2
feeder-circuit = 3%

Step 2: Calculating VD
215-2(d), FPN 2; Table 8, Ch. 9
VD = 2 x R x L x I / 1,000
VD = 2 x 0.0797 x 102' x 178 A / 1,000
VD = 2.89

Step 3: Calculating allowable VD
VD = supply V x 3%
VD = 240 V x 3%
VD = 7.2 V

Step 4: Checking percentage
215-2(d), FPN 2
% = VD / V
% = VD / V
% = 2.89 V / 240 V
% = .0120 or 1.2%

Solution: The voltage drop rating of 2.89 volts is less than 7.2 volts and complies with NEC 215-2(d), FPN 2.

USING DIFFERENT SIZE CONDUCTORS TO LOWER VD

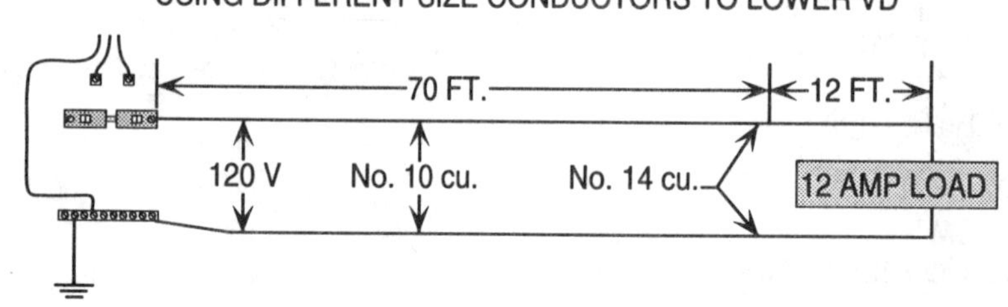

ADDITIONAL HELP 4(a)

• $VD = \frac{2\,RLI}{1{,}000} = \frac{2 \times 1.24 \times 70' \times 12\text{ A}}{1{,}000}$

$= \frac{2{,}083}{1{,}000} = 2.083$

• Check by using 1.24 R per 1,000 ft.

• Total VD = IR = 12 A x .1736$\left(\frac{1.24}{1{,}000} \times 140\right)$

= 2.083 V

• $VD = \frac{2\,RLI}{1{,}000} = \frac{2 \times 3.14 \times 12' \times 12\text{ A}}{1{,}000}$

$= \frac{904.32}{1{,}000} = .9043$

• Check by using 2.57 ohms per 1,000 ft.

• Total VD = IR = 12 A x .07536$\left(\frac{3.14}{1{,}000} \times 24\right)$

= .9043 V

TOTAL VD = 2.083 V + .9043 V = 2.987 V
TOTAL % VD = 2.987 V ÷ 120 V = 2.48% VD

Note: If the VD is to be held to 3 percent on the branch-circuit, the above wiring method complies with the NEC.

Ill. 1-1(c)

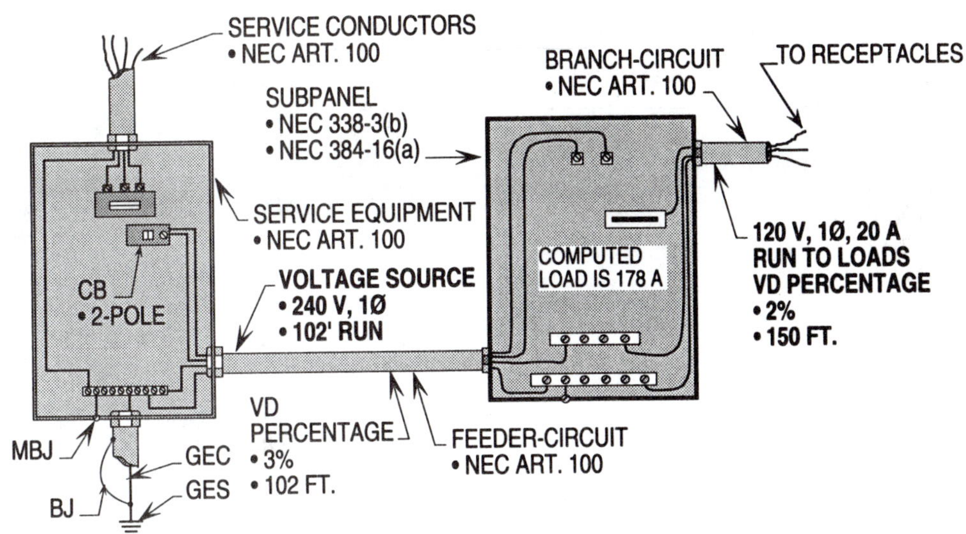

NEC 215-2(d), FPN 2

Branch-circuits
- **NEC 210-19(a), FPN 4**

Feeder-circuits
- **NEC 215-2(d), FPN 2**

Services
- **NEC 230-31(c), FPN**

EGC
- **NEC 250-122(b)**

Grounded Conductor
- **NEC 240-24**

Note: The VD for feeder-circuits are calculated using the same procedure per NEC 215-2(d), FPN 2.

The values that may be used when calculating voltage drop are:

- VD = Voltage drop
- R = Resistivity for conductor material: (Use chapter 9, Table 8, Column 6 or 8 (uncoated ohm/MFT)
- L = One-way length of circuit conductor in feet
- I = Current in conductor in amperage
- CM = Conductor area in circular mills (See Table 8 to Ch. 9)
- 1,000 = Length of conductors based on Table 8 to Ch. 9

Ill. 1-1(d)

GIVEN VALUES

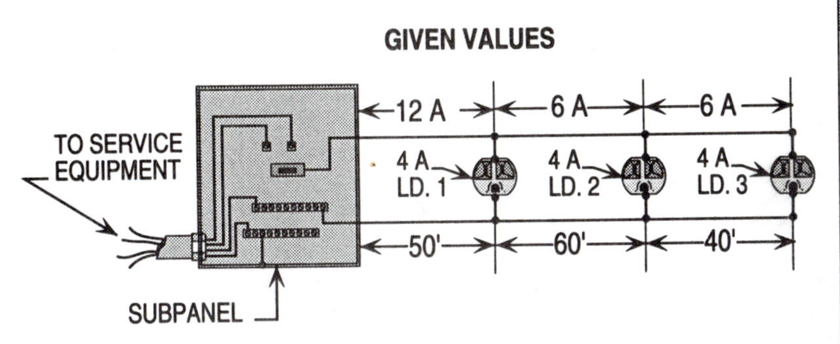

NOTE: VD IN THE (150') BRANCH-CIRCUIT IS GREATER THAN THE 2 PERCENT SHOWN IN THE ILLUSTRATION ABOVE.

ADDITIONAL HELP 4(b)

- VD between panel and load 1
  Table 8 to Ch. 9
  $\frac{100}{1,000}$ x 1.98R x 12 A = 2.376 V
- VD between panel, load 1, and load 2
  .198R x 8 A = 1.584 VD
- VD between panel, load 2, and load 3
  .198R x 4 A = .792 VD
- Total VD
  2.376 VD + 1.584 VD + .792 VD = 4.752 VD

Ill. 1-1(e)

Lighting Load, Etc.
210-19(a); 210-20(a)

The procedure for calculating the elements for a branch-circuit is to compute the noncontinuous load at 100 percent and the continuous load at 125 percent. Add these loads together and the rating in amps or VA is used to size the elements.

Example 1-2. What size OCPD and THWN copper conductors are required for the branch-circuit supplying the loads listed in Illustration 1-2?

Sizing OCPD

Step 1: Calculating OCPD
210-20(a)
12 A x 125% = 15 A
5 A x 100% = 5 A
Total load = 20 A

Step 2: Selecting OCPD
Table 310-16; 240-3(d); 240-3(b); 240-6(a)
20 A requires 20 A OCPD

Solution: A 20 amp OCPD is required.

Sizing Conductors

Step 1: Calculating conductors
210-19(a)
12 A x 125% = 15 A
5 A x 100% = 5 A
Total load = 20 A

Step 2: Selecting conductors
Table 310-16; 240-3(d)
20 A requires #12 THWN

Solution: A No. 12 THWN copper conductor is required.

ADDITIONAL HELP 5:

- Calculating VA when voltage and amps are known.
  - VA = V x Amps
    VA = 120 V x 12 A
    VA = 1,440
  - VA = V x Amps
    VA = 120 V x 5 A
    VA = 600

Calculating total VA
Total VA = 1,440 + 600 = 2,040 VA

NEC 210-3; NEC 210-19(a); NEC 210-20(a); NEC 240-3(d)

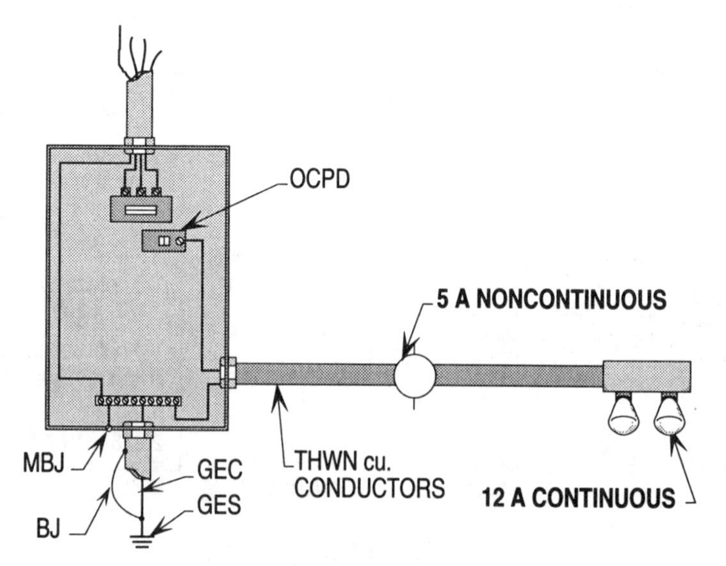

NEC 210-19(a); 210-20(a)

Ill. 1-2

Branch-circuits
210-23(a)

Fixed appliances are allowed to draw up to 50 percent of the rating of a branch-circuit supplying two or more general purpose outlets which serve lighting and receptacle loads.

Example 1-3 . Can the A/C window unit be connected to the existing branch-circuit (BC) in Illustration 1-3 ?

Determining if A/C unit can be added

Step 1: Finding A of BC
210-23(a)
A = 1/2 of 20 A OCPD
A = 10 A

Step 2: Calculating A for A/C unit
210-23(a); 440-62(b); (c); 440-32
8 A x 125% = 10 A

Step 3: Verifying permissive A
210-23(a); 440-62(b); (c); 210-3
20 A OCPD x 80% = 16 A

ADDITIONAL HELP 6:

- Calculating the resistance of 250 ft. of No. 12 THWN cu. run to a small A/C unit.
  - Table 8 to Ch. 9
    #12 cu. = 2.05R
  - 2.05R based on 1,000'
  - $R = \frac{\text{wire ohms x length}}{1{,}000}$

    $R = \frac{2.05R \times 250'}{1{,}000}$

    R = .5125

Solution: **Yes, the A/C window unit rated at 8 amps or 10 amps for sizing elements can be connected to the 20 amp branch-circuit.**

NEC 210-23(a); NEC 210-52(a); NEC 380-8(a); NEC Art. 410, Part C; NEC 440-62(b); (c); NEC 440-63

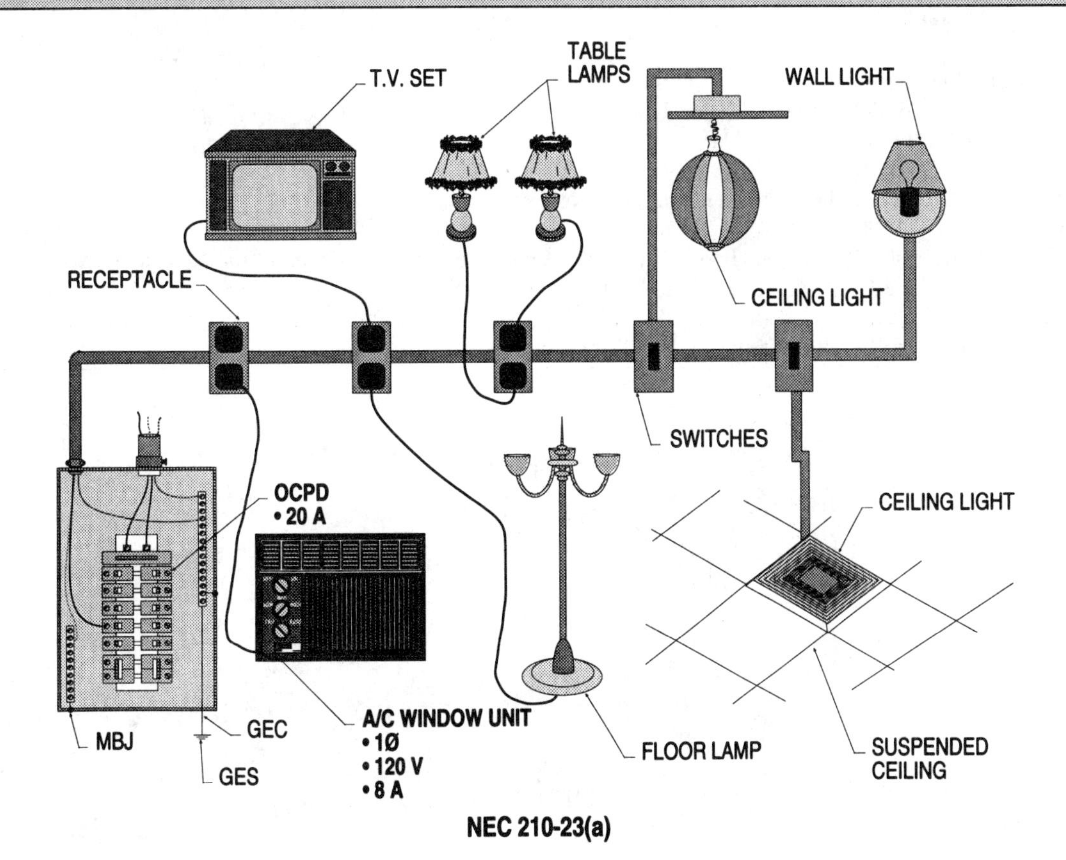

Ill. 1-3

Continuous And Noncontinuous Loads 210-19(a); 210-20(a)

The procedure for calculating the elements for a branch-circuit is to compute the noncontinuous load at 100 percent and the continuous load at 125 percent. Add these loads together and the rating in amps or VA is used to size the elements.

Example 1-4. What size OCPD and THWN copper conductors are required for the branch-circuit supplying the loads listed in Illustration 1-4?

Sizing OCPD

Step 1: Calculating OCPD
210-20(a)

11 A x 125%	= 13.75 A
6 A x 100%	= 6.0 A
Total load	= 19.75 A

Step 2: Selecting OCPD
Table 310-16; 240-3(d); 240-3(b); 240-6(a)
19.75 A requires 20 A OCPD

Solution: A 20 amp OCPD is required.

Sizing Conductors

Step 1: Calculating conductors
210-19(a)

11 A x 125%	= 13.75 A
6 A x 100%	= 6.00 A
Total load	= 19.75 A

Step 2: Selecting conductors
Table 310-16; 240-3(d); 310-10(1) - (4)
19.75 A requires #12 THWN

Solution: A No. 12 THWN copper conductor is required.

ADDITIONAL HELP 7:

Calculating the resistance of a 17 amp electrical circuit that is supplied by a pressure of 120 volts.

- Applying formula

$R = \frac{V}{A}$

$R = \frac{120 V}{17 A}$

R = 7.059

ADDITIONAL HELP 8:

Calculating the resistance of a 17 amp circuit run of 185 ft. with a No. 12 wire.

- Applying Formula

$R = \frac{12 \times \text{length in ft.}}{CM}$

$R = \frac{12 \times 200'}{6,530 CM}$

R = .3675

Note: *The asterisk (*) in Table 310-16 that appears beside* **No.** *14,* **No.** *12, and* **No.** *10 requires* **No.** *12 copper conductors and a 20 amp OCPD to be used. (Also, see Sec. 240-3(d))*

NEC 210-3; NEC 210-19(a); NEC 210-20(a); NEC 240-3(d)

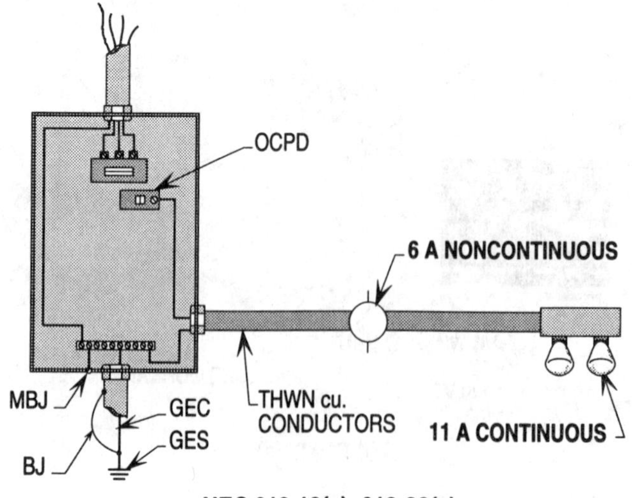

Ill. 1-4 **NEC 210-19(a); 210-20(a)**

Lighting Branch-circuits
220-3(a); Table 220-3(a)

The procedure for calculating the minimum number of branch-circuits is determined by multiplying the square footage per NEC 220-3(a) and Table 220-3(a) and dividing by the size overcurrent protection device times the voltage of the circuit.

Example 1-5(a). How many 15 amp, 2-wire circuits are required to supply power for the general lighting load in Illustration 1-5(a)?

Finding the number of two-wire circuits

Step 1: Finding VA per sq. ft.
220-3(a); Table 220-3(a)
The VA per sq. ft. is 3

Step 2: Calculating total VA
Table 220-3(a)
2,800 sq. ft. x 3 VA = 8,400 VA

Step 3: Calculating the number of circuits
210-11(a)
8,400 VA ÷ 1,800 VA (15 A OCPD x 120 V) = 4.7

Step 4: Selecting number
210-11(a)
4.7 requires 5

Solution: The number of 15 amp, 2-wire circuits is 5.

ADDITIONAL HELP 9:

Calculating amps when VA and Voltage are known.

• Applying formula

$$A = \frac{VA}{V}$$

$$A = \frac{8{,}400\ VA}{120\ V}$$

$$A = 70$$

ADDITIONAL HELP 10:

Calculating voltage when VA and Amps are known.

$$V = \frac{VA}{A}$$

$$V = \frac{8{,}400\ VA}{70\ A}$$

$$V = 120$$

NEC 220-3(a); NEC 210-11(a); NEC 210-11(b); NEC 210-70(a)

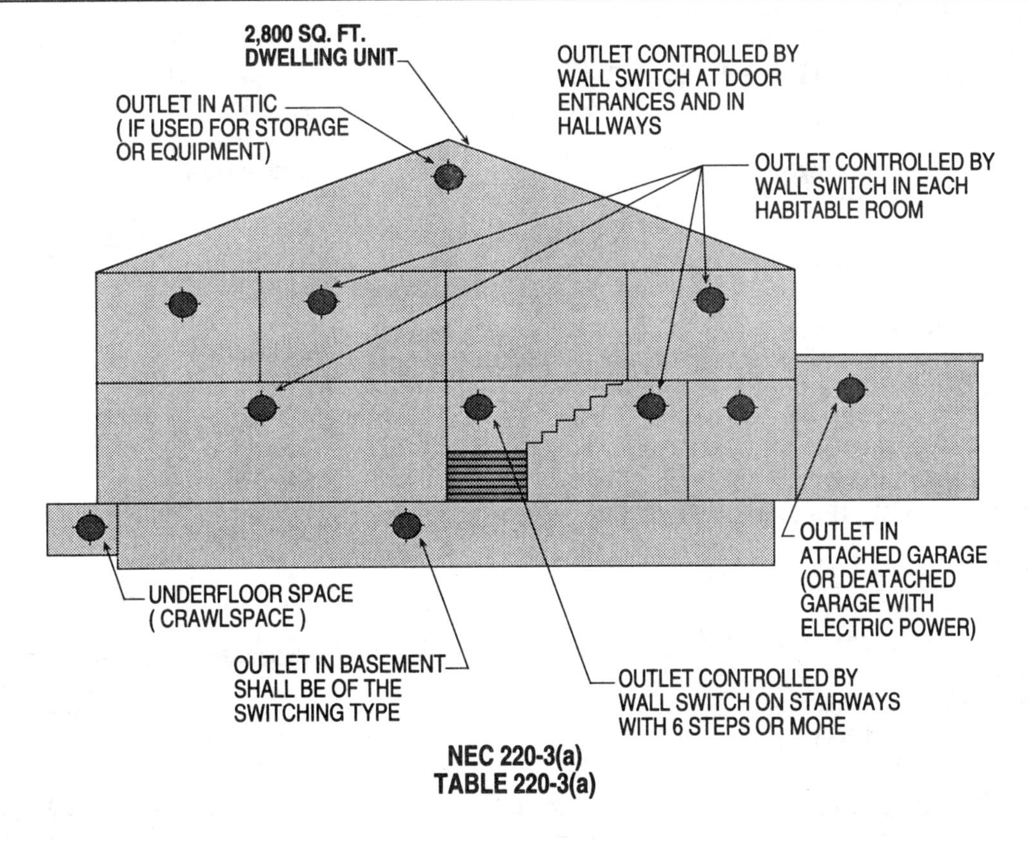

Ill. 1-5(a)

Example 1-5(b). What is the general lighting load in VA for a 6,000 sq. ft. store with 100 ft. of show window, 80 ft. of lighting track, fifty 120 VA outside lighting fixtures, one 1,200 VA sign and 4,800 VA low-voltage lighting load? (See Illustration 1-5(b))

Calculating general lighting load
220-3(a); Table 220-3(a); 210-19(a)

Step 1: 6,000 sq. ft. x 3 VA = 18,000 VA

Step 2: 18,000 VA x 125% = 22,500 VA

Solution: The general lighting load is 22,500 VA.

Calculating show window load
220-12(a); 210-19(a)

Step 1: 100 ft. x 200 VA = 20,000 VA

Step 2: 20,000 VA x 125% = 25,000 VA

Solution: The show window load is 25,000 VA.

Calculating lighting track load
220-12(b)

Step 1: 64 ÷ 2 x 150 VA = 4,800 VA.

Step 2: 4,800 VA x 125% = 6,000 VA

Solution: The lighting track load is 6,000 VA.

Calculating outside lighting load
220-4(b); 210-19(a)

Step 1: 50 x 120 VA = 6,000 VA

Step 2: 6,000 VA x 125% = 7,500 VA

Solution: The outside lighting load is 7,500 VA.

Calculating the sign load
210-19(a); 600-5(b)(3)

Step 1: 1,200 VA x 125% = 1,500 VA

Solution: The sign load is 1,500 VA.

Calculating LVL load
Art. 411; 210-19(a)

Step 1: 4,800 VA x 125% = 6,000 VA

Solution: The low-voltage lighting load is 6,000 VA.

ADDITIONAL HELP 11:

Calculating the number of 20 amp, 2-wire circuits required to supply the lighting load in Ill. 1-5(b)

- **Applying formula 210-11(a)**

$$\# = \frac{\text{VA of LTG. LD.}}{\text{V of cir. x CB rating}}$$

$$\# = \frac{22{,}500\text{ VA}}{120\text{ V x 20 A (2,400 VA)}}$$

$$\# = 9.375$$

Round up to 10

- $$\# = \frac{25{,}000\text{ VA}}{120\text{ V x 20 A (2,400 VA)}}$$

$$\# = 10.417$$

Round up to 11

- $$\# = \frac{6{,}000\text{ VA}}{120\text{ V x 20 A (2,400 VA)}}$$

$$\# = 2.5$$

Round up to 3

- $$\# = \frac{7{,}500\text{ VA}}{120\text{ V x 20 A (2,400 VA)}}$$

$$\# = 3.125$$

Round up to 4

- $$\# = \frac{1{,}500\text{ VA}}{120\text{ V x 20 A (2,400 VA)}}$$

$$\# = .625$$

Round up to 1

- $$\# = \frac{6{,}000\text{ VA}}{120\text{ V x 20 A (2,400 VA)}}$$

$$\# = 2.5$$

Round up to 3

Total # = 10 + 11 + 3 + 4 + 1 + 3 = 32

Solution: The number of 20 amp CB's are 32 to supply the lighting load.

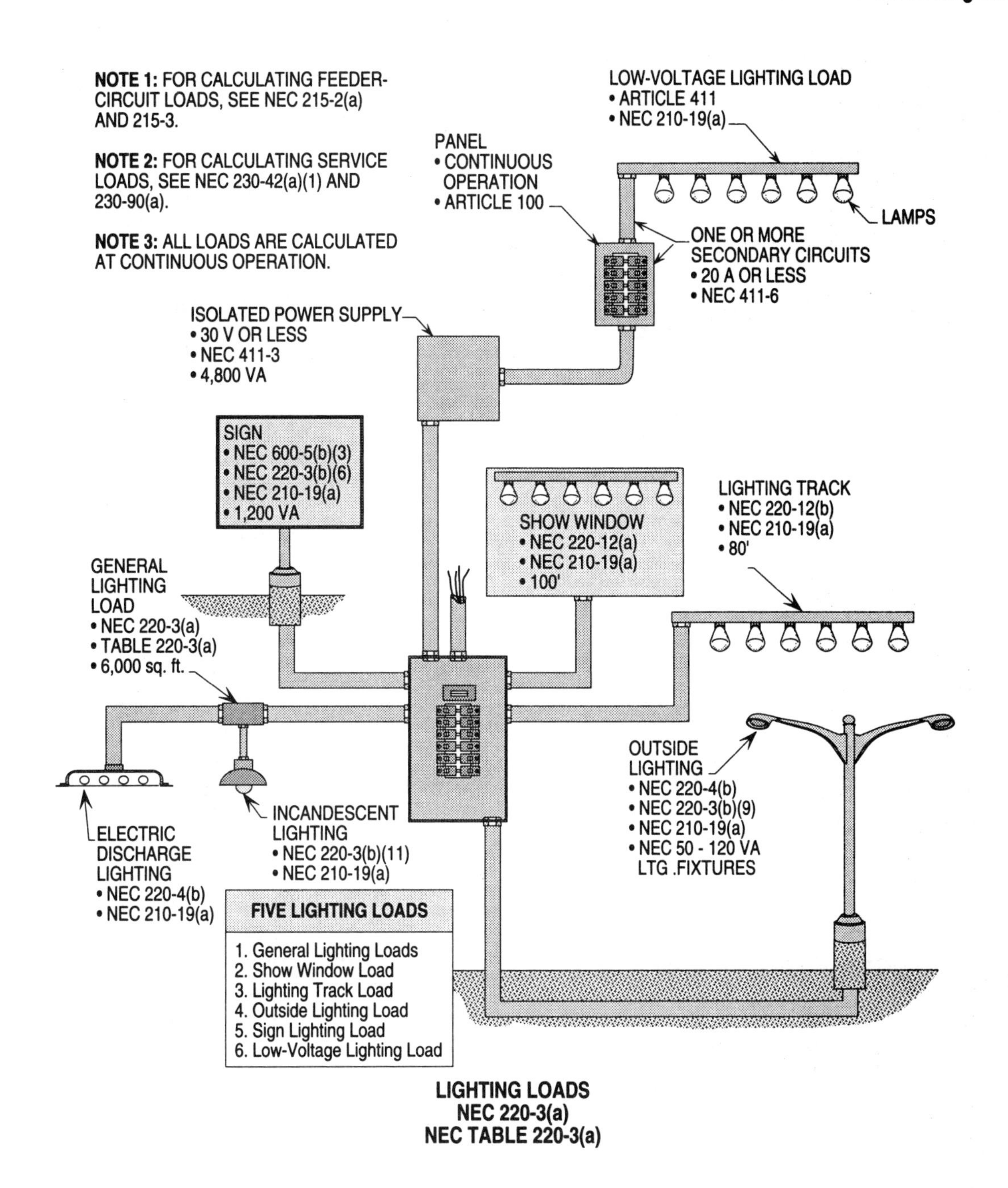

LIGHTING LOADS
NEC 220-3(a)
NEC TABLE 220-3(a)

CALCULATION FACTS

NOTE 1: RECESS LIGHTING FIXTURES MUST BE CALCULATED AT MAXIMUM VA RATING OF THE UNIT PER NEC 220-3(b)(4).

NOTE 2: HEAVY DUTY LAMPHOLDERS MUST BE CALCULATED AT A MINIMUM OF 600 VA PER NEC 220-3(b)(5).

NOTE 3: SHOW WINDOWS MUST BE CALCULATED AT 200 VA PER LINEAR FOOT PER NEC 220-3(b)(7) AND 220-12(a).

NOTE 4: TRACK LIGHTING MUST BE COMPUTED AT 150 VA FOR EVERY 2 FT. PER NEC 220-12(b).

NOTE 5: LIGHTING FIXTURE OUTLETS WITH UNKNOWN VA RATINGS MUST BE CALCULATED AT 180 VA EACH PER NEC 220-3(b)(11).

NOTE 6: FOR DWELLING UNITS, LIGHTING IS CALCULATED PER NEC 220-3(a) AND TABLE 220-3(a).

Ill. 1-5(b)

Commercial Receptacle Load
220-3(b)(9)

The procedure for calculating the commercial receptacle load is determined by 180 volt-amps per outlet. The first 10 kVA is to be calculated at 100 percent, if the receptacles are rated for noncontinuous duty. Continuous loads must be calculated at 180 volt-amps and multiplied by 125 percent to determine the receptacle load.

Example 1-6(a). What is the load for 40 duplex receptacles used for continuous duty and 40 duplex receptacles used for noncontinuous duty in Illustration 1-6(a)?

Calculating continuous loads

Step 1: Finding VA (Continuous)
220-3(b)(9); 210-19(a)
180 VA x 40 x 125% = 9,000 VA

Solution: The continuous duty load is 9,000 VA.

Calculating noncontinuous loads

Step 1: Finding VA (Noncontinuous)
220-3(b)(9); 210-19(a)
180 VA x 40 = 7,200 VA

Step 2: Finding total VA

Continuous	=	9,000 VA
Noncontinuous	=	7,200 VA
Total load	=	16,200 VA

Solution: The total load is 16,200 VA.

Applying demand factors

Step 1: Finding VA (Noncontinuous)
220-3(b)(9); Table 220-13
180 VA x 40 = 7,200 VA

Solution: The noncontinuous duty load is 7,200 VA.

Note: For more information on applying demand factors for general purpose receptacle loads rated over 10,000 VA See pages 2-13 and 2-14 of Chapter 2 of this book..

ADDITIONAL HELP 12:

Based on a 20 amp CB and No. 12 cu. wire, how many outlets should be connected to a branch-circuit?

- Applying formula
  220-3(b)(9)

$$\frac{180 \text{ VA}}{120 \text{ V}} = 1.5 \text{ A}$$

$$\frac{20 \text{ A CB}}{1.5 \text{ A}} = 13 \text{ outlets}$$

Maximum length to the final outlet based on load. (Recommended Procedure)

- 40' or less if such outlet pulls greater than 80 percent of max. load.
- 60' or less if such outlets pulls between 50 and 80 percent of max. load.
- 90' or less if such outlets pulls 50 percent or less of max. load.

Ground-fault circuit-interrupters

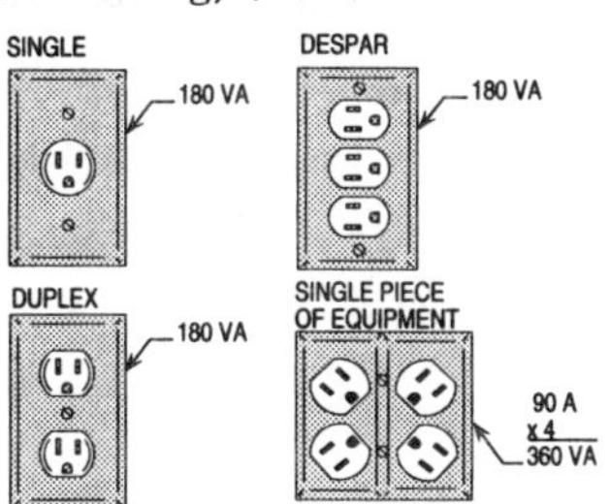

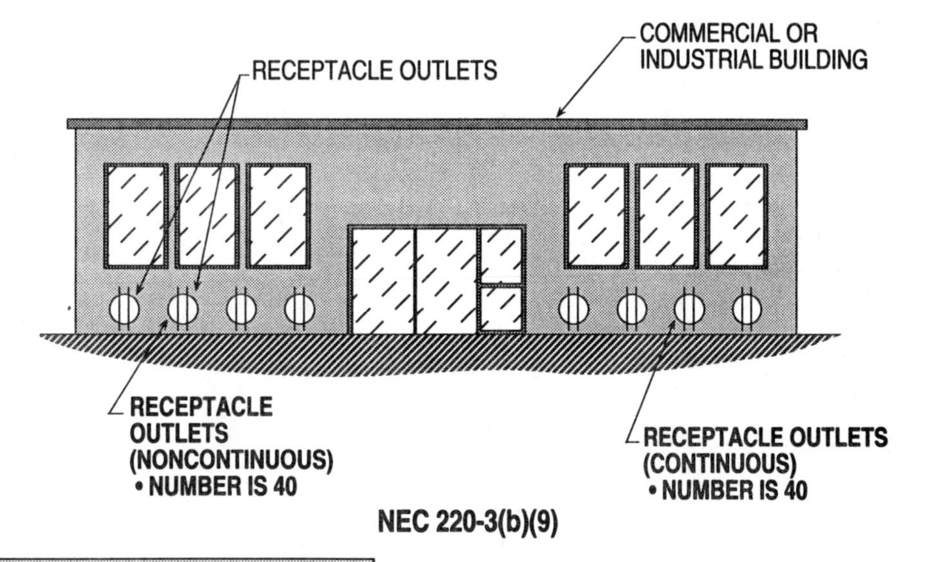

NEC 220-3(b)(9)

Construction Sites	
Services	- NEC 305-4(a)
Feeders	- NEC 305-4(b)
Branch-circuits	- NEC 305-4(c)
Receptacles	- NEC 305-4(d)
GFCI	- NEC 305-6(a)
AEGCP	- NEC 305-6(b)(2)

Installing Receptacles
NEC 210-60
NEC 210-62
NEC 210-63
NEC 410-56(a); (b)

NOTE 1: FOR GFCI-PROTECTION OF 15, 20, AND 30 AMP RECEPTACLE OUTLETS, SEE NEC 305-6(a).

NOTE 2: FOR GFCI-PROTECTION OR AEGC PROGRAM OF RECEPTACLE OUTLETS OTHER THAN 15, 20, AND 30 AMP, SEE NEC 305-6(b).

Ill. 1-6(a)

Multioutlet Assemblies 220-3(b)(8)(a); (b)(8)(b)

The procedure for calculating the load for multioutlet assemblies (not used simultaneously) is to compute the total length by 180 VA and divide by 5. When connected appliances are likely to be used simultaneously, compute the total length by 180 VA to obtain the load rating.

Example 1-6(b). What is the load in VA for the multioutlet assembly in Illustration 1-6(b)?

Step 1: Calculating VA
220-3(b)(8)(a)
VA = assembly length ÷ 5 x 180 VA
VA = 80' ÷ 5 x 180 VA
VA = 2,880

Solution: The multioutlet assembly load is 2,880 VA.

Example 1-6(b). What is the load in VA for such assembly, if the appliances in Illustration 1-6(b) operate simultaneously?

Step 1: Calculating VA
220-3(b)(8)(b)
VA = assembly length x 180 VA
VA = 80' x 180 VA
VA = 14,400

Solution: The multioutlet assembly load is 14,400 VA.

ADDITIONAL HELP 13:

Calculating the demand load for the multioutlet assembly and the receptacle load per Table 220-13.

- Calculating load
  220-3(b)(8)(a); (b); 220-3(b)(9)
  200 x 180 VA = 36,000 VA
  + 2,880 VA
  + 14,000 VA
  + 52,880 VA

- Applying demand factors
  Table 220-13
  First 10,000 VA x 100% = 10,000 VA
  Next 42,880 VA x 50% = 21,440 VA
  31,440 VA

Solution: The demand load is 31,440 VA.

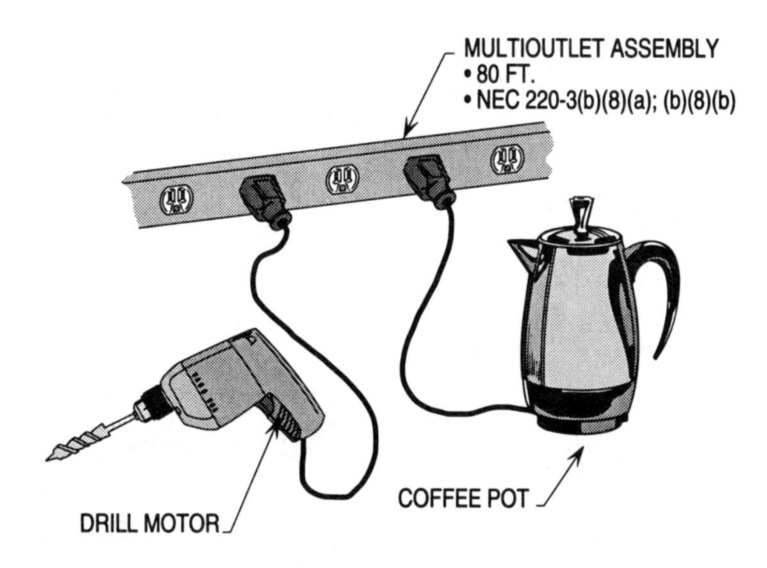

NEC 220-3(b)(8)(a)
NEC 220-3(b)(8)(b)

ADDITIONAL HELP 13:

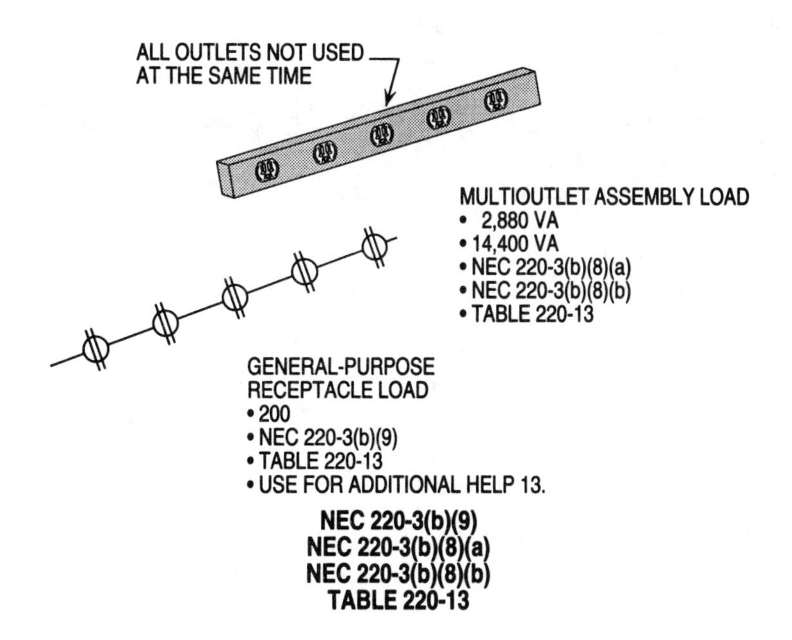

NEC 220-3(b)(9)
NEC 220-3(b)(8)(a)
NEC 220-3(b)(8)(b)
TABLE 220-13

Ill. 1-6(b)

Load For Additions To Existing Installations 220-3(c)(1); (2)

The procedure for adding new circuits or extensions to existing electrical systems in dwelling units of 500 square foot or less is determined by figuring the VA per square foot or the amperes per outlet method. Either method applies to a portion of the dwelling unit that has not been previously wired or an addition which exceeds 500 square foot in area.

Example 1-7(a). What is the lighting load for an addition to an existing dwelling in Illustration 1-7(a)?

Step 1: Calculating load
Table 220-3(a); 220-3(c)(1)
sq. ft. x VA = Ltg. Ld.
576 sq. ft. x 3 VA = 1,728 VA

Solution: The lighting load is 1,728 volt-amps.

Example 1-7(b). Can four new receptacle outlets be added to existing circuit #1 in Illustration 1-7(b)?

Step 1: Calculating number of outlets
220-3(c)(1); 220-3(b)(9)
# of outlets = CB rating - Existing # in A / 1.5 A
# of outlets = 15 A - 9 A = 6
= 6 ÷ 1.5
of outlets = 4

Obtaining Values
180 VA / 120 V = 1.5 A
1.5 A x 6 outlets = 9 A

Solution: Yes, four new receptacle outlets may be added.

Example 1-7(c). What is the load to determine the number of OCPD's for an addition to a existing store building in Illustration 1-7(c)?

Step 1: Calculating VA / sq. ft.
Table 220-3(a); 220-3(c)(2)
6,000 sq. ft. x 3 VA = 18,000 VA

Step 2: Calculating continuous load for OCPD
210-19(a)
18,000 VA x 125% = 22,500 VA

Step 3: Calculating # of outlets
90 x 180 VA = 16,200 VA

Step 4: Calculating continuous load for OCPD
16,200 VA x 125% = 20,250 VA

Solution: The load is 22,500 VA which is based on the larger calculation.

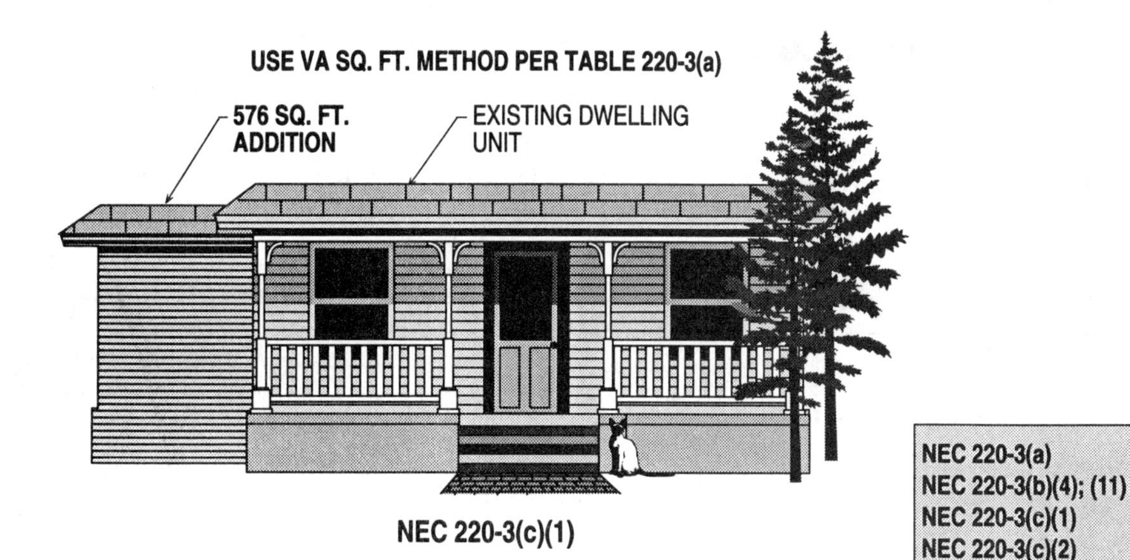

NEC 220-3(a)
NEC 220-3(b)(4); (11)
NEC 220-3(c)(1)
NEC 220-3(c)(2)

Ill. 1-7(a)

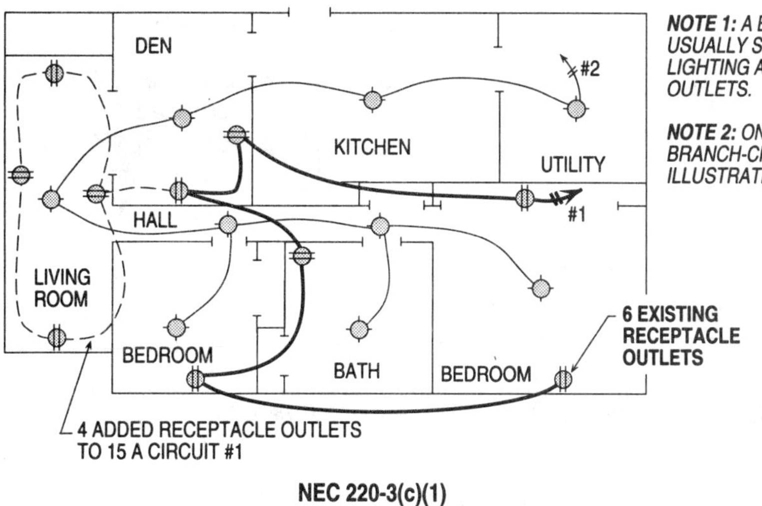

NOTE 1: A BRANCH-CIRCUIT USUALLY SUPPLIES BOTH LIGHTING AND RECEPTACLE OUTLETS.

NOTE 2: ONLY OUTLETS ON BRANCH-CIRCUITS ARE ILLUSTRATED.

NEC 210-52(a)
NEC 220-3(b); (c)
NEC 220-3(a)
TABLE 220-3(a)
NEC 220-3(b)(9)

Ill. 1-7(b)

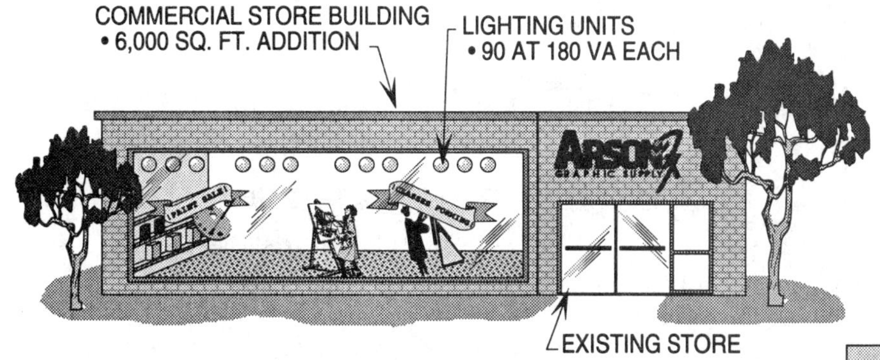

NEC 220-4(b)
NEC 230-42(a)(1)
NEC 220-3(a)
TABLE 220-3(a)

Ill. 1-7(c)

Outside Branch-circuits 225-3(a)

The procedure for calculating the elements for a outside branch-circuit is to compute the noncontinuous load at 100 percent and the continuous load at 125 percent. Add these loads together and the rating in amps or VA is used to size the elements.

Example 1-8. What size OCPD and conductors are required for the branch-circuit in Illustration 1-8?

Finding load

Step 1: Calculating load
225-3(a); 210-19(a)
15.5 A x 125% = 19.375 A

Step 2: Sizing OCPD
225-9; 240-3(b); 240-6(a); 240-3(d)
19.375 A requires 20 A

Solution: The size OCPD is 20 amp.

Sizing conductors

Step 1: Calculating conductors
Table 310-16; 240-3(d)
19.375 A requires #12

Solution: The size conductors are No. 12 THWN copper.

ADDITIONAL HELP 14:

Calculating the amps for a feeder-circuit having a continuous load of 138 amps.

- Calculating amps
  225-3(b); 215-2(a)
  138 A x 125% = 172.5 A
  220-2(b)
  Round up to 175 A

Note: The size THWN cu. conductors are No. 2/0 per Table 310-16 and the size OCPD is 175 amp per NEC 240-3(b) and 240-6(a).

NEC 110-14(c); NEC 240-3(b); NEC 240-6(a); NEC 310-10; TABLE 310-16; NEC 240-3(d)

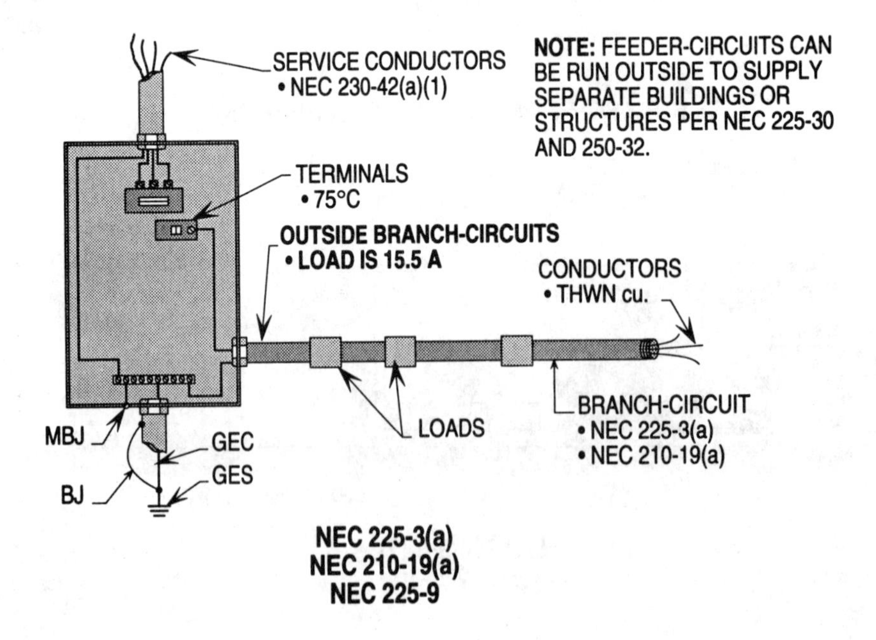

Ill. 1-8

2

Calculating Feeder-Circuit

Feeder-circuits are the circuit conductors between the service equipment or the source of a separately derived system and the final overcurrent protection device protecting the branch-circuit and the equipment.

Service equipment with its elements is the necessary equipment required to supply the needed power for feeder-circuits and branch-circuits which serve all the loads in the electrical system.

Feeder-circuits and service loads are computed based upon their operation. They are computed at continuous or noncontinuous operation or a combination of both.

Overcurrent protection devices, conductors, and other elements are sized and selected based upon these calculations being properly computed by the rules and regulations of the National Electrical Code®.

Quick Reference

Feeder-Circuits Having Continuous And Noncontinuous Loads 215-2(a); 215-3

The procedure for calculating the load for feeder-circuits and service loads is to compute the noncontinuous load at 100 percent and the continuous load at 125 percent. Add these loads together and the rating in amps or VA is used to size the elements.

Example 2-1(a). What size OCPD and conductors are required for the feeder-circuit supplying the loads in Illustration 2-1(a)?

Sizing OCPD

Step 1: Calculating at 125%
215-3

52 A x 100%	= 52 A
76 A x 125%	= 95 A
Total load	= 147 A

Step 2: Sizing OCPD
215-3; 240-3(b); 240-6(a)
147 A requires 150 A

Solution: The size OCPD is 150 amps.

Sizing phase conductors

Step 1: Calculating at 125%
215-2(a)

52 A x 100%	= 52 A
76 A x 125%	= 95 A
Total load	= 147 A

Step 2: Sizing phase conductors
Table 310-16; 310-10(1) - (4)
147 A requires #1/0

Solution: A No. 1/0 THWN copper conductor is required per phase.

Sizing neutral

Step 1: Calculating neutral at 100%
220-22; 310-4(1) - (4); 310-15(b)(4)(a) - (c)

52 A x 100%	= 52 A
76 A x 100%	= 76 A
Total load	= 128 A

Step 2: Sizing neutral conductor
Table 310-16; 250-24(b)(1); (b)(2)
128 A requires #1

Solution: A No. 1 THWN copper conductor is required.

NOTE 1: THE SMALLER SIZE FEEDER-CIRCUIT IS 30 AMPS AND USUALLY SUPPLIED BY No. 10 THWN cu. CONDUCTOR.

NOTE 2: WHEN FEEDER-CIRCUIT CONDUCTORS SUPPLY INDIVIDUAL DWELLING UNITS OR MOBILE HOMES, SUCH FEEDER-CONDUCTORS CAN BE SIZED PER TABLE 310-15(b)(6).

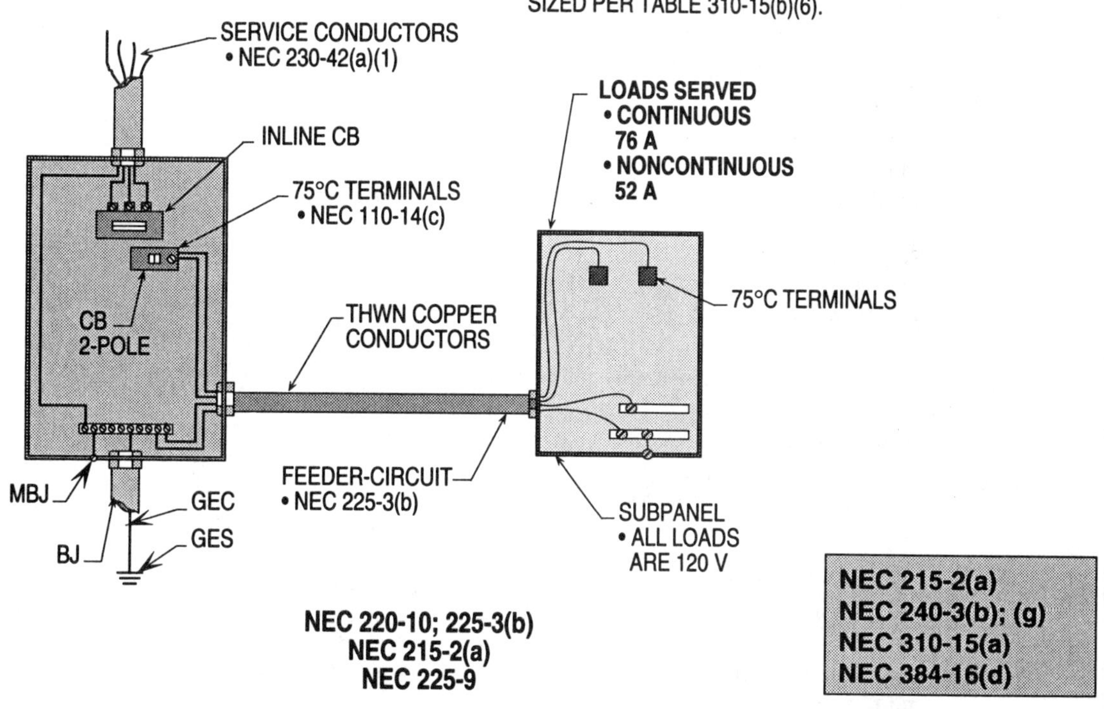

NOTE 3: TO CALCULATE THE LOAD FOR SIZING THE CONDUCTORS FOR A FEEDER-CIRCUIT, SEE NEC 220-10 AND 215-2(a).

NOTE 4: TO CALCULATE THE LOAD FOR SIZING THE OCPD FOR A FEEDER-CIRCUIT, SEE NEC 215-3, 240-2, AND 240-3(a) THROUGH (g).

Ill. 2-1(a)

Continuous And Noncontinuous Loads 220-10; 230-42(a)(1); 230-90(a)

Service loads are calculated at continuous and noncontinuous operation. Where applicable demand factors can be added to the continuous and noncontinuous calculation to derive the total service load.

Example 2-1(b). What size OCPD, phase conductors, and neutral conductors are required for the service equipment supplying the loads in Illustration 2-1(b)?

Sizing OCPD

Step 1: Calculating OCPD at 125%
230-90(a), w/Ex.'s
68 A x 100% = 68.0 A
145 A x 125% = 181.25 A
Total load = 249.25 A

Step 2: Sizing OCPD
240-3(b); 240-6(a)
249.25 A requires 250 A

Solution: The size OCPD is 250 amps.

Sizing phase conductors

Step 1: Calculating phase conductors at 125%
220-10; 230-42(a)(1)
68 A x 100% = 68.0 A
145 A x 125% = 181.25 A
Total load = 249.25 A

Step 2: Sizing phase conductors
Table 310-16; 310-4(1) - (4)
249.25 A requires #250 KCMIL

Solution: A No. 250 KCMIL THWN copper conductor is required per phase.

Sizing neutral

Step 1: Calculating neutral conductors
220-22; 310-4(1) - (4); 310-15(b)(4)(c)
68 A x 100% = 68 A
145 A x 100% = 145 A
Total load = 213 A

Applying demand factors
220-22
First 200 A x 100% = 200 A
Next 13 A x 70% = 9 A
Total load = 209 A

Step 2: Sizing neutral conductor
Table 310-16; 250-24(b)(1); (b)(2)
209 A requires #4/0

Solution: A No. 4/0 THWN copper conductor is required.

NOTE 1: IF THE UNGROUNDED PHASE CONDUCTORS ARE NOT PROTECTED BY THE IN-LINE OCPD THEY MUST BE INCREASED IN SIZE TO COMPLY WITH NEC 230-90(a), Ex. 2 AND NEC 240-3(b). SEE NEC 230-90(a), Ex. 1, 430-62(a), 430-63 AND 240-3(g) WHERE MOTOR LOADS ARE INVOLVED.

SIZING OCPD

- FIRST, SIZE THE OCPD FOR THE SERVICE CONDUCTORS AND EQUIPMENT BASED ON THE AMPACITY OF CONDUCTORS PER NEC 230-90(a), Ex. 1 AND 430-63.
- SECONDLY, SIZE THE OCPD FOR SERVICE CONDUCTORS AND EQUIPMENT BASED ON THE LARGEST OCPD FOR ANY ONE MOTOR AND ADD THIS TOTAL TO ALL THE OTHER COMPUTED LOADS PER NEC 230-90(a), Ex. 1 AND 430-62(a).
- SELECT THE LARGEST OCPD PRODUCED BY THE CALCULATION OF NOTE 1 OR NOTE 2.

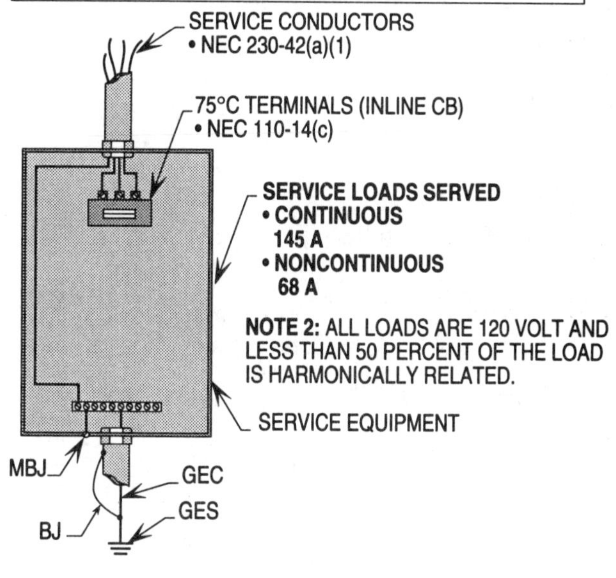

NEC 220-10; 230-42(a)(1)
NEC 230-90(a)

NEC 230-42(a)(1)
NEC 230-90(a), Ex. 2
NEC 310-15(a)
NEC 384-16(d)

Ill. 2-1(b)

Continuous And Noncontinuous Loads 220-10; 230-42(a)(1); 230-90(a), Ex. 2; Ex. 3

The OCPD's do not have to be equal to the ampacity of the service conductors if two to six devices are located in a single enclosure or group of enclosures.

Example 2-1(c). What size conductors are required to supply the loads protected by the OCPD's in Illustration 2-1(c)?

Sizing OCPD

Step 1: Calculating OCPD's at 125%
230-42(a)(1); 230-90(a)
Ld. 1 = 24 A x 125% = 30 A
Ld. 2 = 32 A x 125% = 40 A
Ld. 3 = 40 A x 125% = 50 A
Ld. 4 = 80 A x 125% = 100 A
Ld. 5 = 80 A x 125% = 100 A
Ld. 6 = 80 A x 125% = 100 A

Step 2: Sizing each OCPD
230-90(a), Ex. 2; 240-3(b)
Ld. 1 requires 30 A OCPD
Ld. 2 requires 40 A OCPD
Ld. 3 requires 50 A OCPD
Ld. 4 requires 100 A OCPD
Ld. 5 requires 100 A OCPD
Ld. 6 requires 100 A OCPD

Step 3: Totaling OCPD's
230-90(a), Ex. 2; Ex. 3; 230-80
30 A + 40 A + 50 A + 100 A + 100 A + 100 A = 420 A

Solution: The total rating of the OCPD's is 420 amps.

Note: OCPD may be sized at 400 amps or 450 amps, per NEC 230-90(a) or 230-90(a), Ex. 2 and 240-3(b).

Sizing phase conductor

Step 1: Calculating phase conductors at 100%
220-10
24 A + 32 A + 40 A + 80 A + 80 A + 80 A x 100% = 336 A

Step 2: Increasing phase conductors at 125% to match OCPDs
230-42(a)(1); 310-10(1) - (4)
336 A x 125% = 420 A

Step 3: Selecting phase conductors
Table 310-16
420 A requires #600 KCMIL

Solution: A No. 600 KCMIL THWN copper conductor per phase is required.

Sizing neutral

Step 1: Calculating neutral conductor
250-24(b)(1); Table 250-66
#600 KCMIL requires at least a #1/0

Solution: A No. 1/0 THWN copper conductor is required per phase.

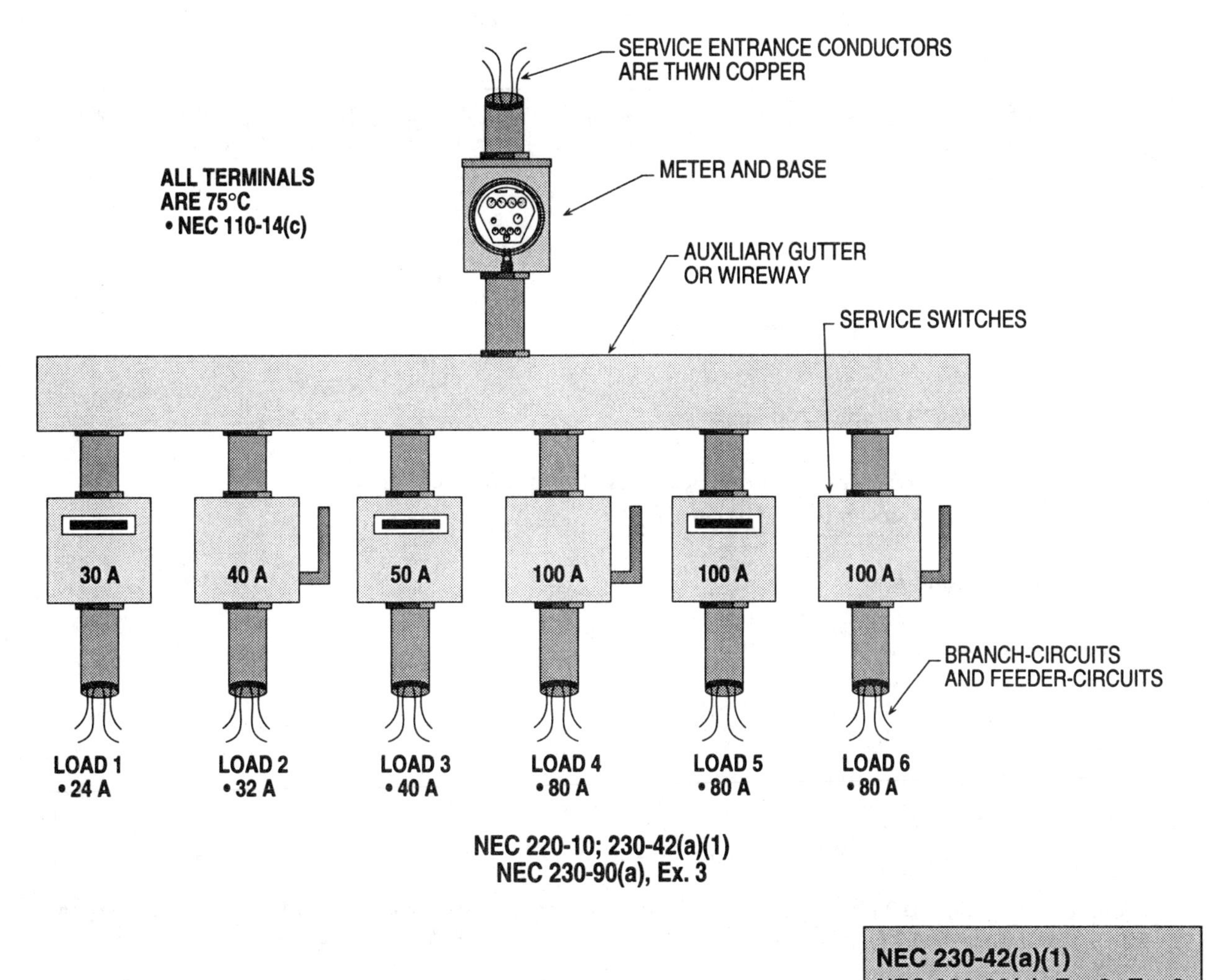

NEC 230-42(a)(1)
NEC 230-90(a), Ex. 2; Ex. 3
NEC 310-15(a)
NEC 384-16(d)
NEC 230-71(a)

Ill. 2-1(c)

NOTE 1: WHERE THE INLINE OCPD IS A SINGLE CB, THE OCPD RATING CAN BE RATED AT 450 AMPS PER NEC 230-90(a), Ex. 2 AND 240-3(b).

NOTE 2: THERE IS NO IN-LINE OCPD IN THE ABOVE ILLUSTRATION WHICH IN MOST CASES WILL ALLOW THE UNGROUNDED SERVICE ENTRANCE CONDUCTORS TO BE SMALLER THAN THE OCPD'S. IN OTHER WORDS, THEY DO NOT HAVE TO BE EQUAL TO THE RATINGS IN AMPS OF THE OCPD'S PER NEC 230-90(a), Ex. 3.

Continuous And Noncontinuous Loads
220-10; 230-42(a)(1); 230-90(a), Ex. 3

OCPD's which are used to allow loads with inrush currents to start and operate can be sized larger than the ampacity of the service conductors.

Example 2-1(d). What size conductors are required to supply the loads protected by the OCPD's in Illustration 2-1(d)?

Sizing OCPD

Step 1: Totaling OCPD's
230-90(a), Ex. 1; 240-3(g)
200 A x 6 = 1,200 A

Solution: Total OCPD's are 1,200 amps.

Note: Amps of service conductors are 900 amps (150 A x 6 = 900 A) per NEC 310-10(1) through (4) and Table 310-16.

Sizing phase conductors

Step 1: Calculating phase conductors
310-4; 220-10
880 A ÷ 6 = 147 A

Step 2: Selecting phase conductors
Table 310-16; 310-4
147 A requires #1/0

Solution: Six No. 1/0 THWN copper conductors are required per phase. Note: Conductors do not have to match the ampere rating of the OCPD's per NEC 230-90(a), Ex. 1.

Sizing neutral

Step 1: Calculating CM for neutral
250-24(b)(2); Table 8, Ch. 9
#1/0 = 105,600 CM x 6 = 633,600 CM
633,600 CM ÷ 1,000 = 633.6 KCMIL

Step 2: Selecting CM for Neutral
250-24(b)(2); Table 8, Ch. 9; Table 250-66
633.6 KCMIL requires #1/0

Step 3: Sizing min. neutral conductor
310-4; 300-5(i)
#1/0 min. allowed for parallel

Solution: Six No. 1/0 THHN copper conductors are required per phase.

NOTE 1: ONE LINE DIAGRAM IS DRAWN FOR SIMPLICITY.

NOTE 2: LOAD SERVED IS CALCULATED AT 880 AMPS.

NOTE 3: THERE IS NO IN-LINE OCPD IN THE ABOVE ILLUSTRATION WHICH IN MOST CASES WILL ALLOW SMALLER UNGROUNDED SERVICE ENTRANCE CONDUCTORS TO BE INSTALLED. IN OTHER WORDS, THEY DO NOT HAVE TO BE EQUAL TO THE RATINGS IN AMPS OF THE OCPD's, WHERE LOADS WITH INRUSH CURRENTS ARE SUPPLIED PER NEC 230-90(a), Ex. 1 AND Ex. 3.

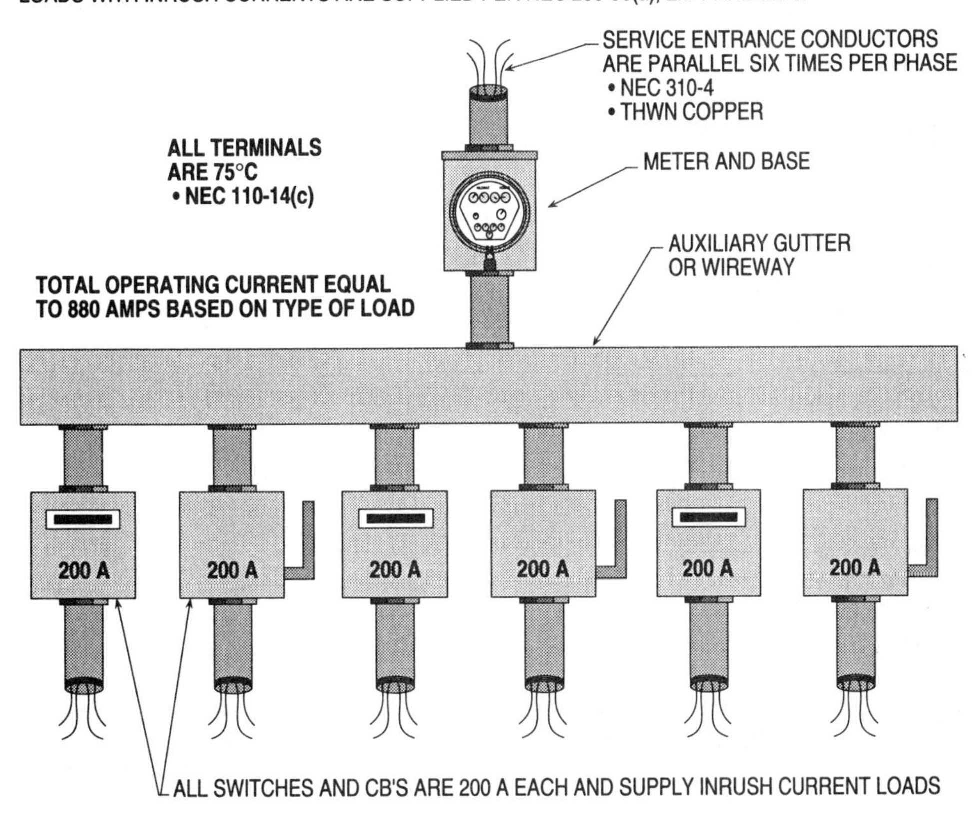

NEC 220-10
NEC 230-42(a)(1)
NEC 230-90(a), Ex. 1
NEC 230-90(a), Ex. 3

NEC 230-42(a)(1)
NEC 230-90(a), Ex.1; Ex. 3
NEC 310-15(a)
NEC 384-16(d)
NEC 230-71(a)

Ill. 2-1(d)

General Lighting
220-11

The procedure for calculating the general lighting load is determined by multiplying the square footage per Table 220-3(a). Demand factors are then applied from Table 220-11.

Example 2-2. What is the general lighting and receptacle load for the dwelling unit in Illustration 2-2?

Finding the loads in VA

Step 1: General lighting load
Table 220-3(a)
2,800 sq. ft. x 3 VA = 8,400 VA

Step 2: Small appliance circuits
220-16(a)
1,500 VA x 2 = 3,000 VA

Step 3: Laundry circuit
220-16(b)
1,500 VA x 1 = 1,500 VA

Step 4: Total load

General lighting	=	8,400 VA
Small appliance	=	3,000 VA
Laundry	=	1,500 VA
Total load	=	12,900 VA

Step 5: Applying demand factors
Table 220-11

First 3,000 x 100%	=	3,000 VA
Next 9,900 x 35%	=	3,465 VA
Total load	=	6,465 VA

Solution: The general lighting and receptacle load is 6,465 VA.

ADDITIONAL HELP 15:

Calculating amps based on 120 volts (2 Hots) or 240 volts (2 Hots) per NEC 220-2(a).

- $A = \frac{VA}{V}$

  $A = \frac{6{,}465 \text{ VA}}{240 \text{ V}}$

  $A = 26.9$

- $A = \frac{VA}{V}$

  $A = \frac{6{,}465 \text{ VA}}{120 \text{ V}}$

  $A = 53.875$

  $A = \frac{53.875 \text{ A}}{2 \text{ Hots}}$

  $A = 26.9$

NEC 220-3(a); NEC 220-11

NOTE 1: *SEE NEC 230-42(b) AND 230-79(c) FOR THE MINIMUM SIZE SERVICE.*

NOTE 2: *SEE NEC 422-12 FOR INDIVIDUAL BC FOR CENTRAL HEATING EQUIPMENT THAT IS NOT FIXED.*

NEC 220-3(a)
TABLE 220-3(a)

Ill. 2-2

Show Window Lighting
220-12(a); 220-3(b)(7)

The procedure for calculating show window lighting is to compute a 180 volt-amps per outlet for medium base lamps and 600 volt-amps for mogul lamps. Continuous loads must have a 125 percent factor applied to the volt-amps per outlet. The show window lighting load must be calculated at 125 percent of the volt-amps per linear foot, if such lighting is considered to be used at continuous duty. **Note:** Example 2-3(a) are loads for service conductors.

Example 2-3(a). What is the lighting load for the show window and the number of lamp outlets in Illustration 2-3(a)?

Finding Load

Step 1: Calculating the load
220-3(b)(7); 220-12(a)
80 ft. x 200 VA = 16,000 VA

Step 2: Calculating continuous load
220-10; 230-42(a)(1)
16,000 VA x 125% = 20,000 VA

Solution: The lighting load is 20,000 volt-amps.

Finding Number of outlets

Step 1: Number of outlets
220-3(b)(9)
71 x 180 VA = 12,780 VA

Step 2: Calculating the load for the OCPD
230-90(a), w/Ex.'s
12,780 VA x 125% = 15,975 VA

Solution: The lighting load is 15,975 VA.

Note: The 20,000 VA computation is used for it is greater than 15,975 VA.

ADDITIONAL HELP 16:

Calculating the amps for the 20,000 VA load when the voltage is 120/208 volt, three-phase, four-wire system per NEC 220-2(a).

- $A = \dfrac{VA}{V \times \sqrt{3}}$

$A = \dfrac{20{,}000 \text{ VA}}{208 \text{ V} \times 1.732}$

$A = 56 \text{ A}$

- $A = \dfrac{VA}{V}$

$A = \dfrac{20{,}000 \text{ VA}}{120 \text{ V}}$

$A = \dfrac{166.7 \text{ A}}{3 \text{ Hots}}$

$A = 56 \text{ A}$

NEC 210-62; NEC 220-3(b)(7); NEC 220-12(a)

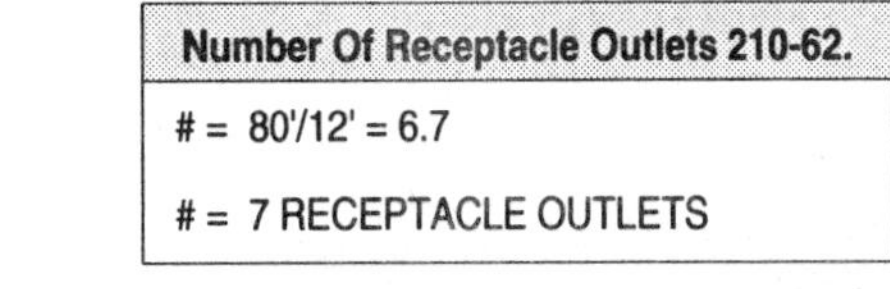

Number Of Receptacle Outlets 210-62.

= 80'/12' = 6.7

= 7 RECEPTACLE OUTLETS

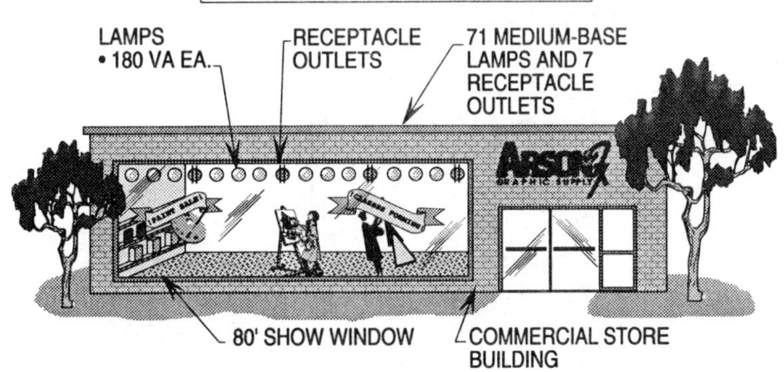

NEC 220-12(a)
NEC 220-3(b)(7)

Ill. 2-3(a)

Lighting Track Load
220-12(b)

The procedure for calculating the lighting track load is to compute the total length by 150 VA and divide by 2.

Example 2-3(b). What is the volt-amp rating of 80 ft. of lighting track in Illustration 2-3(b)?

Finding VA

Step 1: Calculating the load
220-12(b)
VA = track length ÷ 2' x 150 VA
VA = 80' ÷ 2' x 150 VA
VA = 6,000 VA

Solution: **The volt-amp rating for 80 ft. of lighting track is 6,000 VA.**

Note: When dividing VA on circuits, verify procedure with AHJ.

Dividing VA on circuits

For one circuit
Step 1: 150 VA ÷ 1 = 150 VA

For two circuits
Step 1: 150 VA ÷ 2 = 75 VA

For three circuits
Step 1: 150 VA ÷ 3 = 50 VA

Solution: **The VA for one circuit is 150 VA. When dividing on two or three circuits, they are 75 VA and 50 VA respectively.**

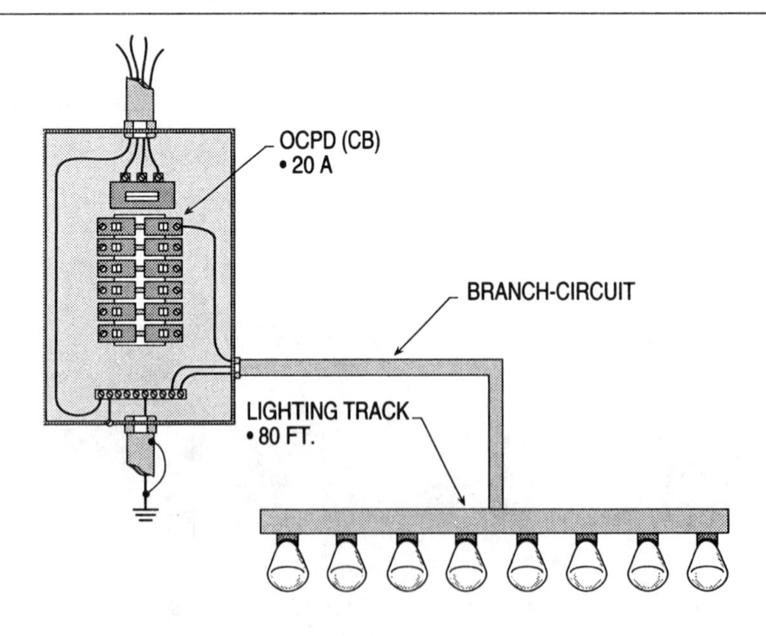

NEC 220-12(b)

Ill. 2-3(b)

Receptacle Loads - Nondwelling Units 220-13

The procedure for calculating receptacle loads in industrial and commercial locations is computed at 100 percent for the first 10 kVA and all remaining volt-amps at 50 percent. This reduction for all receptacles is on the basis that all receptacles are not used at the same time.

Example 2-4(a). What is the receptacle load for the duplex receptacles in Illustration 2-4(a)?

Finding Load

Step 1: Calculating the load
220-3(b)(9)
200 X 180 VA = 36,000 VA

Note: It is permissible to add the VA of multioutlet assemblies to the receptacle outlet load and apply the demand factors of Table 220-11 or Table 220-13.

Step 2: Applying demand factors
Table 220-13
First 10,000 x 100% = 10,000 VA
Next 26,000 x 50% = 13,000 VA
Total load = 23,000 VA

Solution: The receptacle load is 23,000 volt-amps.

NEC 220-3(b)(9); NEC 210-11(a); NEC 220-13; TABLE 220-13

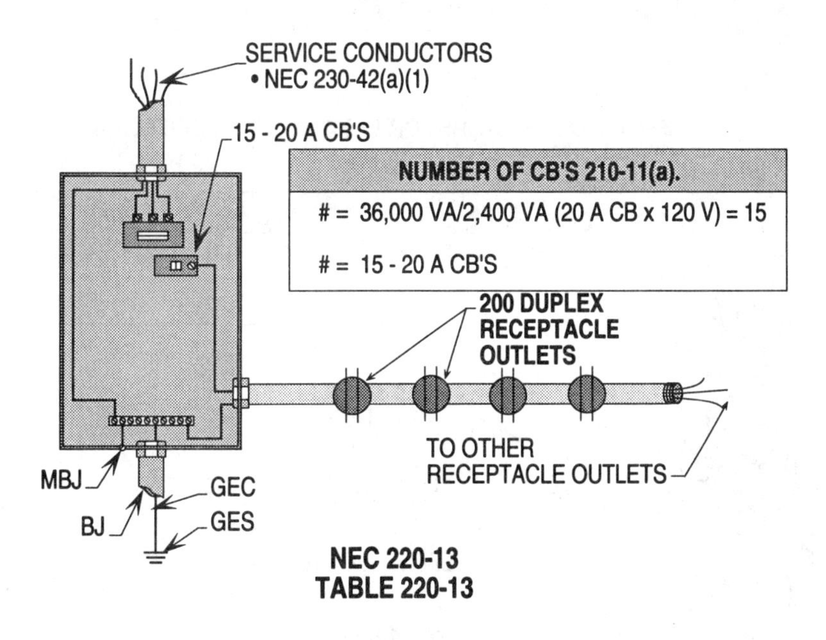

Ill. 2-4(a)

Receptacle Loads - Nondwelling Units 220-13

The receptacle loads in hospitals, hotels, and warehouses may be combined with the lighting load and the demand factors of Table 220-11 applied accordingly.

Example 2-4(b). What is the receptacle load and lighting load in Illustration 2-4(b) where Section 220-11 and Table 220-11 are applied?

Finding Load

Step 1: Calculating receptacle load
220-3(b)(9)
200 x 180 VA = 36,000 VA

Step 2: Calculating lighting load
Table 220-3(a); 220-10
15,000 sq. ft. x 2 VA x 100% = 30,000 VA

Step 3: Applying demand factors
220-13; Table 220-11

Total VA = 36,000 VA + 30,000 VA	= 66,000 VA
First 50,000 VA x 40%	= 20,000 VA
Next 16,000 VA x 20%	= 3,200 VA
Total load	= 23,200 VA

Solution: **The load is 23,200 VA.**

ADDITIONAL HELP 17:

Calculating the load for an 18,000 VA multioutlet assembly load and 10,800 VA receptacle load per NEC 220-3(b)(8), (b)(9), and Table 220-13.

- Adding load
  18,000 VA + 66,000 VA = 84,000 VA

- Applying demand factors
  First 50,000 VA x 40% = 20,000 VA
  Next 34,000 VA x 20% = 6,800 VA

Total VA = 26,800

NEC 220-3(b)(9); NEC 210-11(a); NEC 220-13; TABLE 220-13

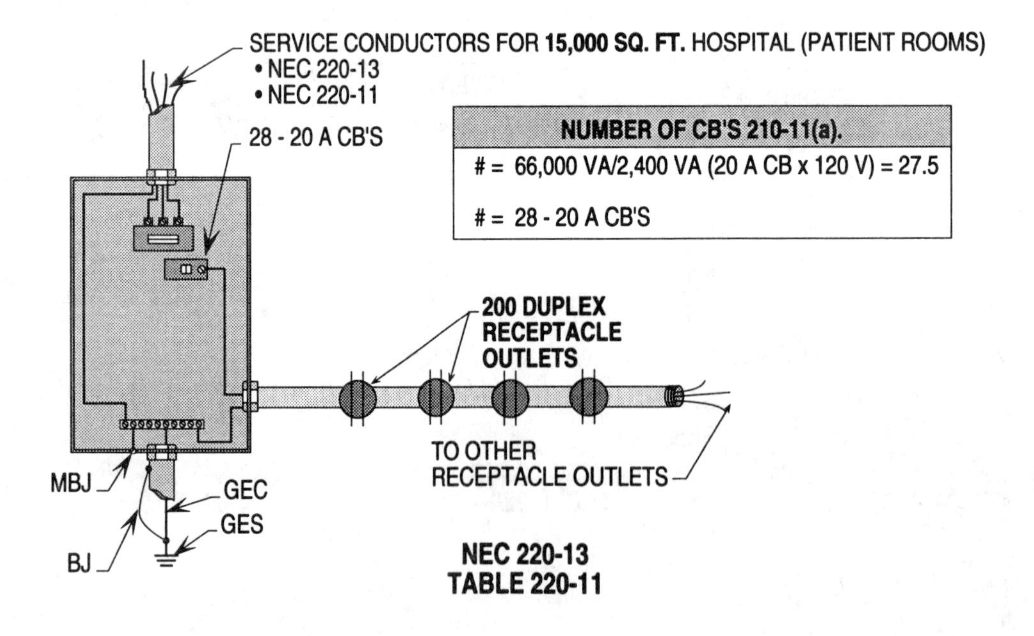

Ill. 2-4(b)

Largest Motor
220-14

The largest motor connected to a feeder or service must be increased 125 percent and added to the remaining motors and other loads per NEC 430-22(a), 430-24, and Part B of Article 220. The largest motor load can also be found by multiplying the largest motor by 25 percent and adding this total to each motor in the group.

Example 2-5. What size conductors, OCPD, overloads and largest motor load are required for the motor in Illustration 2-5?

Sizing conductors

Step 1: Finding FLA of motor
430-6(a)(1); Table 430-148
5 HP = 28 A

Step 2: Calculating load
430-22(a)
28 A x 125% = 35 A

Step 3: Sizing conductors
Table 310-16; 240-3(d)
35 A requires #10

Solution: The size conductors are No. 10 THWN copper.

Sizing overloads

Step 1: Calculating overloads
430-6(a)(2); 430-32(a)(1)
28 A x 115% = 32.2 A

Solution: The size overloads are 32.2 amps.

Sizing OCPD

Step 1: Calculating OCPD
Table 430-152
28 A x 250% = 70 A

Step 2: Sizing OCPD
430-52(c)(1); 240-3(g); 240-6(a)
70 A requires 70 A OCPD

Solution: The size OCPD is 70 amps.

ADDITIONAL HELP 18:

Calculating the largest motor load for an 230 volt, single-phase 1/2 HP, 3/4 HP, 1 HP, 2 HP or 3 HP group of motors on a service or feeder-circuit.

- Selecting motor
  220-14
  5 HP is the largest
- Finding FLA
  Table 430-148
  3 HP = 17 A
- Calculating A
  430-24
  17 A x 25% = 4.25 A

Solution: Largest motor load is 4.25 amps.

Largest Motor Load

Step 1: Calculating load
220-14; 430-24
28 A x 25% = 7 A

Solution: The largest motor load is 7 amps.

Note: This method may be used where there is more than one motor involved on the service or feeder-circuit.

Note: The largest motor load per NEC 220-14 and 430-22(a) is 35 amps per Step 2 in Sizing Conductors and 7 amps for selecting the largest motor load for the service or feeder-circuit.

NOTE 1: TO CALCULATE THE LOAD FOR A SINGLE MOTOR ON A BRANCH-CIRCUIT, FEEDER-CIRCUIT OR SERVICE, SEE NEC 430-22(a).

NOTE 2: TO CALCULATE THE LOAD FOR TWO OR MORE MOTORS ON A FEEDER-CIRCUIT, SEE NEC 430-24.

NOTE 3: TO CALCULATE THE LOAD FOR ONE OR TWO OR MORE MOTORS PLUS OTHER LOADS ON A FEEDER-CIRCUIT OR SERVICE, SEE NEC 430-24.

NOTE 4: TO CALCULATE THE LOAD FOR MOTORS USING DEMAND FACTORS, SEE NEC 430-26.

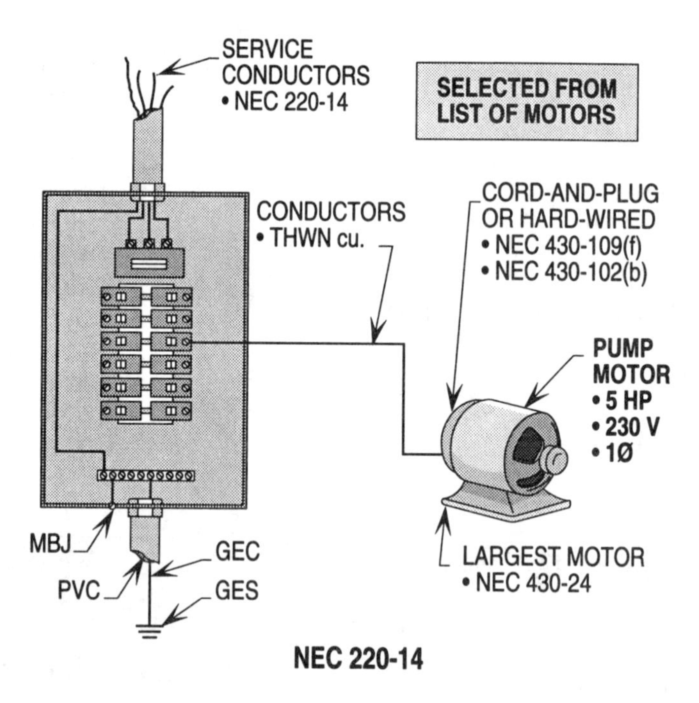

NEC 220-14

NEC 220-3(b)(3)
NEC 220-14
NEC 430-24
NEC 440-34

Ill. 2-5

Small Appliance And Laundry Loads 220-16(a); (b)

The procedure for calculating the small appliance and laundry loads is found by multiplying the 20 amp overcurrent protection device by 120 volts per NEC 210-11(a) and (b) and dividing by 3 VA per sq. ft. from Table 220-3(a).

Example 2-6(a). What is the VA rating for the small appliance circuits in Illustration 2-6(a)?

Finding VA

Step 1: Calculating VA
220-16(a)
1,500 VA x 2 = 3,000 VA

Solution: The rating is 3,000 VA.

Example 2-6(b). How many 20 amp, 2-wire circuits are required for the dwelling unit in Illustration 2-6(b)?

Finding required circuits

Step 1: Finding sq. ft.
210-11(c)(1); (c)(2)

Kitchen	8' x 8'	= 64 sq. ft.
Pantry	6' x 4'	= 24 sq. ft.
Dining	10' x 10'	= 100 sq. ft.
Breakfast	9' x 9'	= 81 sq. ft.
Total		= 269 sq. ft.

Step 2: Finding number of 20 amp small appliance circuits
210-11(a); Table 220-3(a)
20 A OCPD x 120 V ÷ 3 VA = 800 sq. ft.

Step 3: Checking sq. ft. area
210-11(b); Table 220-3(a)

- Two 20 amp small appliance circuits req.
- One 20 amp laundry circuit req. (800 sq. ft.)
- 800 sq. ft. x 2 circuits = 1,600 sq. ft.
- 1,600 sq. ft. is greater than 269 sq. ft.

Solution: Because the 269 sq. ft. area does not exceed 1,600 sq. ft., only two small appliance and one laundry circuit is required.

Note: The method in Example 2-6(b) is used by most designers and installers to determine the number of small appliance circuits required. Check with AHJ to verify the procedure used to determine the proper number required.

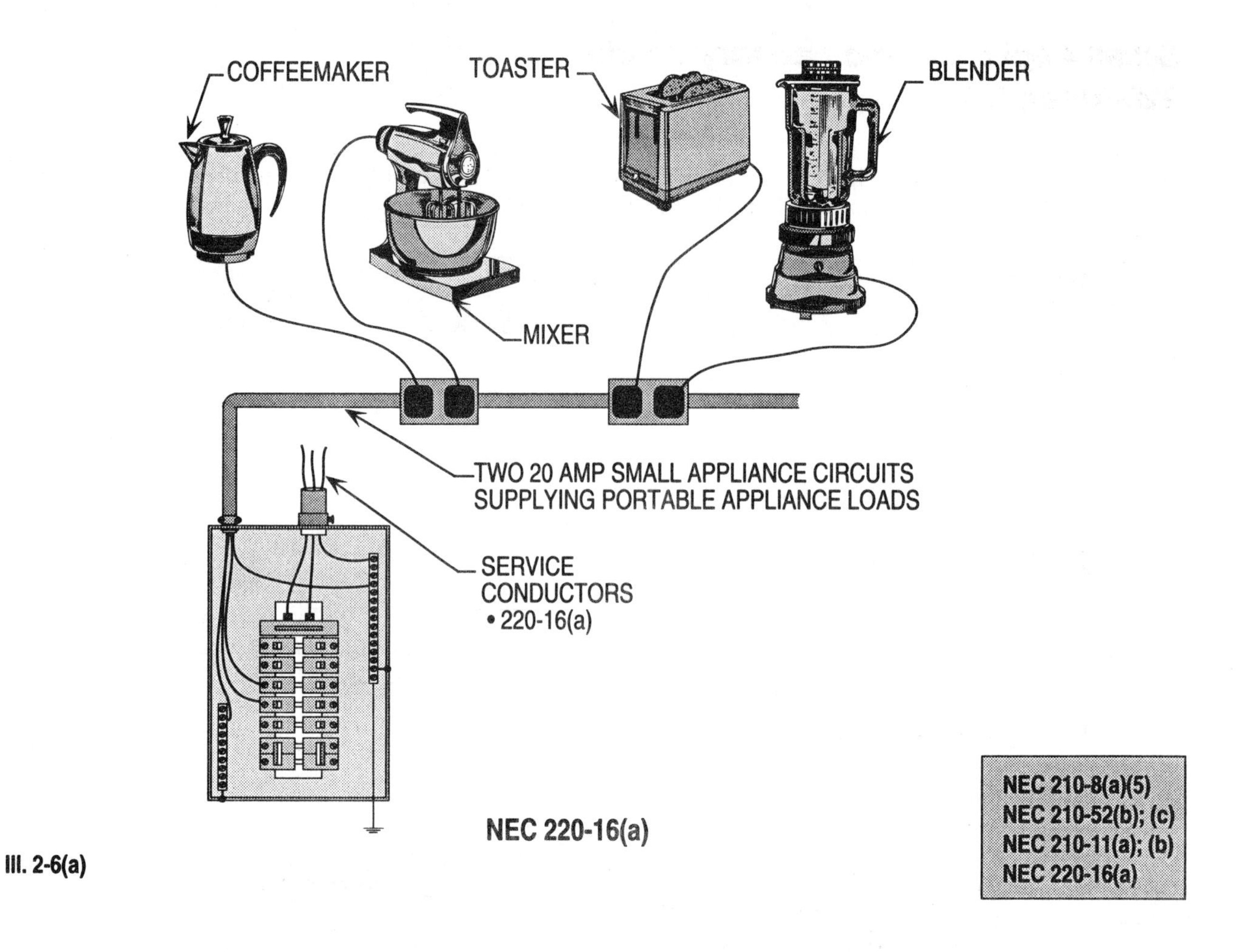

Ill. 2-6(a)

NEC 210-8(a)(5)
NEC 210-52(b); (c)
NEC 210-11(a); (b)
NEC 220-16(a)

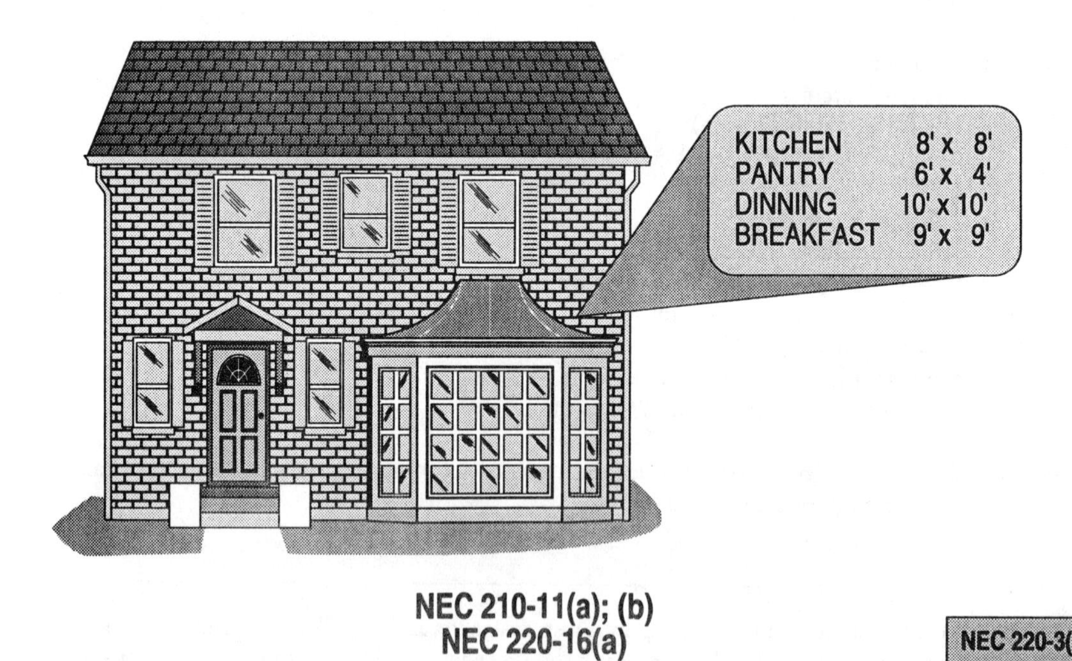

NEC 210-11(a); (b)
NEC 220-16(a)

Ill. 2-6(b)

NEC 220-3(a)
TABLE 220-3(a)
NEC 210-11(a); (b)

Small Appliance And Laundry Loads
220-16(a); (b)

At least one 20 amp small appliance circuit rated at 1,500 VA shall be provided in the calculation for a dwelling unit.

Example 2-6(c). What is the VA load for the laundry circuit in Illustration 2-6(c)?

Finding Load

Step 1: Calculating VA
220-16(b)
1,500 VA x 1 = 1,500 VA

Solution: The rating is 1,500 VA.

ADDITIONAL HELP 19:

Calculating the area that a 20 amp small appliance circuit will serve per Table 220-3(a) and NEC 210-11(a).

- $\frac{20\text{ A CB} \times 120\text{ V}}{3} = 800$

Solution: Area served is 800 sq. ft..

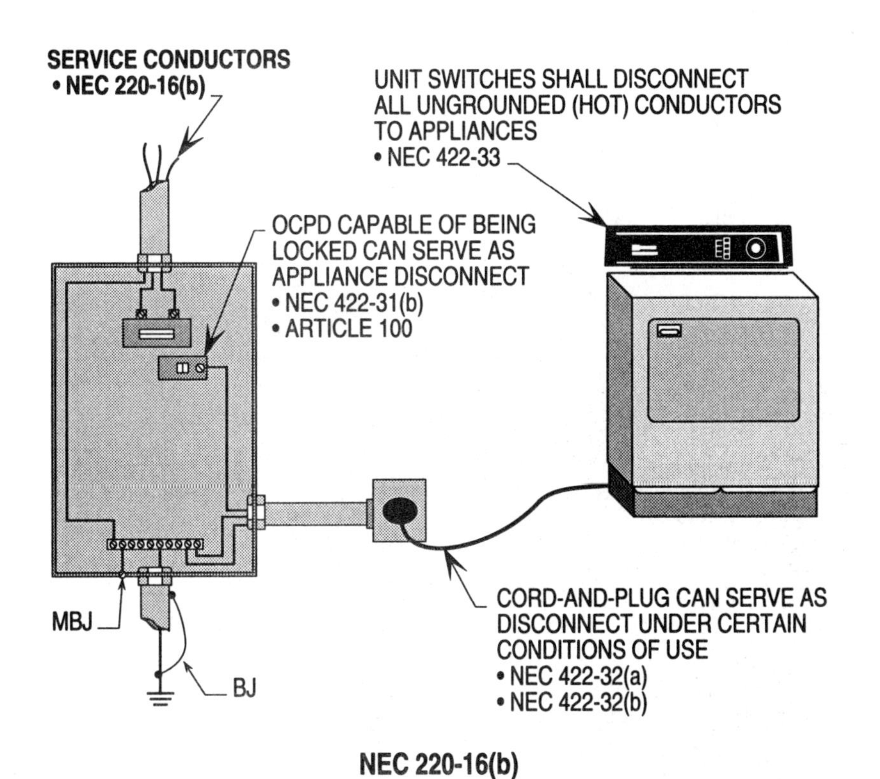

NEC 220-16(b)

Ill. 2-6(c)

Appliance Load - Dwelling Unit(s)
220-17

The procedure for calculating the fixed appliance load is determined by multiplying the VA rating of four or more fixed appliances by 75 percent. Less than three must be figured at 100 percent. NEC 220-17 does not allow heating units, air-conditioner units, dryer or cooking equipment loads to have this demand factor of 75 percent applied.

Example 2-7. What is the fixed appliance load for the appliances in Illustration 2-7?

Finding Load

Step 1: Total all loads
220-17

6.0 kW water heater x 1,000	=	6,000 VA
1.5 kW disposal x 1,000	=	1,500 VA
1.2 kW compactor x 1,000	=	1,200 VA
1.4 kW dishwasher x 1,000	=	1,400 VA
1.5 kW circulating pump x 1,000	=	1,500 VA
Total load	=	11,600 VA

Step 2: Applying demand factors
220-17
11,600 VA x 75% = 8,700 VA

Solution: The fixed appliance load is 8,700 VA.

NEC 210-19(a); NEC 422-16(b)(1)(a), Ex.; NEC 422-16(b)(2)(a), Ex.; NEC 422-16(b)(3); NEC 422-13; NEC 430-22(a); NEC 220-17

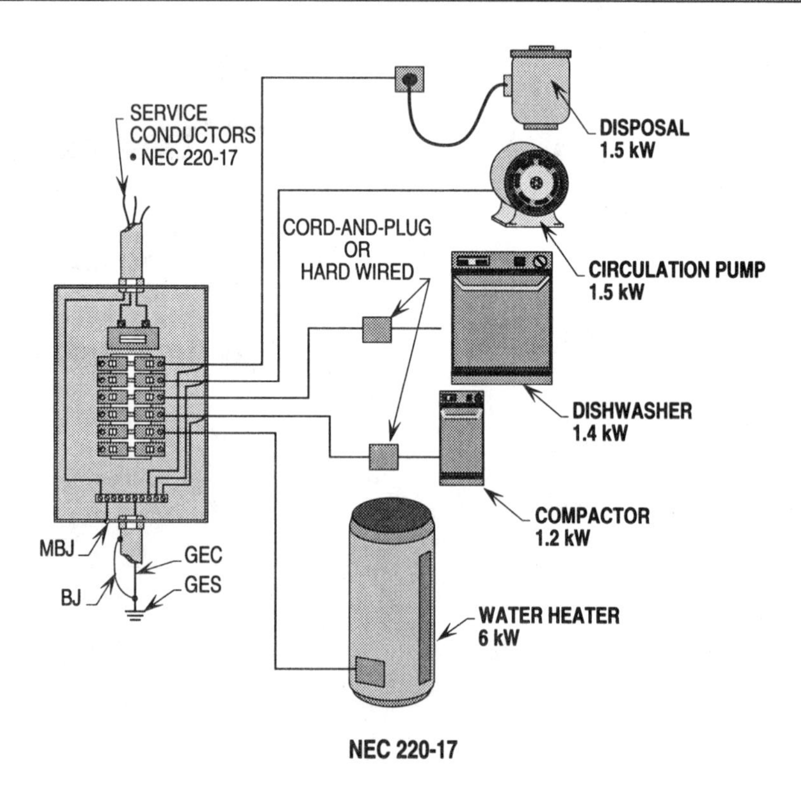

Ill. 2-7

Electric Clothes Dryers
220-18

The procedure for calculating the dryer load is determined by the kW rating of each dryer. Dryers with a kW rating of less than 5,000 W must have their rating increased to 5,000 VA. Dryers with a wattage rating greater than 5,000 VA are computed with the greater rating. Table 220-18 requires four or fewer dryers to be computed at 100 percent of their kW rating.

Example 2-8. What is the dryer load for the dryer in Illustration 2-8?

Finding Load

Step 1: Finding the load
Table 220-18
Four or fewer dryers = 100%

Step 2: 220-18
5 kW = 5 kVA
5 kVA x 1,000 = 5,000 VA

Solution: The dryer load is 5,000 VA.

ADDITIONAL HELP 20:

Calculating load for 3Ø supply.

- Finding dryers per phase
  220-18
  # = 15 dryers ÷ 3Ø
  # = 5
- Finding No. of dryers
  220-18
  # = 5 dryers per phase x 2 (phases)
  # = 10 dryers
- Finding kW rating
  220-18; Table 220-18
  5 kW x 10 x 50% = 25 kW
- Finding VA rating
  220-18
  25 kW x 1,000 = 25,000 VA
- Finding service load
  220-18
  VA = 25,000 VA ÷ 2 (phases) x 3Ø
  VA = 37,500

Solution: Service load is 37,500 VA.

NEC 220-3(b)(2); NEC 220-18; TABLE 220-18

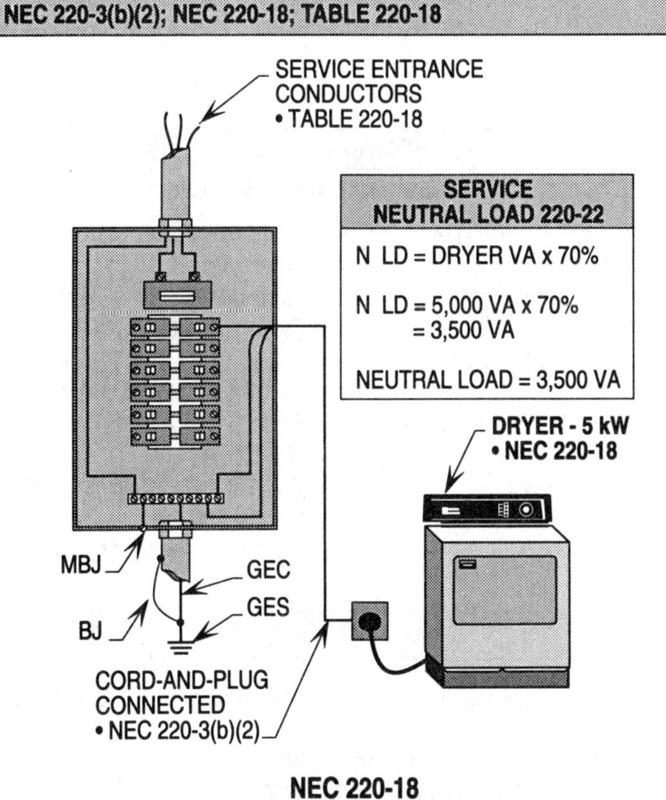

ADDITIONAL HELP 21:

Calculating load for six, 5,000 VA dryers per Table 220-18.

- 5,000 VA x 6 x 70% = 21,000 VA

Solution: The demand load is 21,000 VA

Ill. 2-8

Electric Ranges And Other Cooktop Appliances - Dwelling Unit(s) 220-19

The procedure for calculating the branch-circuit loads for ranges, cooktops and ovens are determined by applying the demand factors in Table 220-19. The demand factors are listed in Column's A, B, and C and are based on the size of the range, cooktop, or oven. When the kW rating exceeds 12, one of the Footnotes is used in conjunction with Table 220-19. Note: For the minimum size neutral for a branch-circuit, see NEC 210-19(c), Ex. 2.

Example 2-9(a). What is the demand load for the range in Illustration 2-9(a)?

Finding Demand Load

Step 1: Calculating the demand load
Table 220-19, Col. C
8.5 kW x 80% = 6.8 kVA

Solution: The demand load is 6.8 kW.

Note: 6.8 kVA x 1,000 = 6,800 VA

NEC 422-32(a); (b); NEC 422-31(b); NEC 422-32(b)

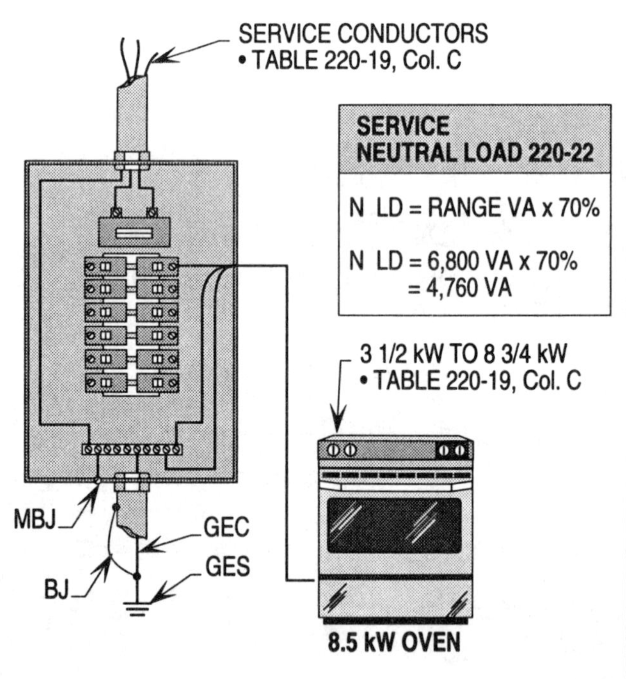

TABLE 220-19, Col. C

ADDITIONAL HELP 22:

Calculating load for 3Ø supply.

- Finding ranges per phase
  220-18
  # = 15 ranges ÷ 3Ø
  # = 5
- Finding No. of ranges
  220-19
  # = 5 ranges per phase x 2 (phases)
  # = 10 ranges
- Finding kW rating
  220-19, Col. A
  10 ranges = 25 kW
- Finding VA rating
  220-19
  25 kW x 1,000 = 25,000 VA
- Finding service load
  220-19
  VA = 25,000 ÷ 2 (phases) x 3Ø
  VA = 37,500 VA

Solution: Service load is 37,500 VA.

ADDITIONAL HELP 23:

Calculating the demand load for 25 - 12 kW ranges per Table 220-19.

- 25 - 12 kW ranges requires 40 kW

Solution: The demand load is 40 kW.

Ill. 2-9(a)

Electric Ranges And Other Cooktop Appliances-Dwelling Unit(s) 220-19

Demand factors may be applied on cooking equipment used in dwelling units by applying Columns A, B, or C based upon kW ratings.

Example 2-9(b). What is the demand load for the oven in Illustration 2-9(b)?

Finding Demand Load

Step 1: Calculating the demand load
Table 220-19, Col. B
2.5 kW x 80% = 2 kW

Solution: The demand load is 2 kVA.

ADDITIONAL HELP 24:
Calculating VA

- 2 kVA x 1,000 = 2,000 VA

Solution: The load is 2,000 VA.

Example 2-9(c). What is the demand load for the range in Illustration 2-9(c)?

Finding Demand Load

Step 1: Calculating the demand load
Table 220-19, Col. A
10.5 kW range = 8 kW

Solution: The demand load is 8 kVA.

ADDITIONAL HELP 25:
Calculating Amps

- $A = \frac{kW \times 1,000}{V}$
- $A = \frac{8 \text{ kW} \times 1,000}{240 \text{ V}} = 33 \text{ A}$

Solution: The load is 33 amps.

Example 2-9(d). What is the demand load for the range in Illustration 2-9(d)?

Finding Demand Load

Step 1: Calculating percentage
Table 220-19, Note 1
18 kW - 12 kW = 6 kW
6 kW x 5% = 30%

Step 2: Calculating demand load
Table 220-19, Note 1
8 kVA x 130% = 10.4 kVA

Solution: The demand load is 10.4 kVA.

ADDITIONAL HELP 26:
Calculating VA and sizing conductors.

- $A = \frac{kVA \times 1,000}{V}$
- $A = \frac{10.4 \text{ kVA} \times 1,000}{240 \text{ V}} = 43 \text{ A}$
- 43 A requires #6 cu. per NEC 338-4(a)

Solution: The size conductors are No. 6 cu..

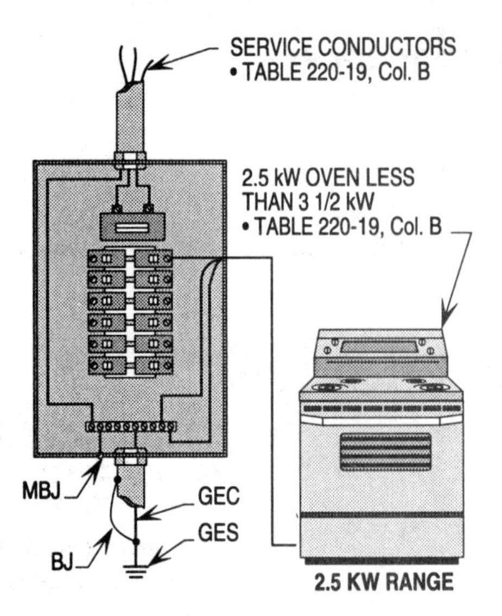

ADDITIONAL HELP 27:
Calculating the demand load for 30 - 10 kW ranges per Table 220-19. • 30 ranges requires 15 kW + 30 kW (1 kW for each range) = 45 kW **Solution: The demand load is 45 kVA.**

NEC 422-32(a); (b)
NEC 422-31(b)
NEC 422-32(b)

Ill. 2-9(b)

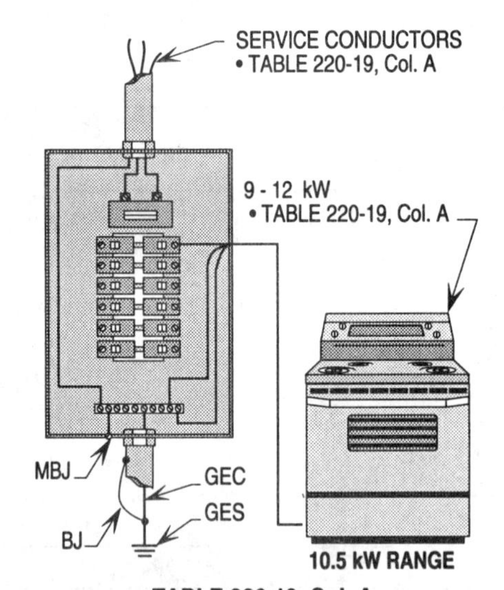

NEC 422-32(a); (b)
NEC 422-31(b)
NEC 422-32(b)

Ill. 2-9(c)

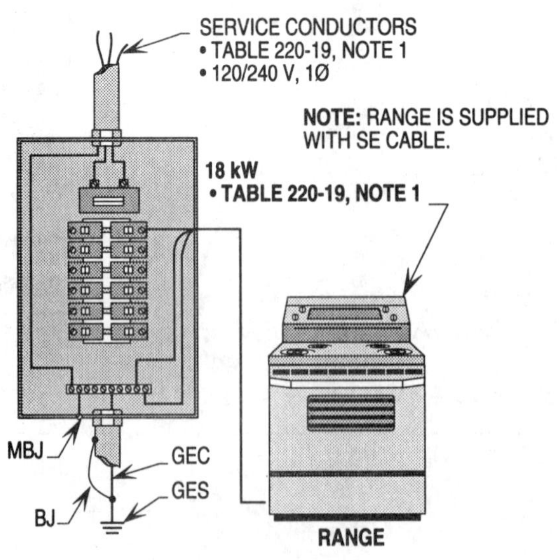

NEC 422-32(a); (b)
NEC 422-31(b)
NEC 422-32(b)

Ill. 2-9(d)

Electric Ranges And Other Cooktop Appliances-
Dwelling Unit(s)
220-19

The demand factors for cooking units exceeding 12 kW are determined by applying one of the ratings to Table 220-19.

Example 2-9(e). What is the demand load for the three ranges in Illustration 2-9(e)?

Finding Demand Load

Step 1: Calculating percentage
Table 220-19, Note 2
12 kW + 18 kW + 24 kW ÷ 3 = 18 kW
18 kW - 12 kW = 6 kW
6 kW x 5% = 30%

Step 2: Calculating demand load
Table 220-19, Note 2
14 kVA x 130% = 18.2 kVA

Solution: The demand load is 18.2 kVA

ADDITIONAL HELP 28:

Calculating the demand load for 50 - 12 kW ranges per Table 220-19.

- 50 ranges requires 25 kW + 37.5 kW (3/4 kW for each range) = 62.5 kW

Solution: The demand load is 62.5 kVA.

Example 2-9(f). What is the demand load for the three pieces of cooking equipment in Illustration 2-9(f) used as one unit?

Finding Demand Load

Step 1: Calculating percentage
Table 220-19, Note 4
8 kW + 10 kW + 12 kW = 30 kW
30 kW - 12 kW = 18 kW
18 kW x 5% = 90%

Step 2: Calculating demand load
Table 220-19, Note 1
8 kVA x 190% = 15.2 kVA

Solution: The demand load is 15.2 kVA.

ADDITIONAL HELP 29:

Calculating the demand load for 10 - 3 kW ranges and 10 - 8 kW ranges per Table 220-19.

- 10 x 3 kW x 49% = 14.7 kW
- 10 x 8 kW x 34% = 27.2 kW
- 14.7 kW + 27.2 kW = 41.9 kVA

Solution: The demand load is 41.9 kVA.

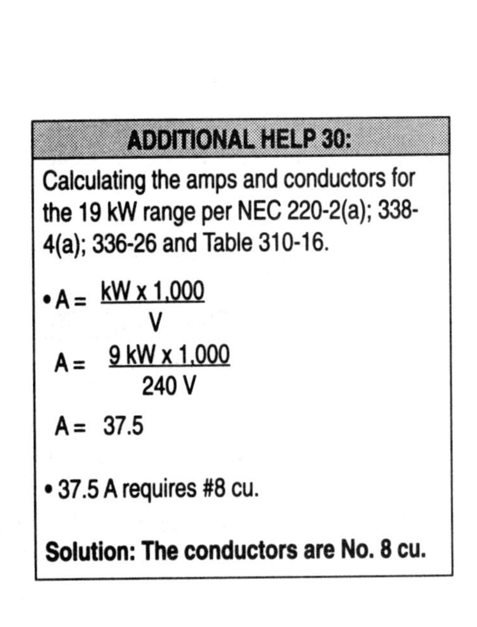

ADDITIONAL HELP 30:

Calculating the amps and conductors for the 19 kW range per NEC 220-2(a); 338-4(a); 336-26 and Table 310-16.

- $A = \frac{kW \times 1,000}{V}$

$A = \frac{9 \text{ kW} \times 1,000}{240 \text{ V}}$

$A = 37.5$

- 37.5 A requires #8 cu.

Solution: The conductors are No. 8 cu.

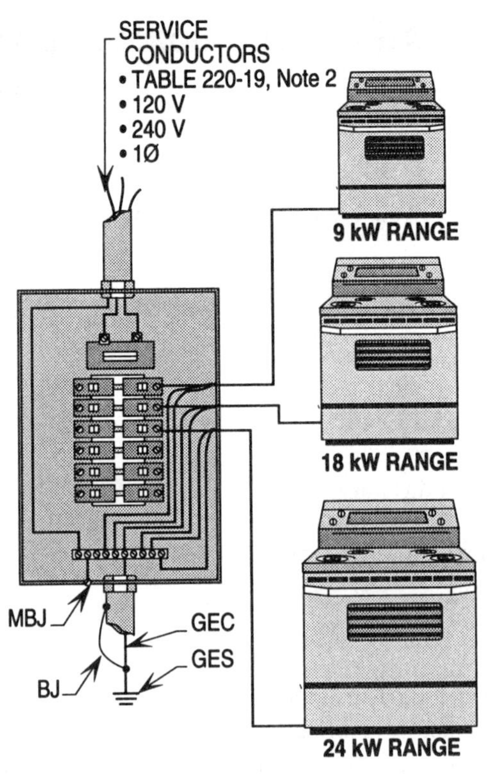

NEC 422-32(a); (b)
NEC 422-31(b)
NEC 422-32(b)

Ill. 2-9(e)

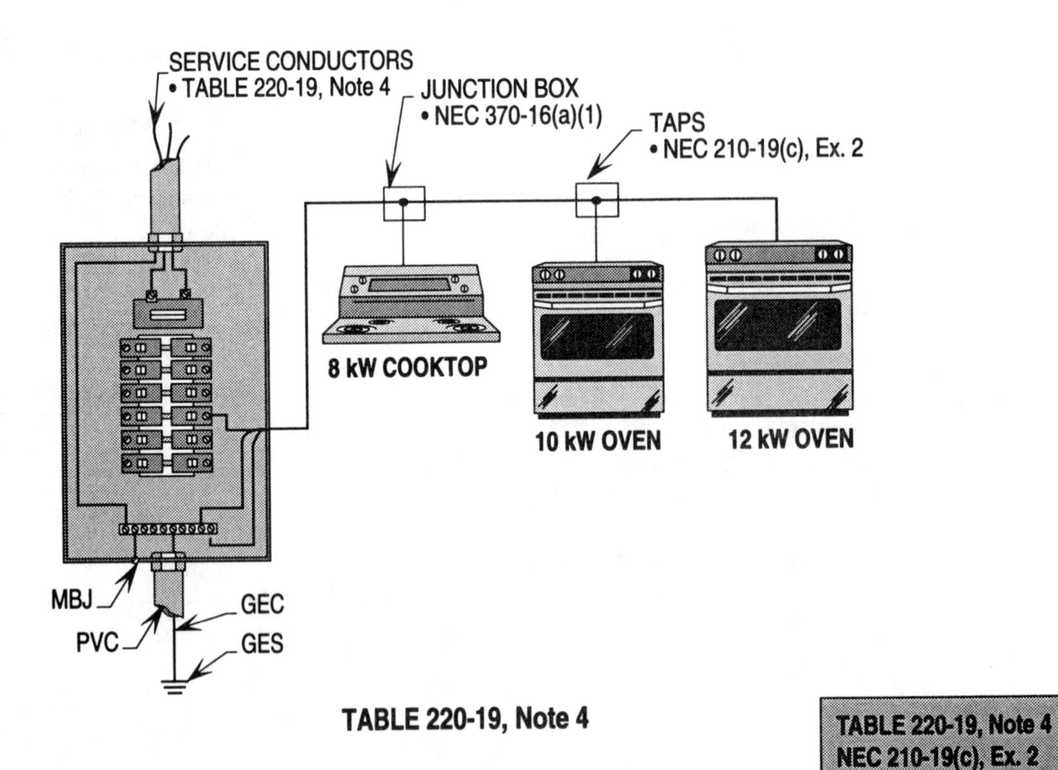

TABLE 220-19, Note 4
NEC 210-19(c), Ex. 2
NEC 220-3(b)(2)

Ill. 2-9(f)

Kitchen Equipment - Other Than Dwelling Units 220-20

The demand factors of Table 220-20 may be used to compute the load for commercial kitchen equipment such as ovens, grills, kettles, fryers, steamers, etc. However, the demand load must be equal to the two largest loads of the kitchen equipment?

Example 2-10. What is the demand load for the pieces of cooking equipment in Illustration 2-10?

Finding Demand Load

Step 1: Calculating percentage
Table 220-20
17 pieces allows 65%

Step 2: Applying demand factors
Table 220-20; 220-20
82 kW x 65% = 53.3 kVA

Solution: The demand load is 53.3 kVA.

Note: 53.3 kVA is greater than the two largest loads of 18 kVA.

ADDITIONAL HELP 31:

Calculating the load in amps for 277/480 volt service.

- $A = \frac{\text{kVA} \times 1{,}000}{\text{V} \times 1.732}$

$A = \frac{53.3 \text{ kVA} \times 1{,}000}{480 \text{ V} \times 1.732}$

$A = 64$

Solution: The load is 64 amps.

NEC 220-20; Table 220-20

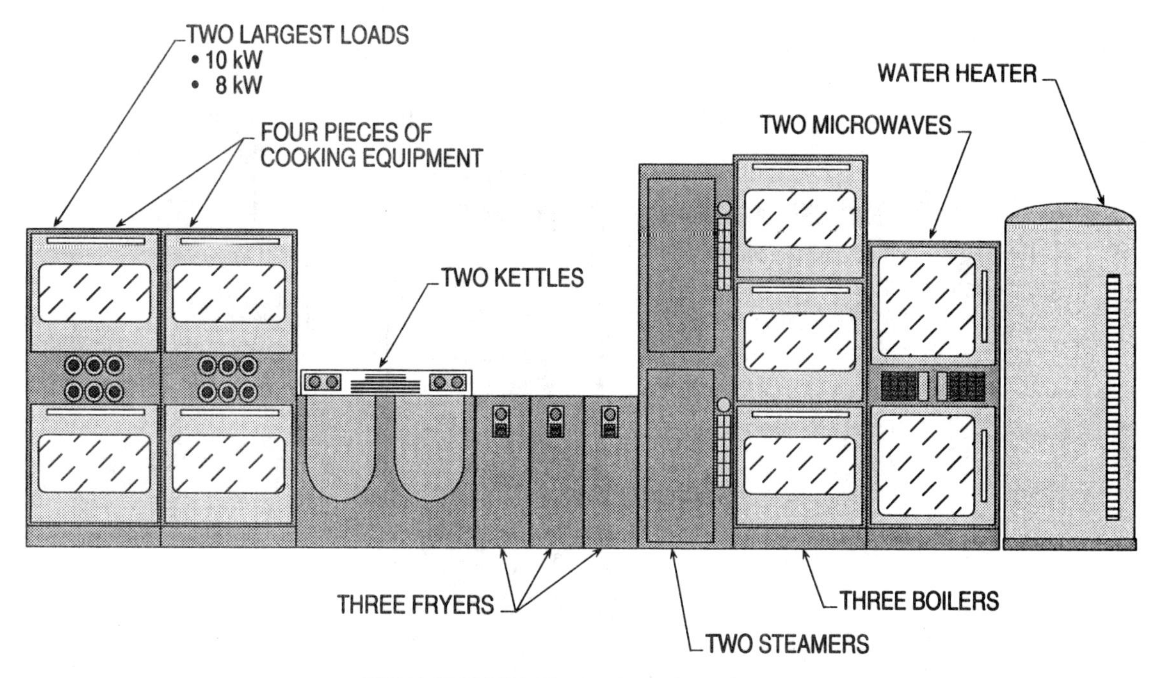

Ill. 2-10

Noncoincident Loads
220-21

The procedure for calculating the noncoincident load is determined between the largest load of the heat and A/C. The heating and A/C loads are computed at 100 percent and the smaller of the two loads is dropped. Heat pumps operating with the heating load is added to the heating load at 100 percent. **Note:** The AHJ may allow the A/C unit to be considered as the largest motor even when it's dropped per NEC 220-21.

Example 2-11. Determine the largest load for the noncoincident load in Illustration 2-11?

Sizing largest HTG. and AC load

Step 1: 220-21
Heating load
10 kW x 100% = 10 kVA
A/C load
5.5 kW x 100% = 5.5 kVA

Note: The 1999 NEC allows more than one load to be dropped per NEC 220-21.

Solution: **Section 220-21 requires 10 kVA heating load to be selected for the largest load between the heating and A/C load.**

NEC 424-3(b); NEC 440-22(a); NEC 440-32

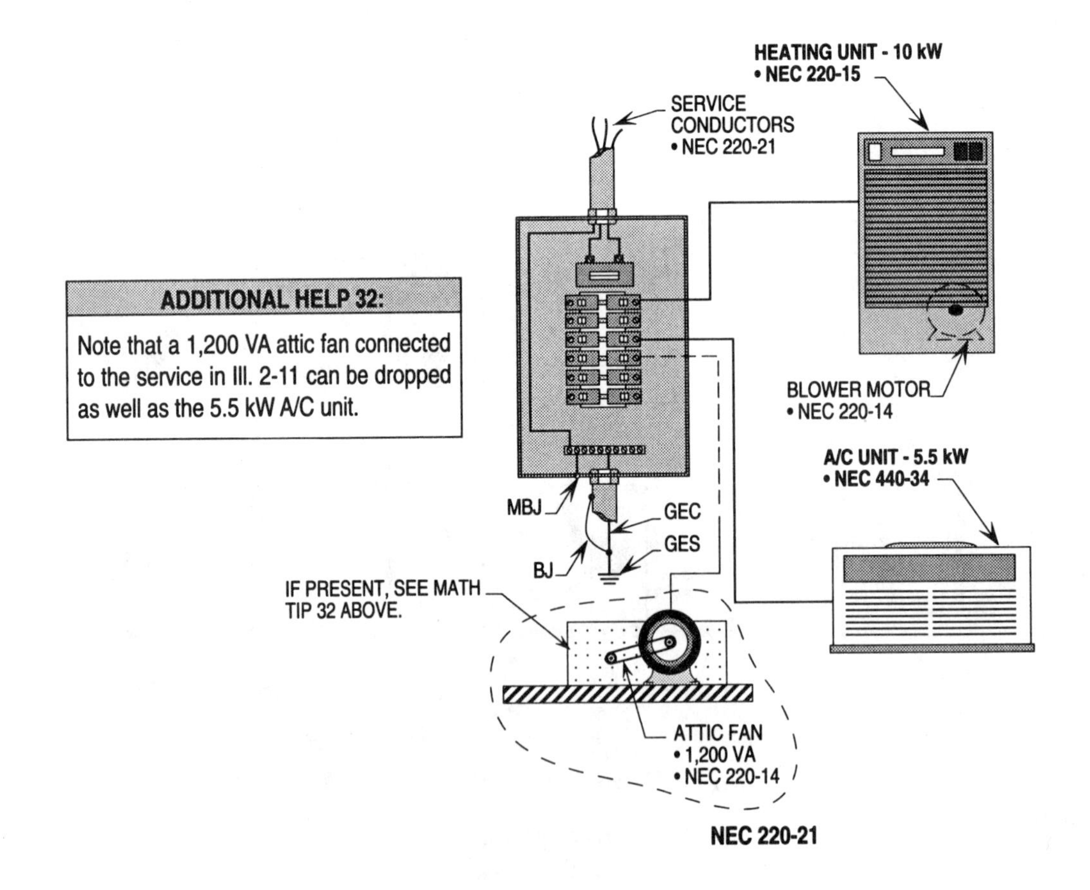

Ill. 2-11

Feeder Neutral Load
220-22; 310-15(b)(4)(a)

The feeder neutral load shall be the maximum unbalanced load connected between the neutral and any one ungrounded conductor. The feeder neutral load shall also be 70 percent of the demand load for cooking equipment or a dryer load. See Figures 2-8 and 2-9(a) for calculating neutral load for cooking equipment and dryers.

Example 2-12(a). Using Illustration 2-12(a), find the following neutral load values?

- Question 12(a)(1) - Neutral load between phases A and B
- Question 12(a)(2) - Neutral load if phase A is lost
- Question 12(a)(3) - Neutral load if 18 amps were removed from phase B

Finding Neutral Load

Question 12(a)(1)

Step 1: Calculating neutral load
220-22
240 V Ld. of 325 A = 0
120 V Ld. of 168 A = 168 A
120 V Ld. of 150 A = 150 A

Step 2: Maximum neutral load
220-22; Table 310-15(b)(4)(a)
Phase B = 168 A
Phase A = 150 A
Total load = 18 A

Solution: The maximum unbalanced neutral load is 18 amps.

Question 12(a)(2)

Step 1: Calculating neutral load
220-22
240 V Ld. of 325 A = 0
120 V Ld. of 168 A = 168 A
120 V Ld. of 150 A = 150 A

Step 2: Maximum neutral load
220-22; Table 310-15(b)(4)(a)
Phase A is lost = 0 A
Phase B = 168 A
Total load = 168 A

Solution: The maximum unbalanced load is 168 amps.

Question 12(a)(3)

Step 1: Calculating neutral load
220-22
240 V Ld. of 325 A = 0
120 V Ld. of 168 A = 168 A
120 V Ld. of 150 A = 150 A

Step 2: Maximum neutral load
220-22
Phase B = (168 - 18) = 150 A
Phase A = 150 A
Total load = 0

Solution: The maximum unbalanced neutral load is zero amps.

NOTE 1: IF THE MAJOR PORTION OF THE NEUTRAL LOAD CONSIST OF NONLINEAR LOADS SUCH AS HARMONIC PRODUCING CURRENTS, THE NEUTRAL ON A FOUR-WIRE, THREE-PHASE WYE SYSTEM IS CONSIDERED CURRENT-CARRYING PER NEC 310-15(b)(4)(c) AND 220-22, FPN 2. (TO CALCULATE SUCH A NEUTRAL, SEE PAGE 2-31)

NOTE 2: THE NEUTRAL WIRE NEVER CARRIES ANY 208 VOLT, 240 VOLT OR 480 VOLT SINGLE-PHASE OR THREE-PHASE LOADS. THEREFORE, CALCULATE ONLY 120 VOLT, 277 VOLT OR ANY OTHER NEUTRAL RELATED LOADS WHEN CALCULATING THE SIZE NEUTRAL LOAD.

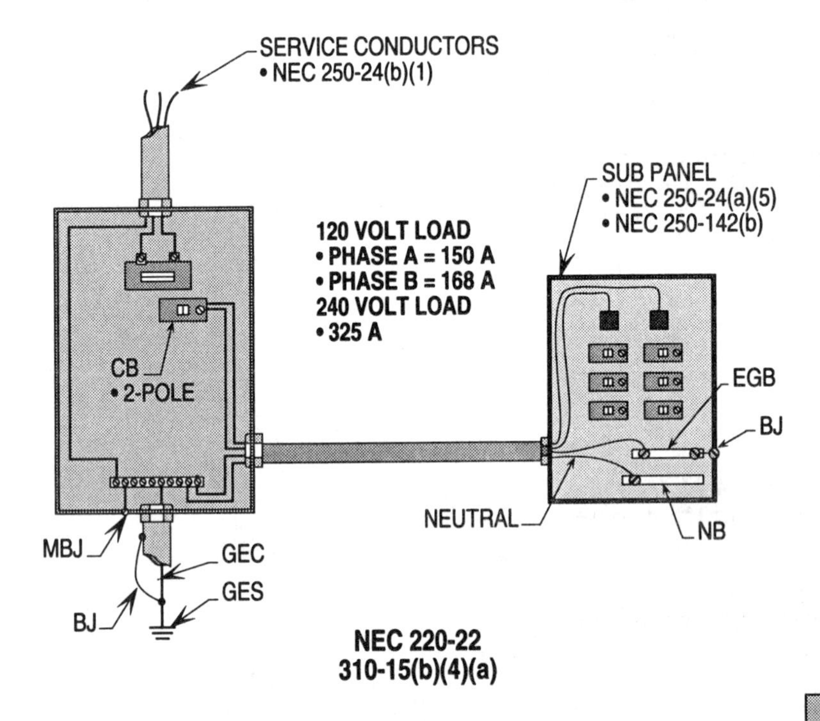

NEC 220-22
NEC 310-15(b)(4)(a)
NEC 250-24(b)

Ill. 2-12(a)

Feeder Neutral Load
220-22

The first 200 amps of neutral current shall be calculated at 100 percent. All resistive loads on the neutral exceeding 200 amps may have a demand factor of 70 percent applied and this value added to the first 200 amps taken at 100 percent.

Example 2-12(b). What is the neutral load in amps in Illustration 2-12(b)?

Finding Neutral Load

Step 1: Calculating neutral load
220-22; 310-15(b)(4)(a)
200 A x 100% = 200 A

Solution: The neutral load is 200 amps.

NEC 220-22; NEC 310-15(b)(4)(a); (b); (c); NEC 250-24(b)

Example 2-12(c). What is the neutral load in amps in Illustration 2-12(c)?

Finding Neutral Load

Step 1: Calculating neutral load
220-22
First 200 A x 100% = 200 A
Next 200 A x 70% = 140 A
Total load = 340 A

Solution: The neutral load is 340 amps.

NEC 220-22; NEC 310-15(b)(4)(a); (b); (c); NEC 250-24(b)

Determine if the neutral is current-carrying by calculating the amps for a four-wire, three-phase wye circuit having a neutral load of 100 amps per phase.

Determine if the neutral is current-carrying

Step 1: Calculating amps
220-22, FPN 2; 310-15(b)(4)(c)
100 A x 51% = 51 amps

Solution: The neutral is considered current-carrying.

Note that the major portion is really any amps over fifty percent. However, fifty-one percent is usually used to get away from applying one-quarter (1/4) percent, one-half (1/2) percent, three quarter (3/4) percent etc. Check with your AHJ for their point of view if this becomes an issue.

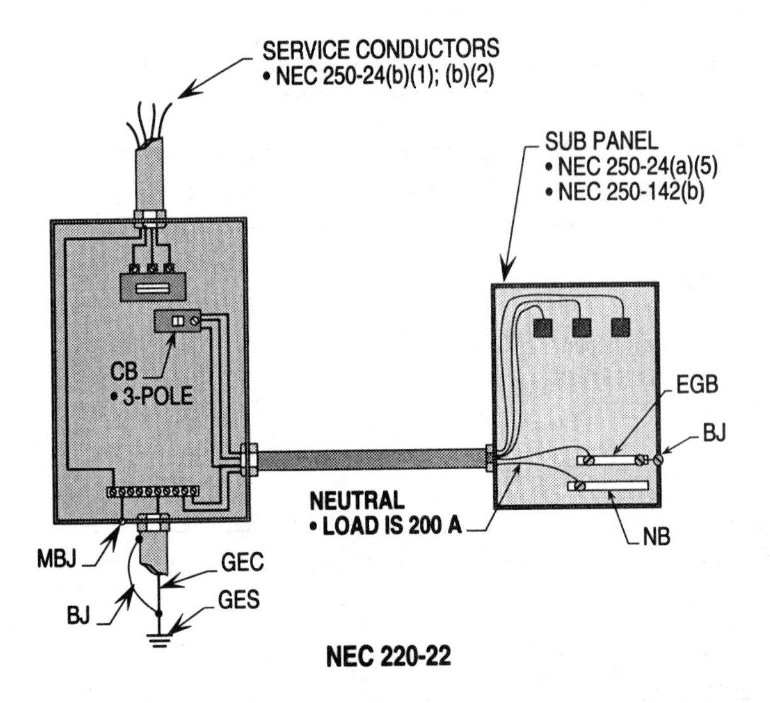

Ill. 2-12(b)

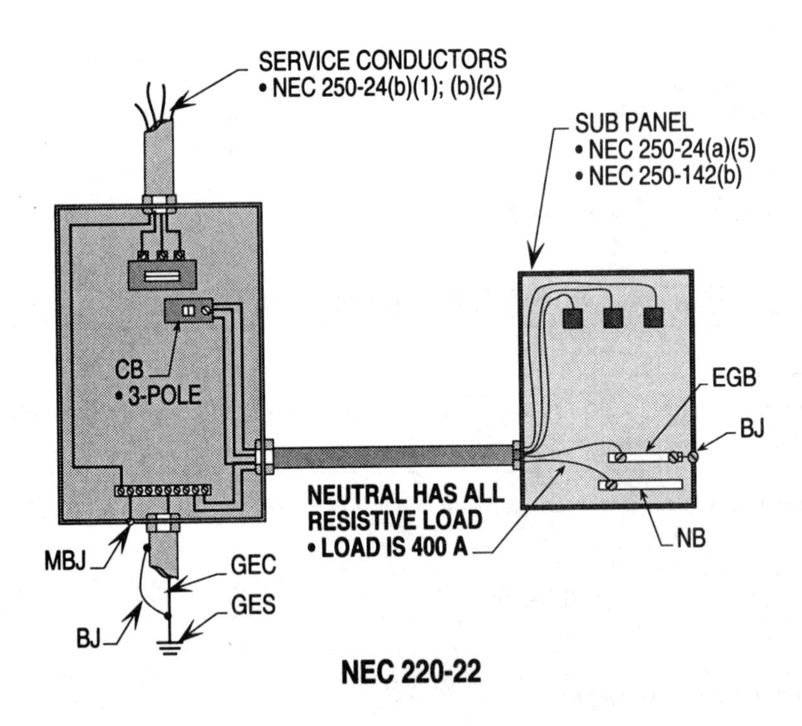

Ill. 2-12(c)

Feeder Neutral Load
220-22

All inductive neutral current shall be computed at 100 percent with no demand factors applied. To determine neutral currents in three-phase systems a special formula must be applied.

Example 2-12(d). What is the neutral load in amps in Illustration 2-12(d)?

Finding Neutral Load

Step 1: Calculating neutral load
220-22; 310-15(b)(4)(c)

First 200 A x 100%	= 200 A
Next 100 A x 100%	= 100 A
Total load	= 300 A

Solution: The neutral load is 300 amps.

Note: There is no reduction of the neutral load due to the predominate inductive load.

Example 2-12(e). What is the maximum size neutral load in amps in Illustration 2-12(e)?

Finding Neutral Load

Step 1: Calculating neutral load
220-22; 310-15(b)(4)(a)

$$A = \sqrt{PH\,A^2 + PH\,B^2 + PH\,C^2 - (PH\,A \times PH\,B) - (PH\,B \times PH\,C) - (PH\,C \times PH\,A)}$$

$$A = \sqrt{125^2 + 90^2 + 110^2 - (125\text{ A} \times 90\text{ A}) - (90\text{ A} \times 110\text{ A}) - (110\text{ A} \times 125\text{ A})}$$

$$A = \sqrt{15{,}625\text{ A} + 8{,}100\text{ A} + 12{,}100\text{ A} - 11{,}250\text{ A} - 9{,}900\text{ A} - 13{,}750\text{ A}}$$

$$A = \sqrt{35{,}825\text{ A} - 34{,}900\text{ A}}$$

$$A = \sqrt{925\text{ A}}$$

ADDITIONAL HELP 33:

Calculating the load on the neutral in Figure 2-12(d) to determine if it is current-carrying.

- Applying Rule of 310-15(b)(4)(c)
  200 A x 51% = 102 A

Solution: The neutral current is equal to 102 amps. However, if the neutral carries 51or more amps of harmonic currents, the neutral is considered current-carrying.

Note 2: Any current in amps that is greater than 50 percent is really a major portion. However, 51 percent is usually used. Check with the AHJ for 51 percent rule.

Solution: The neutral load is 30 amps.

Note 1: If phase A and B were lost, the neutral current would be 110 amps and if phase C were lost the neutral current would be 35 amps (A = 125 A - PH B = 90 A = 35 A).

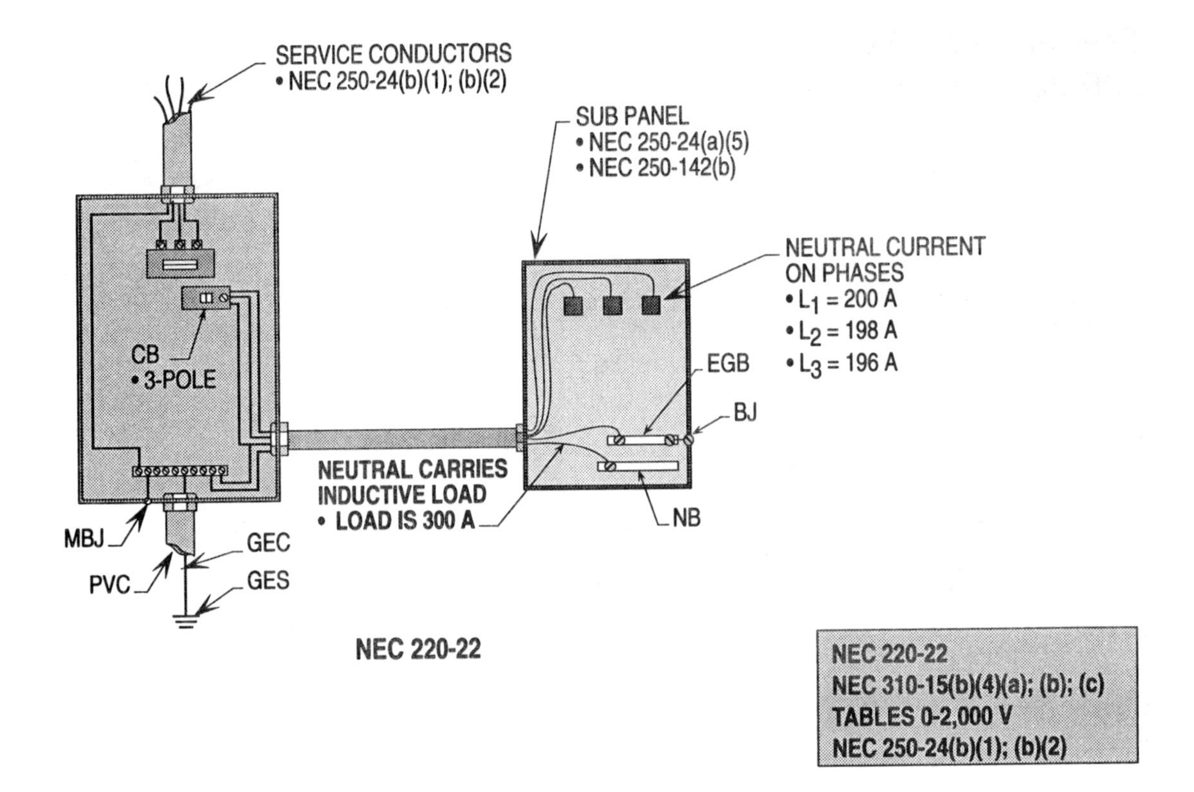

Ill. 2-12(d)

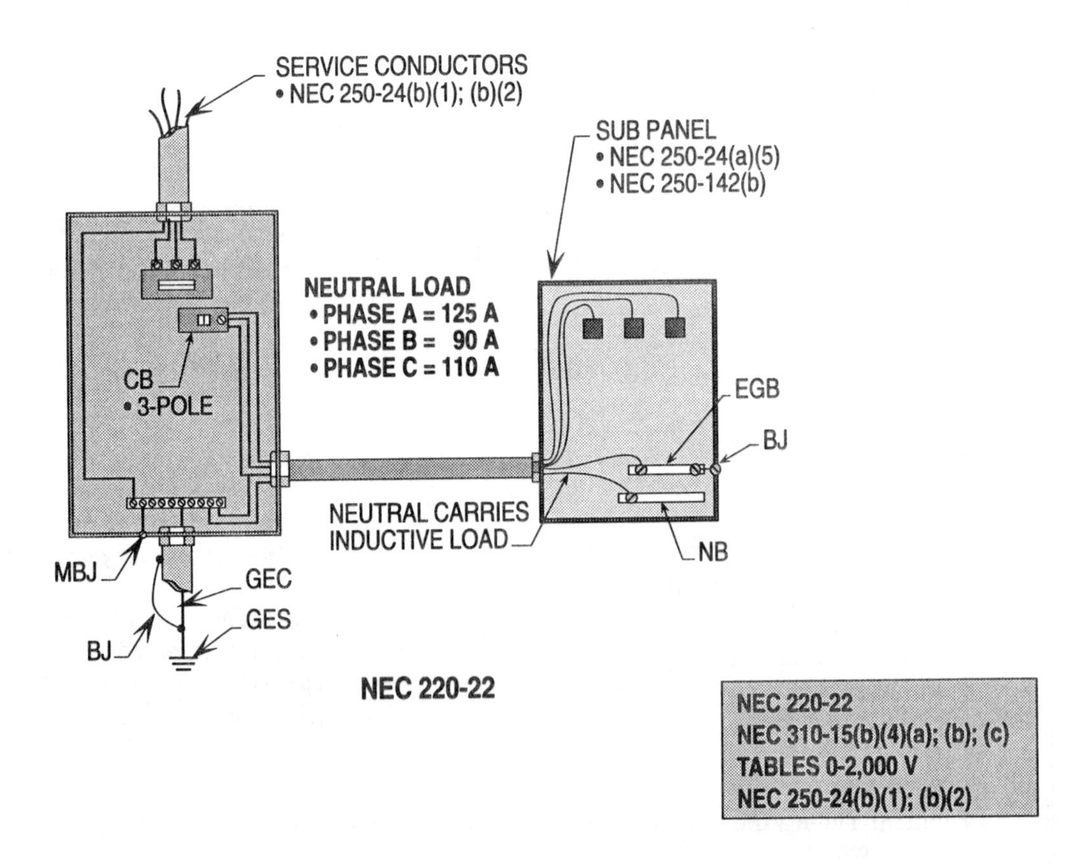

Ill. 2-12(e)

Feeder Neutral Load
220-22; 310-15(b)(4)(c)

Neutral loads which contain harmonic currents add unwanted heating effects to the neutral conductor to components. For this unwanted heating effect larger neutral conductors are sometimes used.

Example 2-12(f). What is the total watts in the neutral conductors in Illustration 2-12(f) that causes the neutral to heat due to harmonic currents?

Finding the watts for one neutral

Step 1: Calculating A per Phase on neutral
220-22; 310-15(b)(4)(c)
Total A per phase = 4.5 x 3 phases
Total A per phase = 13.5 A

Step 2: Calculating watts on N
220-22; 310-15(b)(4)(c)
Watts = I^2 x R
Watts = 13.5 A x 13.5 A x 0.15R
Watts = 27

Solution: The total watts on the single neutral is 27 watts due to harmonic currents.

Finding the watts for each neutral

Step 1: Calculating A per phase on N
220-22; 310-15(b)(4)(c)
Total A per phase = 4.5 A x 3 phases
Total A per phase = 13.5 A

Step 2: Calculating A per N
220-22; 310-15(b)(4)(c)
amps per N = 13.5 A per phase /
3 phases amps per N = 4.5

Step 3: Calculating watts on N's
Watts per N = I^2 x R
Watts per N = 4.5 A x
4.5 A x 0.15R
Watts per N = 3

Solution: The total watts on each neutral is 3 watts due to harmonic currents.

Note: It is the watts flowing in the neutral, due to harmonic currents, that creates extra heat in the neutral by pulling a neutral for each phase, the watts and the heating effect in the neutral are reduced.

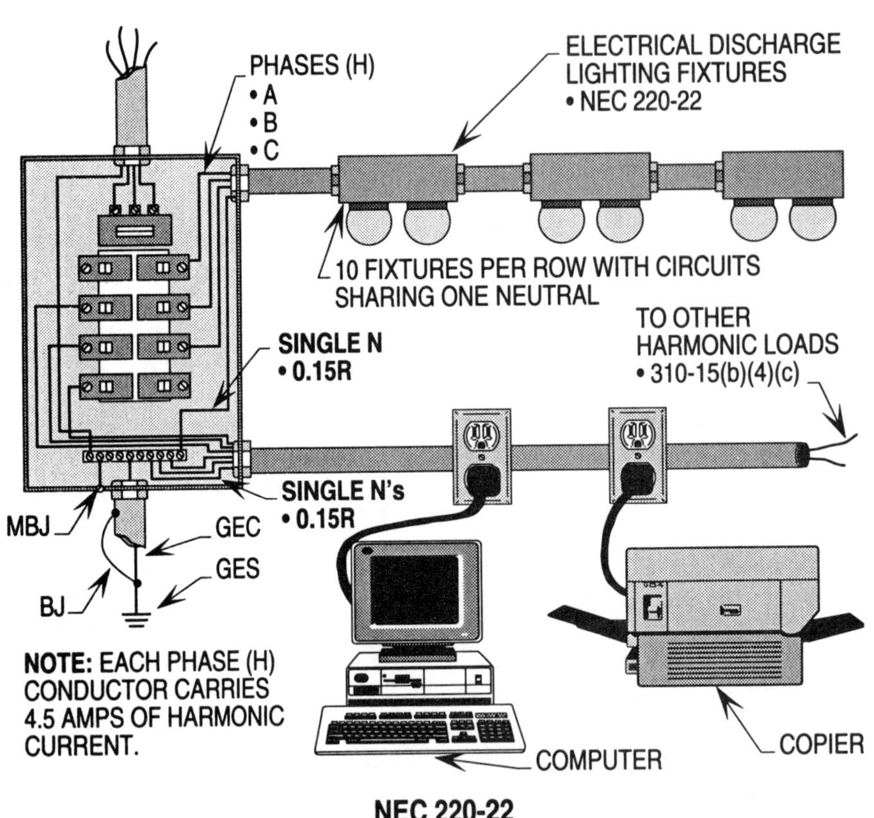
PHASES (H)
• A
• B
• C
ELECTRICAL DISCHARGE
LIGHTING FIXTURES
• NEC 220-22
10 FIXTURES PER ROW WITH CIRCUITS
SHARING ONE NEUTRAL
TO OTHER
HARMONIC LOADS
• 310-15(b)(4)(c)
SINGLE N
• 0.15R
SINGLE N's
• 0.15R
MBJ
GEC
GES
BJ
NOTE: EACH PHASE (H)
CONDUCTOR CARRIES
4.5 AMPS OF HARMONIC
CURRENT.
COMPUTER
COPIER
NEC 220-22
NEC 310-15(b)(4)(c)

Ill. 2-12(f)

Outside Feeder 225-3(b)

The procedure for calculating the load for outside feeders is to compute the noncontinuous load at 100 percent and the continuous load at 125 percent.

Example 2-13. What is the load in amps to size the OCPD and conductors for the feeder-circuit in Illustration 2-13?

Sizing load

Step 1: Calculating load for OCPD
225-3(b); 225-9; 215-3
90 A x 125% = 112.5 A

Solution: The load is 112.5 amps

Sizing OCPD

Step 1: Sizing OCPD
220-10; 240-3(b); 240-6(a)
112.5 A requires 125 A OCPD

Solution: The size OCPD is 125 amps.

Sizing conductors

Step 1: Sizing conductors at 125%
220-10; 215-2(a)
90 A x 125% = 112.5 A

Step 2: Sizing conductors to match OCPD
Table 310-16
112.5 A requires #2

Solution: The size THWN copper conductors are No. 2 (115 A) based on load.

NEC 225-3(a); (b); NEC 225-5; NEC 225-9

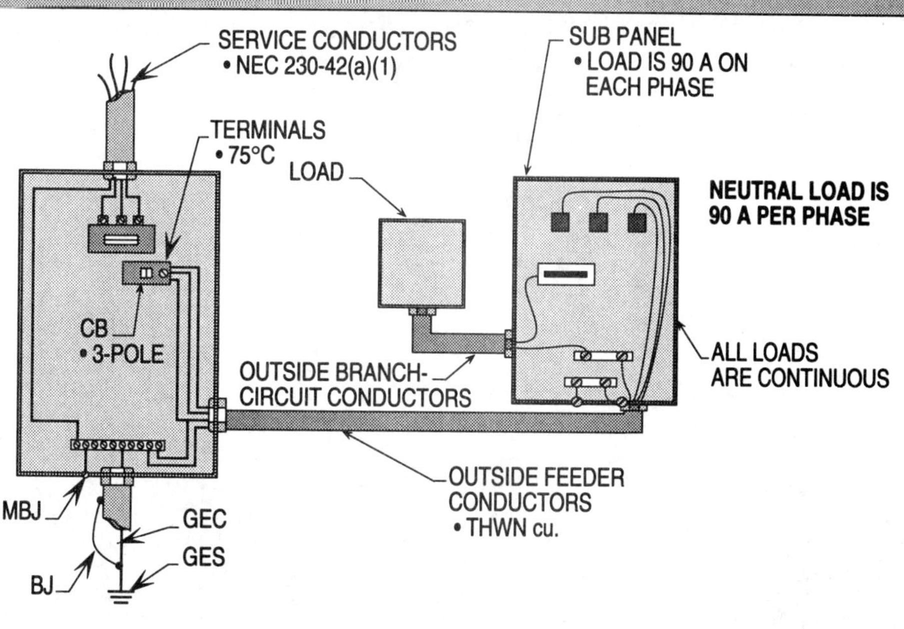

NEC 225-3(b)

Ill. 2-13

Calculating Feeder and Service Loads

3

Feeder and service loads are computed by applying the standard or optional calculation methods. Load computations may also be determined by the application of demand factors. The standard calculation produces greater volt-amp and amp values that require larger elements to be installed, which provides a safe and reliable electrical system.

The optional calculation used under certain conditions of use allows smaller volt-amp and amp values to be utilized in sizing the elements of the electrical system. Demand factors may be applied, based upon the type of loads involved. The application of demand factors produce smaller load values and permits the elements to be selected at ratings smaller than the standard and optional calculation methods.

Quick Reference

Standard Calculation
220-2 thru 220-22

Example 3-1. Determine the following loads of a 2,800 sq. ft. dwelling unit and size the service-entrance conductors required for Phases A and B and the neutral?

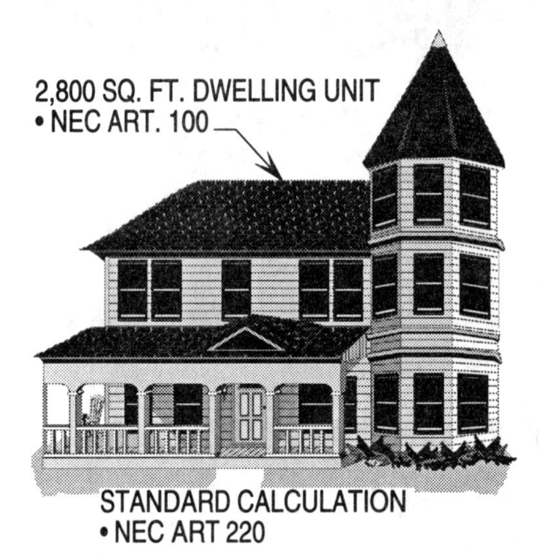

- General lighting and receptacle load
- Small appliance and laundry load
- Special appliance load

• 6,000 VA A/C unit	240 V, single-phase
• 10,000 VA heating unit	240 V, single-phase
• 5,000 VA water heater	240 V, single-phase
• 8,000 VA oven	240 V, single-phase
• 8,500 VA cooktop	240 V, single-phase
• 2,600 VA water pump	120 V, single-phase
• 1,000 VA disposal	120 V, single-phase
• 1,200 VA compactor	120 V, single-phase
• 1,600 VA dishwasher	120 V, single-phase
• 1,000 VA microwave	120 V, single-phase
• 800 VA blower motor	240 V, single-phase
• 5,000 VA dryer	240 V, single-phase

• phase loads
*** neutral loads**

NEC 220-3(a); NEC 220-16(a); (b); NEC 220-19; NEC 220-18; NEC 220-17; NEC 220-21; NEC 220-14

Calculating VA loads (phases)

General lighting and receptacle load

Table 220-3(a)
2,800 sq. ft. x 3 VA = 8,400 VA

Small appliance and laundry load

220-16(a); (b)
1,500 VA x 2 = 3,000 VA
1,500 VA x 1 = 1,500 VA
Total load = 4,500 VA

Applying demand factors

Table 220-11
General lighting load = 8,400 VA
Small appliance and laundry load = 4,500 VA
Total load = 12,900 VA

First 3,000 VA x 100% = 3,000 VA
Next 9,900 VA x 35% = 3,465 VA
Total load = **6,465 VA •**

Cooking equipment loads

Table 220-19, Col. C
Total kW rating

8 kW + 8.5 kW	=	16.5 kW
16.5 kW x 1,000 x 65%	=	**10,725 VA •**

Fixed appliance load

220-17

water heater	=	5,000 VA
water pump	=	2,600 VA *
disposal	=	1,000 VA *
compactor	=	1,200 VA *
dishwasher	=	1,600 VA *
microwave	=	1,000 VA *
blower motor	=	800 VA
Total load	=	13,200 VA
13,200 VA x 75%	=	**9,900 VA •**

Dryer load

220-18; Table 220-18

5,000 VA x 100%	=	**5,000 VA •**

Largest between heating or A/C load

220-21
Heating load

10,000 VA x 100%	=	**10,000 VA •**

Largest motor load

220-14
water pump

2,600 VA x 25%	=	**650 VA •**

Total VA loads (phases)

General lighting load	=	6,465 VA •
Cooking equipment	=	10,725 VA •
Fixed appliance load	=	9,900 VA •
Dryer load	=	5,000 VA •
Heating load	=	10,000 VA •
Largest motor load	=	650 VA •
Total load	=	42,740 VA

Finding amps

I = VA/V
I = 42,740 VA ÷ 240 V
I = 178 A

Calculating VA loads (neutral)

General lighting load

220-22
6,465 VA *

Cooking equipment load

(use 70% of cooking load)
220-22
10,725 VA x 70% = **7,508 VA ***

Fixed appliance load

(use 75% of 120 V (*) fixed appliance load)
220-17
7,400 VA x 75% = **5,550 VA***

Dryer load

(use 70% of dryer load)
220-18; Table 220-18
5,000 x 70% = **3,500 VA ***

Largest motor load

(use 25% of largest motor load)
220-14
2,600 VA x 25% = **650 VA ***

Total VA loads (neutral)

General lighting load	=	6,465 VA *
Cooking equipment	=	7,508 VA *
Dryer load	=	3,500 VA *
Fixed appliance load	=	5,550 VA *
Largest motor load	=	650 VA *
Total VA	=	23,673 VA

Finding Amps

I = VA/V
I = 23,673 VA ÷ 240 V
I = 99 A

Table 310-16 and Table 310-15(b)(6)

Phases A and B - #3/0 THWN copper conductors
Table 310-15(b)(6) allows #2/0 THWN copper conductors
Neutral - #3 THWN copper conductor

Optional Calculation
220-30; Table 220-30

Example 3-2. Determine the following loads of a 2,500 sq. ft. dwelling unit and size the service-entrance conductors required for phases A and B and the neutral?

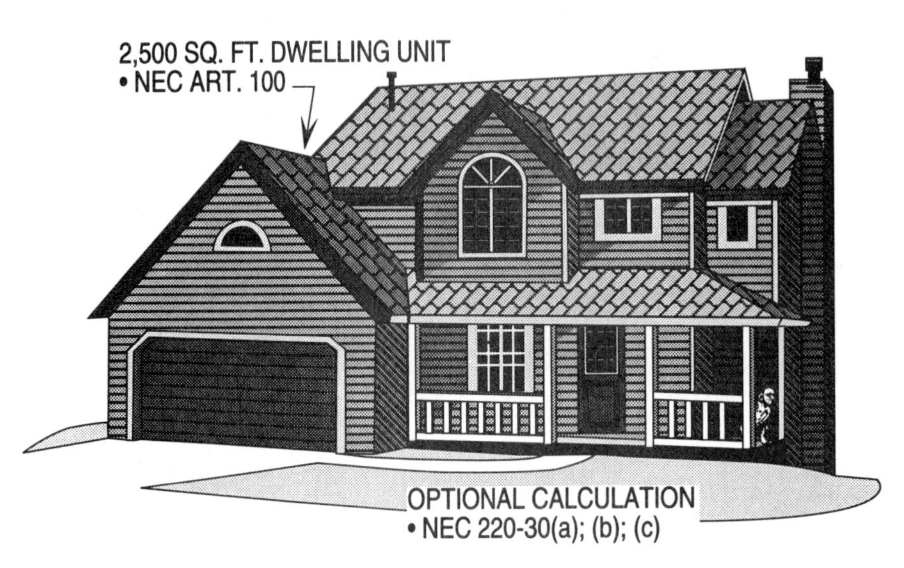

- General lighting and receptacle load
- two small appliance circuits
- one laundry circuit

• 6,000 VA A/C unit	240 V, single-phase
• 20,000 VA heating unit	240 V, single-phase
• 6,000 VA water heater	240 V, single-phase
• 10,000 VA oven	240 V, single-phase
• 9,000 VA cooktop	240 V, single-phase
• 2,600 VA water pump	120 V, single-phase
• 1,000 VA disposal	120 V, single-phase
• 1,200 VA compactor	120 V, single-phase
• 1,600 VA dishwasher	120 V, single-phase
• 1,000 VA pool pump	240 V, single-phase
• 5,000 VA dryer	240 V, single-phase
• 800 VA blower motor	240 V, single-phase

Feeder and Service Load
• NEC 220-30(a)

Other Loads (General)
• NEC 220-30(b)

Demand Factors
• NEC 220-30(b); (c)

• phase loads
* neutral loads

Calculating VA loads (phases)

General lighting loads

220-30(b)(2)
2,500 sq. ft. x 3 VA = 7,500 VA

Small appliance and laundry load

220-30(b)(1)
1500 VA x 2 = 3,000 VA
1500 VA x 1 = 1,500 VA
Total load = 4,500 VA

Cooking equipment loads and appliance loads per 220-30(b)(3); (b)(4)

Cooktop	= 9,000 VA
Oven	= 10,000 VA
Dryer	= 5,000 VA
Water heater	= 6,000 VA
Disposal	= 1,000 VA *
Compactor	= 1,200 VA *
Dishwasher	= 1,600 VA *
Pool pump	= 1,000 VA
Blower motor	= 800 VA
Water pump	= 2,600 VA *
Total load	= 38,200 VA

Applying demand factors

220-30(b)

General lighting load	= 7,500 VA
Small appliance and laundry load	= 4,500 VA
Cooking equip. and appliance load	= 38,200 VA
Total load	= 50,200 VA

First 10,000 VA x 100%	= 10,000 VA
Next 40,200 VA x 40%	= 16,080 VA
Total load	= **26,080 VA •**

Largest load between heating and A/C

220-30(3)(c)(4); (c)(5)
Heating load

20,000 VA x 65%	= **13,000 VA •**

Total VA loads (phases)

Total loads	= 26,080 VA •
Heating load	= 13,000 VA •
Total load	= 39,080 VA

Finding amps

I = VA / V
I = 39,080 VA ÷ 240 V
I = 163 A

Calculating VA loads (neutral)

General lighting load

220-3(a)

2,500 sq. ft. x 3 VA	= 7,500 VA

Small appliance and laundry load

220-16(a); (b)

1,500 VA x 2	= 3,000 VA
1,500 VA x 1	= 1,500 VA
Total load	= 4,500 VA

Applying demand factors

Table 220-11

General lighting load	= 7,500 VA
Small appliance and laundry load	= 4,500 VA
Total load	= 12,000 VA

First 3,000 VA x 100%	= 3,000 VA
Next 9,000 VA x 35%	= 3,150 VA
Total load	= **6,150 VA ***

Cooking equipment load

Table 220-19, Col. A; 220-22
(use 70% of cooking load)

11,000 VA x 70%	= **7,700 VA ***

Fixed appliance load

220-17
(use 75% of fixed appliance load)

disposal	= 1,000 VA
compactor	= 1,200 VA
dishwasher	= 1,600 VA
water pump	= 2,600 VA
Total load	= 6,400 VA
6,400 VA x 75%	= **4,800 VA ***

Dryer load

Table 220-18
(use 70% of dryer load)

5,000 VA x 70%	= **3,500 VA ***

Largest motor load

220-14
(use 25% of largest motor load)

2,600 VA x 25%	= **650 VA ***

Total VA loads (neutral)

General lighting load	= 6,150 VA *
Cooking equipment	= 7,700 VA *
Dryer load	= 3,500 VA *
Fixed appliance load	= 4,800 VA *
Largest motor load	= 650 VA *
Total load	= 22,800 VA

Finding amps

I = VA / V
I = 22,800 VA ÷ 240 V
I = 95 A

Table 310-16 and Table 310-15(b)(6)

Phases A and B - #2/0 THWN copper conductors
Table 310-15(b)(6) allows #1/0 THWN copper conductors
Neutral - #3 THWN copper conductor

Optional Calculation - Existing 220-31

Example 3-3. From the following loads in an existing dwelling unit, it will be determined if a 5,040 VA A/C unit can be added to the service conductors. (The existing service conductors are No. 3 THWN copper conductors).

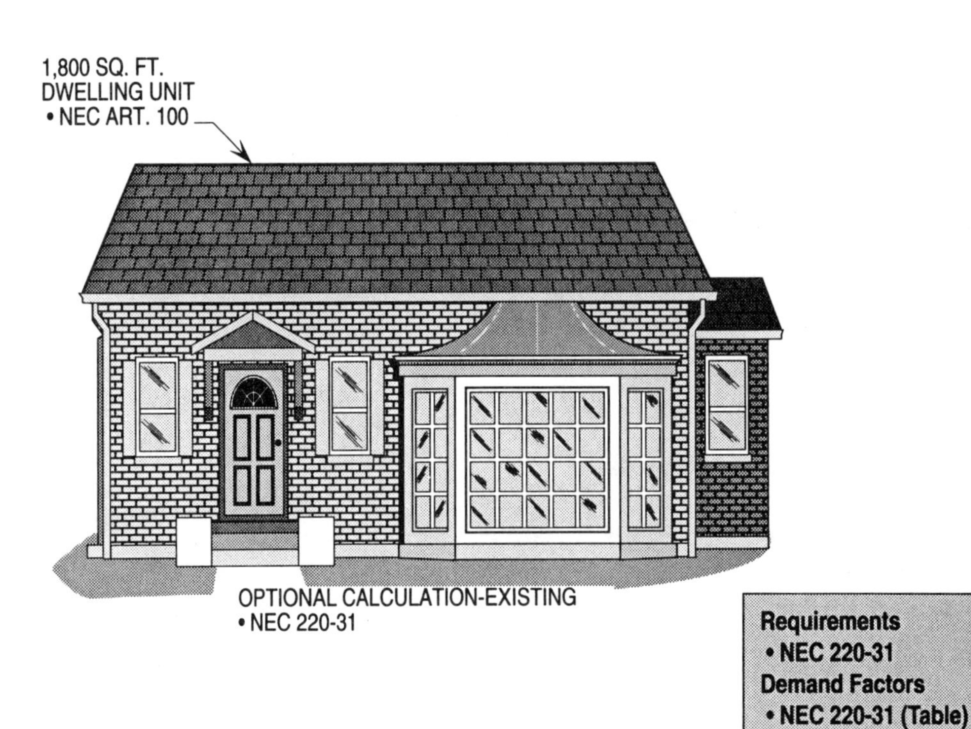

Requirements
- **NEC 220-31**

Demand Factors
- **NEC 220-31 (Table)**

- 1,800 sq. ft. dwelling unit
- 2 small appliance circuits
- 1 laundry circuit

• 12 kW range	240 V, single-phase
• 5 kW dryer	240 V, single-phase
• 1,000 VA compactor	120 V, single-phase
• 900 VA disposal	120 V, single-phase
• 1,200 VA dishwasher	120 V, single-phase

- **phase loads**

Calculating VA loads (phases)

General lighting load

220-31(2)
1,800 sq. ft. x 3 VA = **5,400 VA •**

Small appliance and laundry load

220-31(1)
1,500 VA x 2 = 3,000 VA
1,500 VA x 1 = 1,500 VA
Total load = **4,500 VA •**

Existing load

220-31(3); (4)

Range	= 12,000 VA
Dryer	= 5,000 VA
Compactor	= 1,000 VA
Disposal	= 900 VA
Dishwasher	= 1,200 VA
Total load	= **20,100 VA •**

Demand load

Table to 220-31

General lighting load	= 5,400 VA •
Small appliance and laundry load	= 4,500 VA •
Existing load	= 20,100 VA •
Total load	= 30,000 VA

First 8,000 VA x 100%	= 8,000 VA
Next 22,000 VA x 40%	= 8,800 VA
Total load	= 16,800 VA

Added load

220-31
A/C unit

5,040 VA x 100%	= 5,040 VA
Existing load	= 16,800 VA
Total load	= 21,840 VA

Finding amps

I = VA/V
I = 21,840 VA ÷ 240 V
I = 91 A

Existing service amps

#3 THWN copper conductors = 100 amps
Total load = 91 amps (less than 100 A)

Solution: Yes, the A/C unit rated at 5,040 VA may be added to the existing service without increasing the size of the service conductors.

Note: To calculate the load for sizing the neutral conductor, see the calculation using the asterisk on page 3-5.

Standard Calculation - Multifamily Dwelling
220-2 thru 220-22

Example 3-4. Determine the following loads for 25-1,000 sq. ft. units in a multifamily dwelling and size the service entrance conductors required for Phases A and B and the neutral?

Note: *Parallel service conductors, 6 times per phase.*

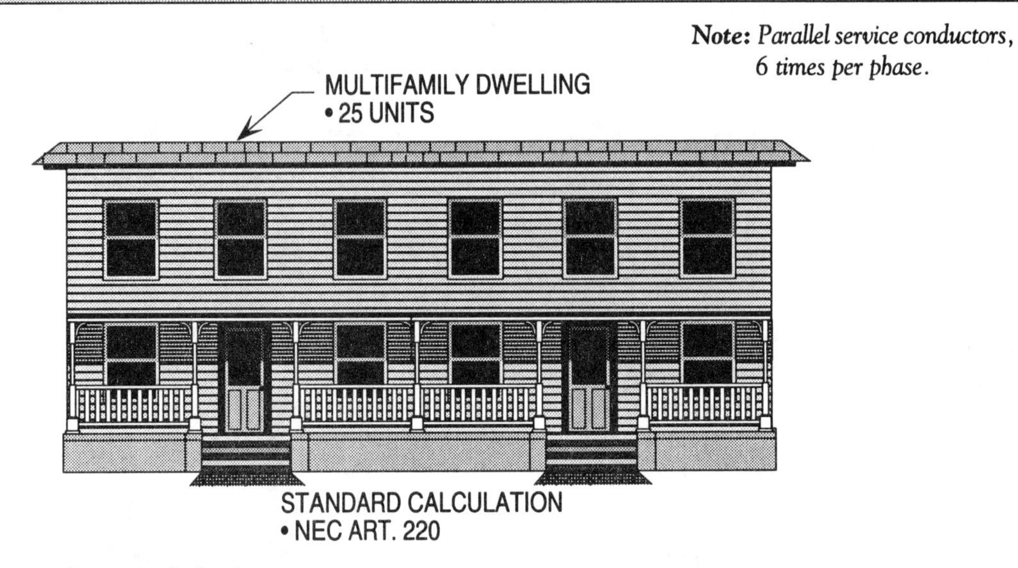

- General lighting and receptacle load
- Two small appliance and one laundry load per unit
- Special appliance load

• 25 - 12 kW ranges	240 V, single-phase
• 25 - 20 kW heating units	240 V, single-phase
• 25 - 1 kW dishwashers	120 V, single-phase
• 25 - 1.2 kW disposals	120 V, single-phase
• 25 - 6 kW water heaters	240 V, single-phase

• phase loads
*** neutral loads**

NEC 220-3(a); NEC 220-16(a); (b); NEC 220-19; NEC 220-18; NEC 220-17; NEC 220-21; NEC 220-14

Calculating VA loads (phases)

General lighting and receptacle load

Table 220-3(a)
1,000 sq. ft. x 3 x 25 = 75,000 VA

Small appliance and laundry load

220-16(a);(b)
1,500 VA x 2 x 25 = 75,000 VA
1,500 VA x 1 x 25 = 37,500 VA
Total load = 187,500 VA

Table 220-11
First 3,000 VA x 100% = 3,000 VA
Next 117,000 VA x 35% = 40,950 VA
Remaining 67,500 VA x 25% = 16,875 VA
Total load = **60,825 VA •**

Cooking equipment load

Table 220-19, Column A
25 kW ranges = **40,000 VA•**

Fixed appliance load

220-17
dishwashers
25 x 1,000 VA x 75% = 18,750 VA
disposals
25 x 1,200 VA x 75% = 22,500 VA
water heaters
25 x 6,000 VA x 75% = 112,500 VA
Total load = **153,750 VA •**

Largest between htg. and A/C load

220-15
heating
20,000 VA x 25 = **500,000 VA •**

Largest motor

220-14
1,200 VA x 25% = **300 VA •**

Total VA loads (phases)

General lighting load	= 60,825 VA •
Range	= 40,000 VA •
Fixed appliances	= 153,750 VA •
Heating	= 500,000 VA •
Largest motor	= 300 VA •
Total loads	= 754,875 VA

Finding Amps

I = VA / V
I = 754,875 VA ÷ 240 V
I = 3,145 A

Calculating VA loads (neutral)

General lighting load

220-22
60,825 VA *

Range

220-22
40,000 VA x 70% = **28,000 VA ***

Fixed appliance load

220-17

dishwashers	= 18,750 VA
disposals	= 22,500 VA
Total load	= **41,250 VA ***

Largest motor

220-14
300 VA *

Total VA loads (neutral)

General lighting load	= 60,825 VA *
Range	= 28,000 VA *
Fixed appliance load	= 41,250 VA *
Largest motor	= 300 VA *
Total load	= 130,375 VA

Finding Amps

I = VA/V
I = 130,375 VA ÷ 240 V
I = 543 A

Table 310-16

Phases A and B
310-4
I = 3,145 A ÷ 6 (no. runs per phase)
I = 524 A

Neutral
220-22
543 A

First 200 A x 100%	= 200 A
Next 343 A x 70%	= 240 A
Total load	= 440 A

310-4
I = 440 A ÷ 6 (no. runs per phase)
I = 73 A

Phases A and B - #1,000 KCMIL THWN copper conductors
Neutral - #1/0 THWN copper conductor

Note: Sections 310-4 and 250-24(b)(2) requires a No. 1/0 (N) and larger conductor to be installed when in parallel.

Optional Calculation - Multifamily Dwelling 220-32; Table 220-32

Example 3-5. Determine the following loads for 25-1,000 sq. ft. units in a multifamily dwelling and size the service entrance conductors required for Phases A and B and the neutral?

Note: *Parallel service conductors, 6 times per phase.*

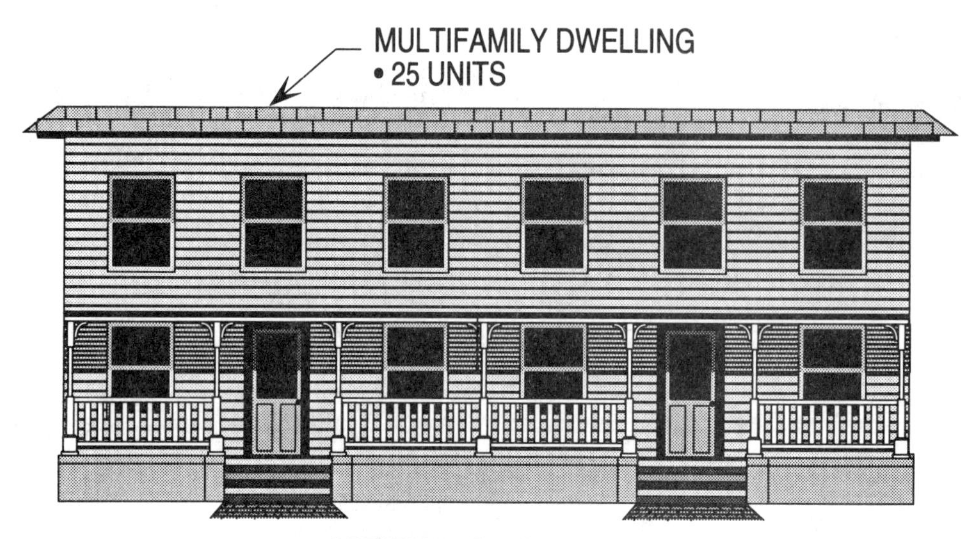

OPTIONAL CALCULATION
- NEC 220-32
- TABLE 220-32

- **phase loads**

* **neutral loads**

- General lighting and receptacle load
- Two small appliance and one laundry load per unit
- Special appliance load

• 25 - 12	kW ranges	240 V, single-phase
• 25 - 20	kW heating units	120 V, single-phase
• 25 - 1	kW dishwashers	120 V, single-phase
• 25 - 1.2	kW disposals	240 V, single-phase
• 25 - 6	kW water heaters	240 V, single-phase

Feeder or Service Load
- **NEC 220-32(a)**

House Loads
- **NEC 220-32(b)**

Connected Loads
- **NEC 220-32(c)**

Demand Factors
- **Table 220-32**

Calculating VA loads (phases)

General lighting and receptacle load

220-32(c)(2)
1,000 sq. ft. x 3 x 25 = **75,000 VA •**

Small appliance and laundry load

220-32(c)(1)
1,500 VA x 2 x 25 = 75,000 VA
1,500 VA x 1 x 25 = 37,500 VA
Total load = **112,500 VA •**

Range loads

220-32(c)(3)
12,000 VA x 25 = **300,000 VA •**

Heating loads

220-32(c)(5)
20,000 VA x 25 = **500,000 VA •**

Fixed appliance load

220-32(c)(3)
dishwasher load
1,000 VA x 25 = 25,000 VA
disposal load
1,200 VA x 25 = 30,000 VA
water heater load
6,000 VA x 25 = 150,000 VA
Total load = **205,000 VA •**

Total VA loads (phases)

General lighting load and receptacle load = 75,000 VA •
Small appliance and laundry load = 112,500 VA •
Range loads = 300,000 VA •
Heating loads = 500,000 VA •
Fixed appliance loads = 205,000 VA •
Total loads = 1,192,500 VA

Applying demand factors

Table 220-32
1,192,500 VA x 35% = 417,375 VA

Finding amps

I = VA / V
I = 417,375 VA ÷ 240 V
I = 1,739 A

Table 310-16

Phases A and B
310-4
I = 1,739 A ÷ 6 (no. runs per phase)
I = 290 A

Phases A and B - #350 THWN copper conductors
Neutral - #1/0 THWN copper conductor

Note: See Example #3-4 Standard Calculation for neutral load and conductor size. (Reference Columns 1 and 2 of page 3-9)

Standard Calculation - Store And Warehouse
220-2 Thru 220-22

Example 3-6. Determine the following loads in VA and amps for a store and warehouse with 40,000 sq. ft. of store space and 20,000 sq. ft. warehouse area.

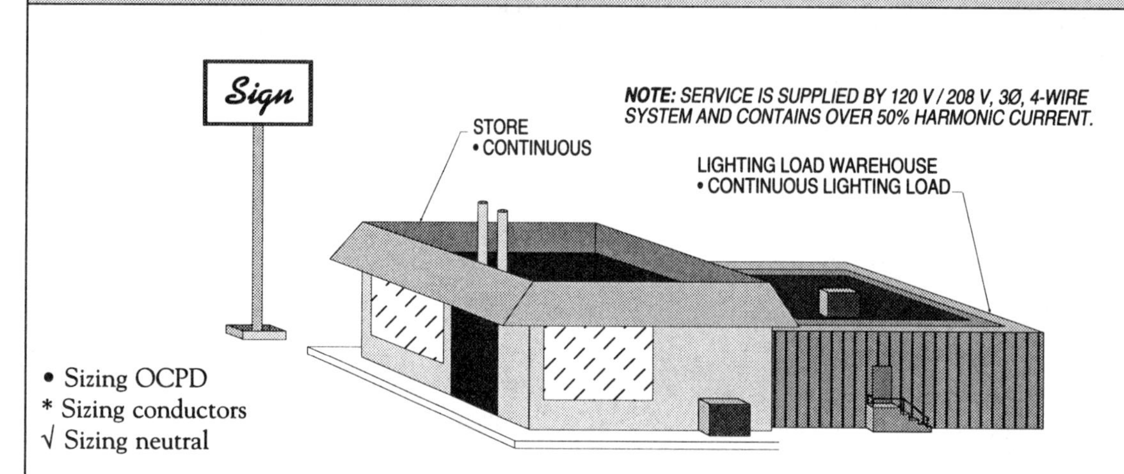

• Sizing OCPD
* Sizing conductors
√ Sizing neutral

120 V, Single-phase loads

- 80 linear feet of show window (noncontinuous)
- 144' - 0" lighting track (noncontinuous)
- 30 - 180 VA ballasts outside lighting (continuous duty)
- 3,600 VA sign lighting (continuous duty)
- 65 receptacles (noncontinuous duty)
- 28 receptacles (continuous duty)
- 80' - 0" multioutlet assembly (heavy-duty)

208 V, Three-phase loads

- 7,380 VA freezer
- 5,580 VA ice cream boxes
- 1,296 VA exhaust fan
- 10,000 VA water heater
- 9,540 VA walk-in cooler
- 50,000 VA heating unit
- 22,320 VA A/C unit
- 3,801 VA water pump

NEC 220-4(b)	NEC 600-5(b)(3)	NEC 220-3(b)(8)	NEC 430-24
NEC 220-12(a)	NEC 220-3(b)(9)	NEC 230-42(a)(1)	NEC 220-14
NEC 220-12(b)	NEC 220-13	NEC 220-21	NEC 220-22

Calculating lighting load

General lighting load

Table 220-3(a); 230-42(a)(1)
40,000 sq. ft. x 3 VA = 120,000 VA √
120,000 VA x 125% = 150,000 VA *

20,000 sq. ft. x 1/4 = 5,000 VA √
5,000 VA x 125% = 6,250 VA *

Show window load

220-12(a)
80 ft. x 200 = 16,000 VA * √

Track lighting load

220-12(b)
144 ft. ÷ 2 x 150 VA = 10,800 VA * √

Outside lighting load

220-3(b)(11); 230-42(a)(1)
30 x 180 VA = 5,400 VA √
5,400 VA x 125% = 6,750 VA *

Sign lighting load

220-3(b)(6); 600-5(b)(3); 230-42(a)(1)
3,600 VA x 100% = 3,600 VA √
3,600 VA x 125% = 4,500 VA *

Total lighting load

General lighting load	= 150,000 VA
Warehouse	= 6,250 VA
Show window load	= 16,000 VA
Track lighting load	= 10,800 VA
Outside lighting load	= 6,750 VA
Sign lighting load	= 4,500 VA
Total loads	= **194,300 VA** •

Calculating receptacle load

Noncontinuous duty

220-3(b)(9); 220-3(b)(8)(b)

65 x 180 VA	= 11,700 VA
80 ft. x 180 VA	= 14,400 VA
Total load	= 26,100 VA
Table 220-13	
First 10,000 VA x 100%	= 10,000 VA
Remaining 16,100 VA x 50%	= 8,050 VA
Total load	= 18,050 VA * √

Continuous duty

220-3(b)(9); 230-42(a)(1)

28 x 180 VA	= 5,040 VA √
5,040 VA x 125%	= 6,300 VA *

Total receptacle load

Noncontinuous load	= 18,050 VA
Continuous load	= 6,300 VA
Total load	= **24,350 VA •**

Calculating special loads

Freezer

230-42(a)(1)

7,380 VA x 100% = 7,380 VA *

Ice cream boxes

5,580 VA x 100% = 5,580 VA *

Water heater

10,000 VA x 100% = 10,000 VA*

Walk-in cooler

9,540 VA x 100% = 9,540 VA*

Total loads = **32,500 VA •**

Calculating heating or A/C load

220-21

50,000 VA x 100% = **50,000 VA * •**

Calculating motor load

Water pump

430-24

3,801 VA x 100% = 3,801 VA *

Exhaust fan

1,296 VA x 100%	= 1,296 VA *
Total load	= **5,097 VA •**

Calculating largest motor

Walk-in cooler

220-14; 430-24

9,540 VA x 25% = **2,385 VA * •**

Calculating OCPD

230-90(a); 220-10

Lighting load	= 194,300 VA •
Receptacle loads	= 24,350 VA •
Special loads	= 32,500 VA •
Heating load	= 50,000 VA •
Motor load	= 5,097 VA •
Largest motor load	= 2,385 VA •
Total load	= 308,632 VA

Finding amps for OCPD

I = VA / V x √3
I = 308,632 VA / 208 V x 1.732 (360 V)
I = 857 A

Calculating phase conductors

220-10; 230-42(a)(1)

General lighting	
(store)	= 150,000 VA *
(warehouse)	= 6,250 VA *
Show window	= 16,000 VA *
Track lighting	= 10,800 VA *
Outside lighting	= 6,750 VA *
Sign lighting	= 4,500 VA *
Receptacle (noncontinuous)	= 18,050 VA *
Receptacle (continuous)	= 6,300 VA *
Freezer	= 7,380 VA *
Ice cream boxes	= 5,580 VA *
Water heater	= 10,000 VA *
Walk-in cooler	= 9,540 VA *
Heating	= 50,000 VA *
Water pump	= 3,801 VA *
Exhaust fan	= 1,296 VA *
Largest motor	= 2,385 VA *
Total load	= 308,632 VA

Finding amps for phase conductors

I = VA / V x √3
I = 308,632 VA / 208 V x 1.732 (360 V)
I = 857 A

Calculating neutral

220-22 (add all loads marked)

Lighting load	= 160,800 VA √
Receptacle load	= 23,090 VA √
Total load	= 183,890 VA

Finding amps for neutral conductors

I = VA / V x √3
I = 183,890 VA / 208 V x 1.732 (360 V)
I = 511 A

Standard Calculation - Office Building
220-2 thru 220-22

Example 3-7. Determine the following loads in VA and amps for a office building with an area of 150,000 sq. ft. General lighting load is 277 volt, single-phase.

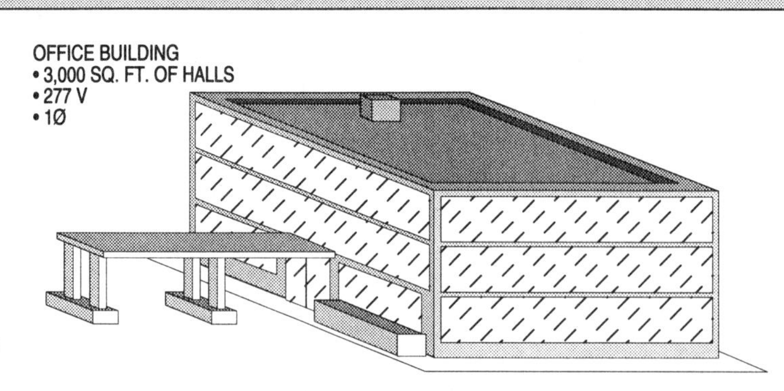

• Sizing OCPD
* Sizing conductors
√ Sizing neutral

NEC 220-2 thru 220-22

120 V, Single-phase loads

- 200' - 0" multioutlet assembly (used at same time)
- 60' - 0" lighting track (noncontinuous)
- 20 - 180 VA ballasts outside lighting (continuous-duty)
- 4,800 VA sign lighting (continuous-duty)
- 182 receptacles (noncontinuous-duty)
- 121 receptacles (continuous-duty)

480 V, Three-phase loads

- 40 HP elevator (15 minute intermitten duty)
- 40 kW heating unit
- 12,000 VA A/C unit

208 V, Three-phase loads

- 5 - 1,450 VA copying machines
- 8,500 VA water heater
- 25 - 225 VA data processors
- 10 - 175 VA word processors
- 4 - 1,200 VA printers

NEC 220-4(b)	NEC 600-5(b)(3)	NEC 220-3(b)(8)(b)	NEC 430-24
NEC 220-12(a)	NEC 220-3(b)(9)	NEC 230-42(a)(1)	NEC 220-14
NEC 220-12(b)	NEC 220-13	NEC 220-21	NEC 220-20

Calculating lighting load

General lighting load

Table 220-3(a); 220-10; 230-42(a)(1)
(office)
150,000 sq. ft. x 3.5 VA = 525,000 VA √
525,000 VA x 125% = 656,250 VA *
(halls)
3,000 sq. ft. x 1/2 VA = 1,500 VA √
1,500 VA x 125% = 1,875 VA *

Lighting track load

220-12(b)
60 ft. ÷ 2 x 150 VA = 4,500 VA *

Outside lighting load

230-42(a)(1)
180 VA x 20 = 3,600 VA
3,600 VA x 125% = 4,500 VA *

Sign lighting load

220-3(b)(6); 230-42(a)(1); 600-5(b)(3)
4,800 VA x 100% = 4,800 VA
4,800 VA x 125% = 6,000 VA *

Total lighting load

Office lighting load = 656,250 VA
Hall lighting load = 1,875 VA
Lighting track lighting load = 4,500 VA
Outside lighting load = 4,500 VA
Sign lighting load = 6,000 VA
Total load = **673,125 VA •**

Calculating receptacle load

Noncontinuous load

220-3(b)(8)(b); (b)(9); Table 220-13
182 x 180 VA = 32,760 VA
200' x 180 VA = 36,000 VA
Total load = 68,760 VA

Applying demand factors

Table 220-13
First 10,000 VA x 100% = 10,000 VA
Remaining 58,760 VA x 50% = 29,380 VA
Total load = 39,380 VA *

Continuous duty

220-3(b)(9); 220-10; 230-42(a)(1)

121 x 180 VA	= 21,780	VA
21,780 VA x 125%	= 27,225	VA *

Total receptacle load

Noncontinuous duty	= 39,380	VA
Continuous duty	= 27,225	VA
Total load	= **66,605**	VA •

Calculating special loads

Copying machines

220-10; 230-42(a)(1)

1,450 VA x 5	= 7,250	VA
7,250 VA x 125%	= 9,063	VA *

Water heater

220-10; 422-13

8,500 VA x 100%	= 8,500	VA *

Data processors

225 VA x 25	= 5,625	VA
5,625 VA x 125%	= 7,031	VA *

Word processors

220-10; 230-42(a)(1)

175 VA x 10	= 1,750	VA
1,750 VA x 125%	= 2,188	VA *

Printers

220-10; 230-42(a)(1)

1,200 VA x 4	= 4,800	VA
4,800 VA x 125%	= 6,000	VA *

Total special loads

Copying machines	= 9,063	VA
Water heater	= 8,500	VA
Data processors	= 7,031	VA
Word processor	= 2,188	VA
Printers	= 6,000	VA
Total loads	= **32,782**	VA •

Calculating heating or A/C load

220-21; 220-15

40,000 VA x 100%	= **40,000**	VA * •

Calculating motor loads

430-24; 430-22(b); Table 430-22(b)

52 A x 480 V x 1.732 x 85%	= **36,746**	VA * •

Calculating largest motor load

220-14; 430-24

36,746 VA x 25%	= **9,187**	VA * •

Calculating OCPD

230-90(a); 220-10

General lighting load	= 673,125	VA •
Receptacle load	= 66,605	VA •
Special loads	= 32,782	VA •
Heating load	= 40,000	VA •
Motor load	= 36,746	VA •
Largest motor load	= 9,187	VA •
Total load	= 858,445	VA

Finding amps for phase conductors

I = VA / V x √3
I = 858,445 VA / 480 V x 1.732 (831 V)
I = 1,033 A

Calculating phase conductors

220-10(b); 230-42(a)(1)

Office lighting load	= 656,250	VA *
Hall lighting load	= 1,875	VA *
Lighting track load	= 4,500	VA *
Outside lighting load	= 4,500	VA *
Sign lighting load	= 6,000	VA *
(noncontinuous)		
Receptacle load	= 39,380	VA *
(continuous)		
Receptacle load	= 27,225	VA *
Copying machine load	= 9,063	VA *
Water heater load	= 8,500	VA *
Data processor load	= 7,031	VA *
Word processor load	= 2,188	VA *
Printers load	= 6,000	VA *
Heating load	= 40,000	VA *
Motor load	= 36,746	VA *
Largest motor load	= 9,187	VA *
Total loads	= 858,445	VA

Finding amps for phase conductors

I = VA / V x √3
I = 858,445 VA / 480 V x 1.732 (831 V)
I = 1,033 A

Calculating Neutral

220-22

General lighting load	
(office building)	= 525,000 VA √
(halls)	= 1,500 VA √
Total load	= 526,500 VA

Finding amps for neutral

I = VA / V x √3
I = 526,500 VA / 480 V x 1.732 (831 V)
I = 634 A

Standard Calculations - Welding Shop
220-2 Thru 220-22

Example 3-8. Determine the following loads in VA and amps for a welding shop with an 120 V / 208 volt, three-phase, four-wire service.

• Sizing OCPD
* Sizing conductors
√ Sizing neutral

NEC 220-2 thru 220-22

120 V, Single-phase loads

- 9,000 VA inside lighting load (continuous duty)
- 6 - 180 VA outside lighting loads (continuous duty)
- 1,200 VA sign lighting load (noncontinuous duty)
- 50 receptacles (noncontinuous duty)
- 20 receptacles (continuous duty)

208 V, Three-phase loads

- 2 - 10 kW heating units
- 5,400 VA A/C units
- 7.5 HP air compressor
- 2 - 3/4 HP grinders
- Welders - Resistance (50% duty cycle)
- 12 kW
- 8 kW
- Welders - Motor-Generator Arc (90% duty cycle)
- 14 kW
- 12 kW
- Welders - Arc Welder without motor (80% duty cycle)
- 13 kW
- 9 kW

NEC 220-4(b) **NEC 600-5(b)(3)** **NEC 220-3(b)(8)** **NEC 430-24** **NEC 220-12(b)** **NEC 220-3(b)(9)**
NEC 220-10 **NEC 220-14** **NEC 220-12(a)** **NEC 220-13** **NEC 220-21** **NEC 220-22**

Calculating lighting load

Inside lighting load

220-10; 230-42(a)(1)
9,000 VA x 100% = 9,000 VA √
9,000 VA x 125% = 11,250 VA *

Outside lighting load

220-10; 230-42(a)(1)
6 x 180 VA = 1,080 VA √
1,080 VA x 125% = 1,350 VA *

Sign lighting load

220-3(b)(6); 600-5(b)(3)
1,200 VA x 100% = 1,200 VA * √

Total lighting load

Inside lighting load	= 11,250 VA
Outside lighting load	= 1,350 VA
Sign lighting load	= 1,200 VA
Total loads	= **13,800 VA** •

Calculating receptacle load

Noncontinuous duty

220-3(b)(9); 220-10; 230-42(a)(1)
50 x 180 VA = 9,000 VA * √

Continuous duty

220-3(b)(9); 220-10; 230-42(a)(1)

20 x 180 VA = 3,600 VA √
3,600 VA x 125% = 4,500 VA *

Total receptacle load

Noncontinuous duty	= 9,000 VA
Continuous duty	= 4,500 VA
Total loads	= **13,500 VA •**

Calculating special loads

Welders - resistance

630-11(a); (b)
12,000 VA x 71% = 8,520 VA *
8,000 VA x 71% x 60% = 3,408 VA *

Welders - Motor-Generator arc

630-11(a); (b)
14,000 VA x 96% = 13,440 VA *
12,000 VA x 96% = 11,520 VA *

Welders - Arc Welder without motor

630-11(a); (b)
13,000 VA x 89% = 11,570 VA *
9,000 VA x 89% = 8,010 VA *

Total special loads

Resistance welders	= 11,928 VA
Motor-Generator arc welders	= 24,960 VA
Arc Welder without motor	= 19,580 VA
Total loads	= **56,468 VA •**

Calculating heating or AC load

220-21; 220-15
20,000 VA x 100% = **20,000 VA * •**

Calculating motor loads

Air compressor

430-24
24.2 A x 100% x 208 V x 1.732 = 8,718 VA *

Grinders

430-24
7 A x 100% x 208 V x 1.732 = 2,522 VA *

Total motor loads

Air compressor	= 8,718 VA
Grinders	= 2,522 VA
Total loads	= **11,240 VA * •**

Calculating largest motor load

220-14; 430-24
24.2 A x 100% x 208 V x 1.732 x 25%
= **2,180 VA * •**

Calculating OCPD

230-90(a); 220-10; 230-42(a)(1)

Lighting loads	= 13,800 VA •
Receptacle loads	= 13,500 VA •
Special loads	= 56,468 VA •
Heating loads	= 20,000 VA •
Motor loads	= 11,240 VA •
Largest motor load	= 2,180 VA •
Total loads	= 117,188 VA

Finding amps for OCPD

I = VA / V x √3
I = 117,188 VA / 208 V x 1.732 (360 V)
I = 326 A

Calculating Phase conductors

220-10; 230-42(a)(1)

Lighting loads	= 13,800 VA *
Receptacle loads	= 13,500 VA *
Special loads	= 56,468 VA *
Heating load	= 20,000 VA *
Motor loads	= 11,240 VA *
Largest motor load	= 2,180 VA *
Total loads	= 117,188 VA

Finding amps for phase conductors

I = VA / V x √3
I = 117,188 VA / 208 V x 1.732 (360 V)
I = 326 A

Calculating neutral

220-22

Lighting loads	= 11,280 VA √
Receptacle loads	= 12,600 VA √
Total loads	= 23,880 VA

Finding amps for neutral

I = VA / V x 1.732
I = 23,880 VA / 208 V x 1.732 (360 V)
I = 66 A

Standard Calculations - Schools
220-2 thru 220-22

Example 3-9. Determine the following loads in VA and amps for a school with a 36,000 sq. ft. classroom area, 5,000 sq. ft. auditorium area, 3,000 sq. ft. cafeteria area, and 1,000 sq. ft. assembly hall area and size the service entrance conductors required for Phases A and B and the neutral. (The school is supplied by a 277/480 volt, three-phase, four-wire service)?

Note 1: Parallel service conductors, 6 times per phase.

Note 2: Cafeteria is also used for study hall activities.

OPTIONAL CALCULATION
- NEC 220-20
- TABLE 220-20

- **Sizing OCPD**

* **Sizing Conductors**

√ **Sizing Neutral**

NEC 220-4(b)	NEC 600-5(b)(3)	NEC 220-3(b)(8)(a); (b)	NEC 430-24	NEC 220-12(b)	NEC 220-3(b)(9)
NEC 220-10	NEC 220-14	NEC 220-12(a)	NEC 220-13	NEC 220-21	NEC 220-22

Lighting loads

- General lighting load
  277 V, single-phase
- 40 linear feet show window (noncontinuous)
  120 V, single-phase
- 200 VA lighting track (noncontinuous)
  120 V, single-phase
- 40 - 225 VA outside lighting loads (continuous duty)
  277 V, single-phase
- 1,200 VA sign lighting loads (continuous duty)
  120 V, single-phase

Receptacle loads

- 200 receptacles (noncontinuous duty)
  120 V, single-phase
- 50 receptacles (continuous duty)
  120 V, single-phase
- 200' - 0" multioutlet assembly (heavy-duty)
  120 V, single-phase

Special loads

- 40 - 10 kW heating units
  480 V, three-phase
- 40 - 6 kVA A/C units
  480 V, three-phase
- 10 - 1,500 VA soft drink boxes
  120 V, single-phase (noncontinuous duty)
- 4 - 1,000 VA copying machines
  120 V, single-phase (noncontinuous duty)

Motor loads

- 4 - 1 HP hood fans
  208 V, single-phase
- 3 - 3/4 HP grill vent fans
  208 V, single-phase
- 20 - 1 HP exhaust fans
  480 V, three-phase

Kitchen equipment (cafeteria)
(all units equipped with thermostatic controls)

- 4 - 12 kW ranges
  208 V, single-phase
- 3 - 9 kW ovens
  208 V, single-phase
- 4 - 4 kW fryers
  208 V, single-phase
- 6 kW booster heater
  208 V, single-phase
- 3 - 3 kW heaters
  208 V, single-phase
- 2 - 1 kW toaster
  120 V, single-phase
- 6 kW water heater
  208 V, single-phase
- 4 - 1.5 kW refrigerators
  120 V, single-phase
- 3 - 1.5 kW freezers
  120 V, single-phase

Calculating lighting load

Table 220-3(a); 220-10; 230-42(a)(1)

36,000 sq. ft. x 3	= 108,000 VA
5,000 sq. ft. x 1	= 5,000 VA
1,000 sq. ft. x 1	= 1,000 VA
3,000 sq. ft. x 1	= 3,000 VA
Total	= 117,000 VA √
117,000 VA x 125%	= 146,250 VA *

Show window load

220-12(a)

40 ft. x 200 VA	= 8,000 VA *

Track lighting load

220-12(b)

200 ÷ 2 x 150 VA	= 15,000 VA *

Outside lighting load

220-10; 230-42(a)(1)

225 VA x 40	= 9,000 VA √
9,000 VA x 125%	= 11,250 VA *

Sign lighting load

220-3(b)(6); 600-5(b)(3)

1,200 VA x 125%	= 1,500 VA *

Total lighting load

General lighting load	= 146,250 VA
Show window load	= 8,000 VA
Track lighting load	= 15,000 VA
Outside lighting load	= 11,250 VA
Sign lighting load	= 1,500 VA
Total loads	= **182,000 VA** •

Calculating receptacle loads

Noncontinuous duty

220-3(b)(9); 220-10; Table 220-13

200 x 180 VA	= 36,000 VA
200' x 180 VA	= 36,000 VA
Total load	= 72,000 VA

Applying demand factors

Table 220-13

First 10,000 VA x 100%	= 10,000 VA
Remaining 62,000 VA x 50%	= 31,000 VA
Total load	= 41,000 VA *

Continuous duty

220-3(b)(9); 220-10; 230-42(a)(1)

50 x 180 VA	= 9,000 VA
9,000 VA x 125%	= 11,250 VA *

Total receptacle load

Noncontinuous load	= 41,000 VA
Continuous load	= 11,250 VA
Total load	= **52,250 VA** •

Calculating special loads

Soft drink boxes

220-10; 230-42(a)(1)

10 x 1,500 VA	= 15,000 VA
15,000 VA x 100%	= 15,000 VA *

Copying machines

220-10; 230-42(a)(1)

4 x 1,000 VA	= 4,000 VA
4,000 VA x 100%	= 4,000 VA *

Kitchen equipment 220-20

Ranges	
4 x 12 kW x 1,000	= 48,000 VA
Ovens	
3 x 9 kW x 1,000	= 27,000 VA
Fryers	
4 x 4 kW x 1,000	= 16,000 VA
Booster heater	
1 x 6 kW x 1,000	= 6,000 VA
Heaters	
3 x 3 kW x 1,000	= 9,000 VA
Toasters	
2 x 1 kW x 1,000	= 2,000 VA

Water heater
1 x 6 kW x 1,000 = 6,000 VA
Refrigerators
4 x 1.5 kW x 1,000 = 6,000 VA
Freezers
3 x 1.5 kW x 1,000 = 4,500 VA
Total load = 124,500 VA

Applying demand factors

Table 220-20
124,500 VA x 65% = 80,925 VA *

Total special load

Soft drink boxes = 15,000 VA
Copying machines = 4,000 VA
Kitchen equipment = 80,925 VA
Total load = **99,925 VA •**

Calculating Heating or A/C load

220-21
40 x 10 kW x 1,000 = 400,000 VA *
400,000 VA x 100% = **400,000 VA •**

Calculating motor loads Tables 430-150; 430-148

Exhaust fans

42 A x 100% x 480 V = 20,160 VA
20,160 VA x √3 = 34,917 VA *
(2.1 A x 20 = 42 A)

Hood fans

35.2 A x 100% x 208 V = 7,322 VA *
(8.8 A x 4 = 35.2 A)

Grill vent fans

22.8 A x 100% x 208 V = 4,742 VA *
(7.6 A x 3 = 22.8 A)

Total motor loads

Exhaust fans = 34,917 VA
Hood fans = 7,322 VA
Grill vent fans = 4,742 VA
Total load = **46,981 VA •**

Calculating Largest motor load

220-14
8.8 A x 100% x 208 V = 1,830 VA
1,830 VA x 25% = **458 VA *•**

Calculating OCPD

230-90(a); 220-10(b)
Lighting load = 182,000 VA •
Receptacle load = 52,250 VA •
Special load = 99,925 VA •
Heating or A/C load = 400,000 VA •
Motor load = 46,981 VA •
Largest motor load = 458 VA •
Total load = 781,614 VA

Finding amps for OCPD

I = VA/V x √3
I = 781,614 VA / 480 x 1.732 (831 V)
I = 941 A

Calculating phase conductors

220-10; 230-42(a)(1)
General lighting = 146,250 VA *
Show window = 8,000 VA *
Track lighting = 15,000 VA *
Outside lighting = 11,250 VA *
Sign lighting = 1,500 VA *
Receptacle (noncontinuous) = 41,000 VA *
Receptacle (continuous) = 11,250 VA *
Soft drink boxes = 15,000 VA *
Copying machines = 4,000 VA *
Kitchen equipment = 80,925 VA *
Heating or A/C = 400,000 VA *
Exhaust fans = 34,917 VA *
Hood fans = 7,322 VA *
Grill vent fans = 4,742 VA *
Largest motor = 458 VA *
Total load = 781,614 VA

Finding amps phase for conductors

I = VA/A x √ 3
I = 781,614 VA / 480 V x 1.732 (831 V)
I = 941 A

Calculating neutral

220-22
General lighting load = 117,000 VA √
Outside lighting load = 9,000 VA √
Total load = 126,000 VA

Finding amps for neutral conductors

I = VA/V x √3
I = 126,000 VA ÷ 480 V x 1.732 (831 V)
I = 152 A

Sizing OCPD at 125%

Step 1: Selecting OCPD
230-90(a); 240-3(c); 240-6(a)
941 A requires 900 A

Step 2: Selecting OCPD based on six parallel runs at 125%
310-4; 300-5(i)
I = 941 A ÷ 6
I = 157 A

Step 3: Selecting conductors at 125%
310-4; Table 310-16
A requires #2/0 THWN cu.

Step 4: Calculating OCPD based on conductors at 125%
#2/0 THWN = 175 A x 6 = 1,050 A

Solution: **The OCPD is required to be selected at 1,000 amp.**

Sizing conductors based at 100% of load

Step 1: Selecting conductors based on six parallel runs at 125%
230-42(a)(1); 220-10
I = 941 A ÷ 6
I = 157 A

Step 2: Selecting conductors at 125%
157 A requires #2/0 THWN cu.

Step 3: Calculating OCPD based on conductors at 100%
#2/0 THWN = 175 A x 6 =1,050 A

Solution: **The OCPD is required to be selected at 1,000 amps.**

Sizing OCPD and conductors

Step 1: Selecting OCPD at 125%
230-90(a); 240-3(c); 240-6(a)
OCPD must be 1,000 A and conductors must be #2/0 THWN copper.

Solution: **Conductors must be increased to 1,000 amp in rating to be considered protected by the 1,000 amp OCPD computed at 125 percent.**

Note: See Steps 4 and 3 for selecting OCPD based upon 125 percent or 100 percent for continuous loads.

Sizing neutral at 100%

Step 1: Parallel neutral six times in parallel
220-22; 310-4
I = 152 A ÷ 6
I = 25 A

Solution: **A 25 amp conductor cannot be connected in parallel.**

Selecting neutral conductors

Step 1: Selecting neutral six times in parallel
250-24(b)(2); Table 8, Ch. 9
152 A requires # 2/0
#2/0 = 133,100

Step 2: Selecting a single neutral, if needed
250-24(b)(2); Table 250-94
133.1 KCMIL requires #2/0

Step 3: Sizing min. neutral conductors in parallel
310-4; 300-5(i)
#1/0 min. allowed for parallel

Solution: **Six No. 1/0 THWN copper conductors are required per phase.**

Note: For neutral load of 152 amps, see Columns 1 and 2 of page 3-20.

Optional Calculation - Schools
220-34; Table 220-34

Example 3-10. Determine the following loads in VA and amps for a school with a 36,000 sq. ft. classroom area, 5,000 sq. ft. auditorium area, 3,000 sq. ft. cafeteria area, and 1,000 sq. ft. assembly hall area and size the service entrance conductors required for Phases A, B and C and the Neutral. (The school is supplied by a 277/480 volt, three-phase, four-wire service that is paralleled 4 times per phase.)?

OPTIONAL CALCULATION
- NEC 220-20
- TABLE 220-20

- **Sizing OCPD**
- * **Sizing Conductors**
- √ **Sizing Neutral**

Demand Factors	Requirements
• Table 220-20	• NEC 220-20

Lighting loads

(See CALC. LTG. LD pg. 3-19)
- 117,000 VA inside lighting load √
  277 V, single-phase
- 8,000 VA show window load
  120 V, single-phase
- 15,000 VA track lighting load
  20 V, single-phase
- 9,000 VA outside lighting load √
  277 V, single-phase
- 1,200 VA sign lighting load
  120 V, single-phase

Receptacle loads

- 36,000 VA receptacles (noncontinuous)
  120 V, single-phase
- 9,000 VA receptacles (continuous)
  120 V, single-phase
- 18,000 VA multioutlet assembly
  120 V, single-phase

Special loads

- 15,000 VA soft drink boxes
  120 V, single-phase
- 4,000 VA copying machines
  120 V, single-phase
- 400,000 VA heating units
  480 V, three-phase
- 240,000 VA A/C units
  480 V, three-phase

Motor loads

- 34,917 VA exhaust fans
  480 V, three-phase
- 7,322 VA hood fans
  208 V, single-phase
- 4,742 VA grill vent fans
  208 V, single-phase

Kitchen equipment

- 48,000 VA ranges
  208 V, single-phase
- 27,000 VA ovens
  208 V, single-phase
- 16,000 VA fryers
  208 V, single-phase
- 6,000 VA booster heater
  208 V, single-phase
- 9,000 VA heaters
  208 V, single-phase
- 2,000 VA toasters
  120 V, single-phase
- 6,000 VA water heater
  208 V, single-phase
- 6,000 VA refrigerators
  120 V, single-phase
- 4,500 VA freezers
  120 V, single-phase

Total connected load

Table 220-34

Inside lighting load	=	117,000 VA √
Show window load	=	8,000 VA
Track lighting load	=	15,000 VA
Outside lighting load	=	9,000 VA √
Sign lighting load	=	1,200 VA
Receptacles (noncontinuous)	=	36,000 VA
Receptacles (continuous)	=	9,000 VA
Multioutlet assembly	=	18,000 VA
Soft drink boxes	=	15,000 VA
Copying machines	=	4,000 VA
Heating units	=	400,000 VA
Exhaust fans	=	34,917 VA
Hood fans	=	7,322 VA
Grill vent fans	=	4,742 VA
Ranges	=	48,000 VA
Ovens	=	27,000 VA
Fryers	=	16,000 VA
Booster heater	=	6,000 VA
Heaters	=	9,000 VA
Toasters	=	2,000 VA
Water heater	=	6,000 VA
Refrigerators	=	6,000 VA
Freezers	=	4,500 VA
Total load	=	803,681 VA

Total square footage

Table 220-34

Classroom area	=	36,000 sq. ft.
Auditorium area	=	5,000 sq. ft.
Cafeteria area	=	3,000 sq. ft.
Hall area	=	1,000 sq. ft.
Total area	=	45,000 sq. ft.

VA per sq. ft.

Table 220-34
VA per sq. ft. = Total VA / Total sq. ft.
VA per sq. ft. = 803,681 VA ÷ 45,000 sq. ft.
VA per sq. ft. = 17.859578

Applying demand factors

Table 220-34
First 3 VA @ 100%, Next 14.859578 VA @ 75%
3 VA per sq. ft. x 45,000 sq. ft.
x 100% = 135,000.00 VA
14.859578 VA per sq. ft. x 39,000 sq. ft.
x 75% = 434,642.65 VA
Total VA = 569,642.65 VA

Finding amps

Total Amps	Amps in parallel
I = VA / V x √3	A = 685 A ÷ 4
I = 569,642.65 VA ÷ 831 V	A = 171
I = 685	

Total VA loads (neutral 277 V)

General lighting load	=	117,000 VA √
Outside lighting load	=	9,000 VA √
Total load	=	126,000 VA

Finding amps

Total Amps	Amps in parallel
I = VA / V x √3	A = 152 A ÷ 4
I = 126,000 VA ÷ 831 V	A = 38 A
I = 152 A	

Sizing conductors

Phases A, B, and C (310-4)
171 A = 4 #2/0 THHN cu.
Neutral (310-4); (250-24(b)(2))
38 A requires 4 #1/0 in parallel

Optional Calculations For Additional Loads To Existing Installations 220-35 or 220-35(1), Ex.

Example 3-11. Determine if the existing service can have a 15.1 kVA load added which is supplied by four No. 400 KCMIL THW copper conductors?

Step 1: Finding demand
220-35
Maximum demand = 78.4 kVA

Step 2: Calculating existing demand
78.4 kVA x 125% = 98 kVA

Step 3: Calculating existing and added load
98 kVA + 15.1 kVA = 113.1 kVA

Step 4: Calculating amperage
Table 310-16
#400 KCMIL THW copper = 335 A
113.1 kVA x 1,000 = 113,100 VA
113,100 VA ÷ 208 x 1.732 (360 V) = 314 A
314 A is less than 335 A

Solution: The 15.1 kVA load can be applied to the existing service.

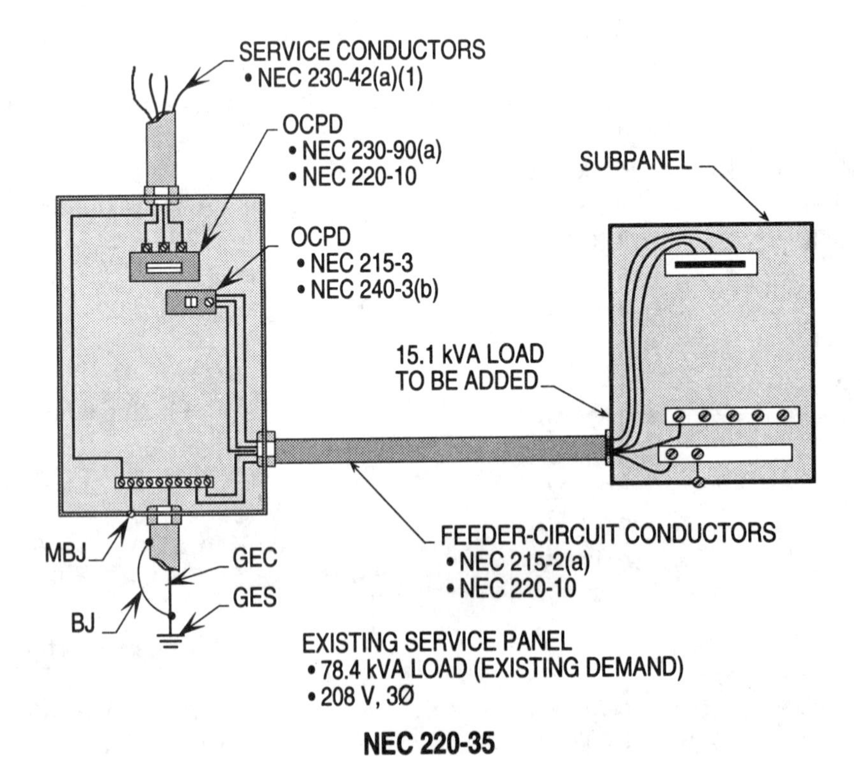

Ill. 3-11

Standard Calculation - Restaurants
220-2 thru 220-22

Example 3-12(a). Determine the following loads in VA and amps for a 70 ft. x 80 ft. restaurant and size the service entrance conductors required for Phases A, B and C and the neutral. **Note:** The lighting load is 51 percent or more incandescent lighting. (The restaurant is supplied by 120/208 volt, three-phase, four-wire service)?

- • Sizing OCPD
- * Sizing Conductors
- √ **Sizing Neutral**

Lighting load

30'-0" lighting track
 120 V, single-phase
10-180 VA outside lighting (continuous duty)
 120 V, single-phase
1,200 VA sign lighting (continuous duty)
 120 V, single-phase

Receptacle load

35 receptacles (noncontinuous duty)
 120 V, single-phase
25 receptacles (continuous duty)
 120 V, single-phase
25'-0" multioutlet assembly (heavy duty)
 120 V, single-phase

Note 1: Do not apply demand factors of Table 220-13 to multioutlet assembly load.

Special loads

2 - 20 kW heating units
 208 V, three-phase
2 - 8,650 VA A/C units
 208 V, three-phase
1 - 5 HP water pumps
 208 V, three-phase
1 - 1/2 HP exhaust fans
 208 V, three-phase

Kitchen equipment

3,800 VA boiler
 208 V, three-phase
2 - 2,700 deep fat fryers
 208 V, three-phase

Note 2: The demand factors of Table 220-13 are permitted (not required) to be applied to the multioutlet assembly load. Application of demand factors do not reduce the load that much.

20 A walk in cooler
208 V, three-phase
6,000 VA water heater
208 V, three-phase
3,650 VA ice cream box
208 V, single-phase
13 A freezer
208 V, single-phase
11,000 VA cooktop
208 V, single-phase
2 - 9,000 VA ovens
208 V, single-phase
12,000 range
208 V, single-phase
14 A refrigerator
208 V, single-phase
18 A refrigerator
208 V, single-phase
2 - 1,500 VA dishwashers
120 V, single-phase
3 - 1,000 VA toasters
120 V, single-phase
820 VA mixer
120 V, single-phase
2 - 1,650 VA dishwashers
120 V, single-phase
1,000 VA dishwasher (booster heater)
120 V, single-phase
6 A bun warmer
120 V, single-phase
2 - 1,820 VA coffee makers
120 V, single-phase
1,200 VA grinder
120 V, single-phase
16 A soft drink box
120 V, single-phase

Calculating lighting load

General lighting load

Table 220-3(a); 220-10; 230-42(a)(1)

5,600 sq. ft. x 2 VA	= 11,200 VA √
11,200 VA x 125%	= 14,000 VA *

Track lighting load

220-12(b)

30 ÷ 2 x 150 VA	= 2,250 VA *√

Outside lighting load

220-10; 230-42(a)(1)

180 VA x 10 units	= 1,800 VA √
1,800 VA x 125%	= 2,250 VA *

Sign lighting load

220-3(b)(6); 600-5(b)(3)

1,200 VA x 100%	= 1,200 VA √
1,200 VA x 125%	= 1,500 VA *

Total lighting load

General lighting load	= 14,000 VA
Track lighting load	= 2,250 VA
Outside lighting load	= 2,250 VA
Sign lighting load	= 1,500 VA
Total load	= **20,000 VA** •

Calculating receptacle load

Receptacle load (noncontinuous)

220-10; 220-3(b)(9)

35 x 180 VA	= 6,300 VA *√

Receptacle load (continuous)

220-10; 230-42(a)(2); 220-3(b)(9)

25 x 180 VA	= 4,500 VA √
4,500 VA x 125%	= 5,625 VA *

Multioutlet assembly

220-3(b)(8)(b)

25' x 180 VA	= 4,500 VA *√

Total receptacle load

Receptacle load (noncontinuous)	= 6,300 VA
Receptacle load (continuous)	= 5,625 VA
Multioutlet assembly	= 4,500 VA
Total load	= **16,425 VA** •

Calculating special load

(kitchen equipment)

Boiler	= 3,800 VA
Deep fat fryers	= 5,400 VA
Walk in cooler	= 7,200 VA
Water heater	= 6,000 VA
Ice cream box	= 3,650 VA
Freezer	= 2,704 VA
Cooktop	= 11,000 VA
Ovens	= 18,000 VA
Range	= 12,000 VA
Refrigerator	= 2,912 VA
Refrigerator	= 3,744 VA
Dishwashers	= 3,000 VA √
Toasters	= 3,000 VA √
Mixer	= 820 VA √
Dishwashers	= 3,300 VA √
Dishwasher (booster heater)	= 1,000 VA √
Bun warmer	= 720 VA √
Coffee makers	= 3,640 VA √
Grinder	= 1,200 VA √
Soft drink box	= 1,920 VA √
Total load	= 95,010 VA

Applying demand factor
Table 220-20

95,010 VA x 65%	= **61,757 VA** • *

Calculating heating or A/C load

220-21

20 kW x 2	= 40,000 VA • *

Calculating motor loads

430-22(a); 430-24: 430-25

Water pump	
16.7 A x 100% x 208 V	= 3,473.6 VA
3,473.6 VA x 1.732	= 6,016.3 VA
Exhaust fans	
2.4 A x 100% x 208 V	= 499.2 VA
499.2 VA x 1.732	= 864.6 VA

Total motor load (See special loads)

Water pump	= 6,016.3 VA
Exhaust fans	= 864.6 VA
Total load	= 6,881 VA • *

Calculating largest motor load

220-14; 430-24

Walk in cooler (Phases)	
20 A x 100% x 208 V	
4,160 VA x 1.732 x 25%	= 1,801 VA • *

Total VA loads (Phases - OCPD)

230-90(a); 220-10

General lighting load	= 20,000 VA •
Receptacle load	= 16,425 VA •
Special load	= 61,757 VA •
Heating or A/C load	= 40,000 VA •
Motor load	= 6,881 VA •
Largest motor load	= 1,801 VA •
Total load	= 146,864 VA

Finding amps

I = VA / V x √3
I = 146,864 VA ÷ 360 V (208 V x 1.732)
I = 408 A

Calculating lighting load (neutral)

220-3(a)

Lighting load	= 16,450 VA √

Calculating Receptacle Load (neutral)

220-3(b)(9)

Receptacle load	= 15,300 VA √

Calculating special load (neutral)

220-20

Dishwashers	= 3,000 VA
Toasters	= 3,000 VA
Mixer	= 820 VA
Dishwashers	= 3,300 VA
Dishwasher (booster heater)	= 1,000 VA
Bun warmer	= 720 VA
Coffee maker	= 3,640 VA
Grinder	= 1,200 VA
Soft drink box	= 1,920 VA
Total load	= 18,600 VA
18,600 VA x 65%	= 12,090 VA √

Calculating largest motor load (neutral)

220-14; 430-24

Soft drink box	
16 A x 100% x 120 V	
1,920 VA x 25%	= 480 VA √

Total VA loads (Neutral)

220-22

Lighting load	= 16,450 VA √
Receptacle load	= 15,300 VA √
Special load	= 12,090 VA √
Largest motor load	= 480 VA √
Total load	= 44,320 VA

Finding amps

I = VA / V x √3
I = 44,320 VA ÷ 360 V (208 V x 1.732)
I = 123 A

Conductors (Phases A, B, and C) (at 100% and 125% of loads)

220-10; 230-42(a)(1)

General lighting load	= 20,000 VA *
Receptacle load	= 16,425 VA *
Special load	= 61,757 VA *
Heating or A/C load	= 40,000 VA *
Motor load	= 6,881 VA *
Largest motor load	= 1,801 VA *
Total load	= 146,864 VA

Finding amps

I = VA / V x √3
I = 146,864 VA ÷ 360 V (208 V x 1.732)
I = 408 A

Parallel 310-4
I = A / no. runs per phase
I = 408 A ÷ 2 (no. runs per phase)
I = 204 A
I = A/ No runs per Neutral
I = 123 A ÷ 2 (no. runs per phase)
I = 62 A

Sizing conductors

Table 310-16; 310-4; 250-24(b)(2)
Phases - 204 A = # 4/0 THWN copper
Neutral - 62 A = # 1/0 THWN copper

Optional Calculation - Restaurants
220-36; Table 220-36

Example 3-12(b). Determine the following loads in VA and amps for a 70 ft. x 80 ft. restaurant and size the service entrance conductors required for Phases A and B and the neutral. **Note:** The lighting load is 51 percent or more incandescent lighting. (The restaurant is supplied by 120/208 volt, three-phase, four-wire service)?

OPTIONAL CALCULATION
- NEC 220-36
- TABLE 220-36
- USE LOADS IN EXAMPLE PROBLEM 3-12(a)

• Phase Loads
√ Neutral Loads

Calculating lighting load

General lighting load

Table 220-3(a)
5,600 sq. ft. x 2 VA = 11,200 VA

Track lighting load

220-12(b)
30' ÷ 2 x 150 VA = 2,250 VA

Outside lighting load

220-10
180 VA x 10 units = 1,800 VA

Sign lighting load

220-3(b)(6); 600-5(b)(3)
1,200 VA x 100% = 1,200 VA

Total lighting load

General lighting load = 11,200 VA
Track lighting load = 2,250 VA
Outside lighting load = 1,800 VA
Sign lighting load = 1,200 VA
Total load = **16,450 VA** • √

Calculating receptacle load

Receptacle load (noncontinuous)

220-3(b)(9)
35 x 180 VA = 6,300 VA

Receptacle load (continuous)

220-10
25 x 180 VA = 4,500 VA

Multioutlet assembly

220-3(b)(8)(b)
25' x 180 VA = 4,500 VA

Total receptacle load

Receptacle load (noncontinuous) = 6,300 VA
Receptacle load (continuous) = 4,500 VA

Multioutlet load	= 4,500 VA
Total load	= **15,300 VA** •√

Calculating special loads

Special loads (kitchen equipment)

Boiler	= 3,800 VA
Deep fat fryers	= 5,400 VA
Walk in cooler	= 7,200 VA
Water heater	= 6,000 VA
Ice cream box	= 3,650 VA
Freezer	= 2,704 VA
Cooktop	= 11,000 VA
Ovens	= 18,000 VA
Range	= 12,000 VA
Refrigerator	= 2,912 VA
Refrigerator	= 3,744 VA
Dishwashers	= 3,000 VA √
Toasters	= 3,000 VA √
Mixer	= 820 VA √
Dishwashers	= 3,300 VA √
Dishwasher (booster heater)	= 1,000 VA √
Bun warmer	= 720 VA √
Coffee makers	= 3,640 VA √
Grinder	= 1,200 VA √
Soft drink box	= 1,920 VA √
Total load	= **95,010 VA** •

Calculating heating or A/C load

220-21

20 kW x 2	= **40,000 VA** •

Calculating motor loads

430-22(a); 430-24; 430-25

Water pump	
16.7 A x 100% x 208 V	= 3,473.6 VA
3,473.6 VA x 1.732	= 6,016.3 VA
Exhaust fans	
2.4 A x 100% x 208 V	= 499.2 VA
499.2 VA x 1.732	= 864.6 VA

Total motor loads

Water pump	= 6,016.3 VA
Exhaust fans	= 864.6 VA
Total load	= **6,881 VA** •

Total VA loads (phases)

Lighting loads	= 16,450 VA •
Receptacle loads	= 15,300 VA •
Special loads	= 95,010 VA •
Heating or A/C loads	= 40,000 VA •
Motor loads	= 6,881 VA •
Total loads	=173,641 VA

Applying demand factors
Table 220-36
Total load = VA x demand factor
Total load = 173,641 VA x 80%
Total load = 138,913 VA

Finding amps

I = VA / V x √3
I = 138,913 VA ÷ 360 V (208 V x 1.732)
I = 386 A

Parallel
I = A ÷ 3 (no. runs per phase)
I = 386 A ÷ 3 (no. runs per phase)
I = 129 A

Total VA loads (neutral)

Lighting load	= 16,450 VA √
Receptacle load	= 15,300 VA √
Special load	= 18,600 VA √
Total loads	= 50,350 VA

Finding amps

I = VA / V x √3
I = 50,350 A ÷ 360 V (208 V x 1.732)
I = 140 A

Parallel 310-4
I = A ÷ 3 (no. runs per phase)
I = 140 A ÷ 3 (no. runs per phase)
I = 47 A

Sizing conductors

Table 310-16; 310-4
Phases - 129 A = #1/0 THWN copper (parallel)
250-24(b)(2)
Neutral - 47 A = #1/0 THWN copper (parallel)

Farm Loads
220-40; 220-41

Calculation Tip Where the dwelling unit has electric heat and one of the farm loads is equipped with an electric grain dryer system, the load for the dwelling unit must be computed using the standard calculation.

Example 3-13(a). Calculate the load in amps for phases A and B for the dwelling unit in Illustration 3-13(a) using the standard calculation?

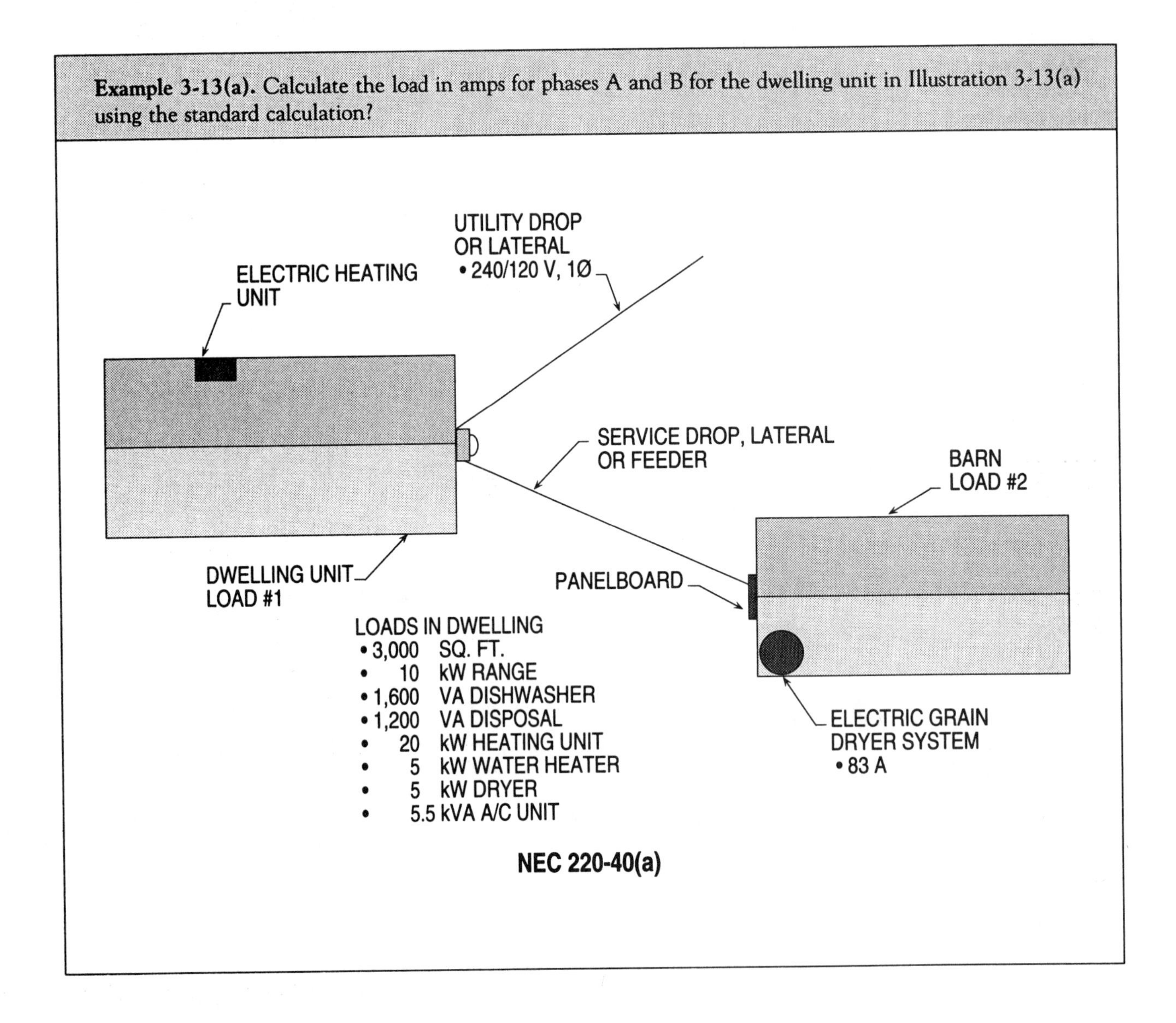

NEC 220-40(a)

Calculating VA loads (phases)

General lighting load

Table 220-3(a)
3,000 sq. ft. x 3 VA = 9,000 VA

Small appliance load

220-16(a); (b)
1,500 VA x 3 = 4,500 VA
Total load = 13,500 VA

Applying demand factors

Table 220-11
First 3,000 VA @ 100% = 3,000 VA
Next 10,500 VA @ 35% = 3,675 VA
Total load = **6,675 VA**

Range load

Table 220-19, Col. A
10 kW = 8,000 VA

Dryer load

Table 220-18
5 kW = 5,000 VA

Fixed appliance load

220-17
1,600 VA x 100% = 1,600 VA
1,200 VA x 100% = 1,200 VA
5,000 VA x 100% = 5,000 VA
Total load = **7,800 VA**

Heating unit load

220-15; 220-21
20 kW = **20,000 VA**

Largest motor load

220-14; 430-24
1,200 VA x 25% = **300 VA**

Total VA loads (phases)

Lighting load	=	6,675 VA
Cooking load	=	8,000 VA
Dryer load	=	5,000 VA
Fixed appliance	=	7,800 VA
Heating or A/C load	=	20,000 VA
Motor load	=	300 VA
Total load	=	47,775 VA

Calculating amps for phases A and B

I = VA / V
I = 47,775 VA / 240 V
I = 199 A

Solution: The load for phases A and B is 199 amps and is determined by the standard calculation and not the optional per NEC 220-40(a). (The electric grain dryer system of 83 amps must be added to the 199 amp computed load.)

Note: The neutral is sized per NEC 220-22 using the standard calculation.

Total amps for phases A & B

Dwelling unit load	=	199 A
Electric grain dryer load	=	83 A
Total load	=	282 A

Farm Loads 220-40

NEC 220-40 and Table 220-40 allows demand factors to be applied to the amps of farm loads which produces smaller values to select service elements.

Example 3-13(b). What is the load in amps for farm loads 2, 3, and 4 in Illustration 3-13(b)?

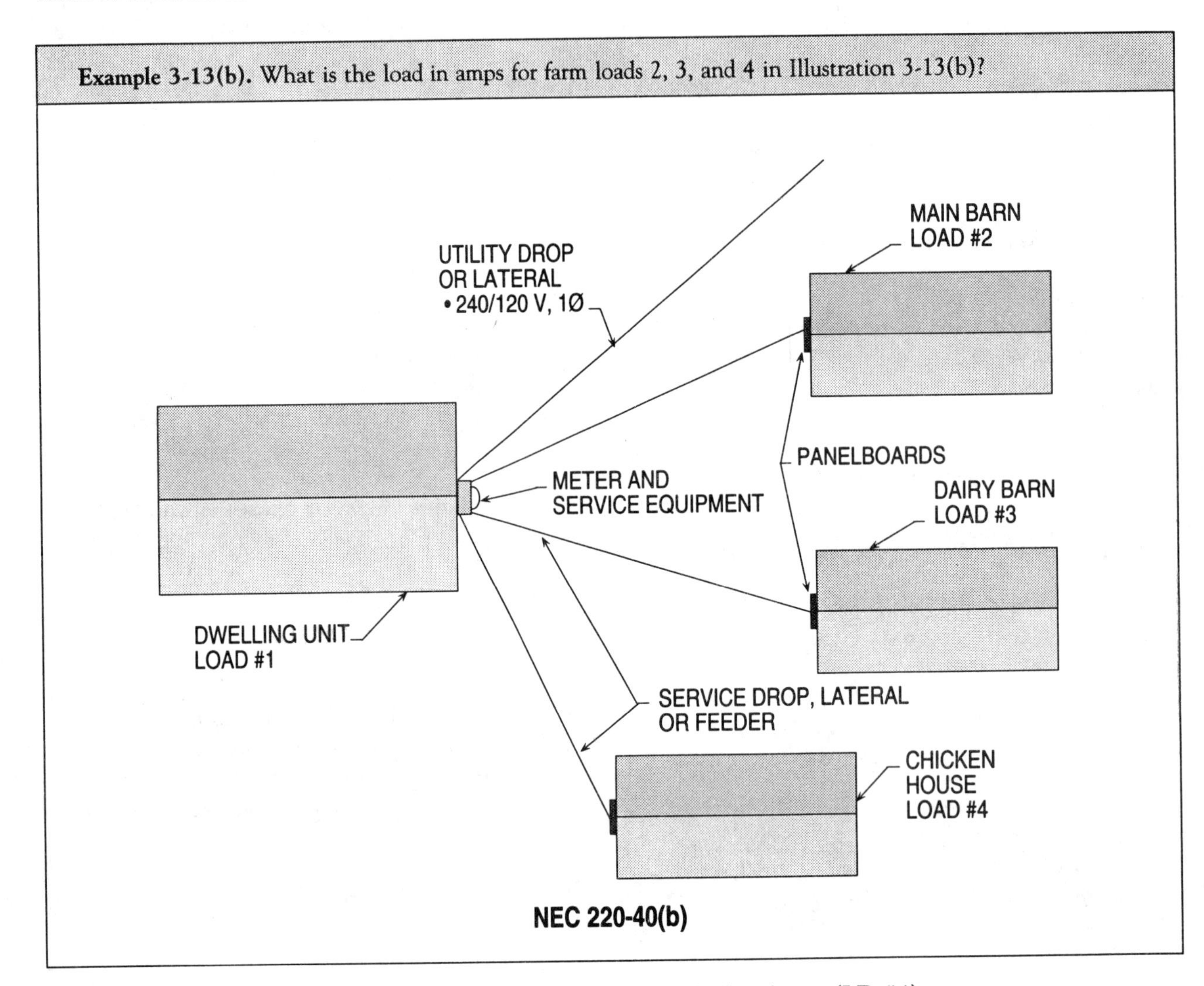

NEC 220-40(b)

Main barn loads (LD #2)

Feed grinder = 6,440 VA (5 HP, 230, 1Ø)
- Auger = 2,300 VA (1 1/2 HP, 230 V, 1Ø)
- Lighting = 2,500 VA
- Receptacles = 4,500 VA

Chicken house (LD #4)

- Brooder = 3,500 VA
- Lighting = 600 VA
- Receptacle = 1,800 VA

Smaller type - (loads)

Dairy barn loads (LD #3)

- Milker = 2,300 VA (1 1/2 HP, 230 V, 1Ø)
- Water Heater = 5,000 VA
- Cooler = 1,840 VA (1 HP, 230 V, 1Ø)
- A/C Unit and htg. pump = 3,910 VA (3 HP, 230 V, 1Ø)
- Lighting = 3,000 VA
- Receptacles = 5,400 VA

Total Amps (phases)

Main barn load #2
220-40(b)

• Feeder grinder (5 HP motor)	=	6,440 VA
• Auger (1 1/2 HP)	=	2,300 VA
• Lighting	=	2,500 VA
• Receptacles	=	4,500 VA
Total load	=	15,740 VA

Load #2 in amps

A = VA / V
A = 15,740 VA / 240 V
A = 66 A

Dairy barn load #3
220-40(b)

• Milker (1 1/2 HP)	=	2,300 VA
• Water heater	=	5,000 VA
• Cooler (1 HP)	=	1,840 VA
• A/C unit and Htg. pump (3 HP)	=	3,910 VA
• Lighting	=	3,000 VA
• Receptacles	=	5,400 VA
Total load	=	21,450 VA

Load #3 in amps

A = VA / V
A = 21,450 VA / 240 V
A = 89 A

Chicken house load #4
220-40(b)

• Brooder	=	3,500 VA
• Lighting	=	600 VA
• Receptacles	=	1,800 VA
Total load	=	5,900 VA

Load #4 in amps

A = VA / V
A = 5,900 VA / 240 V
A = 25 A

Applying demand factors
Table 220-40

Load #2 = 28 A x 25% + 66 A	=	73 A
Load #3	=	89 A
Load #4	=	25 A
Total load	=	187 A

First 60 A @ 100%	=	60 A
Next 60 A @ 50%	=	30 A
Remainder 67 A @ 25%	=	17 A
Total load	=	107 A

Solution: The demand load for the farm load is 107 amps and must be added to the dwelling unit load.

Note: The largest motor load is the 5 HP motor in the main barn #2.

Farm Loads
220-41

NEC 220-41 and Table 220-41 permits demand factors to be applied based upon a largest load sequence which produces smaller amps to select service elements.

Example 3-13(c). What is the total connected load in amps for the farm loads listed in Illustration 3-13(c)?

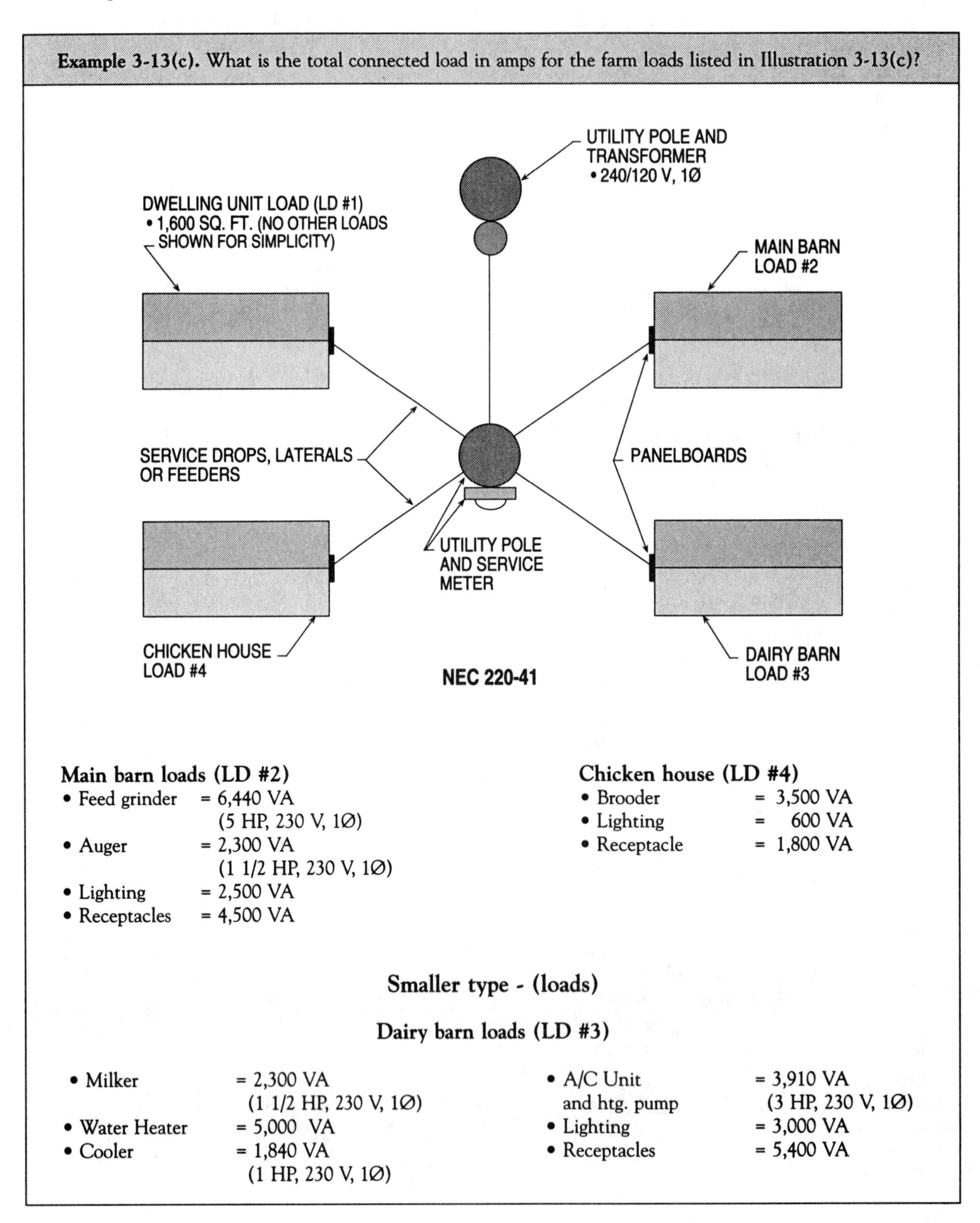

Calculating VA load (phases)

Dwelling unit load #1

• Gen. ltg. and receptacle load
Table 220-3(a)
1,600 sq. ft. x 3 = 4,800 VA

• Small appliance and laundry load
220-16(a);(b)
1,500 VA x 3 = 4,500 VA

• Applying demand factor
Table 220-11
Gen. ltg. load = 4,800 VA
Small appliance load = 4,500 VA
Total load = 9,300 VA

First 3,000 VA @ 100% = 3,000 VA
Next 6,300 VA @ 35% = 2,205 VA
Total load = 5,205 VA

Load #1 in amps

A = VA / V
A = 5,205 VA / 240 V
A = **22 A**

Main barn load #2

• Feeder grinder = 6,440 VA
(5 HP motor)
• Auger = 2,300 VA
(1 1/2 HP)
• Lighting = 2,500 VA
• Receptacles = 4,500 VA
Total load = 15,740 VA

Load #2 in amps

A = VA / V
A = 15,740 VA / 240 V
A = 66 A

Dairy barn load #3

• Milker = 2,300 VA
(1 1/2 HP)
• Water heater = 5,000 VA
• Cooler = 1,840 VA
(1 HP)
• A/C unit and Htg. pump = 3,910 VA
(3 HP)
• Lighting = 3,000 VA
• Receptacles = 5,400 VA
Total load = 21,450 VA

Load #3 in amps

A = VA / V
A = 2,1450 VA / 240 V
A = 89 A

Chicken house load #4

• Brooder = 3,500 VA
• Lighting = 600 VA
• Receptacles = 1,800 VA
Total load = 5,900 VA

Load #4 in amps

A = VA / V
A = 5,900 VA / 240 V
A = 25 A

Applying demand factor Table 220-41

Largest demand load #1 - Ld. 3
89 A x 100% = 89 A
Second largest load #2 - Ld. 2
28 A x 25% + 66 A x 75% = 55 A
Third largest load #3 - Ld. 4
25 A x 65% = 16 A
Balance of load #4 - Ld. 1
22 A x 50% = 11 A
Total load = 171 A

Solution: The total load in amps for the farm loads is 171 amps.

Note: *The largest motor load is the 5 HP motor in the main barn #2.*

Available Short-circuit Current 110-9; 110-10; 250-2(d)

The interrupting capacity of OCPD's shall be equal to the short-circuit current (SCC) that may be imposed upon it, and all other current-carrying elements of the service equipment shall be designed to withstand such currents without being damaged. There are basically two methods used for easy application in finding the SCC.

Example 3-14. What is the available short-circuit current (ASCC) at the terminals of the service equipment in Illustration 3-14?

Calculating ASCC at service equipment using method 1

Step 1: Calculating FLC of transformer
FLC = kVA x 1,000 ÷ V x √3
FLC = 200 kVA x 1,000 ÷ 480 V x 1.732
FLC = 241 A

Step 2: Calculating ASCC at transformer
ASCC = TFLC ÷ Z
ASCC = 241 A ÷ .015
ASCC = 16,067 A

Step 3: Calculating F value using manufacturer Table of C values
Table B of appendix
F = 1.732 x L x I(ASCC) ÷ C x V(L to L)
F = 1.732 x 100' x 16,067 ÷ 16,483 x 480
F = 2,782,804 ÷ 7,911,840
F = .352

Step 4: Calculating M value
M = 1 ÷ 1 + F
M = 1 ÷ 1 + .352
M = .740

Step 5: Calculating line-to-line ASCC at service equipment
110-9; 110-10; 250-2(d)
ASCC = ASCC x M
ASCC = 16,067 x .740
ASCC = 11,890 A

Solution: **The ASCC at the terminals of the service equipment is 11,890 amps.**

Calculating ASCC at service equipment using method 2

Step 1: Calculating ASCC at transformer;
Step 2 above
ASCC = TFLC / Z
ASCC = 241 A / .015
ASCC = 16,067 A

Step 2: Calculating R of 100' of #250 KCMIL
Table A and C of appendix
R = R x L / 1,000
R = .0535 x (100 / 1,000) .10
R = .0054 (rounded up)

Step 3: Calculating R of transformer
R = V to Grd. / ASCC
R = 277 V / 16,067 A
R = .01724

Step 4: Calculating total R
R of #250 KCMIL= .0054
R of transformer = .01724
Total R = .02264

Step 5: Calculating ASCC at service equip.
ASCC = V to Grd. / Total R
ASCC = 277 V / .02264 R
ASCC = 12,235 A

Solution: **The ASCC at the terminals of the service equipment is 12,235 amps.**

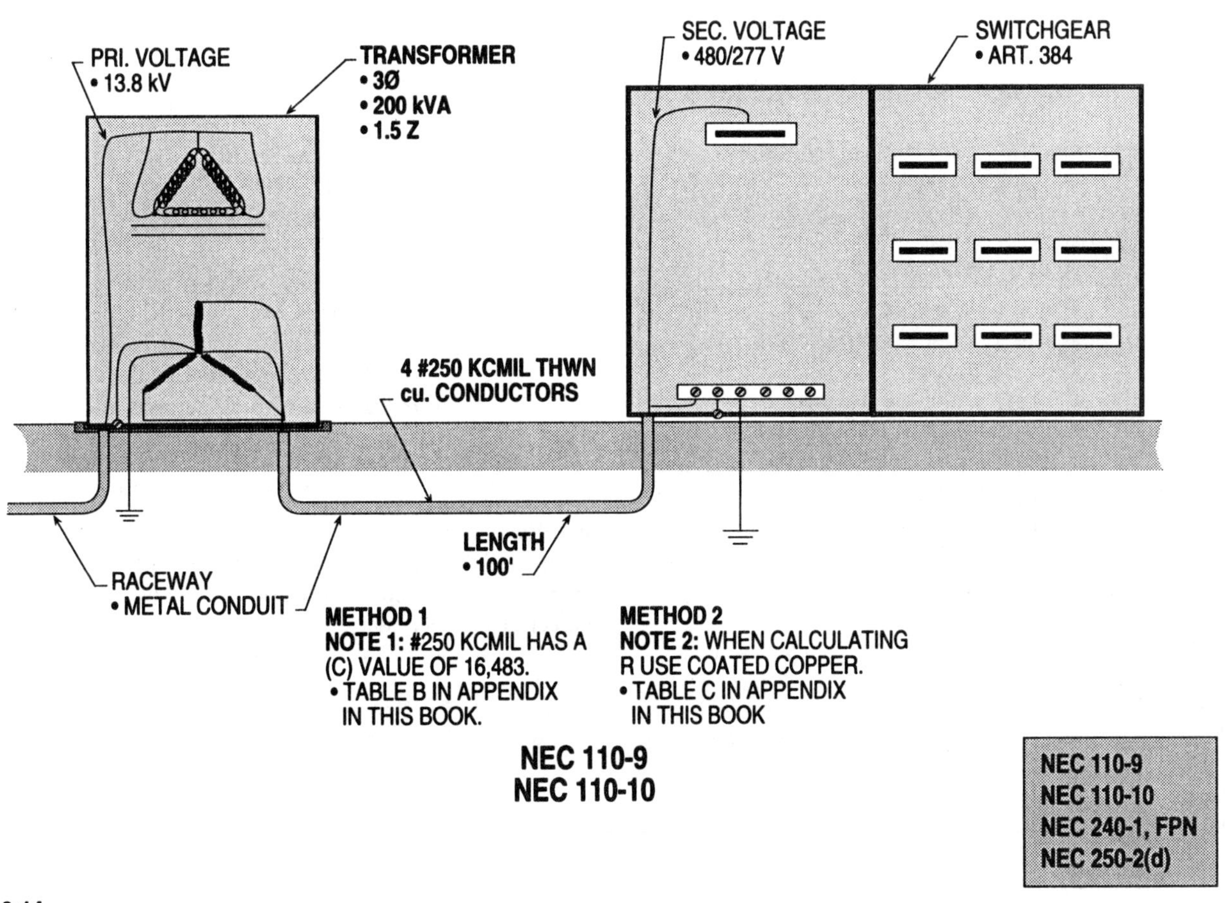
PRI. VOLTAGE
• 13.8 kV
TRANSFORMER
• 3Ø
• 200 kVA
• 1.5 Z
SEC. VOLTAGE
• 480/277 V
SWITCHGEAR
• ART. 384
4 #250 KCMIL THWN
cu. CONDUCTORS
LENGTH
• 100'
RACEWAY
• METAL CONDUIT
METHOD 1
NOTE 1: #250 KCMIL HAS A
(C) VALUE OF 16,483.
• TABLE B IN APPENDIX
IN THIS BOOK.
METHOD 2
NOTE 2: WHEN CALCULATING
R USE COATED COPPER.
• TABLE C IN APPENDIX
IN THIS BOOK
NEC 110-9
NEC 110-10
NEC 110-9
NEC 110-10
NEC 240-1, FPN
NEC 250-2(d)

Ill. 3-14

Ground-fault Protection Of Equipment 230-95

Ground fault protection of equipment is required for grounded wye connected services of more than 150 volts-to-ground and OCPD's rated 1,000 amps or greater. This rule applies mainly for 480/277 volt wye services. Calculations shall be made to find the time-delay in cycles and the setting of the ground-fault sensor.

Example 3-15. What is the maximum time-delay and trip setting of the ground-fault sensor at the main disconnect in Illustration 3-15?

Calculating max. time delay

Step 1: Selecting cycles of max. allowable energy level (MEL)
MEL = 200,000 A cycles

Step 2: Calculating FLA of transformer
FLA = kVA x ,1,000 / V x $\sqrt{3}$
FLA = 1,000 x 1,000 / 480 V x 1.732
FLA = 1,203 A

Step 3: Calculating max. ground fault A (MGFA)
MGFA = transformer FLA x 6
MGFA = 1,203 FLA x 6 = 7,218 A

Step 4: Calculating time delay cycles (TDC)
TDC = step 1 / step 3
TDC = 200,000 A cycles / 7,218 A
TDC = 27.7 cycles

Solution: **The maximum time delay in cycles is 27.7 cycles.**

Calculating the trip setting (TS) 230-95

Step 1: Calculating the setting
TS = OCPD x 20%
TS = 1,200 A x 20% = 240 A

Solution: **The trip setting of the ground-fault sensor is 240 amp.**

Note: The trip setting may be calculated from 10 percent to 50 percent of OCPD as long as 400 amp setting is not exceeded.

For Example: For the 1,200 amp device, the maximum setting is 399.6 amps (1,200 A x 33.3% = 399.6 A). However, most designers start with 20 percent and increase percentage as necessary.

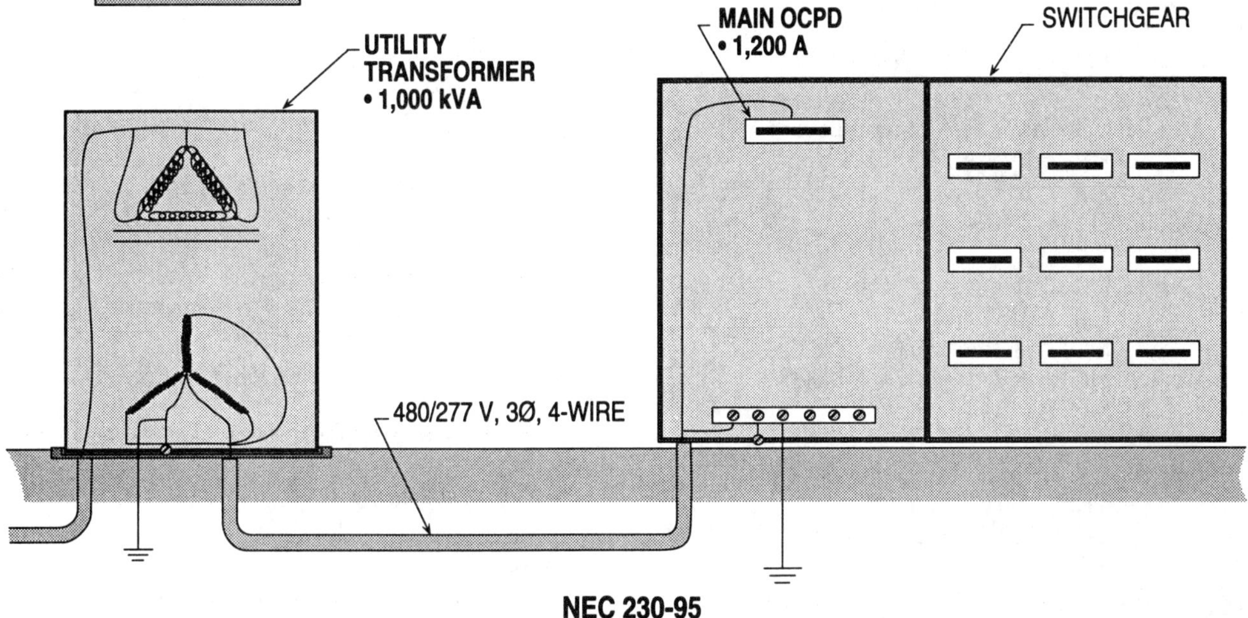

NEC 230-95

NEC 215-10
NEC 230-95
NEC 240-13

Ill. 3-15

Selective Coordination Of Short-circuit Protection 230-95; 240-12

Selective coordination ensures that certain OCPD's in an electrical system will open and by choice, localize a fault condition to restricted outages and equipment affected.

Example 3-16(a). What is the maximum time delay in cycles and the recommended trip setting for the subpanel CB's in Illustration 3-16(a)?

Sizing max. time delay

Step 1: Finding cycles of max. allowable energy level (MEL)
MEL = 200,000 A cycles

Step 2: Calculating FLA of transformer
FLA = kVA x 1,000 / V x $\sqrt{3}$
FLA = 2,000 x 1,000 / 480 V x 1.732
FLA = 2,407

Step 3: Calculating max. ground fault A (MGFA)
MGFA = Trans. FLA x 6
MGFA = 2,407 A x 6 = 14,442 A
MGFA = 14,442 A

Step 4: Calculating time delay cycles (TDC)
TDC = Step 1 / Step 3
TDC = 200,000 A cycles / 14,442 A
TDC = 13.85 cycles

Solution: **The maximum time-delay setting for the main in the service equipment is 13.85 cycles.**

Sizing trip setting (TS) for OCPD's

Step 1: Calculating setting for service OCPD
230-95; 240-13
TS = OCPD x 10%
TS = 2,000 A x 10%
TS = 200 A

Step 2: Calculating setting for FC 1
215-10; 240-13
TS = OCPD x 10%
TS = 1,000 A x 10%
TS = 100 A

Step 3: Calculating setting for FC 2
215-10; 240-13
TS = OCPD x 10%
TS = 400 A x 10%
TS = 40 A

Solution: **The recommended trip settings are 200 amps for OCPD at the service equipment, 100 amps for feeder-cir cuit 1 and 40 amps for feeder 2.**

Note: OCPD in service equipment and feeder-circuits, can't be achieved, ground-fault limiters may have to be used.

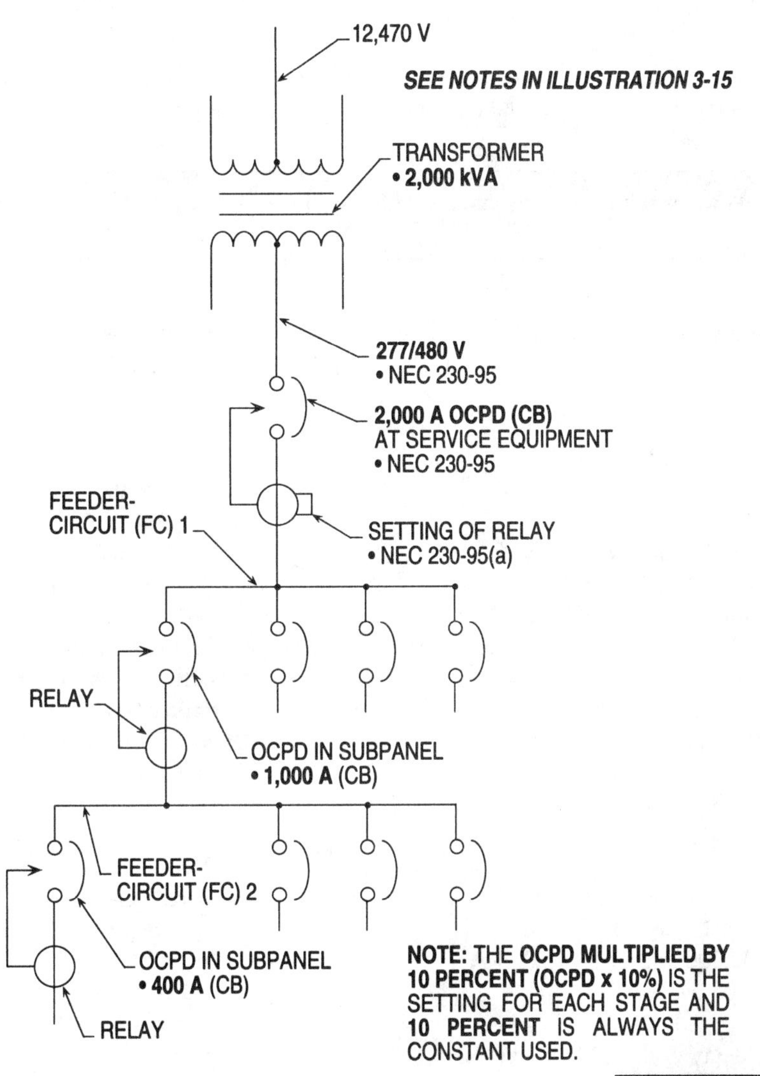

NEC 230-95
NEC 240-12

Services	Feeders
• NEC 230-95	• NEC 215-10
Setting	GFPE
• NEC 230-95(a)	• NEC 240-13
Testing	Hospitals
• NEC 230-95(c)	• NEC 517-17

Ill. 3-16(a)

Selective Coordination Of Short-circuit Protection 230-95; 240-12

The use of a ground-fault limiter will more adequately cause the selected OCPD in the electrical system to trip open, which will minimize personnel hazards, damaged equipment, and provides an orderly shutdown of only the affected area.

Example 3-16(b). What is the time delay cycles (TDC) of the 2,500 amp CB and ground-fault limiter (GFL) in Illustration 3-16(b)?

Sizing the time delay cycles without the GFL

Step 1: Finding cycles of max. allowable Energy level (MEL)
MEL = 200,000 A cycles

Step 2: Calculating FLA of transformer
FLA = kVA x 1,000 / V x √3
FLA = 2,500 x 1,000 / 480 V x 1.732
FLA = 3,008 A

Step 3: Calculating max. ground fault A (MGFA)
MGFA = Trans. FLA x 6
MGFA = 3,008 A x 6
MGFA = 18,048 A

Step 4: Calculating time delay cycles (TDC)
TDC = Step 1 / Step 3
TDC = 200,000 A cycles / 18,048 A
TDC = 11.08 cycles

Solution: The time delay cycles (11.08 cycles) without the use of a ground-fault limiter does not coordinate with the downstream settings of 30 and 15 cycles.

Sizing the time delay cycles with the ground-fault limiter

Step 1: Calculating TDC for GFL
TDC = 200,000 A cycles / GFL (trip setting)
TDC = 200,000 A cycles / 2,500 A
TDC = 80 cycles

Solution: The time delay cycles (80 cycles) with the use of the ground-fault limiter has a longer setting and will coordinate with the down stream settings of 30 and 15 cycles.

Note: A time-delay setting of 60 cycles instead of 80 cycles for the service equipment may be used. Use 30 cycles (60 ÷ 2 = 30) for FC 1 and 15 cycles (30 ÷ 2 = 15) for FC 2. This procedure provides a better coordination of settings in cycles.

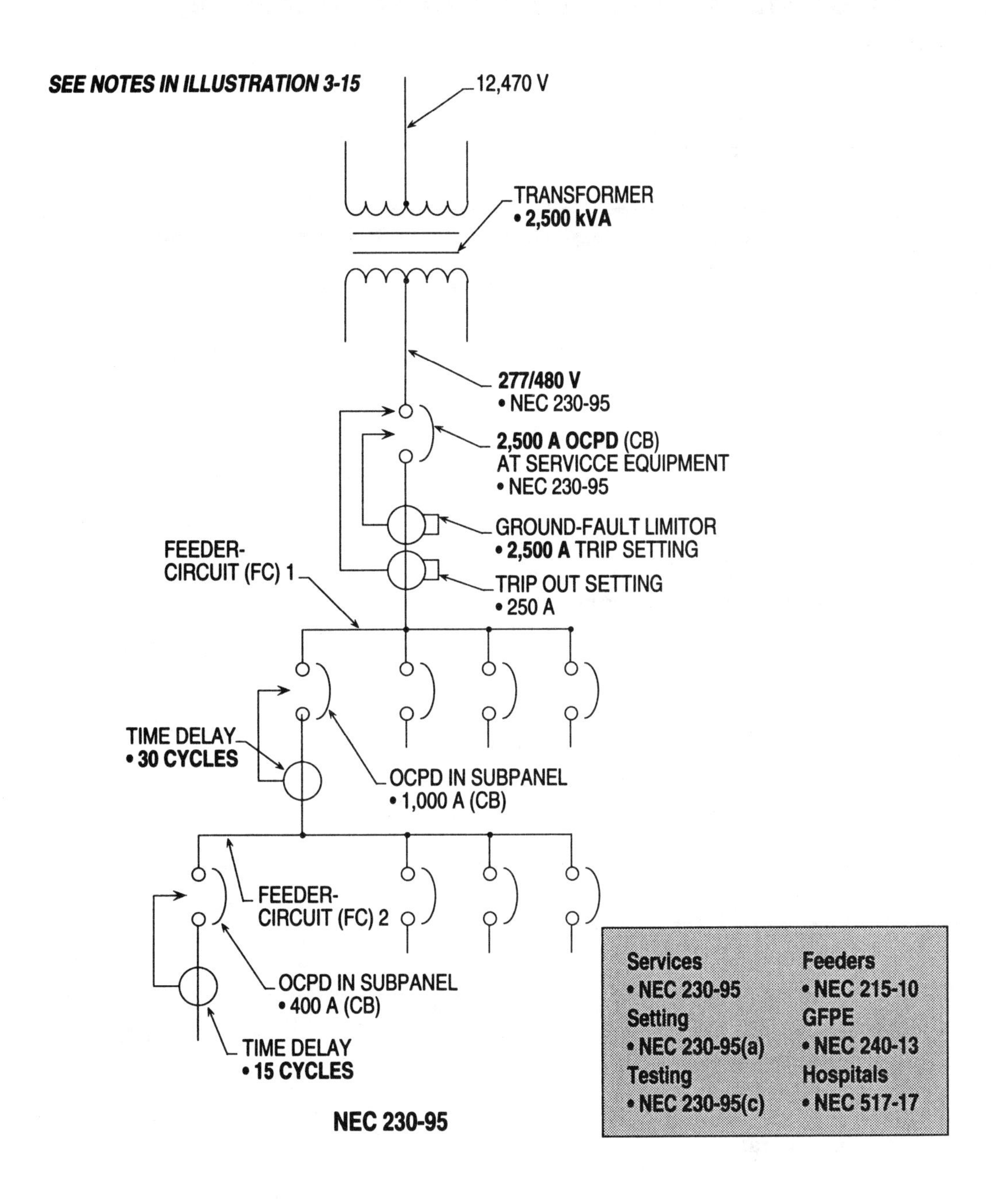
SEE NOTES IN ILLUSTRATION 3-15
12,470 V
TRANSFORMER
• 2,500 kVA
277/480 V
• NEC 230-95
2,500 A OCPD (CB)
AT SERVICCE EQUIPMENT
• NEC 230-95
GROUND-FAULT LIMITOR
• 2,500 A TRIP SETTING
FEEDER-
CIRCUIT (FC) 1
TRIP OUT SETTING
• 250 A
TIME DELAY
• 30 CYCLES
OCPD IN SUBPANEL
• 1,000 A (CB)
FEEDER-
CIRCUIT (FC) 2
OCPD IN SUBPANEL
• 400 A (CB)
TIME DELAY
• 15 CYCLES
NEC 230-95
Services
• NEC 230-95
Setting
• NEC 230-95(a)
Testing
• NEC 230-95(c)
Feeders
• NEC 215-10
GFPE
• NEC 240-13
Hospitals
• NEC 517-17

Ill. 3-16(b)

Overcurrent Protection For Services (Over 600 V) 230-208

The OCPD's for conductors supplying systems over 600 volts can be sized above the ampacity of the conductors to allow loads with high inrush currents to operate.

Example 3-17(a). What is the maximum setting for a circuit breaker (CB) used as a main OCPD in Illustration 3-17(a)?

Finding the setting

Step 1: Selecting percentage
230-208
CB = 6 times (600%)

Step 2: Calculating OCPD setting
230-208
200 A x 6 = 1,200 A

Solution: The maximum setting for the CB is 1,200 amps.

Example 3-17(b). What is the maximum rating of the fuses in Illustration 3-17(b)?

Finding the rating

Step 1: Selecting percentages
230-208
Fuses = 3 times (300%)

Step 2: Calculating rating
230-208
200 A x 3 = 600 A

Solution: The rating of each fuse is 600 amps.

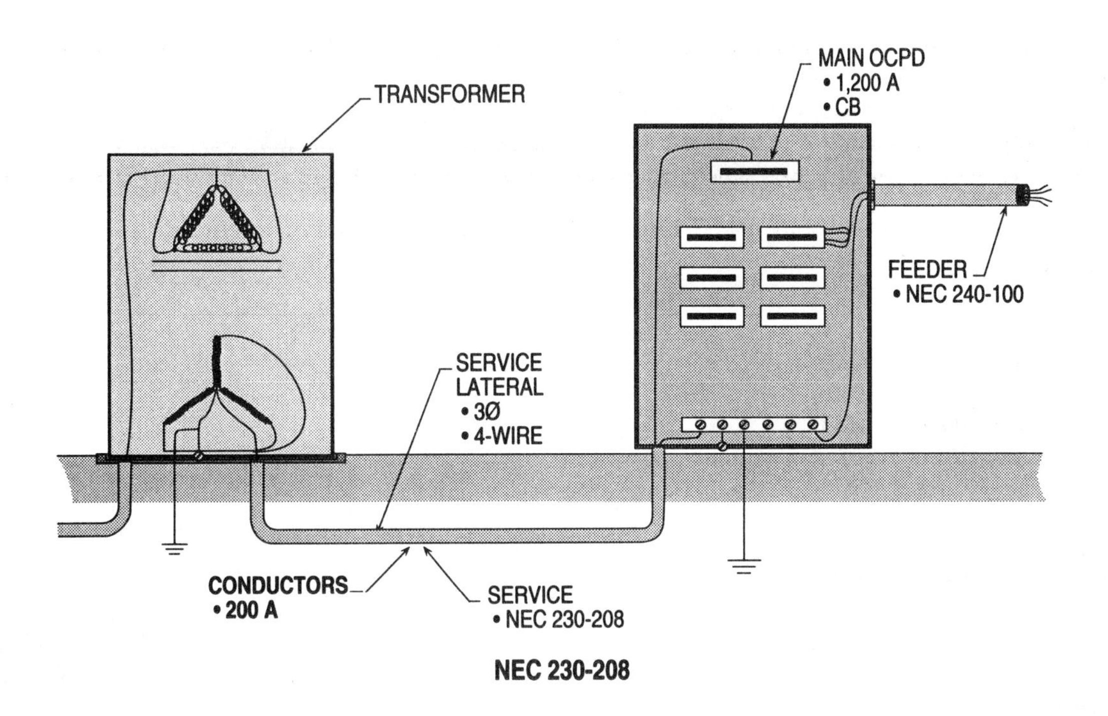

NEC 230-208
NEC 240-100; 101
NEC 490-20

Ill. 3-17(a)

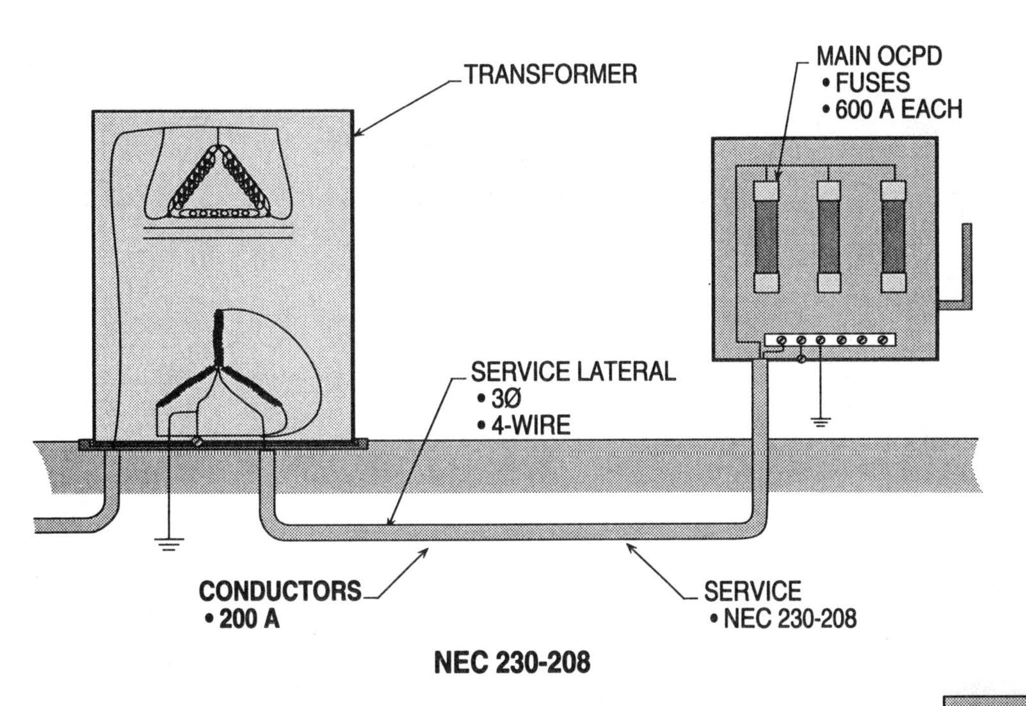

NEC 230-208
NEC 240-100; 101
NEC 490-20

Ill. 3-17(b)

Overcurrent Protection For Feeders (Over 600 V) 230-208(b)

OCPD's supplying power to systems rated over 600 volts are not required to be derated 80 percent for continuous operation.

Example 3-18. What size OCPD is required to supply the connected load in Illustration 3-18?

Finding the size

Step 1: Selecting percentage
Art. 100; 230-208(b)
Over 600 V = 100%
(For continuous or noncontinuous operation)

Step 2: Calculating OCPD
230-208(b)
200 A x 100% = 200 A

Step 3: Selecting OCPD
240-6(a)
200 A requires 200 A

Solution: The size OCPD is 200 amps.

NEC 384-16(d), Ex.
• 600 V OR LESS
NEC 230-208(b)
• OVER 600 V
NEC 240-100; 101

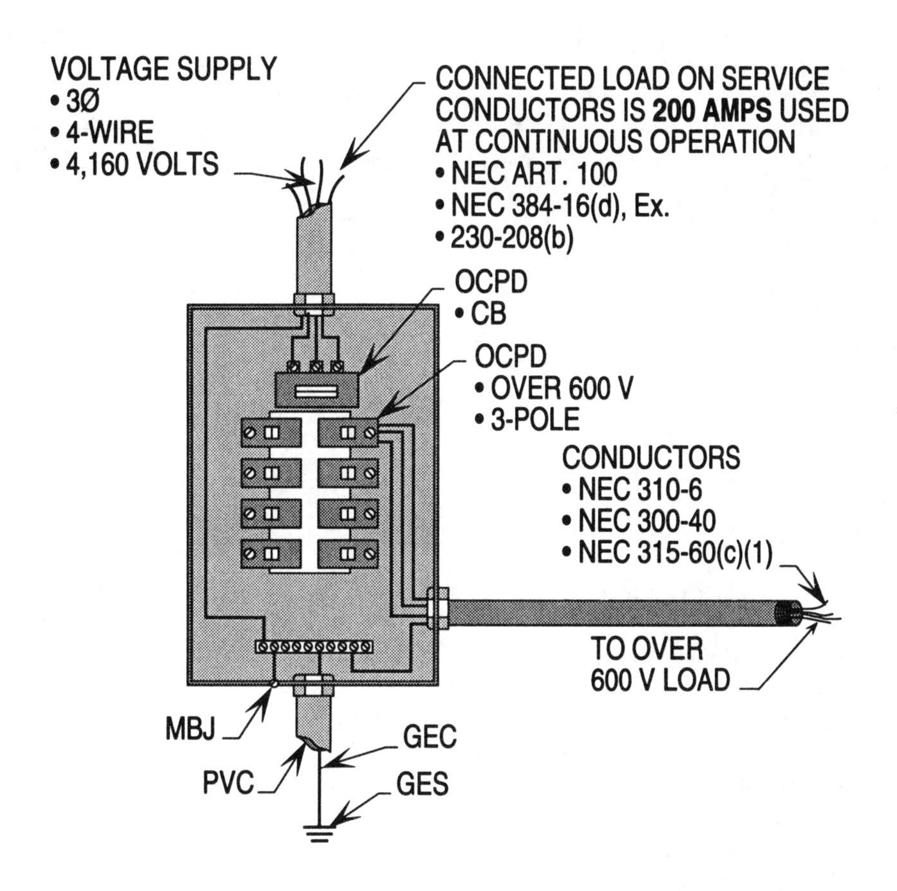

NEC 230-208(b)

Ill. 3-18

Calculating Working Clearances And Flash Protection Pertaining To Equipment
Table 110-34(a); NFPA 70E, CH. 2-3.3.3.2

Certain working clearances must be provided in front of electrical equipment, for the safety of personnel. Calculations to verify if flash protection is needed should also be performed.

Example 3-19. What is the working clearance and flash protection requirements for the electrical equipment in Illustration 3-19?

Finding Clearance
Table 110-34(a)
and 110-32

Step 1: Min. clearance
Table 110-34(a)
4,160 V requires 3 ft.

Step 2 : Min. width
110-32
4,160 V requires 3 ft.

Step 3 : Min. height
110-32
6' 6" vertically

Solution: **The minimum working clearance in front of the electrical equipment is 3 ft., the width is 3 ft., and the height is 6 ft. 6 in..**

Calculating Flash Protection

Step 1 : Finding MVA
MVA = V x 1.732 x SCA X 10^{-6}
MVA = 4,160 V x 1.732 x 2,024 A x 10^{-6}
MVA = 14.6 or 15 rounded up.

Step 2 : Calculating distance
for curable burn

Applying simple formula
FPB = $\sqrt{53 \times mVA \times T}$
FPB = $\sqrt{53 \times 15 \times .1}$
FPB = 8.9 ft.

Solution: **The flash protection boundary distance is 8.9 ft. from the arc to the employee.**

NOTE: WHEN QUALIFIED PERSONNEL WORK IN THE FLASH PROTECTION BOUNDARY, IF THE EQUIPMENT IS ENERGIZED WITH EXPOSED PARTS, THEY MUST WEAR PROTECTIVE CLOTHING AND USE INSULATED HAND TOOLS.

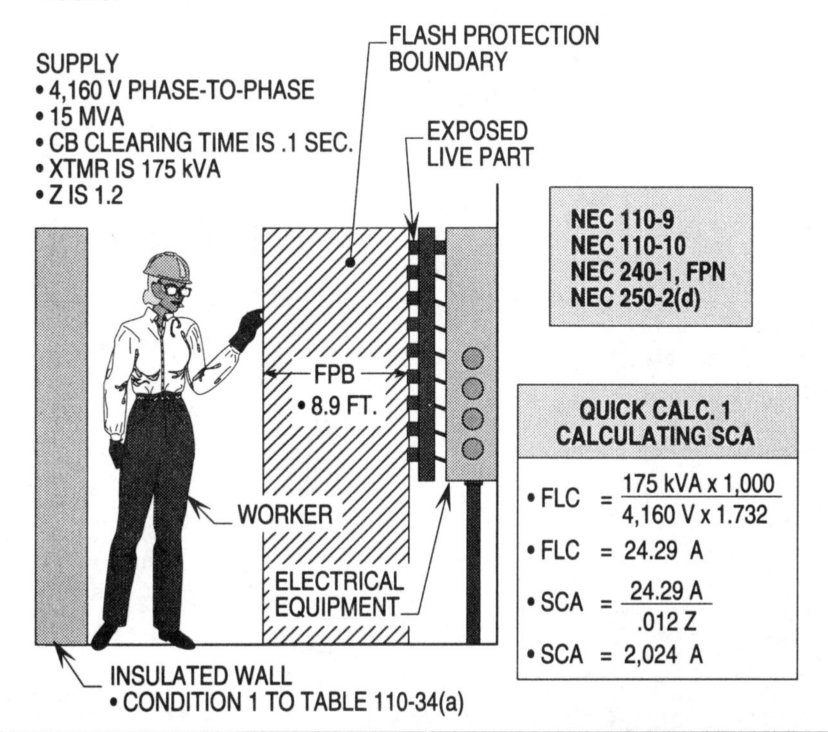

QUICK CALC. 3 CALCULATING POWER	QUICK CALC. 2 CALCULATING MVA
• P = 1.732 x V x SCA x 10^{-6} x $.707^{-2}$	• MVA = V x 1.732 x ISC x 10^{-6}
• P = 1.732 x 4,160 V x 2,024 A x 10^{-6} x $.707^{-2}$	• MVA = 4,160 V x 1.732 x 2,024 A x 10^{-6}
• P = 7.28 MW OR 7,200,000 W	• MVA = 14.5 OR 15 ROUNDED UP

TABLE 110-34(a)
NFPA 70 E, CH. 2.3.3.3.2

Ill. 3-19

4

Sizing Overcurrent Protection Devices

Overcurrent protection devices (OCPD's) must be sized to keep the current flow in a circuit at a level that prevents overheating of terminals, conductors and equipment. Excessive current flowing in a circuit generates heat, which raises the temperature of the wire and circuit elements. High temperature of the wire can cause the ignition of insulation protecting conductors and equipment. Loads that are properly calculated and overcurrent protection devices selected which are based upon these computations provides the necessary protection for elements of electrical systems. There are two parts in any installation to be protected and they are the circuit conductors and elements of the equipment.

Quick Reference

Parallel Conductors
310-4

Example 4-1(a). What size THWN copper conductors are required for the parallel connections in Illustration 4-1(a)?

Step 1: Calculating load
220-10; 230-42(a)(1)
600 A x 125% = 750 A

Step 2: Calculating number of conductors
Table 310-16; 310-4
No. of conductors = A of load / A of conductor used
No. of conductors = 750 A / 255 A (#250 KCMIL)
No. of conductors = 2.9

Step 3: Calculating total A
310-4
Total A = A of conductors x # in parallel
Total A = 255 A x 3
Total A = 765 A

Solution: **Each conductor connected in parallel must be No. 250 KCMIL copper conductor rated at 255 amps each.**

Note: Three No. 250 KCMIL cu. coppers (255 A x 3 = 765 A) will serve a load of 750 amps.

Example 4-1(b). What size THWN copper conductors are required for the parallel connections in Illustration 4-1(b)?

Step 1: Calculating load
220-10; 230-42(a)(1)
960 A x 125% = 1,200 A

Step 2: Calculating number of conductors
Table 310-16; 310-4
No. of conductors = A of load / A of conductor used
No. of conductors = 1,200 A / 420 A (#600 KCMIL)
No. of conductors = 2.8

Step 3: Calculating total A
310-4
Total A = A of conductors x # in parallel
Total A = 420 A x 3
Total A = 1,260 A

Solution: **Each conductor connected in parallel must be No. 600 KCMIL copper conductor rated at 420 amps each.**

Note: Three No. 600 KCMIL cu. coppers (420 A x 3 = 1,260 A) will serve a load of 1,200 amps.

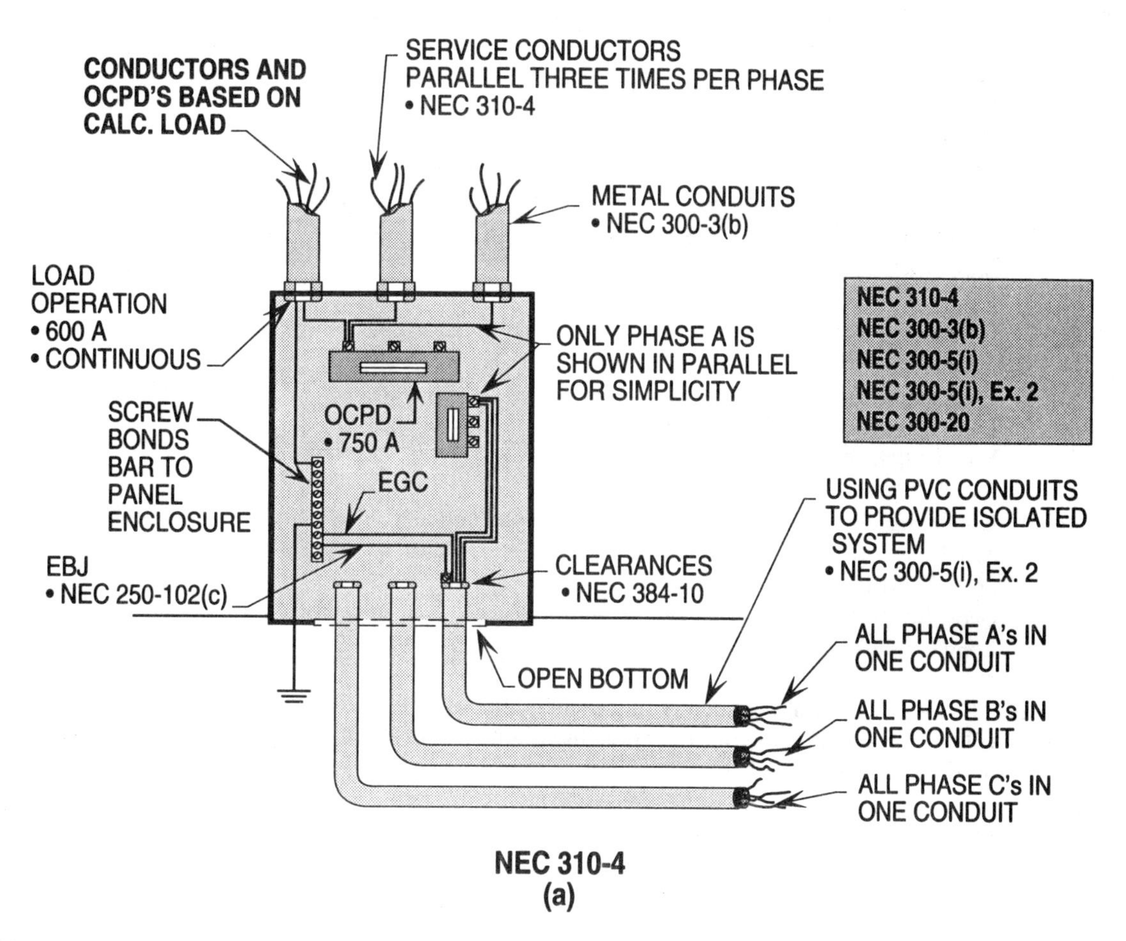

NEC 310-4
(a)

Ill. 4-1(a)

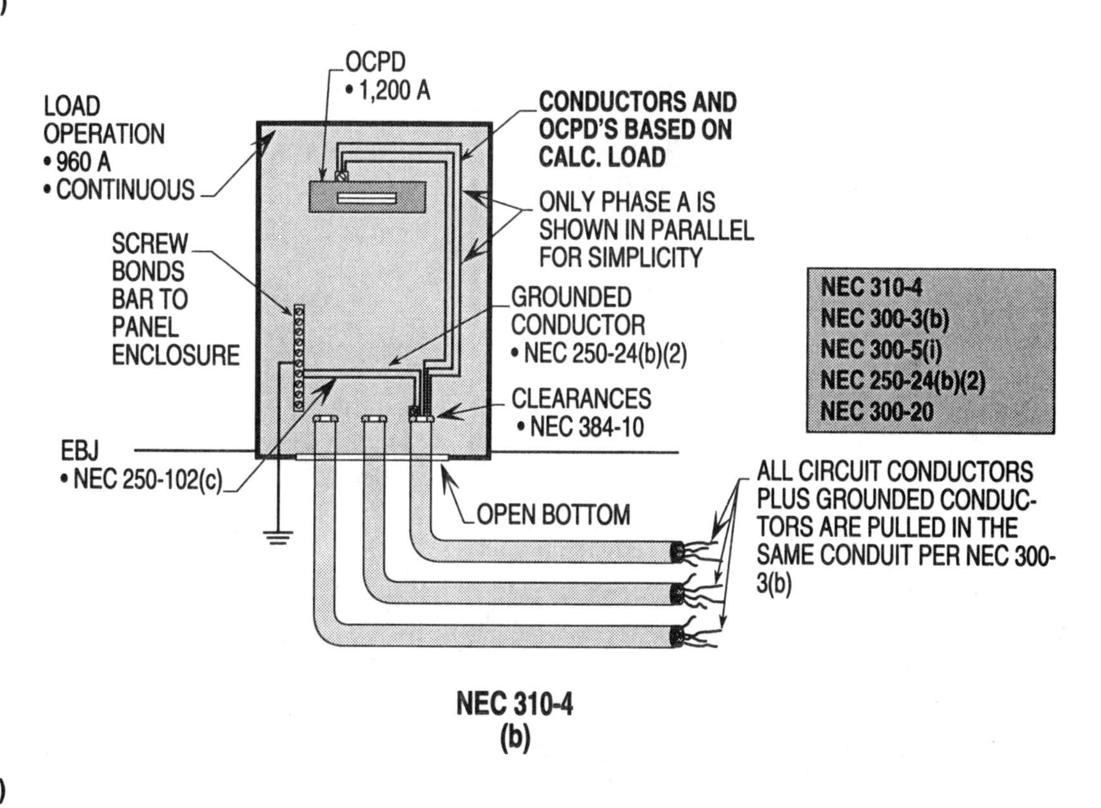

NEC 310-4
(b)

Ill. 4-1(b)

Example 4-1(c). What is the ampacity of the short length of conductors connected in parallel in Illustration 4-1(c)?

Step 1: Calculating short length A
310-4
Short length A = long length / total length x A of OCPD
Total length = 25' + 30' = 55'
Short length A = 30' / 55' x 800 A
Short length A = 436 A

Step 2: Calculating long length A
310-4
Long length A = short length / total length x A of OCPD
Long length A = 25' / 55' x 800 A
Long length A = 363.6 A

Step 3: Comparing A of the two lengths to A of parallel conductor
310-4
#600 KCMIL cu. = 420 A
Short length = 436 A
Long length = 363.6 A

Solution: **The short length of conductors in parallel will carry excessive current and must be adjusted in length to prevent overloading the No. 600 KCMIL THWN copper conductors.**

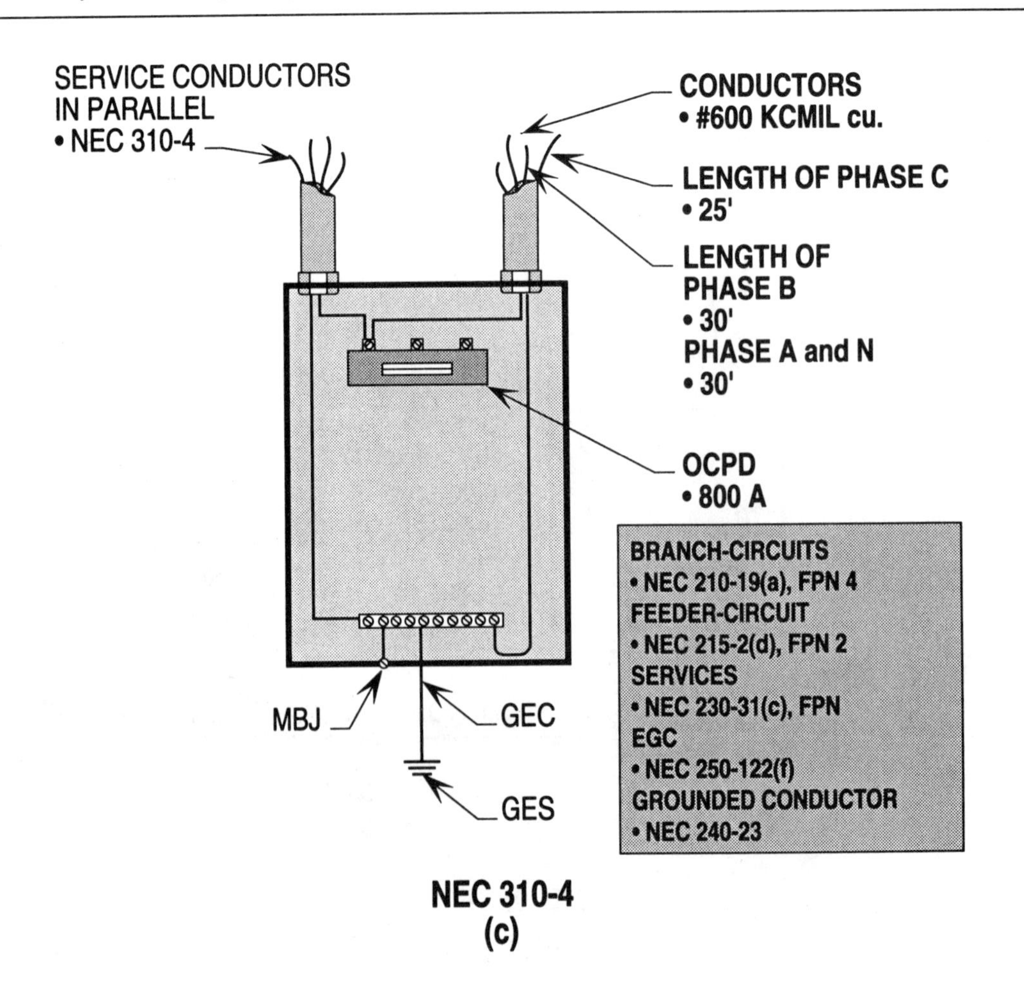

Ill. 4-1(c)

Protection Of Equipment
240-2

There are two parts in any electrical installation that must be protected and they are the circuit conductors and elements of the equipment supplied. The NEC requires the OCPD to be of a rating small enough in size to protect both conductors and equipment or a second stage of protection must be provided such as the overloads in a motor circuit.

Example 4-2. What size OCPD is required to protect the conductors and equipment in Illustration 4-2?

Sizing OCPD

Step 1: Calculating load
384-16(d); 210-19(a)
80 A x 125% = 100 A

Step 2: Sizing OCPD
240-2; 240-6(a)
100 A requires 100 A

Solution: The size OCPD is 100 amps.

Sizing Conductors

Step 1: Sizing the conductors
Table 310-16
100 A requires #3

Solution: The size THWN copper conductors are No. 3 based on OCPD and equipment.

NEC 384-16(d); NEC 210-19(a); NEC 215-2(a); NEC 230-42(a)(1)

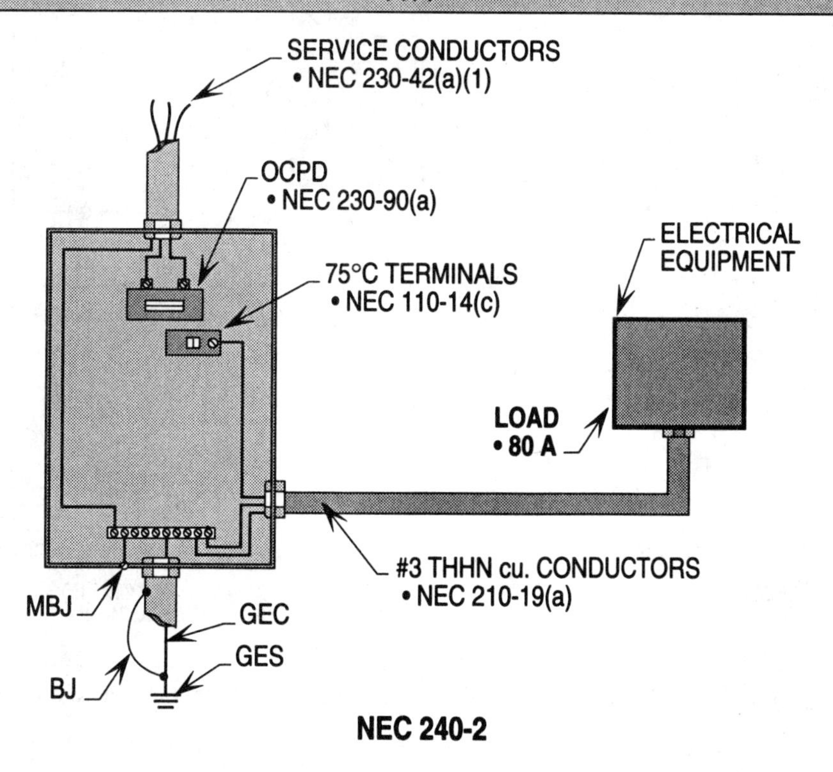

Ill. 4-2

Transformer Secondary Conductors 240-3(f)

Under certain conditions, conductors supplied by the secondary side of a single-phase or three-phase transformer, having a two-wire or three-wire secondary, is considered protected by the primary OCPD.

Examples 4-3(a) and (b). Are the secondary conductors for the transformers in Illustrations 4-3(a) and (b) considered protected by the primary OCPD?

Single-phase, two-wire system

Step 1: Finding ratio A
240 V ÷ 480 V x 100 A = 50 A

Solution: The 50 amp OCPD on the primary can protect the secondary.

Three-phase, three-wire system

Step 1: Finding ratio A
240 V ÷ 480 V x 420 A = 210 A

Solution: The 200 amp OCPD on the primary can protect the secondary.

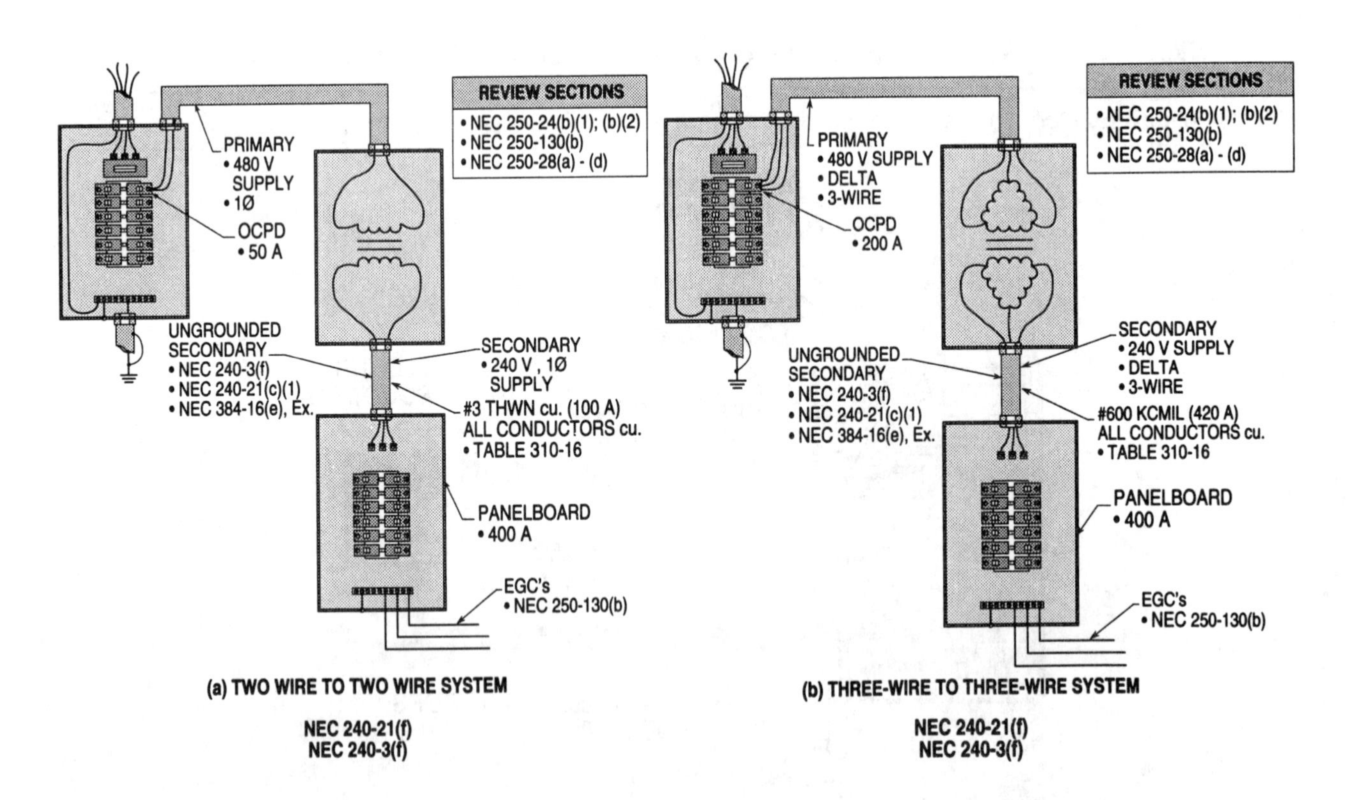

Note: CB's in secondary panelboards are nonautomatic and are utilized as disconnect switches only.

Ill.'s 4-3(a) and (b)

Protection Of Flexible Cords 240-4(b)(1)

Flexible cords approved for use with specific listed appliances or portable lamps connected to a branch-circuit can be smaller than the power supply conductors and protecting OCPD.

Example 4-4. What is the smallest size flexible cord conductor connected to a branch-circuit and considered protected by the OCPD in Illustration 4-4?

Sizing Flexible Cord

Step 1: Sizing flexible cord
240-4(b)(1)
20 A OCPD allows #18

Solution: The size conductor in the flexible cord is No. 18.

NEC 210-19(d), Ex. 2; NEC 210-19(a); NEC 210-21

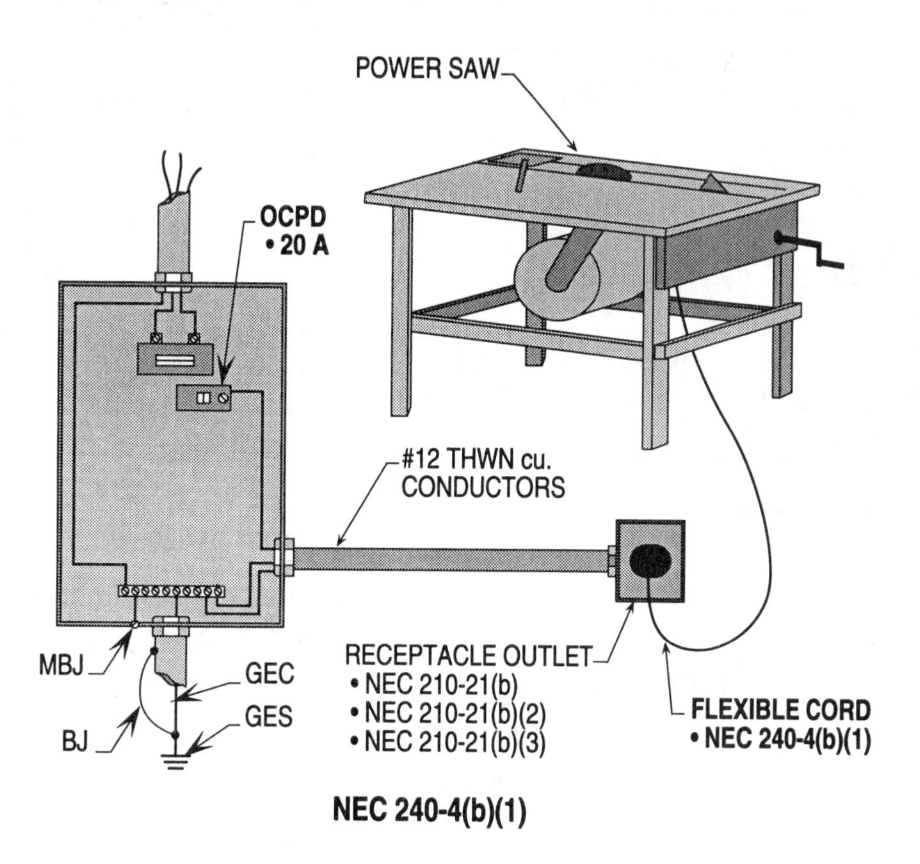

Ill. 4-4

Protection Of Fixture Wires 240-4(b)(2)

Fixture wires connected to 120 volt or higher branch-circuits can be smaller in size than the power supply conductors and protecting OCPD.

Example 4-5(a). What is the smallest size fixture wire permitted in a stem and connected to the branch-circuit in Illustration 4-5(a)?

Sizing Fixture Wire

Step 1: Sizing fixture wire
240-4(b)(2)
40 A OCPD allows #12

Solution: The size fixture wire allowed is No. 12.

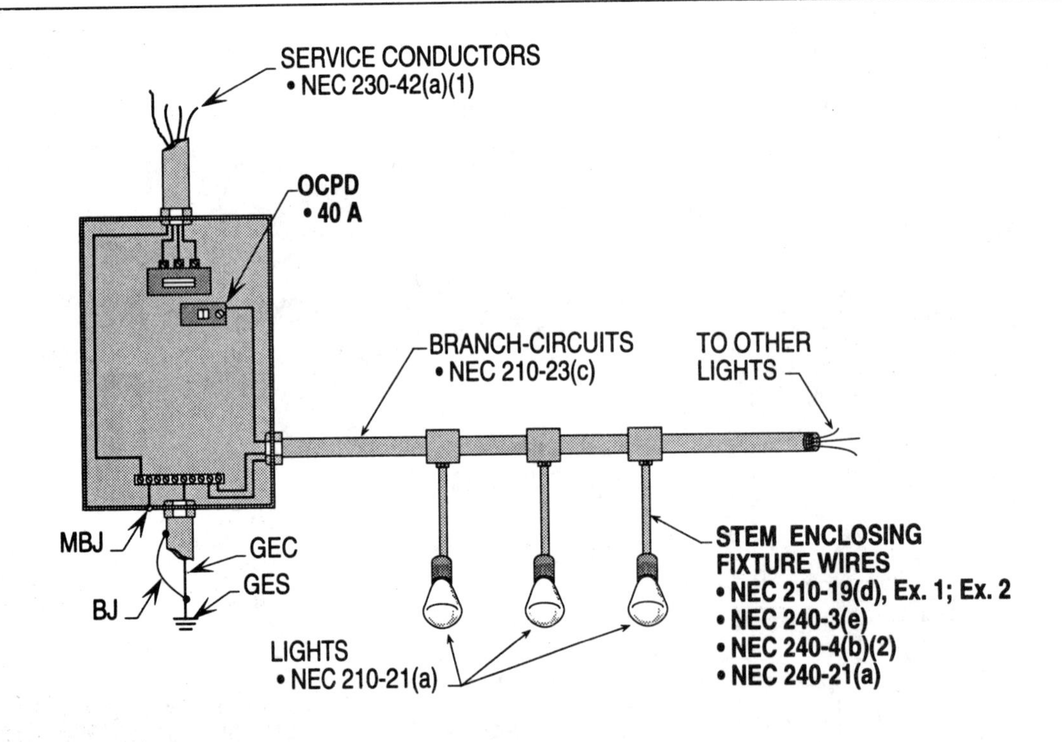

NEC 210-19(d), Ex. 1
NEC 210-21(a)
NEC 210-23(a) - (d)

Ill. 4-5(a)

Example 4-5(b). What is the smallest size fixture wire permitted in a fixture whip connected between a 20 amp branch-circuit and a fluorescent lighting fixture in Illustration 4-5(b)?

Sizing Fixture Wire

Step 1: Sizing fixture wire
240-4(b)(2)
20 A OCPD allows #18

Solution: **The size fixture wire allowed is No. 18.**

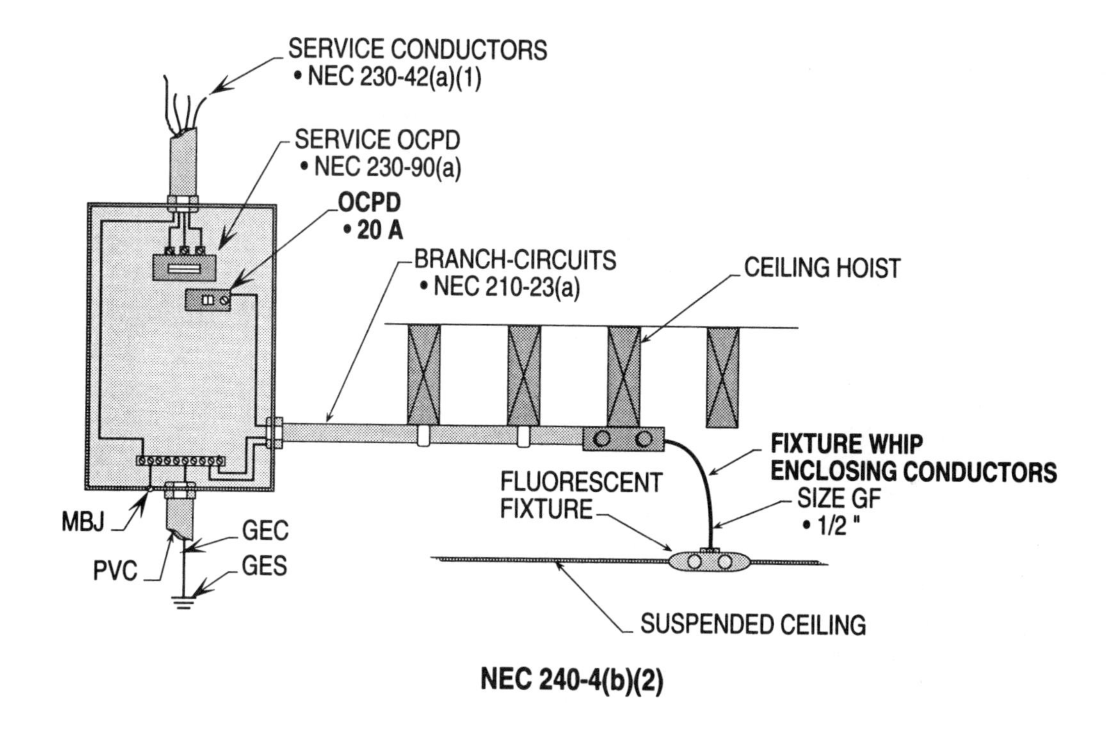

NEC 240-4(b)(2)

Ill. 4-5(b)

NEC 410-67(c)
NEC 250-118(6)
NEC 350-14
NEC 410-16(c)

Sizing Taps 240-21(a) - (g)

The general rule requires an OCPD to be placed at the source of the circuit to which it is connected. If a branch-circuit taps to a feeder-circuit, an OCPD protecting the conductors and elements of the circuit must be installed at the point where the tap is made to the feeder-conductors. However, there are Subparts (a) through (g) to NEC 240-21 which permits taps to be made without OCP at the point of the tap.

10 ft. Tap Rule 240-21(b)(1)

The conductors used in conjunction with the 10 ft. tap must be equal (OCPD ÷ 10) to the OCPD protecting the feeder-circuit conductors or the OCPD must not exceed 10 times the ampacity of the tapped conductors.

Calculation Tip A main at the end of the tap is only required if the panelboard is a lighting and appliance panel per NEC 384-14(a) and 16(a).

Example 4-6(a). What is the minimum size conductors and OCPD required for the tap in Illustration 4-6(a)?

Sizing conductors

Step 1: Calculating min. size tap
240-21(b)(1)(d)
A = 600 A OCPD ÷ 10
A = 60

Step 2: Sizing conductors
Table 310-16
#6 THWN cu. = 65 A

Step 3: Verifying size
240-21(b)(1)(d)
65 A is greater than 60 A

Solution: The size THWN copper conductors is No. 6.

Sizing OCPD

Step 1: Calculating OCPD
240-3(e); 240-21(b)(1)(a); 240-3(b); 240-6(a)
65 A conductors allows a 60 A
OCPD to protect conductors from
overload

Solution: The size OCPD is allowed to be 60 amps.

Note 1: Another method is to multiply the No. 6 THWN copper conductors at 65 amps by 10 (65 A x 10 = 650 A), which produces 650 amps. This total exceeds the rating of the 600 amp OCPD and complies with NEC 240-21(b)(1)(d).

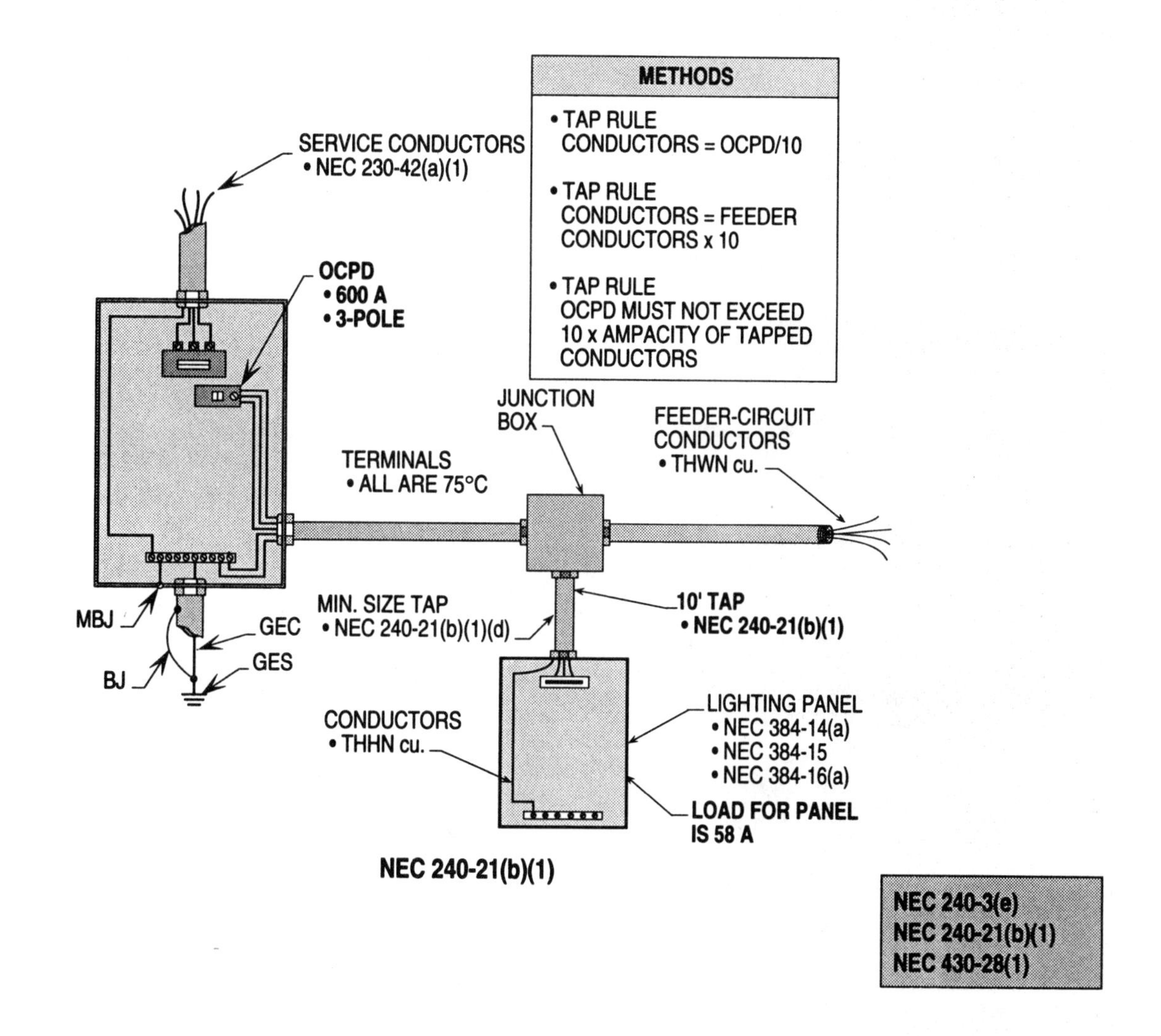

NEC 240-3(e)
NEC 240-21(b)(1)
NEC 430-28(1)

Ill. 4-6(a)

Calculation Tip Dividing the feeder OCPD by 10 derives minimum ampacity of the tap conductors. The ampacity of the tap conductors can be multiplied by 10 and if this new ampacity is equal to or greater than the feeder OCPD, the tap is sized correctly. Another method is to multiply the ampacity of the tapped conductors by 10 and the OCPD ahead of the feeder-circuit conductors must not exceed this rating in amps. **Note:** Either of these procedures produces the same results and provides the proper sized conductors.

25 ft. Tap Rule
240-21(b)(2)

The 25 ft. tap rule allows smaller tap conductors to be made to larger sized conductors and extended to a distance of over 10 ft. to 25 ft. provided the current-carrying capacity of the tap is at least 1/3 that of the feeder OCPD.

Calculation Tip Tap conductors must terminate in an OCPD to protect the elements of the tap.

Example 4-6(b). What is the minimum size conductors and OCPD required for the tap in Illustration 4-6(b)?

Sizing tap conductor

Step 1: Calculating min. size tap
240-21(b)(2)(a)
1/3 x 600 A OCPD = 200 A

Step 2: Selecting conductors
Table 310-16
200 A requires #3/0

Solution: The size THWN copper conductors is No. 3/0.

Sizing OCPD

Step 1: Selecting OCPD
240-3(e); 240-21(b)(2)(b); 240-3(b); 240-6(a)
#3/0 requires a 200 A

Solution: The size OCPD is 200 amps.

NEC 240-3(e); NEC 240-21(b)(2); NEC 430-28(2)

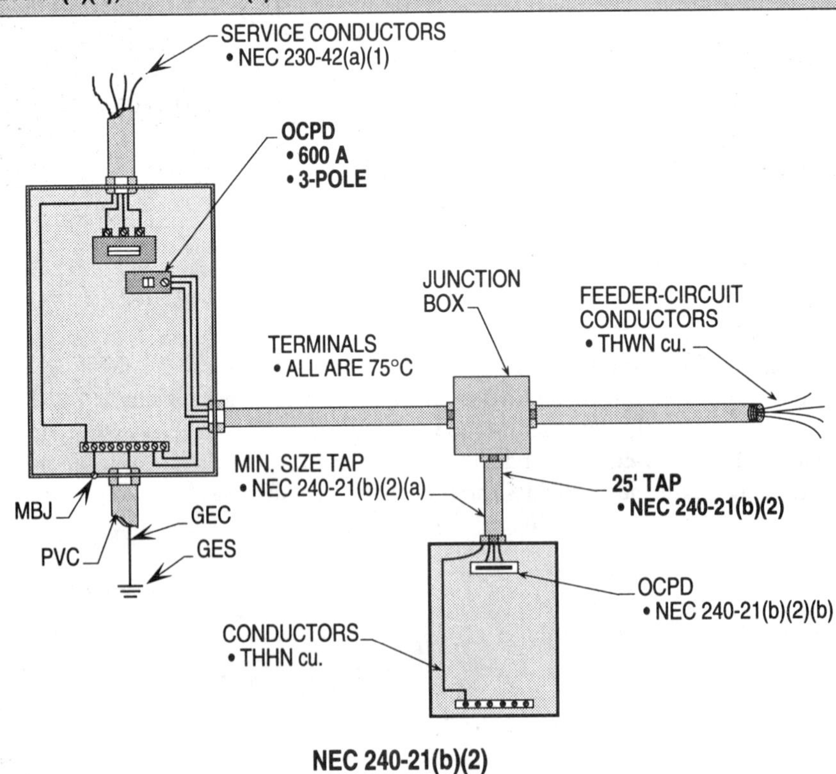

Ill. 4-6(b) **NEC 240-21(b)(2)**

Transformer Feeder Tap Plus Primary And Secondary Not Over 25 ft. Long 240-21(b)(3)

The primary tap for this tap rule must be at least 1/3 of the OCPD device protecting the larger feeder-circuit conductors. The secondary tap conductors must be at least 1/3 of the OCPD protecting the feeder-circuit conductors based on the primary-secondary transformer ratio.

Example 4-6(c). What are the minimum size conductors required for the primary and secondary including OCPD in Illustration 4-6(c)?

Sizing pri. tap conductors

Step 1: Calculating primary tap
240-21(b)(3)(a)
1/3 of 250 A = 83 A

Step 2: Selecting conductors
Table 310-16
83 A requires #4

Solution: The size THWN copper conductors are No. 4.

Sizing sec. tap conductors

Step 1: Calculating secondary tap
240-21(b)(3)(b)
480 V / 208 V x 1/3 x 250 A = 192 A

Step 2: Selecting conductors
Table 310-16
192 A requires #3/0

Solution: The size THWN copper conductors are No. 3/0.

Sizing sec. tap OCPD

Step 1: Selecting OCPD in secondary tap
240-3(e); 240-21(b)(3)(e); 240-6(a)
200 A (#3/0) requires 200 A

Solution: The size OCPD is 200 amps.

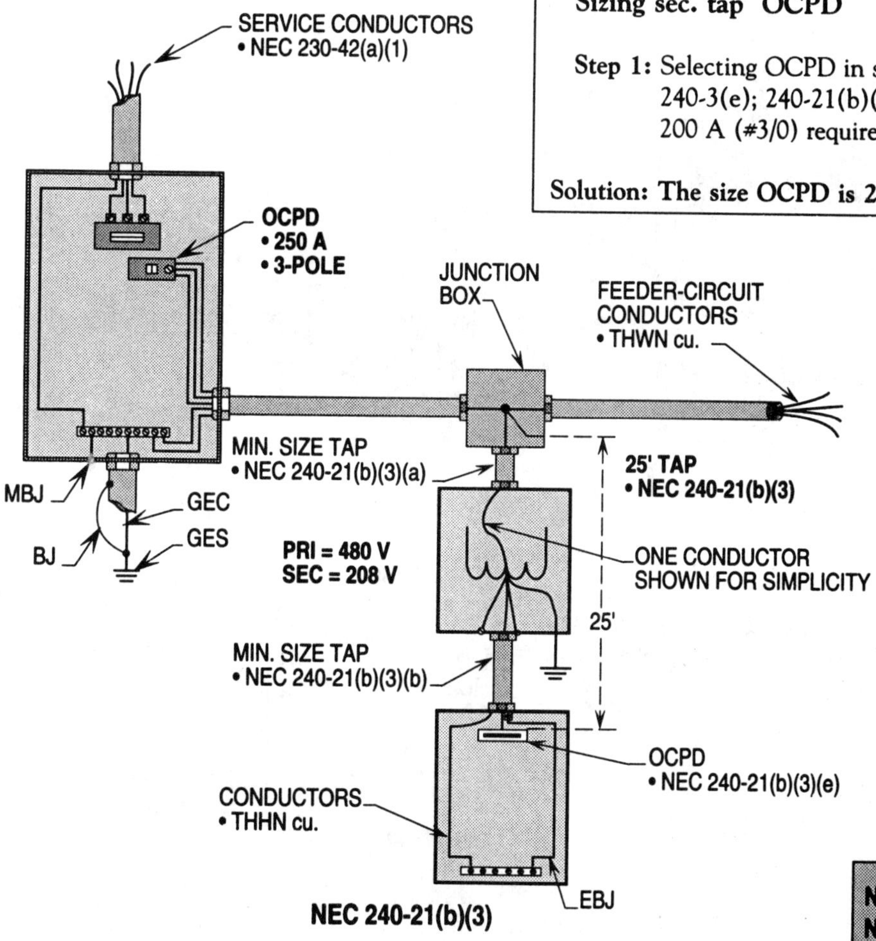

SERVICE CONDUCTORS
• NEC 230-42(a)(1)
OCPD
• 250 A
• 3-POLE
JUNCTION BOX
FEEDER-CIRCUIT CONDUCTORS
• THWN cu.
MIN. SIZE TAP
• NEC 240-21(b)(3)(a)
25' TAP
• NEC 240-21(b)(3)
MBJ
GEC
GES
BJ
PRI = 480 V
SEC = 208 V
ONE CONDUCTOR SHOWN FOR SIMPLICITY
25'
MIN. SIZE TAP
• NEC 240-21(b)(3)(b)
OCPD
• NEC 240-21(b)(3)(e)
CONDUCTORS
• THHN cu.
EBJ
NEC 240-21(b)(3)

NEC 240-3(e)
NEC 240-21(b)(3)
NEC 430-28(2)

Ill. 4-6(c)

100 ft. Tap Rule 240-21(b)(4)

It is permissible to apply the 100 ft. tap rule in high bay manufacturing buildings where the height is over 35 ft. The ampacity of the tap conductors must be at least 1/3 of the rating of the OCPD protecting the larger feeder-circuit conductors. Junction box must be installed at a height of at least 30 ft.

Example 4-6(d). What are the minimum size OCPD and conductors required for the tap in Illustration 4-6(d)?

Sizing conductor

Step 1: Calculating minimum size tap
240-21(b)(4)(c)
1/3 of 350 A = 116.7 A

Step 2: Selecting conductors
Table 310-16
116.7 requires #1

Solution: **The size THWN copper conductors is No. 1.**

Sizing OCPD

Step 1: Selecting OCPD
240-3(e); 240-21(b)(4)(d); 240-3; 240-6(a)
130 A (#1 THWN cu.) requires 125

Solution: **The size OCPD is 125 amps.**

NEC 240-3(e), NEC 240-21(b)(4), NEC 430-28, Ex.

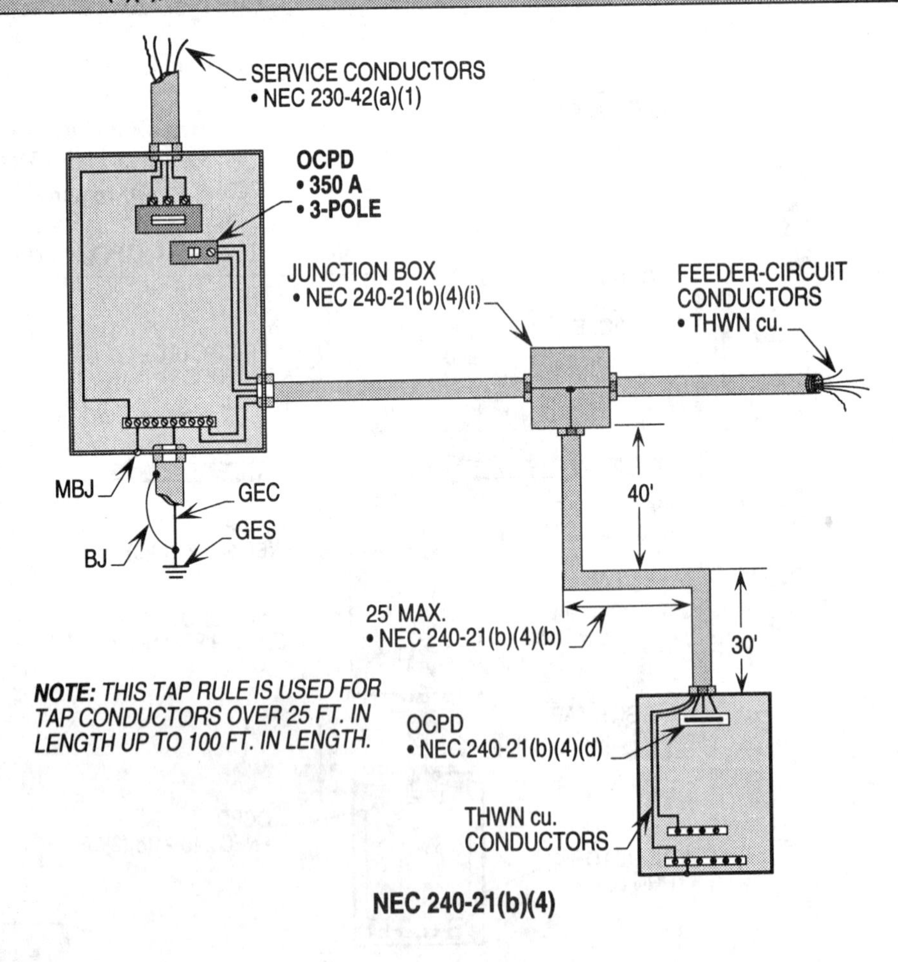

Ill. 4-6(d)

XTMR Secondary Conductors Not Over 10 ft. Long 240-21(c)(2)

If certain rules and conditions are followed, conductors in 10 ft. or less lengths can be connected to the secondary side of transformers.

Example 4-7. What size OCPD and THWN copper conductors are required to be connected from the secondary side of the transformer in Illustration 4-7?

Sizing Conductors

Step 1: Calculating min. size
240-21(c)(2)
Computed load is 148 A

Step 2: Sizing Conductors
Table 310-16
#1/0 THWN cu. = 150 A

Step 3: Verifying size
240-21(c)(2)
150 A is greater than 148 A

Solution: The size THWN copper conductors are No. 1/0 rated at 150 amps.

Sizing OCPD

Step 1: Calculating OCPD
240-3(e); 240-21(c)(2); 240-3(b); 240-6(a)
#1/0 cu. = 150 A
OCPD rated at 150 A protects conductors from overload

Solution: The size OCPD is allowed to be 150 amps.

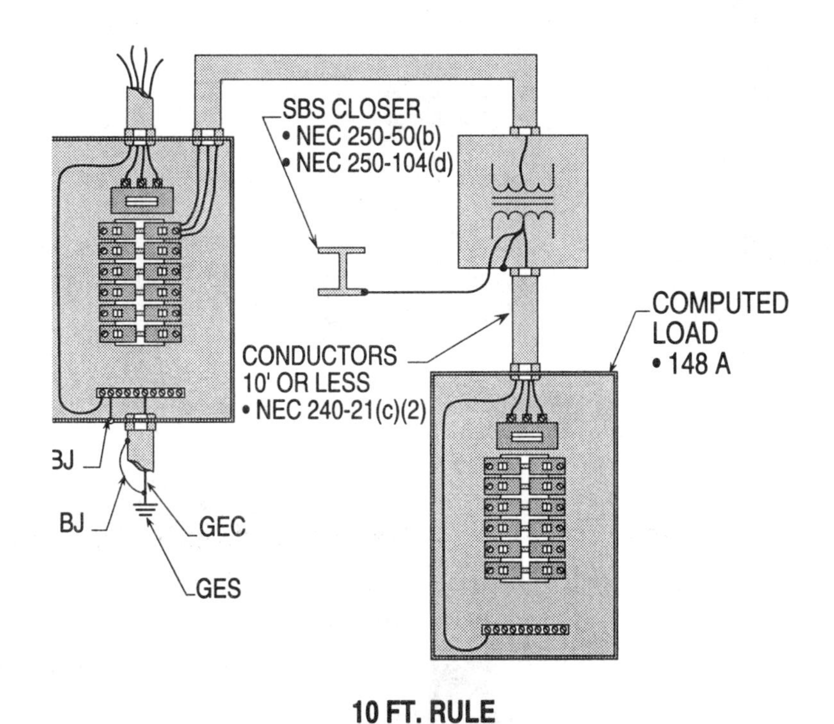

Ill. 4-7

XTMR Secondary Conductors Not Over 25 ft. Long 240-21(c)(3)

If certain rules and regulations are followed, conductors over 10 ft. long 25 ft. or less in length can be connected to the secondary side of transformers.

Example 4-8. What size OCPD is required to protect the conductors that are connected to the secondary side of the transformer in Illustration 4-8?

Sizing OCPD

Step 1: Calculating FLA (Sec.)
FLA = 200 kVA x 1,000 ÷ 208 V x 1.732
FLA = 555.6 A

Step 2: Sizing secondary OCPD
240-21(c)(3)(b)
500 A secondary OCPD is less than the 555.6 A output

Solution: The secondary OCPD rated at 500 amps protects the secondary output and the 4 No. 1/0 parallel per phase is equal to and greater than the output.

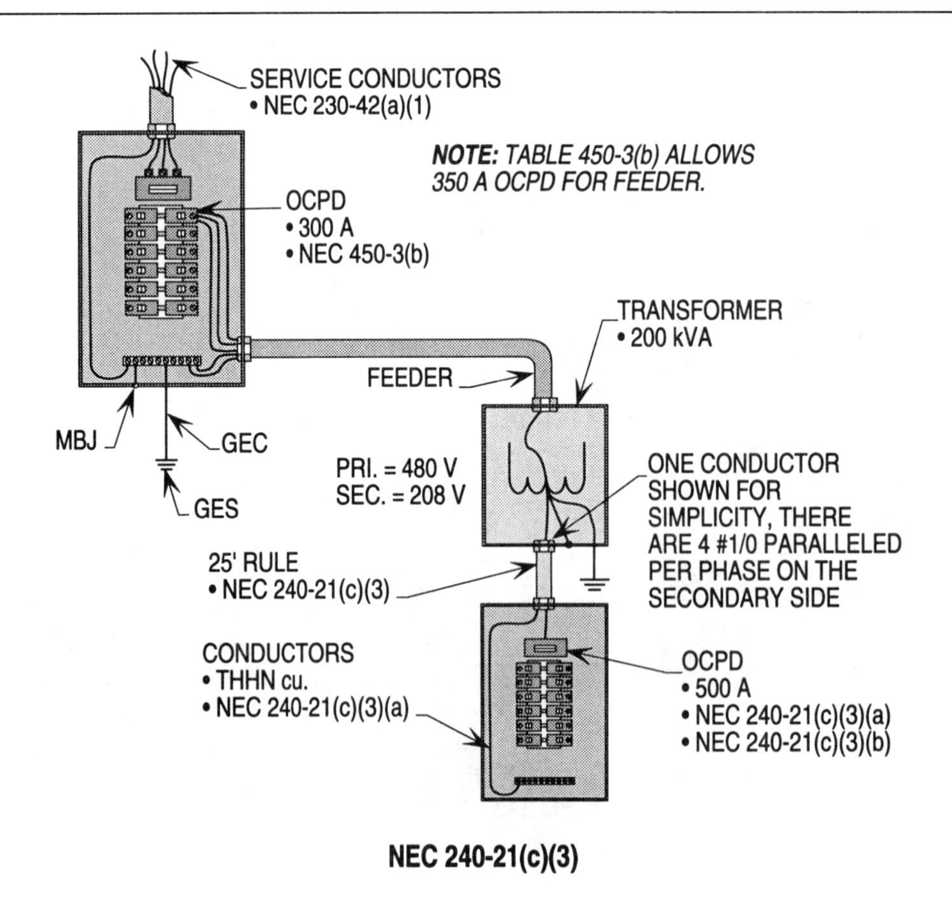

NEC 240-21(c)(3)

Ill. 4-8

Outside Secondary Conductors 240-21(c)(4)

Outside conductors are allowed to be connected at the transformer secondary, without installing an overcurrent protection device at the point of connection.

Example 4-9. What size OCPD and THWN copper conductors are required for the secondary connection from the outside transformer in Illustration 4-9?

Sizing OCPD and Conductors

Step 1: Calculating FLA (Sec.)
FLA = 500 kVA x 1,000 ÷ 480 V x 1.732
FLA = 601.7 A

Step 2: Sizing Secondary OCPD
240-21(c)(4)
600 A secondary OCPD is less than the 601.7 A output

Solution: **The 600 amp OCPD protects the secondary of transformers and the 4 No. 2/0 THWN conductors per phase (175 A x 4 = 700 A) are equal to and greater than the output.**

Note: The AHJ may allow 4 No. 1/0 per phase (150 A x 4 = 600 A).

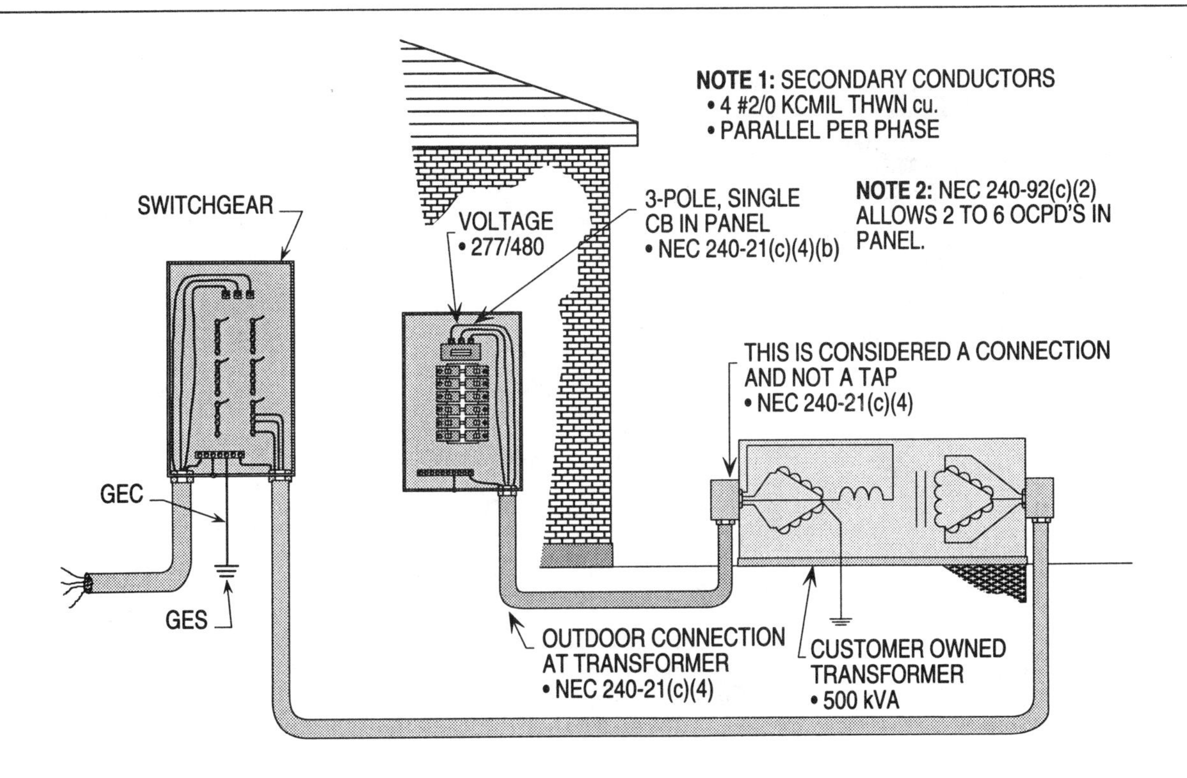

NEC 240-21(c)(4)

Ill. 4-9

Conductors Over 25 ft. up to 50 ft. in length 240-92(b)(1)(a)

In supervised industrial installations, conductors can be connected over 25 ft. up to 50 ft. in length from the secondary side of the transformer.

Example 4-10. What size OCPD is required in the primary to allow a 50 ft. connection of conductors from the secondary side of the transformer in Illustration 4-10?

Sizing Primary OCPD

Step 1: Finding amps
240-92(b)(1)(a)
A = 480 V ÷ 4,160 V (.1153) x 600 A x 150% = 103.8 A

Solution: The size OCPD in primary is 100 amps or less.

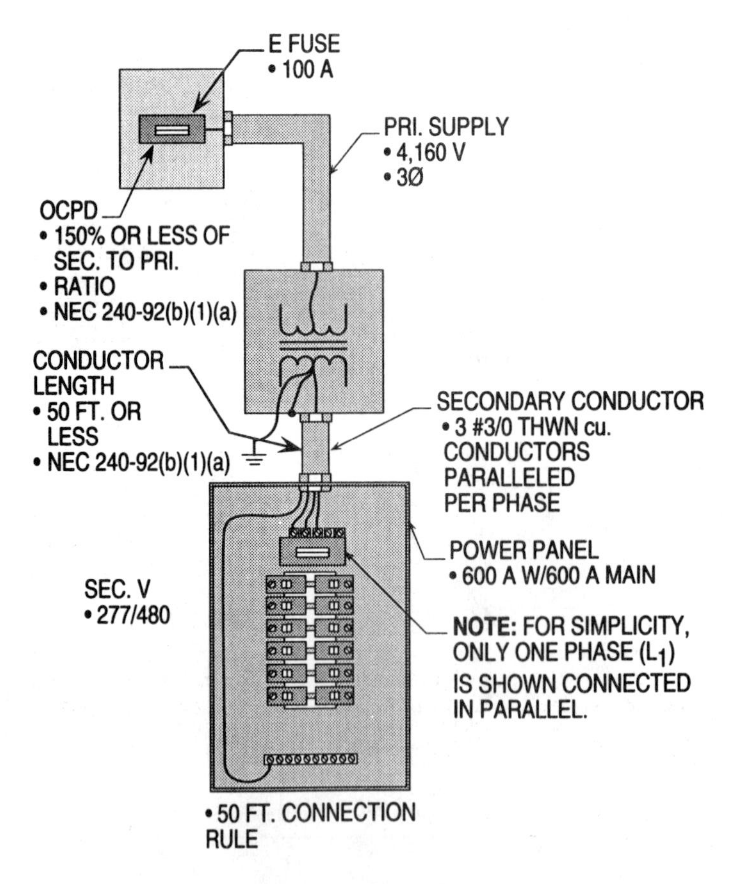

NEC 240-92(b)(1); (b)(1)(a)

Ill. 4-10

Conductors Over 50 ft. up to 75 ft. in Length 240-92(b)(1)(b); (b)(1)(c)

In supervised industrial installations, conductors can be connected over 50 ft. up to 75 ft. in length from the secondary side of transformers.

Example 4-11. Will the available fault current in amps in the secondary be great enough to clear the primary OCPD and allow a 75 ft. connection in the secondary side of the transformer in Illustration 4-11?

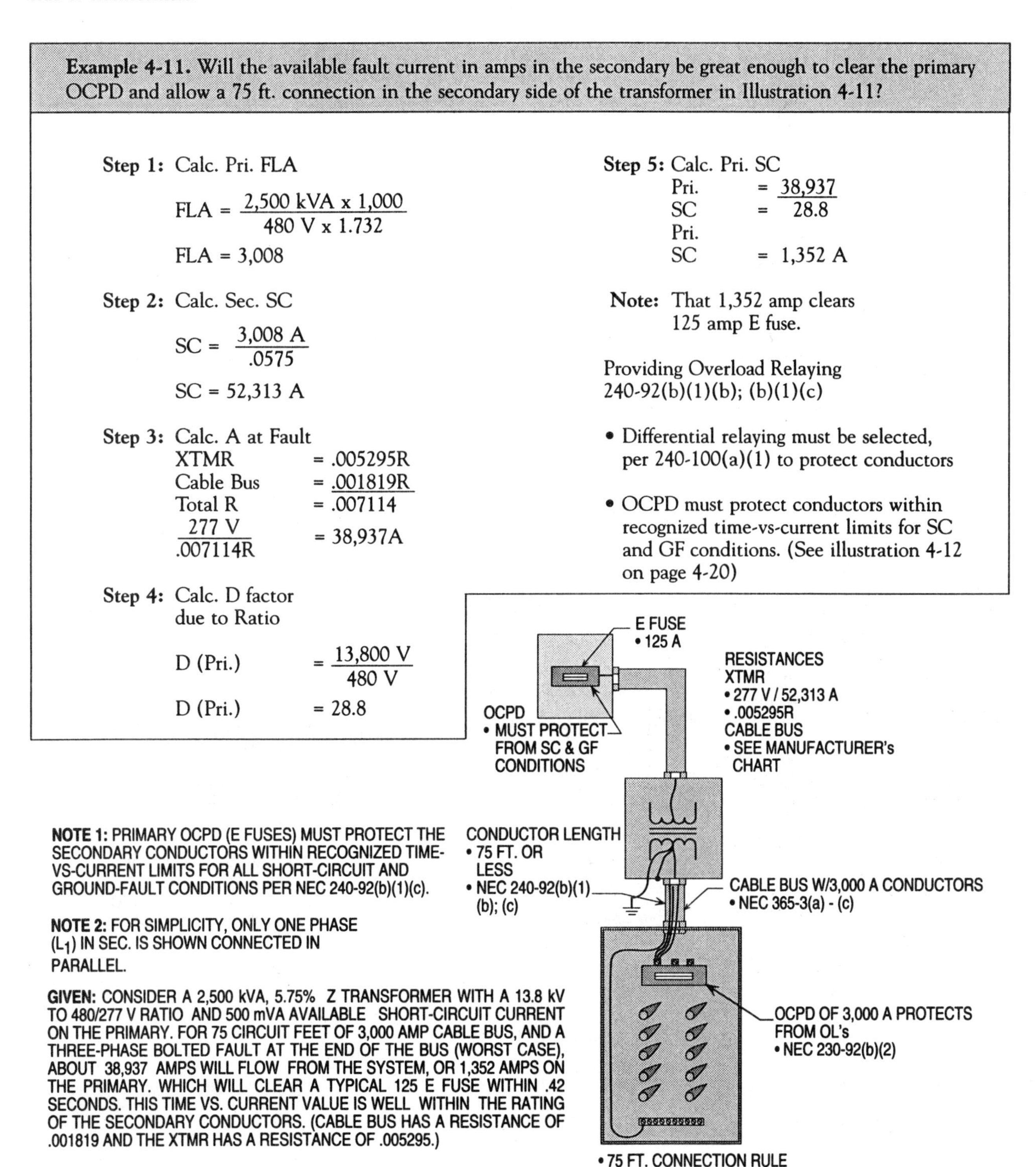

Step 1: Calc. Pri. FLA

$$FLA = \frac{2{,}500 \text{ kVA} \times 1{,}000}{480 \text{ V} \times 1.732}$$

FLA = 3,008

Step 2: Calc. Sec. SC

$$SC = \frac{3{,}008 \text{ A}}{.0575}$$

SC = 52,313 A

Step 3: Calc. A at Fault

XTMR = .005295R
Cable Bus = .001819R
Total R = .007114

$$\frac{277 \text{ V}}{.007114\text{R}} = 38{,}937\text{A}$$

Step 4: Calc. D factor due to Ratio

$$D \text{ (Pri.)} = \frac{13{,}800 \text{ V}}{480 \text{ V}}$$

D (Pri.) = 28.8

Step 5: Calc. Pri. SC

$$\text{Pri. SC} = \frac{38{,}937}{28.8}$$

Pri. SC = 1,352 A

Note: That 1,352 amp clears 125 amp E fuse.

Providing Overload Relaying
240-92(b)(1)(b); (b)(1)(c)

- Differential relaying must be selected, per 240-100(a)(1) to protect conductors
- OCPD must protect conductors within recognized time-vs-current limits for SC and GF conditions. (See illustration 4-12 on page 4-20)

NOTE 1: PRIMARY OCPD (E FUSES) MUST PROTECT THE SECONDARY CONDUCTORS WITHIN RECOGNIZED TIME-VS-CURRENT LIMITS FOR ALL SHORT-CIRCUIT AND GROUND-FAULT CONDITIONS PER NEC 240-92(b)(1)(c).

NOTE 2: FOR SIMPLICITY, ONLY ONE PHASE (L_1) IN SEC. IS SHOWN CONNECTED IN PARALLEL.

GIVEN: CONSIDER A 2,500 kVA, 5.75% Z TRANSFORMER WITH A 13.8 kV TO 480/277 V RATIO AND 500 mVA AVAILABLE SHORT-CIRCUIT CURRENT ON THE PRIMARY. FOR 75 CIRCUIT FEET OF 3,000 AMP CABLE BUS, AND A THREE-PHASE BOLTED FAULT AT THE END OF THE BUS (WORST CASE), ABOUT 38,937 AMPS WILL FLOW FROM THE SYSTEM, OR 1,352 AMPS ON THE PRIMARY. WHICH WILL CLEAR A TYPICAL 125 E FUSE WITHIN .42 SECONDS. THIS TIME VS. CURRENT VALUE IS WELL WITHIN THE RATING OF THE SECONDARY CONDUCTORS. (CABLE BUS HAS A RESISTANCE OF .001819 AND THE XTMR HAS A RESISTANCE OF .005295.)

NEC 240-92(b)(1)(b); (b)(1)(c)

Ill. 4-11

Conductors 50 ft. up to 75 ft. in length 240-92(b)(2)

In supervised industrial installations, conductors can be connected 50 ft. up to 75 ft. **in length from the secondary side** of the transformer.

Example 4-12. What size OCPD is required at the switchgear and switch to **protect the secondary conductors** and output of the transformer in Illustration 4-12?

Step 1: Size OCPD at Switchgear 1
240-92(b)(2)(a)

Solution: OCPD must be 3,000 amp or less in rating.

Step 2: Size relaying at Switchgear 2
240-92(b)(2)(c); 240-100(a)(1)

Solution: Relaying must be equal to or less than conductor ampacity of 3,000 amps.

Step 3: Size fuses at switch
240-90(b)(2)(a)

Solution: Fuses must be 3,000 amp or less in rating.

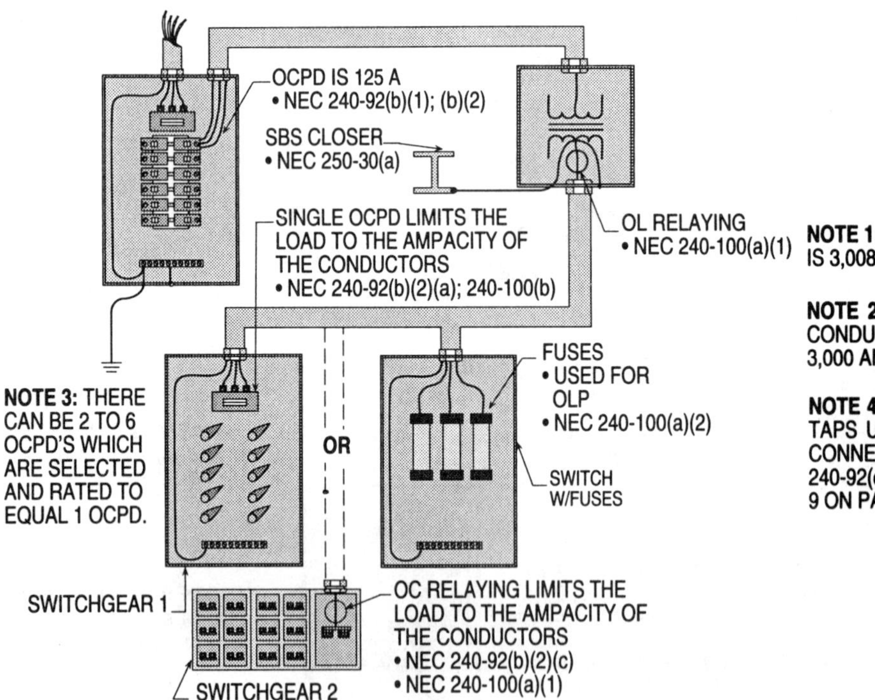

NEC 240-92(b)(2)

Ill. 4-12

OCPD's For Feeders
240-100; 240-101

For short-circuit protection of high-voltage feeders, the fuse rating may be increased up to three times the conductor ampacity. Circuit breaker setting may be increased up to six times the conductor ampacity. **Note:** Overload protection must be provided.

Example 4-13. What is the setting for the CB in the switch gear and the fuses in the switch in Illustration 4-13?

Sizing OCPD

Step 1: Finding cable ampacity
Table 310-74
#3/0 Alu. = 200 A

Step 2: Calculating CB rating
240-100; 240-101
200 A x 600% = 1,200 A

Solution: The CB setting is 1,200 amps.

Sizing fuses

Step 1: Calculating fuse rating
240-100; 240-101
200 A x 300% = 600 A

Solution: The fuse rating is 600 amps.

NEC 230-208; NEC 240-100; NEC 490-21(a); NEC 490-21(b)

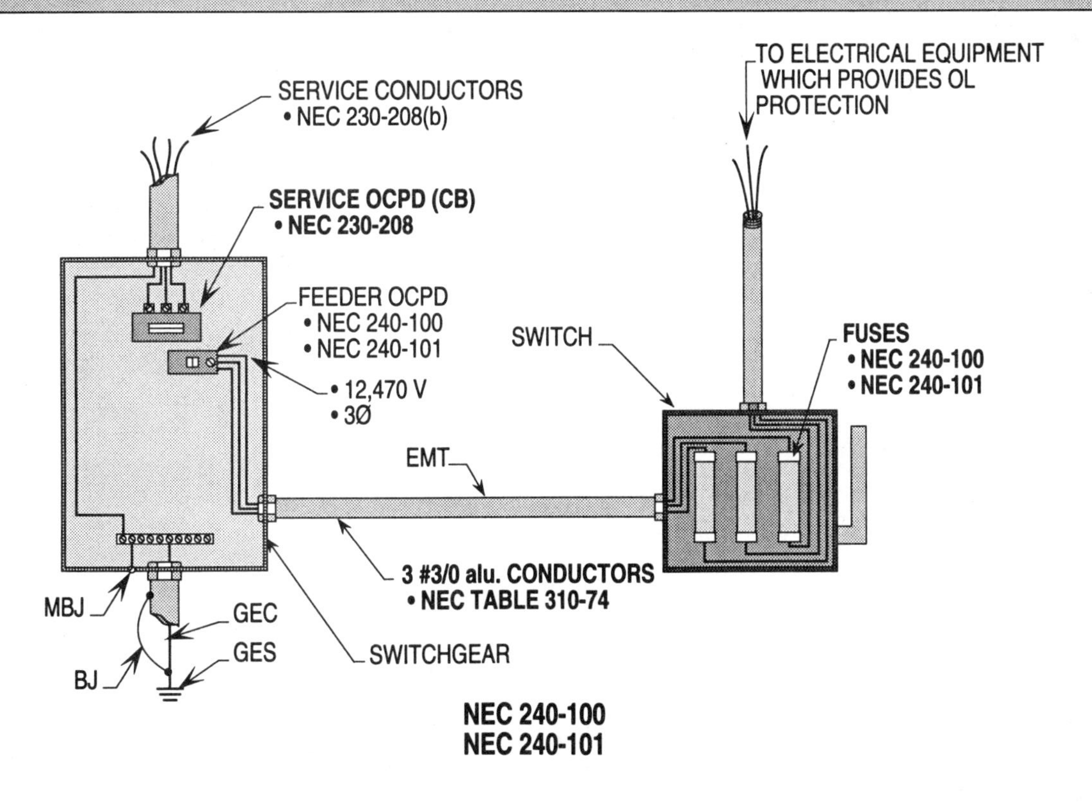

Ill. 4-13

5 Grounding and Bonding

Electrical systems are circuits and systems grounded to drain off to ground lightning surges, regulate voltage-to-ground and facilitate the operation of the OCPD's. The noncurrent-carrying metal parts of equipment are grounded to limit the voltage-to-ground. Grounding electrode conductors (GEC), Equipment grounding conductors (EGC) and bonding jumpers (BJ) must be sized properly to ensure a well designed and safe grounding scheme.

The GEC connects the EGC's, GSC, grounded bus bar and neutral conductors to earth ground. The EGC's interconnects all noncurrent-carrying metal parts to the grounded busbar in the service equipment enclosure. The BJ's ensure that all metal enclosures and raceways enclosing elements and conductors are connected to the EGC's making up the grounding system.

Quick Reference

Direct-current Systems
250-162(b); 250-166

Direct-current (DC) systems are not grounded by the same procedure as alternating-current (AC) systems. AC systems are grounded by the provisions of NEC 250-66 and Table 250-66. DC systems are grounded in accordance with NEC 250-166(a) or (b) based upon phase or neutral conductors.

Example 5-1(a). What is the size copper GEC connected to the electrode in Illustration 5-1(a)?

Sizing GEC

Step 1: Calculating size GEC
250-166(a)
#3/0 neutral requires #3/0 cu.

Solution: The size GEC is No. 3/0 cu.

Note: For grounding electrodes (direct-current systems), other than for a metal water pipe or building structure steel, see NEC 250-166(c) through (e).

ADDITIONAL HELP 34:

Calculating the amps of a DC circuit when volts and VA are known.

• Calc. Amps

$$A = \frac{kVA \times 1{,}000}{V}$$

$$A = \frac{80.4\ kVA \times 1{,}000}{240\ V}$$

$$A = 335$$

Solution: Amps is equal to 335.

ADDITIONAL HELP 35:

Calculating the Watts when volts and amps are known.

• Calc. W

$$W = E \times I$$

$$W = 240\ V \times 330\ A$$

$$W = 79{,}200$$

Solution: Watts is equal to 79,200.

Example 5-1(b). What is the size GEC connected to the electrode in Illustration 5-1(b)?

Sizing GEC

Step 1: Calculating size GEC
250-166(b)
#600 KCMIL requires #600 KCMIL cu.

Solution: The size GEC is No. 600 KCMIL cu.

Note: For grounding electrodes (direct-current systems), other than for a metal water pipe or building structure steel, see NEC 250-166(c) through (e).

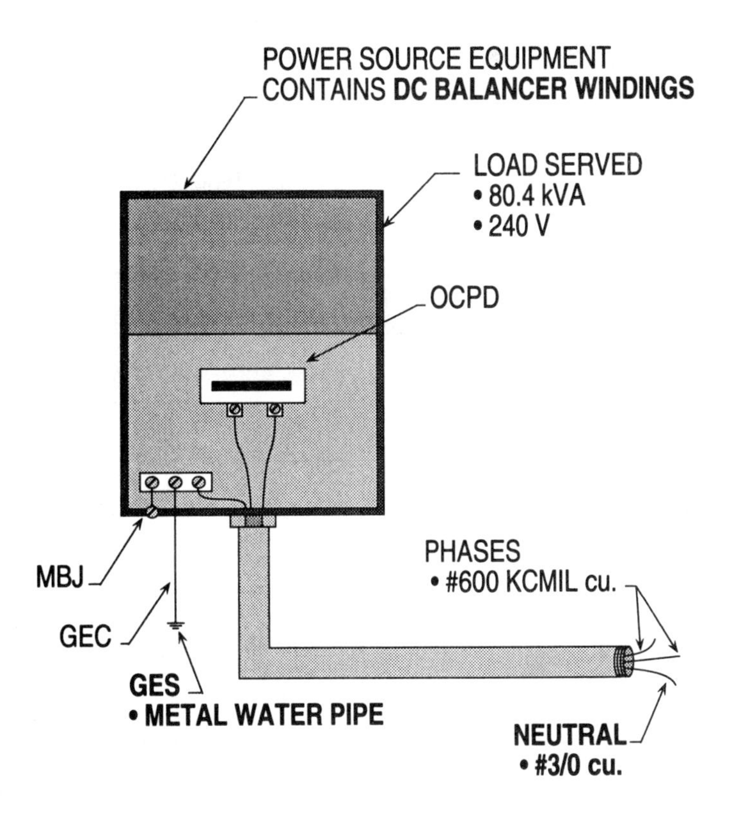

Ill. 5-1(a)

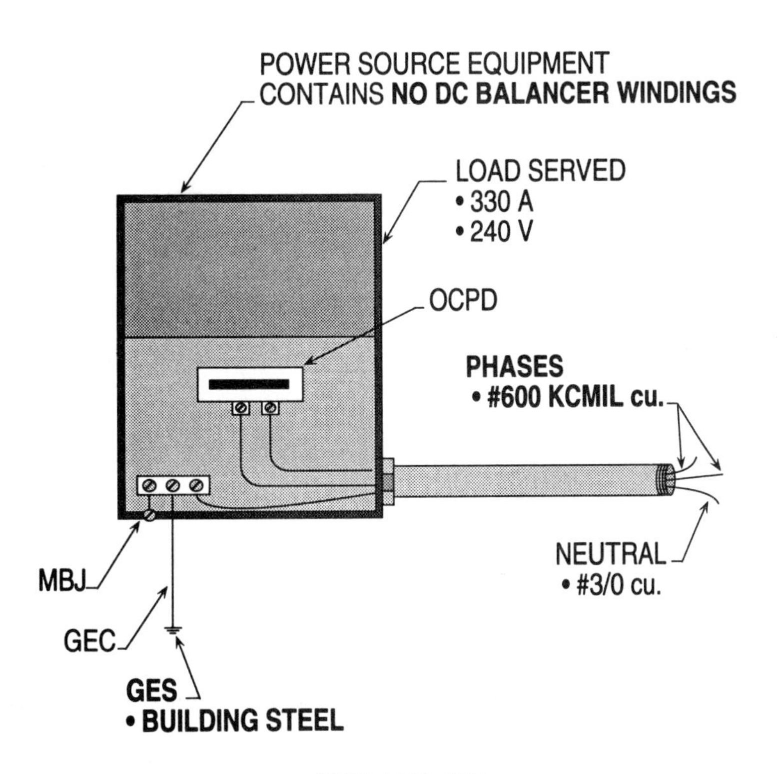

Ill. 5-1(b)

Alternating Current Systems
250-20(b); 250-66

Alternating current (AC) systems are grounded per NEC 250-66 and Table 250-66 based upon the size phase conductors or the type of grounding electrode per NEC 250-50(a) through (d) or 250-52(c) and (d).

Example 5-2(a). What are the sizes of the GEC's connected to the electrodes in Illustration 5-2(a)?

Sizing GEC to water pipe

Step 1: Calculating size GEC to water pipe
250-66; Table 250-66; 250-104(a); 250-50(a)
#350 KCMIL requires #2 cu.

Solution: The size GEC is No. 2 copper.

Sizing EGC to CEE

Step 1: Calculating size GEC to CEE
250-66(b); 250-50(c)
#350 KCMIL requires #4 cu.

Solution: The size GEC is No. 4 cu.

Sizing GEC to building steel

Step 1: Calculating size GEC to building steel
250-66; Table 250-66; 250-50(b); 250-104(d)
#350 KCMIL requires #2 cu.

Solution: The size GEC is No. 2 cu.

Sizing GEC to ground ring

Step 1: Calculating size GEC to ground ring
250-66(c); 250-50(d)
#350 KCMIL requires # 2 cu.

Solution: The size GEC is No. 2 cu.

Note: The size GEC to the driven rod is No. 6 cu. per 250-66(a).

ADDITIONAL HELP 36:

Calculating the amps to ground on each grounding electrode in Illustration 5-2(a).

- Calc. A on MWP
250-50(a)

$$A = \frac{V}{R}$$
$$A = \frac{120\ V}{3R}$$
$$A = 40$$

- Calc. A on GR
250-50(d)

$$A = \frac{V}{R}$$
$$A = \frac{120\ V}{10R}$$
$$A = 12$$

- Calc. A on BS
250-50(b)

$$A = \frac{V}{R}$$
$$A = \frac{120\ V}{5R}$$
$$A = 24$$

- Calc. A on DR

$$A = \frac{V}{R}$$
$$A = \frac{120\ V}{20R}$$
$$A = 6$$

- Calc. A to CCE
250-50(c)

$$A = \frac{V}{R}$$
$$A = \frac{120\ V}{8R}$$
$$A = 15$$

Solution: The amps which can flow over the MWP is 40, over the BS is 24, over the CCE is 15, over the GR is 12, and the DR is 6.

Note: The types of grounding electrodes are found in NEC 250-50(a) through (d). The procedure for sizing the GEC's is determined by the requirements listed in NEC 250-66 and Table 250-66 for AC systems. Section 250-166 is used to size the GEC's for DC systems.

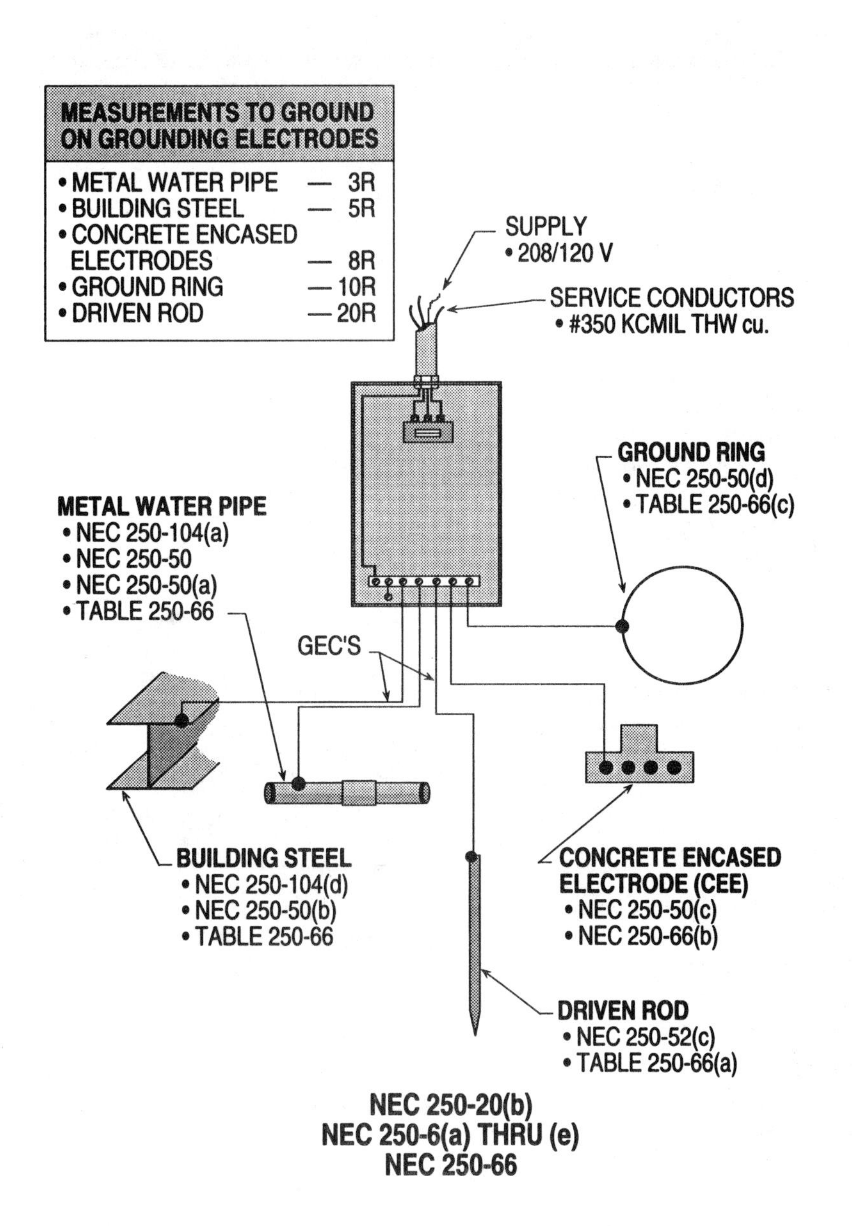
MEASUREMENTS TO GROUND ON GROUNDING ELECTRODES
• METAL WATER PIPE — 3R
• BUILDING STEEL — 5R
• CONCRETE ENCASED ELECTRODES — 8R
• GROUND RING — 10R
• DRIVEN ROD — 20R
SUPPLY
• 208/120 V
SERVICE CONDUCTORS
• #350 KCMIL THW cu.
GROUND RING
• NEC 250-50(d)
• TABLE 250-66(c)
METAL WATER PIPE
• NEC 250-104(a)
• NEC 250-50
• NEC 250-50(a)
• TABLE 250-66
GEC'S
BUILDING STEEL
• NEC 250-104(d)
• NEC 250-50(b)
• TABLE 250-66
CONCRETE ENCASED ELECTRODE (CEE)
• NEC 250-50(c)
• NEC 250-66(b)
DRIVEN ROD
• NEC 250-52(c)
• TABLE 250-66(a)
NEC 250-20(b)
NEC 250-6(a) THRU (e)
NEC 250-66

Ill. 5-2(a)

Alternating Current Systems
250-20(b); 250-66

If the grounding electrode system has two or more electrodes available on the premises, they can be bonded together and used collectively as the grounding electrodes.

Example 5-2(b). What size bonding conductor is required to loop and connect to the electrodes in Illustration 5-2(b)?

Sizing bonding jumper

Step 1: Calculating bonding conductor
250-50(a); (b); Table 250-66
- Building steel requires #2 cu.
- Metal water pipe requires #2 cu.

Solution: The bonding conductor must be No. 2 cu. based upon the largest bonding jumper required per NEC 250-50(a) and (b).

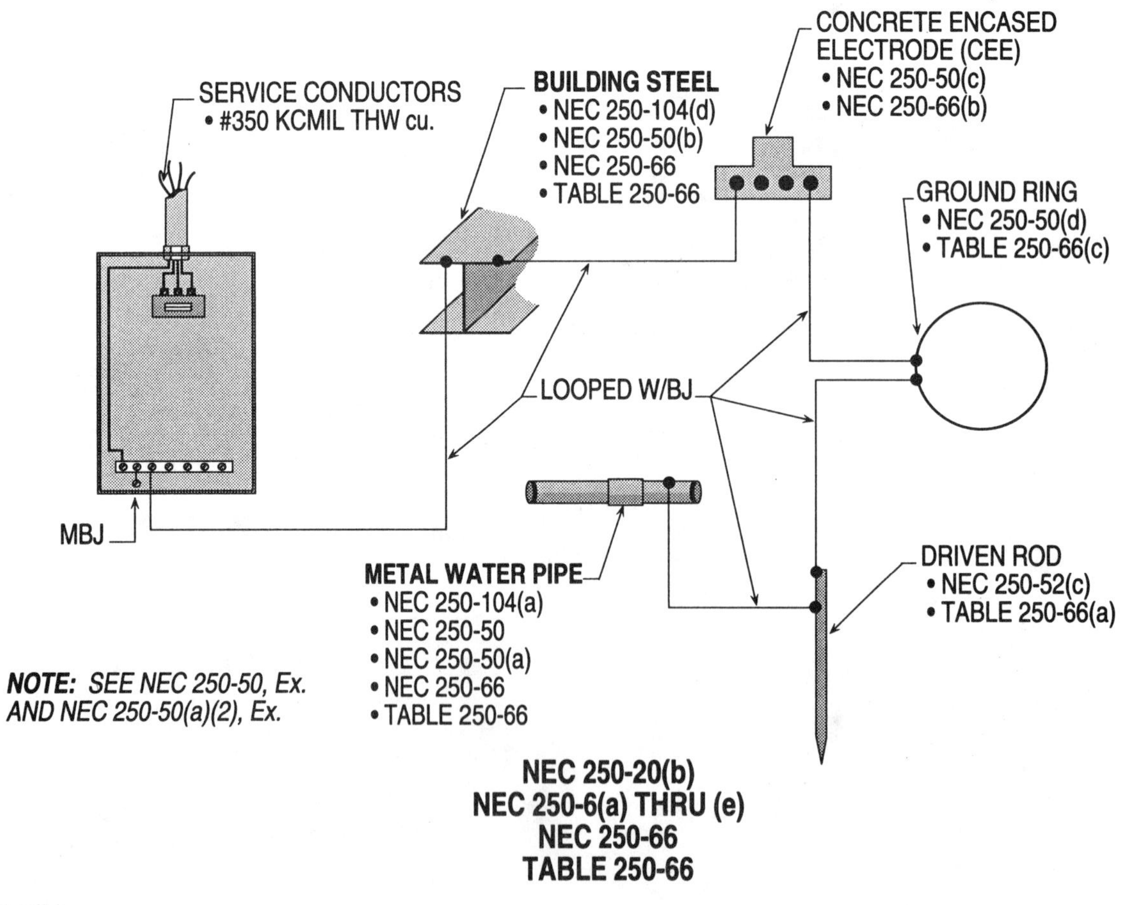

Ill. 5-2(b)

Grounded Service Conductor 250-24(b)(1)

The grounded service conductor (GSC) must be sized first to carry the maximum unbalanced neutral current per NEC 220-22. In addition, the grounded conductor has to be sized to carry the maximum available fault current that it may be called upon to carry per NEC 250-24(b)(1). The condition of use of the GSC is found in 310-15(b)(4)(a) and (4)(c) to ampacity Tables 0-2,000 volts.

Example 5-3(a). What size grounded service conductor is required to carry the maximum fault current based upon the phase conductors in Illustration 5-3(a)?

Sizing grounded service conductor

Step 1: Calculating min. size grounded conductor
250-24(b)(1); Table 250-66
#350 KCMIL cu. requires #2 cu.

Solution: The size grounded service conductor is No. 2 THHN copper.

NEC 220-22; NEC 250-24(b)(1); Table 310-15(b) to Tables 0-2000 V

RULE OF THUMB METHOD FOR SIZING GROUNDED CONDUCTOR

- CACULATING CM BASED ON OCPD TIMES MULTIPLIER
  CM = OCPD x 159
  CM = 350 A x 159 = 55,650 CM

SOLUTION: THE SIZE GROUNDED CONDUCTOR IS NO. 2 COPPER.

NOTE 1: WHEN USING THE RULE OF THUMB METHOD, ALWAYS ROUND OCPD UP TO THE NEXT SIZE ABOVE AMPACITY OF CONDUCTOR.

NOTE 2: WHEN APPLYING THE RULE OF THUMB METHOD, 159 IS A CONSTANT IN THE FORMULA THAT MUST BE USED EVERYTIME THE FORMULA IS USED TO DETERMINE THE SIZE OF THE GROUNDED CONDUCTORS.

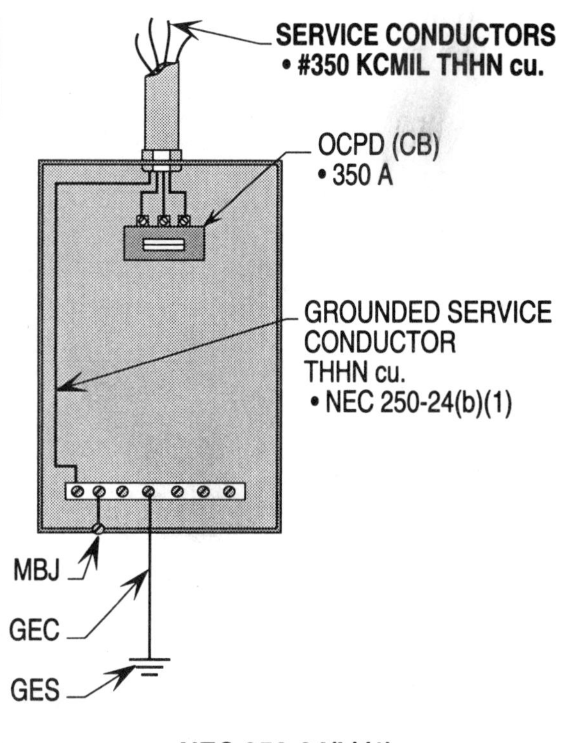

Ill. 5-3(a)

Grounded Service Conductor 250-24(b)(2)

The grounded conductors shall be brought to each disconnecting means which makes up the service equipment.

Example 5-3(b). What size grounded service conductor is required for each conduit in Illustration 5-3(b)?

Sizing grounded service conductor

Step 1: Calculating grounded conductor
250-24(b)(1); (b)(2)
#600 KCMIL x 3 per phase =
1,800 KCMIL

Step 2: Calculating total KCMIL
250-24(b)(1); (b)(2)
1,800 KCMIL x .125 = 225 KCMIL

Step 3: Calculating KCMIL for each conduit
250-24(b)(2); 310-4
225 KCMIL / 3 per phase = 75 KCMIL

Step 4: Calculating each conductor
250-24(b)(2); 310-4
75 KCMIL x 1,000 = 75,000 CM

Step 5: Sizing each conductor
250-24(b)(2); 310-4; Table 8, Ch. 9
75,000 CM requires #1

Solution: **Section 310-4 requires a No. 1/0 THHN copper conductor in each conduit.**

NEC 250-24(b)(2); NEC 310-4, NEC 300-5(i); NEC 300-20(a) and (b)

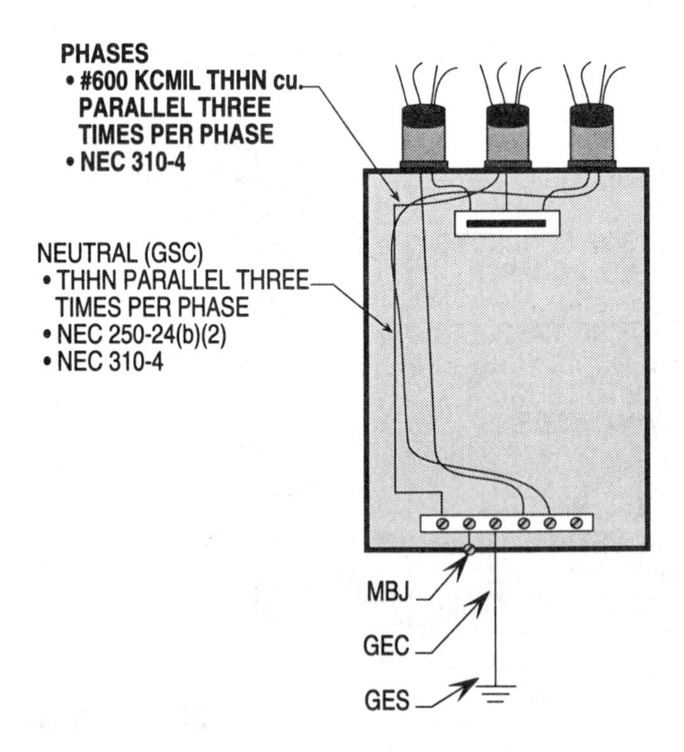

Ill. 5-3(b)

Grounding Separately-Derived Systems
250-20(d); 250-30(a)(1) through (a)(4); ART. 100

Separately derived systems are grounded per NEC 250-30(a)(1) through (a)(4) based upon the largest phase conductor per Table 250-66 or the type electrode per 250-30(a)(3)(c). **Note:** Only one electrode is to be utilized per 250-24(a)(5), FPN and 250-30(a)(3).

Example 5-4(a). What size GEC and BJ is required to ground the separately derived system in Illustration 5-4(a)? (If any one of the grounding electrodes (GE's) is available).

Sizing GEC for water pipe or building steel

Step 1: Calculating GEC for water pipe or building steel
250-30(a)(2); 250-30(a)(3)(a); (b);
Table 250-66
#3/0 cu. requires #4 cu.

Solution: The size GEC is No. 4 cu.

Sizing GEC to driven rod

Step 1: Calculating size GEC to driven rod
250-30(a)(3)(c); 250-66(a)
#3/0 cu. requires #6 cu.

Solution: The size GEC is No. 6 cu.

Sizing bonding jumper

Step 1: Calculating size BJ
250-30(a)(2); 250-28(d); Table 250-66
#3/0 cu. requires #4 cu.

Solution: The size BJ is No. 4 cu.

NEC 250-20(d); NEC 250-24(a)(5), FPN; NEC 250-66; NEC 250-50(a) through (d); Table 250-66

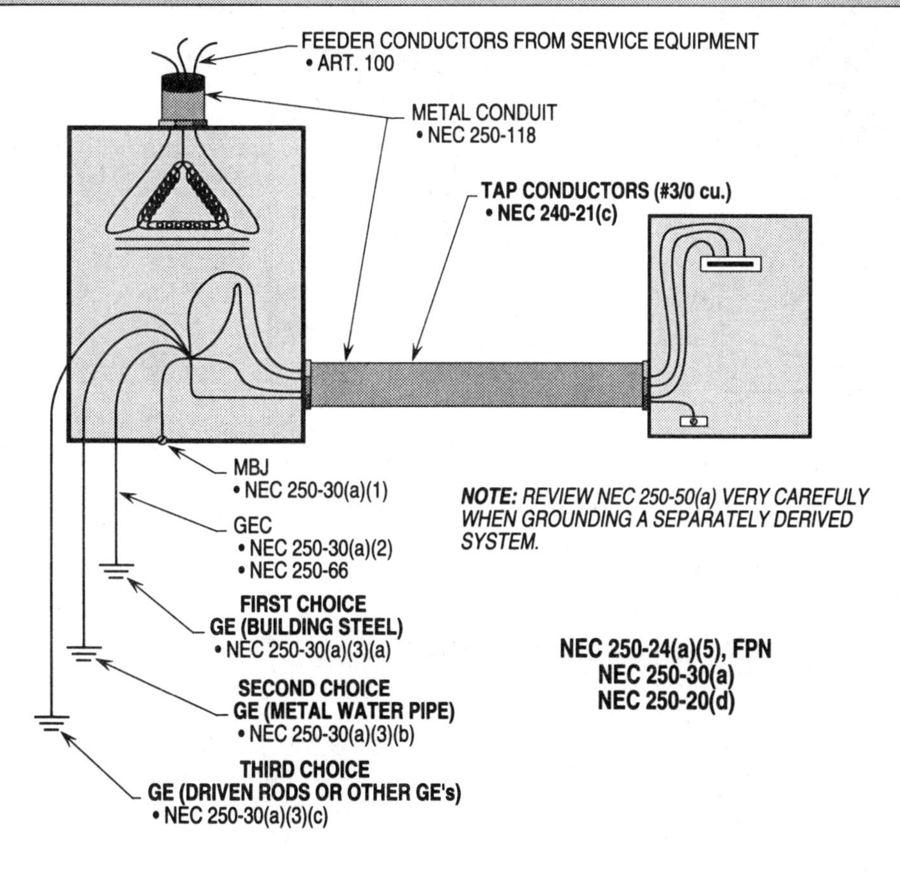

Ill. 5-4(a)

Grounding Separately-derived Systems
250-20(d); 250-30(a)(1) through (a)(4); ART. 100

The grounding of the separately derived system may be done at the transformer or equipment serviced per NEC 250-30(a)(2) and NEC 250-24(a)(5), FPN. **Note:** There is a priority list of electrodes that must be used if they are available.

Example 5-4(b). What size GEC is required to ground at the panelboard in Illustration 5-4(b)?

Sizing GEC

Step 1: Calculating GEC
250-30(a)(2); Table 250-66
#3/0 cu. requires #4 cu.

Solution: The size GEC is No. 4 cu.

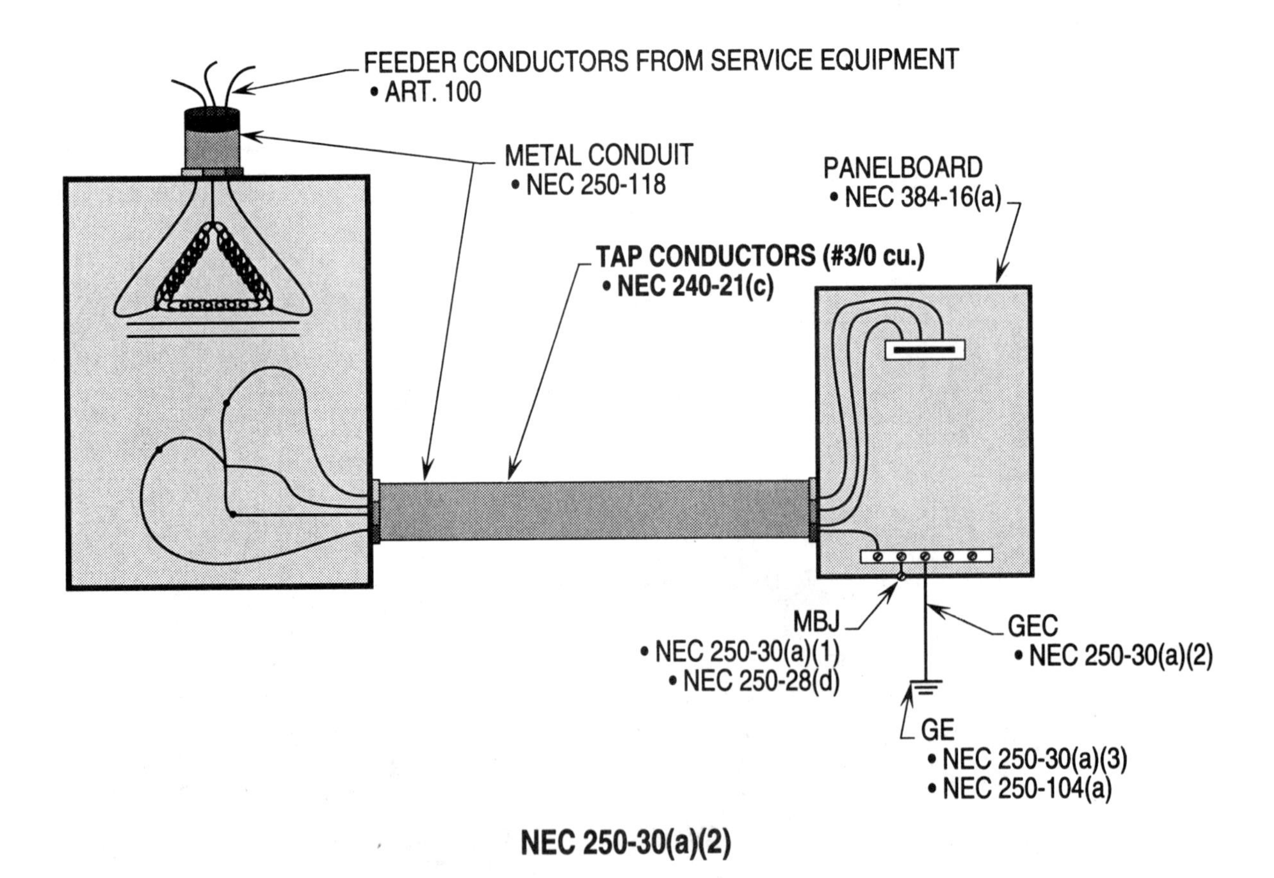

Ill. 5-4(b)

Bonding Supply Side 250-28(d)

Main bonding jumpers (MBJ's) are sized from the provisions listed in NEC 250-28(d) which requires the MBJ's to be sized per Table 250-66 for phase conductors rated at 1,100 KCMIL or less for cu. and 1,750 KCMIL or less for Alu.. For KCMIL ratings exceeding those listed above, the KCMIL ratings must be multiplied by 12 1/2 percent (.125) and the MBJ selected from Table 8, Ch. 9 based upon this computation.

Example 5-5(a). What size main bonding jumper (MBJ) is required to bond the busbars in Illustration 5-5(a)?

Sizing main bonding jumper

Step 1: Calculating MBJ
250-28(d); Table 250-66
#350 KCMIL cu. requires #2 cu.

Solution: The size MBJ is No. 2 cu.

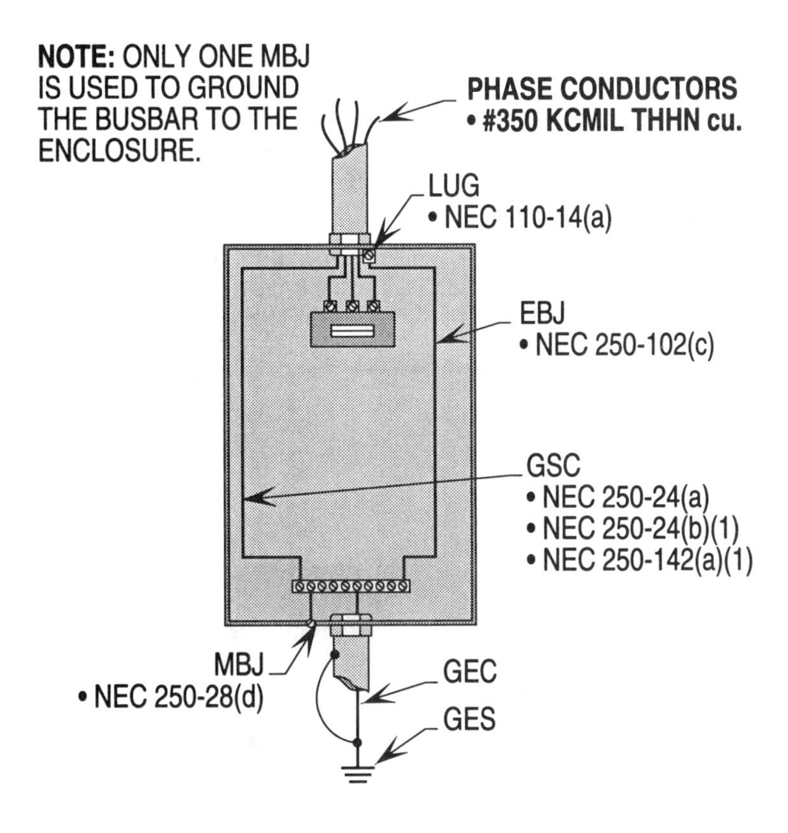

Ill. 5-5(a)

Bonding Supply Side
250-28(d)

Table 250-66 is used to size main bonding jumpers (MBJ's) on the supply side because there is no OCPD between the windings of the transformers and the loads supplied.

Example 5-5(b). What size MBJ is required to bond the busbar in Illustration 5-5(b)?

Sizing main bonding jumper

Step 1: Calculating MBJ
250-28(d)
#700 KCMIL x 3 = 2,100 KCMIL

Step 2: Calculating total KCMIL
250-28(d)
2,100 KCMIL x .125 = 262.5 KCMIL

Step 3: Finding MBJ
Table 250-66; Table 8, Ch. 9
262.5 KCMIL requires #300 KCMIL cu.

Solution: The size EBJ is No. 300 KCMIL cu.

Note: Only one MBJ is used to connect the busbar to the enclosure.

NEC 250-90; NEC 250-92(a); NEC 250-94; NEC 250-28(d)

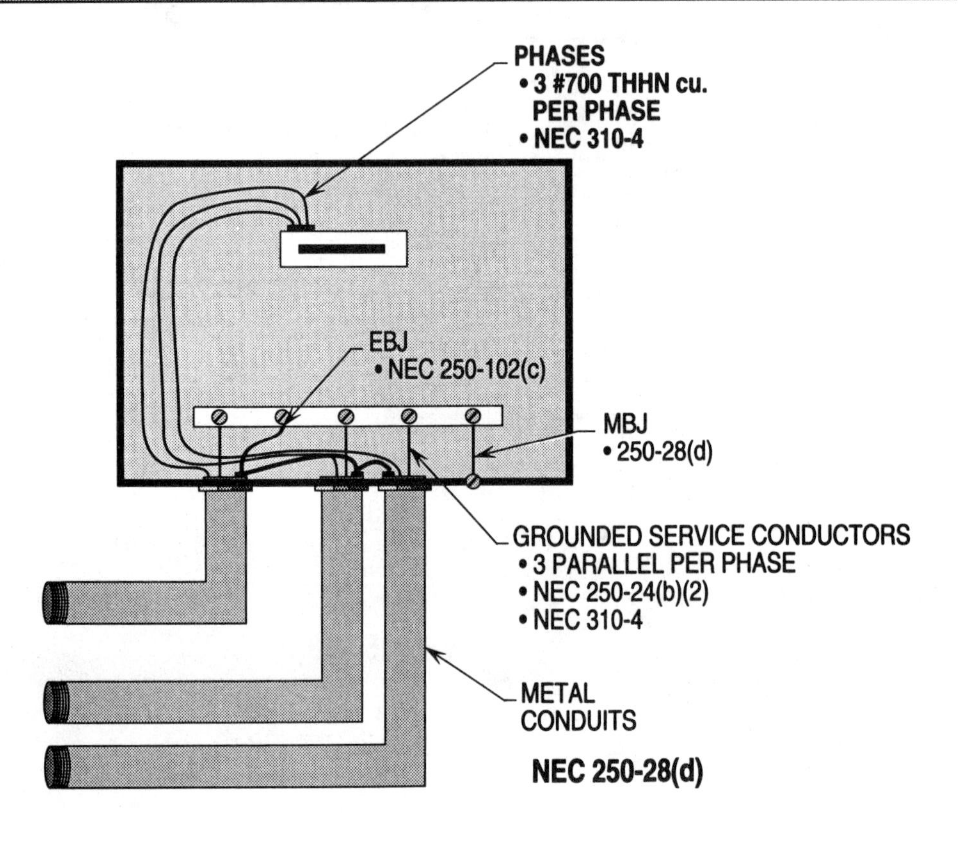

Ill. 5-5(b)

Bonding Supply Side
250-28(d)

Table 250-66 is used to size main bonding jumpers (MBJ's) on the supply side because there is no OCPD between the windings of the transformers and the load supplied.

Example 5-5(c). What size MBJ is required to bond the busbar in Illustration 5-5(c)?

Sizing main bonding jumper

Step 1: Calculating MBJ
250-28(d); Table 250-66
#700 KCMIL requires #2/0 cu.

Solution: **The size MBJ is No. 2/0 cu.**

NEC 250-90; NEC 250-92(a); NEC 250-96(a); NEC 250-28(d)

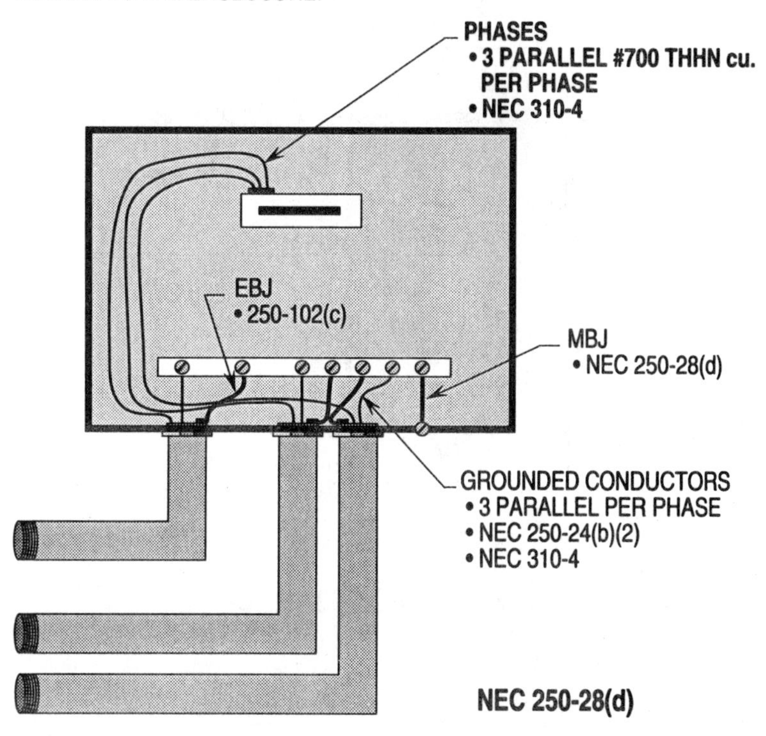

Ill. 5-5(c)

Bonding Supply Side 250-102(c)

Equipment bonding jumpers (EBJ's) are sized from the provisions listed in NEC 250-102(c) which requires the EBJ's to be sized per Table 250-66 for phase conductors rated at 1,100 KCMIL or less for cu. and 1,750 KCMIL or less for Alu.. For KCMIL ratings exceeding those listed above, the KCMIL ratings must be multiplied by 12 1/2 percent (.125) and the EBJ selected from Table 8, Ch. 9 based upon this computation.

Example 5-5(d). What size bonding jumpers (EBJ's) are required to bond the conduits in Illustration 5-5(d)?

Sizing Equipment bonding jumper

Step 1: Calculating EBJ
250-102(c); Table 250-66
#350 KCMIL cu. requires #2 cu.

Solution: The size EBJ is No. 2 cu.

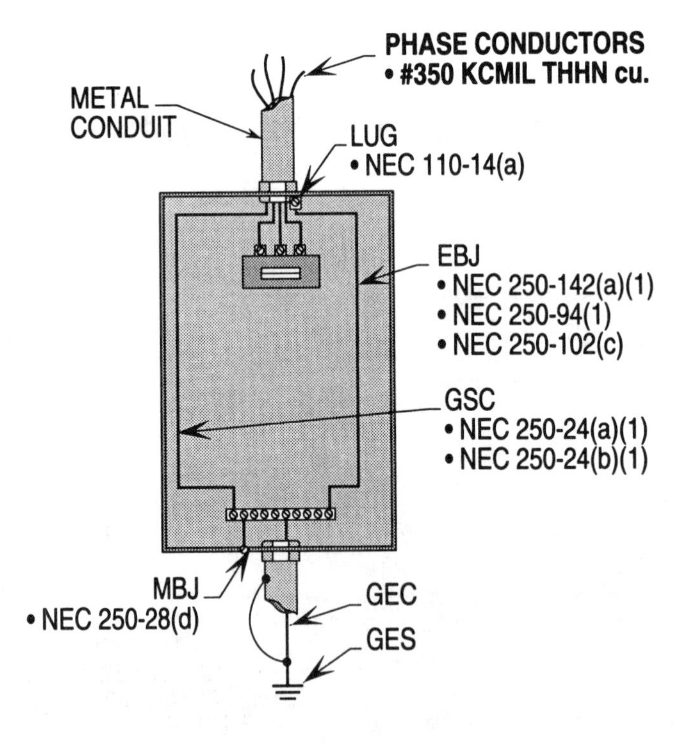

Ill. 5-5(d)

Bonding Supply Side
250-102(c)

Table 250-66 is used to size bonding jumpers (EBJ's) on the supply side because there is no OCPD between the windings of the transformers and the loads supplied.

Example 5-5(e). What size EBJ is required to bond the conduits in Illustration 5-5(e)?

Sizing equipment bonding jumper

Step 1: Calculating EBJ
250-28(d)
#700 KCMIL x 3 = 2,100 KCMIL

Step 2: Calculating total KCMIL
250-28(d)
2,100 KCMIL x .125 = 262.5 KCMIL

Step 3: Finding EBJ
Table 250-66; Table 8, Ch. 9
262.5 KCMIL requires #300 KCMIL cu.

Solution: The size EBJ is No. 300 KCMIL cu.

Note: Only one EBJ is used to connect the three service conduits.

NEC 250-90; NEC 250-94; NEC 250-96(a); NEC 250-102(c)

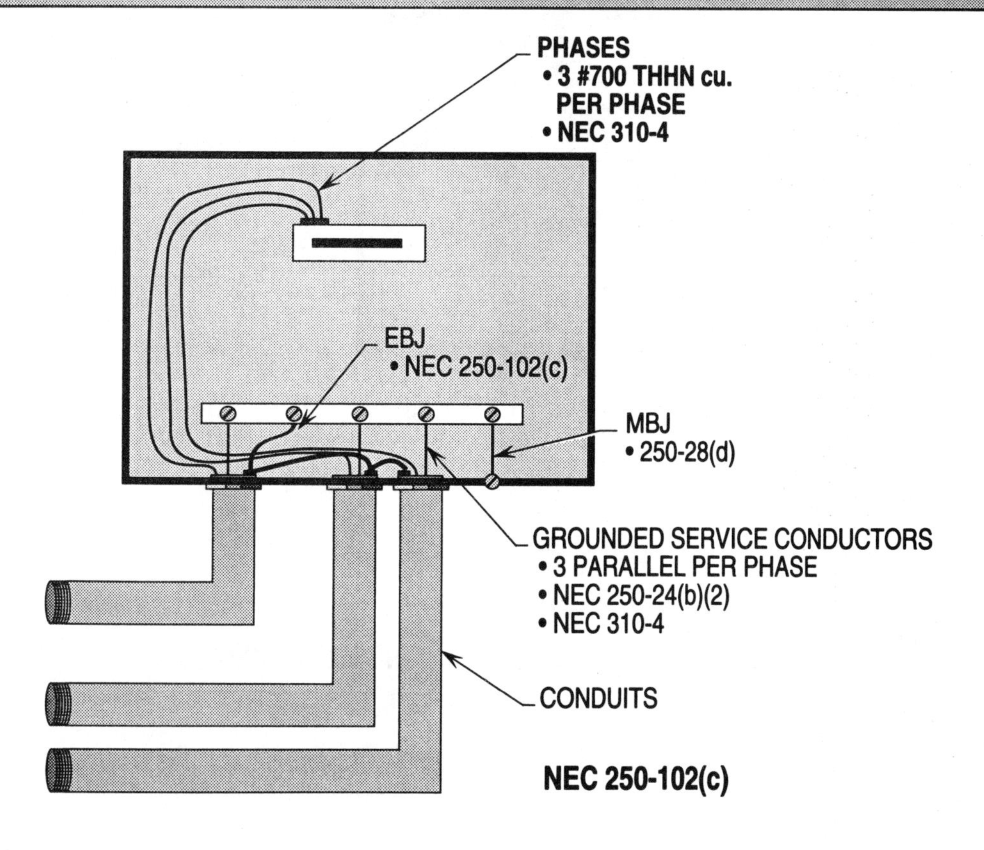

Ill. 5-5(e)

Bonding Supply Side
250-102(c)

Table 250-66 is used to size equipment bonding jumpers (EBJ's) on the supply side because there is no OCPD between the windings of the transformers and the load supplied.

Example 5-5(f). What size EBJ's are required to bond the busbar in Illustration 5-5(f)?

Sizing bonding jumper

Step 1: Calculating EBJ's
250-102(c); Table 250-66
#700 KCMIL requires #2/0 cu.

Solution: **The size EBJ's for each conduit is No. 2/0 cu.**

NEC 250-90; NEC 250-94; NEC 250-96(a); NEC 250-102(c)

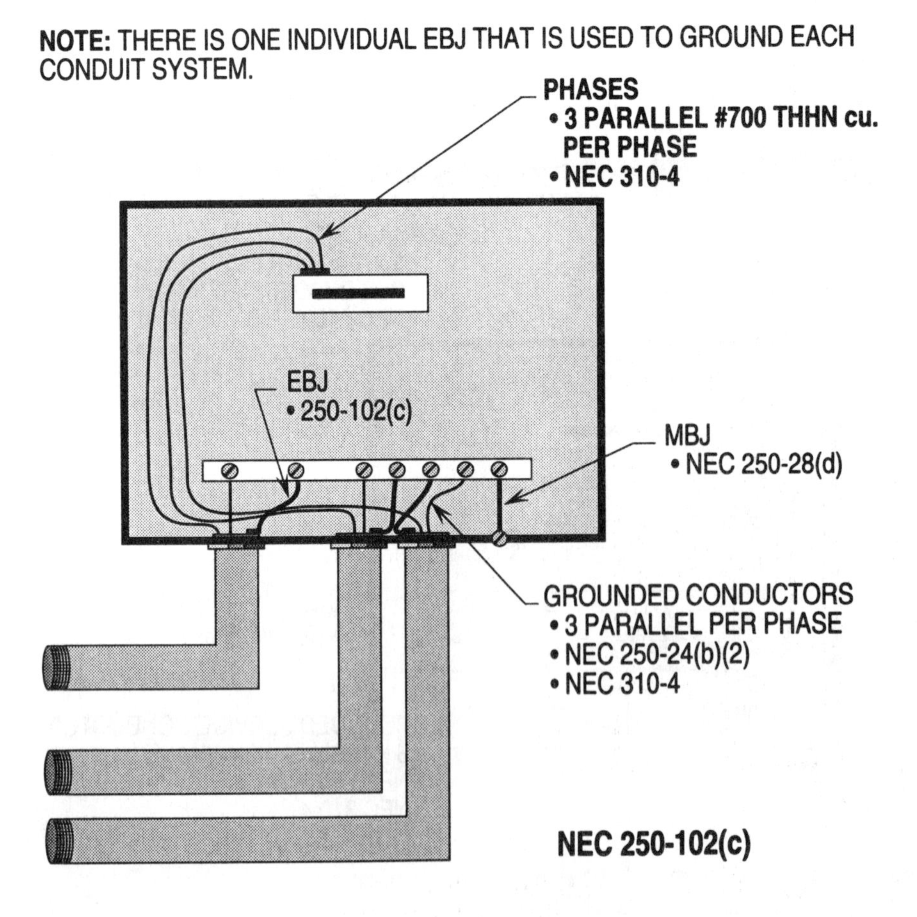

Ill. 5-5(f)

Bonding Load Side
250-102(d)

Table 250-122 is used to size equipment bonding jumpers (EBJ's) on the load side because there is an OCPD between the conductors and the load supplied.

Equipment bonding jumpers (EBJ's) are used to bond the metal of equipment together to create equipotentials. For circuits, other than service loads of more than 250 volts-to-ground where oversized concentric or eccentric knockouts are encountered a bonding jumper must be used.

Example 5-6. What size EBJ is required to bond the conduit in Illustration 5-6?

Sizing EBJ

Step 1: Calculating EBJ
250-102(c); Table 250-122
70 A OCPD requires #8 cu.

Solution: **The size EBJ is No. 8 cu.**

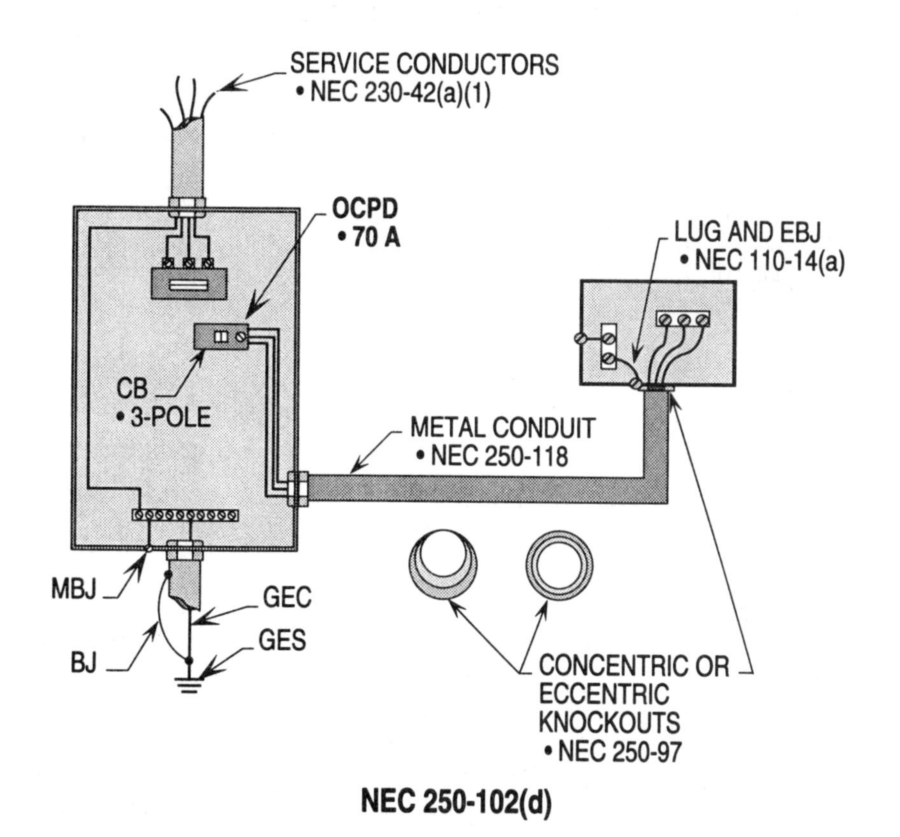

Ill. 5-6

Grounding Equipment
250-122(a); Table 250-122

Equipment grounding conductors (EGC's) are sized from Table 250-122 based upon the size OCPD protecting the branch-circuit conductors and equipment.

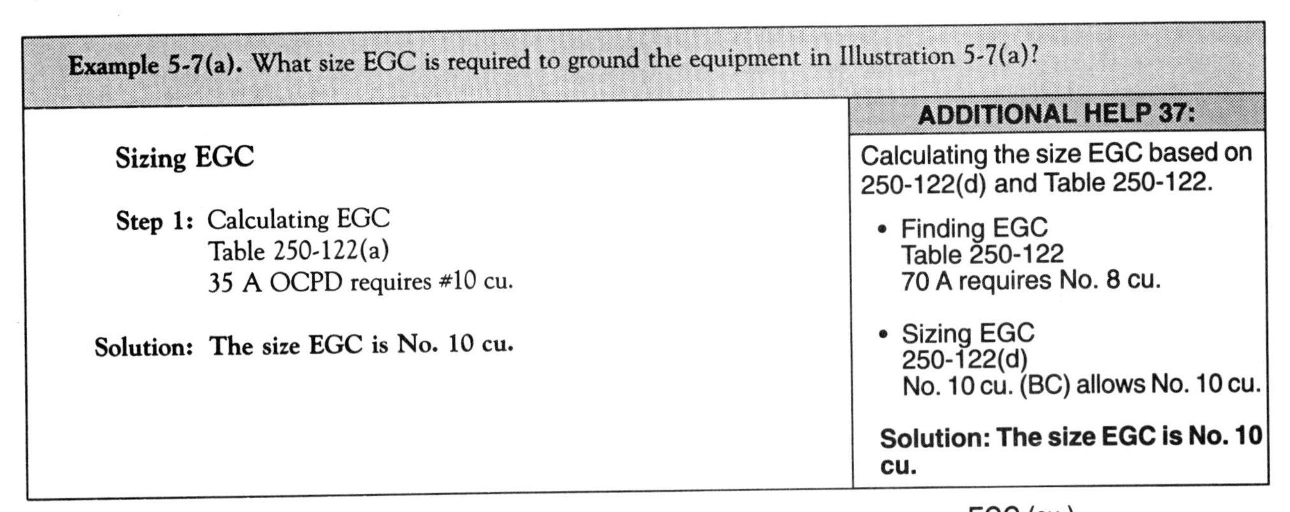

Example 5-7(a). What size EGC is required to ground the equipment in Illustration 5-7(a)?

Sizing EGC

Step 1: Calculating EGC
Table 250-122(a)
35 A OCPD requires #10 cu.

Solution: The size EGC is No. 10 cu.

ADDITIONAL HELP 37:

Calculating the size EGC based on 250-122(d) and Table 250-122.

- Finding EGC
  Table 250-122
  70 A requires No. 8 cu.
- Sizing EGC
  250-122(d)
  No. 10 cu. (BC) allows No. 10 cu.

Solution: The size EGC is No. 10 cu.

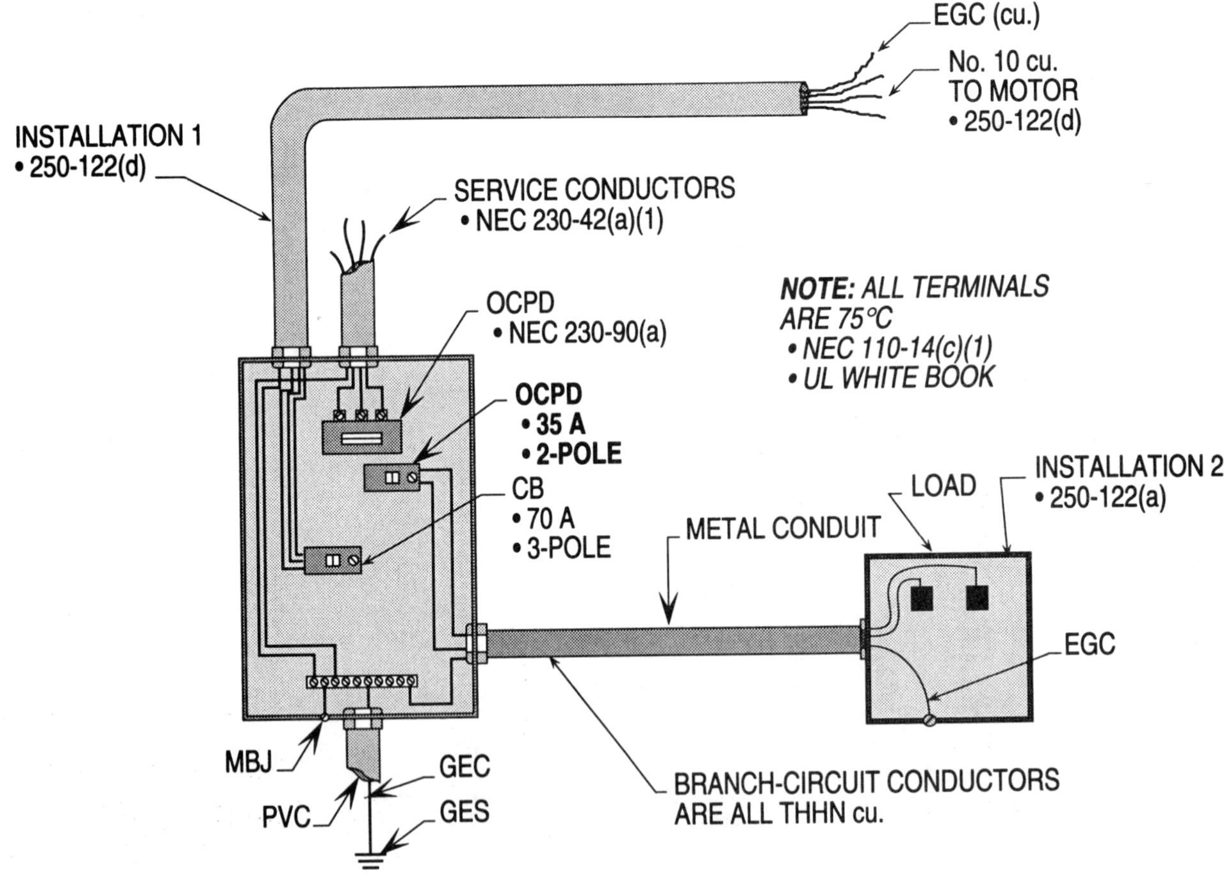

Ill. 5-7(a)

Grounding Equipment
250-122(a); Table 250-122

Equipment grounding conductors (EGC's) are sized from Table 250-122 based upon the OCPD protecting the feeder-circuit conductors and equipment.

Example 5-7(b). What size EGC is required to ground the subpanel in Illustration 5-7(b)?

Sizing EGC

Step 1: Calculating EGC
Table 250-122
200 A OCPD requires #6 cu.

Solution: The size EGC is No. 6 cu.

ADDITIONAL HELP 38:

Calculating the size and number of EGC's required to serve as a ground for the circuits in installation No. 1.

- Finding EGC
  250-122(c); Table 250-122

Solution: The size EGC is No. 10 cu.

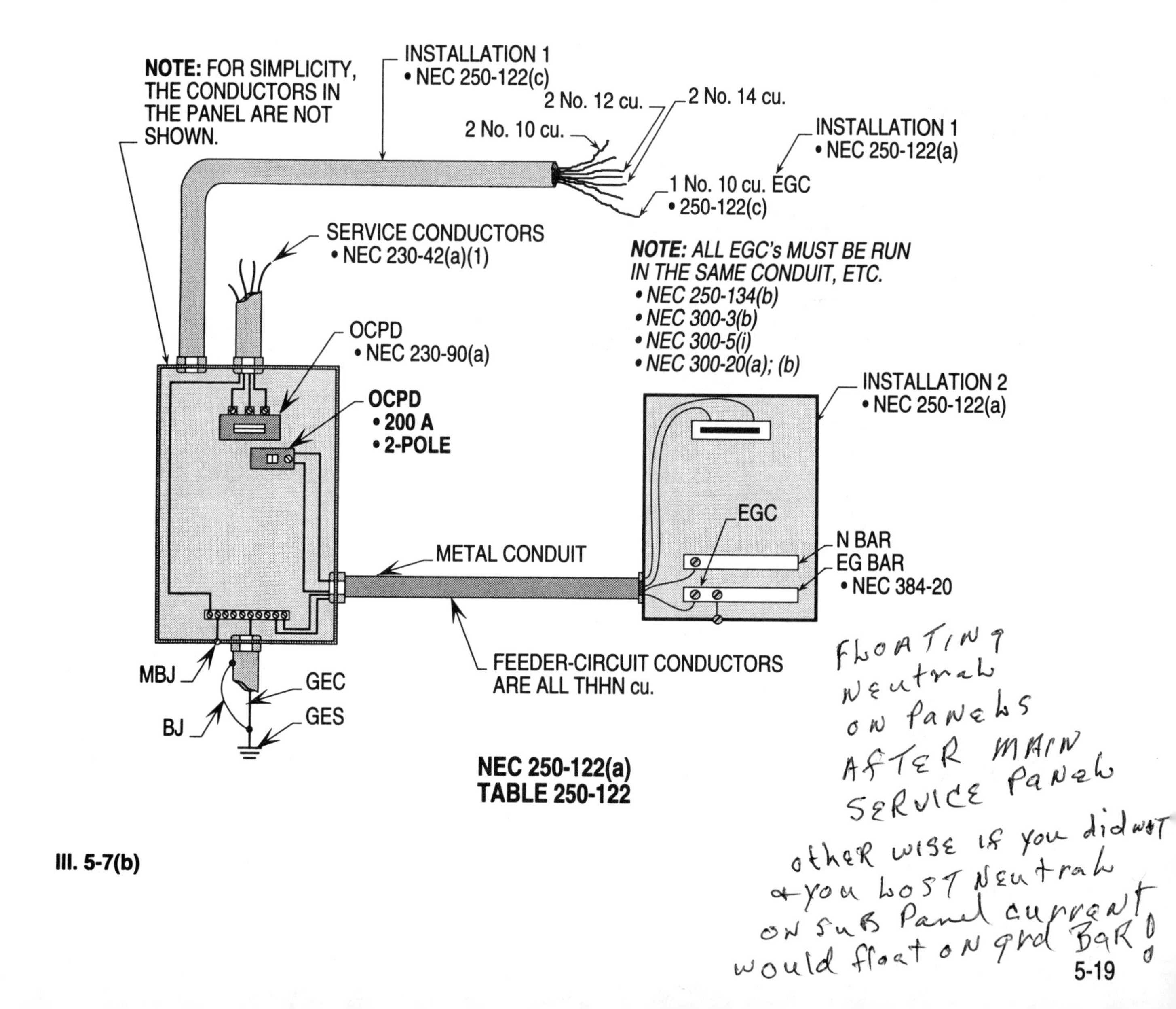

NEC 250-122(a)
TABLE 250-122

Ill. 5-7(b)

Grounding Equipment
250-122(b); Table 250-122

The equipment grounding conductors (EGC's) shall be increased in size based upon the phase conductors being oversized due to poor voltage drop (VD) in extremely long runs.

Example 5-7(c). What size equipment grounding conductor (EGC) is required for the circuit in Illustration 5-7(c) due to voltage drop (VD)?

Finding VD in circuit

Step 1: Calculating EGC
250-122(b); Table 250-122
175 A requires #6 cu.

Step 2: Calculating feeder-circuit conductors
215-2(a); 220-10; 215-3
140 A x 125% = 175 A

Step 3: Selecting conductors based on OCPD
Table 310-16
175 A requires #2/0
Increased to #3/0 due to VD

Step 4: Finding CM of conductors
250-122(b); Table 8, Ch. 9
#2/0 (175 A CB) = 133,100 CM
#3/0 (due to VD) = 167,800 CM
#6 (See step 1) = 26,240 CM

Step 5: Finding multiplier based on VD
250-122(b)
Multiplier = 167,800 CM / 133,100 CM
Multiplier = 1.2607062

Step 6: Calculating EGC based on VD
EGC = 26,240 CM x 1.2607062
EGC = 33,081 CM

Step 7: Selecting EGC based on VD
Table 8 to Ch. 9
33,081 CM requires #4 cu.
(41,740 CM)

Solution: **The size EGC based upon VD is No. 4 cu. and not No. 6 cu.**

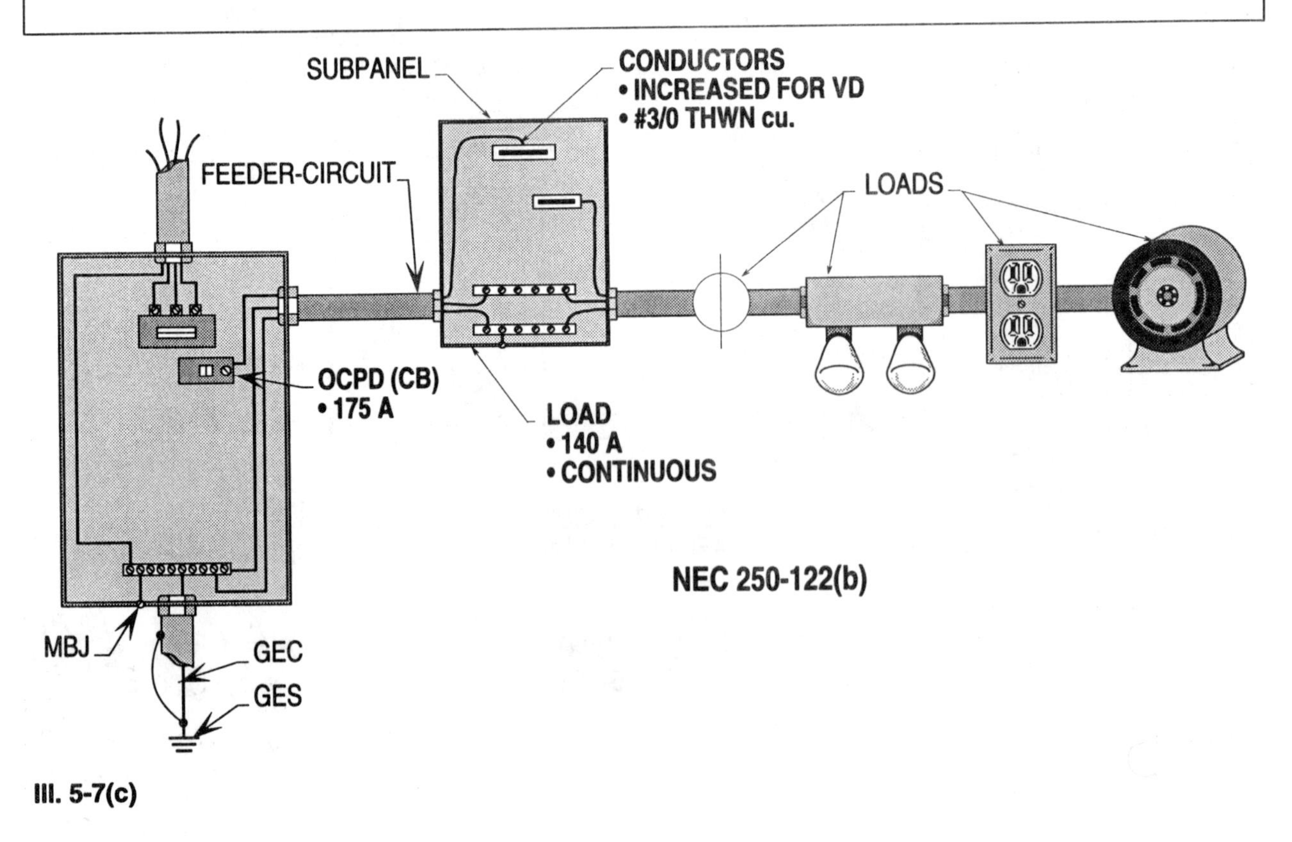

Ill. 5-7(c)

Grounding Equipment
250-122(c); Table 250-122

Where multiple circuits are routed in a conduit, the EGC for the largest circuit can be used to serve as the EGC for the smaller circuits.

Example 5-7(d). What size EGC is required to serve as the EGC for the circuits in illustration 5-7(d)?

Finding the size EGC

Step 1: Calculating EGC
250-122(c); Table 250-122
30 A requires #10 cu.

Solution: **One No. 10 copper EGC can be used to serve as the EGC for the No. 10, No. 12, and No. 14 copper circuit conductors.**

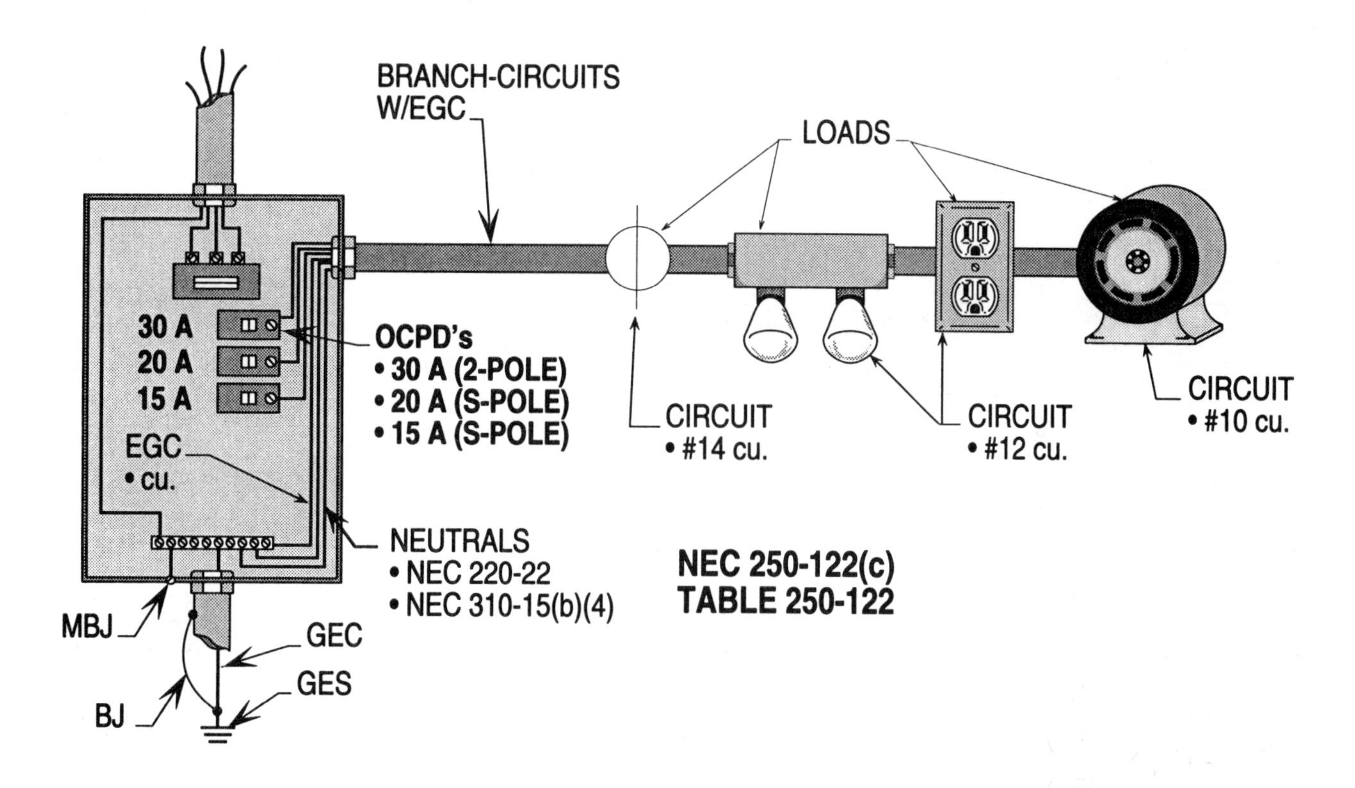

Ill. 5-7(d)

Grounding Equipment
250-122(d); Table 250-122

Under certain conditions of use, the selection of the minimum size EGC can be made from Table 250-122 based on the trip rating of motor-circuit overload device.

Example 5-7(e). What size EGC is required for the motor circuit in Illustration 5-7(e) based on the motor-circuit overload device?

Finding the Size EGC

Step 1: Calculating EGC based on MOPD
250-122(d); Table 250-122
91 A MOPD requires #8 cu.

Step 2: Calculating EGC based on CB
250-122(d); Table 250-122(d)
175 A CB requires #6 cu.

Solution: **The size EGC, based on the CB in the service equipment panel is No. 6 copper.**

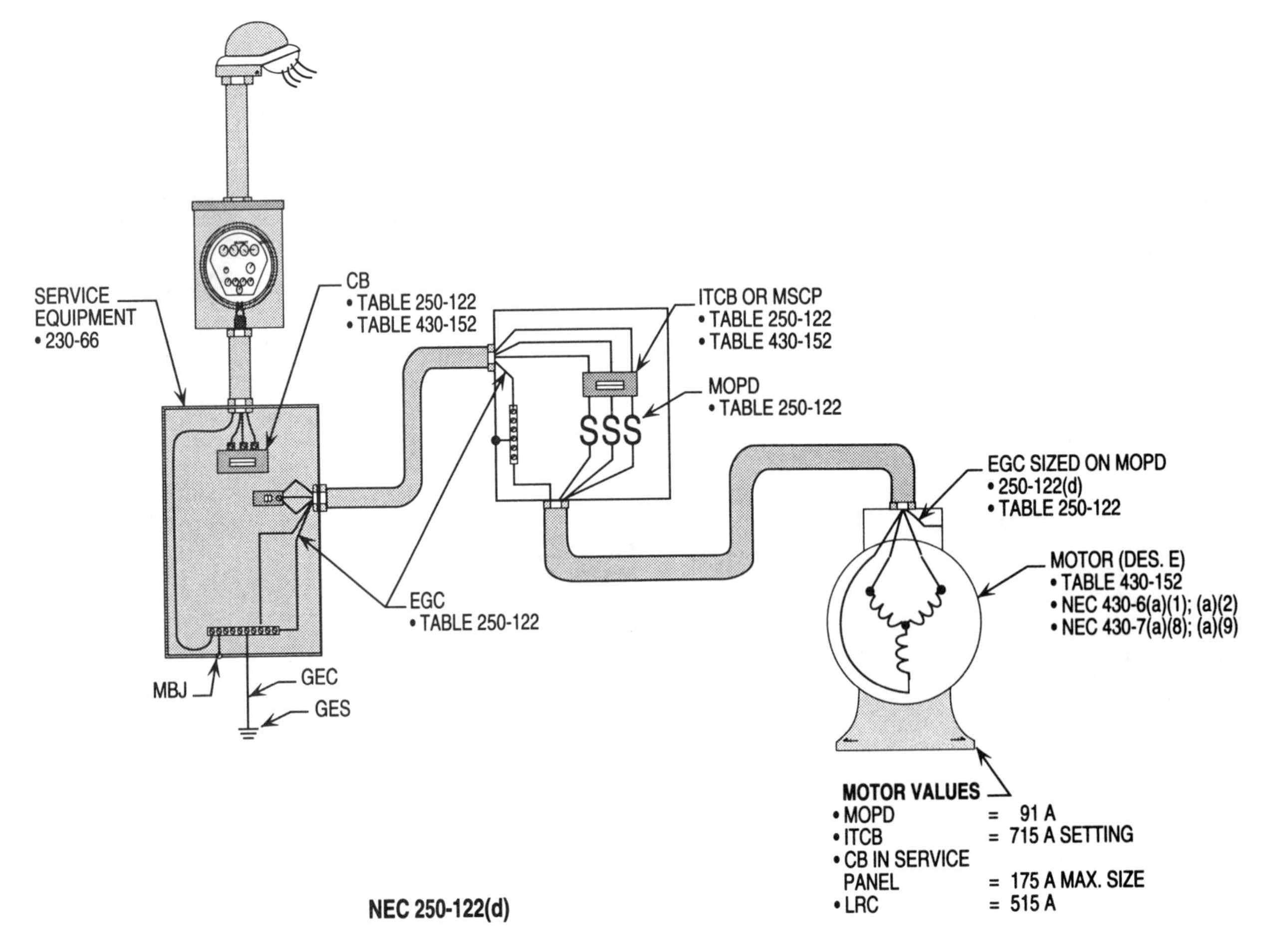

Ill. 5-7(e)

Grounding Equipment
250-122(f); Table 250-122

Full sized equipment grounding conductors (EGC's) shall be pulled in each conduit in a parallel connected system.

Example 5-7(f). What size copper EGC is required in each conduit in Illustration 5-7(f)?

Sizing EGC

Step 1: Calculating EGC
250-122(f); Table 250-122
600 A OCPD requires #1 cu.

Solution: The size EGC in each conduit is No. 1 cu.

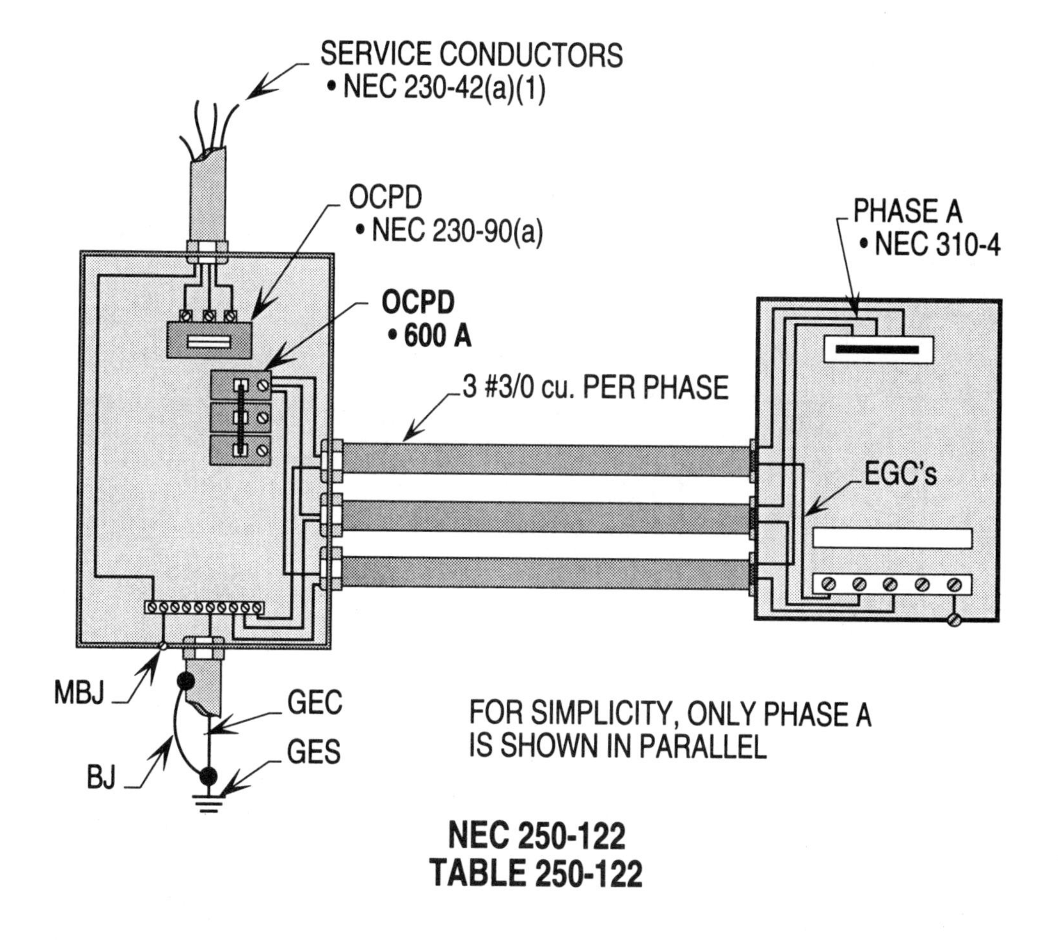

Ill. 5-7(f)

Grounding Equipment
250-122; Table 250-122

Properly sized equipment grounding conductors (EGC's) will provide a safe touch voltage (TV) at the enclosure of electrical equipment.

Example 5-7(g). What is the touch voltage at the enclosure of the equipment in Illustration 5-7(g)?

Finding SCC at enclosure

Step 1: Calculating SCC
240-1, FPN; 250-2(d)
SCC = V to Gd. / Total R
SCC = 120 V / .4R + .045R
SCC = 269.7 A

Finding touch V at enclosure

Step 1: Calculating tough V
240-1, FPN; 250-2(d)
TV = SCC x R of EGC
TV = 269.7 A x .045R
TV = 12 V

Solution: The touch voltage at the enclosure of the equipment is 12 volts.

NEC 240-1, FPN; NEC 110-9; NEC 110-10

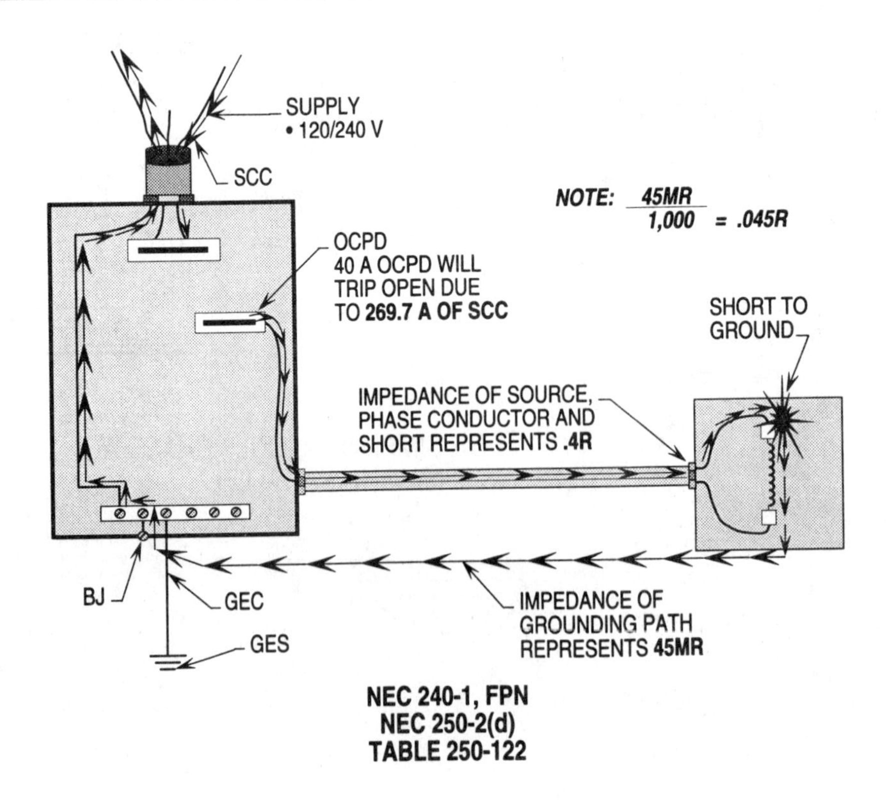

Ill. 5-7(g)

Grounding Equipment
250-122; Table 250-122

Poor connections between conduit and fittings can cause the metal conduit system to react as a heating element which cause electrical shock or fire.

Example 5-7(h). What is the short-circuit current (SCC) and heating element effect in VA (watts) for the circuit in Illustration 5-7(h)?

Finding SCC in circuit

Step 1: Calculating SCC
240-1, FPN; 250-2(d)
SCC = V to Gd. / R
SCC = 120 V / 4.5R
SCC = 26.7 A

Solution: The short-circuit current due to the loose connection is 26.7 amps.

Finding heating element effect

Step 1: Calculating VA
240-1, FPN; 250-2(d)
VA = V x A
VA = 120 V x 26.7 A
VA = 3,204

Solution: The volt-amps due to the short-circuit has a heating element effect of 3,204 VA (watts).

NEC 240-1, FPN; NEC 110-9; NEC 110-10

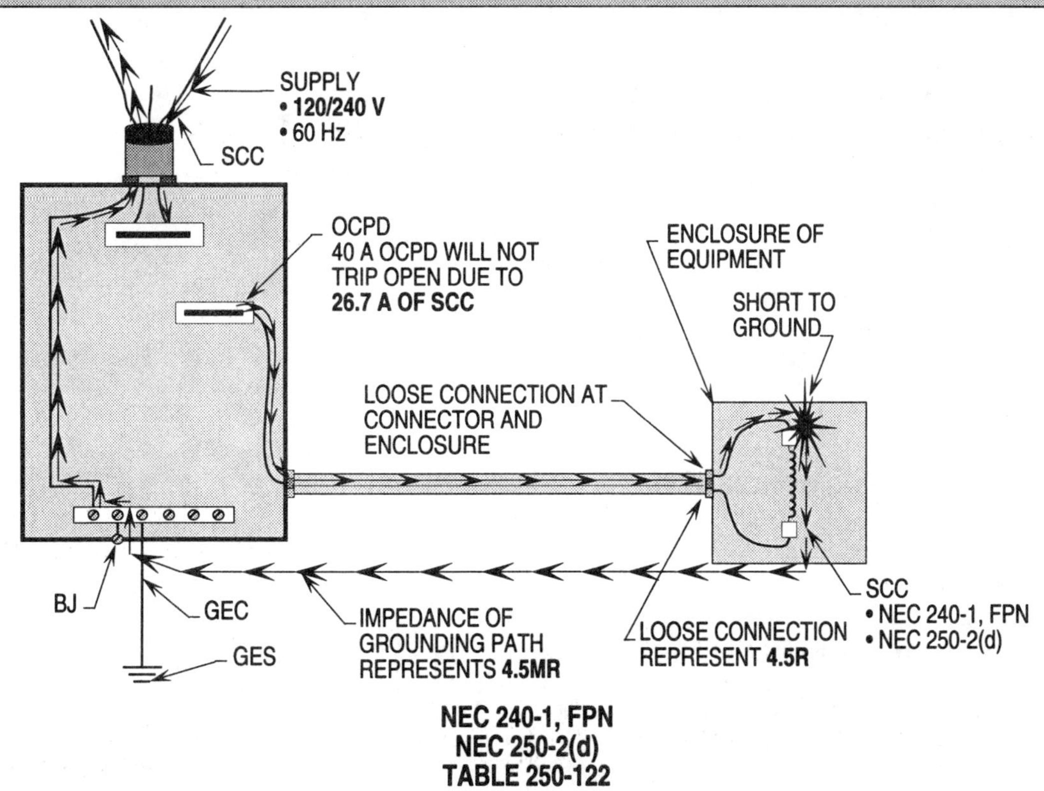

Ill. 5-7(h)

GROUNDING DC SYSTEMS
250-162

DC systems are grounded by the rules listed in NEC 250-166(a) and (b). **See Examples 5-1(a) and (b)** for a step by step procedure on how to ground DC systems.

GROUNDING AC SYSTEMS
250-66

AC systems are grounded by the rules listed in NEC 250-66 and Table 250-66. The GEC's are sized based upon the largest phase conductor or the type of grounding electrode utilized. **See Examples 5-2(a) and (b)** for a step by step procedure on how to ground AC systems.

DO ALL SYSTEMS HAVE TO BE GROUNDED
250-20(b); 250-24(a); 250-130(a); (b)

The answer is no, not all AC electrical systems have to be grounded. Section 250-24(a) states plainly "is grounded" and NEC 250-130(a) and (b) addresses both grounded and ungrounded systems and the exact methods of grounding and bonding are discussed. These Sections of the NEC makes it very clear that not all AC electrical systems have to be grounded. Care must be exercised when applying "shall be grounded", for it only applies if the AC systems are grounded and bonded. Check with the authority having jurisdiction for the requirements concerning grounding AC electrical systems.

GROUNDED SERVICE CONDUCTORS
250-24(a); (b); 250-2(d)

The grounded service conductor(s) (GEC's) must carry the maximum normal neutral current and the maximum short-circuit current the transformer can deliver through the phase conductors. **(See Examples 5-3(a) and (b) for a step by step procedure on how to size the GSC).**

EQUIPMENT GROUNDING CONDUCTOR
250-122

The equipment grounding conductor(s) (EGC's) must be sized large enough to carry any ground-fault current that may occur between any ungrounded phase conductor and metal enclosure housing or enclosing such conductors. **(See Examples 5-7(a) through (h) for the procedures for sizing EGC's).**

MAIN AND EQUIPMENT BONDING JUMPER
250-28(d); 102(c); (d)

The bonding jumpers (BJ's) are used to bond metal parts of the electrical grounding system together. In other words, the BJ's ensure the grounding system is continuous from point to point. **(See Examples 5-5(a) through (f) and 5-6 for a step by step procedure for sizing bonding jumpers on the supply and load side of the electrical system).**

GROUNDING ELECTRODE CONDUCTOR
250-66

The grounding electrode conductors (GEC's) are used to connect the grounding system to an electrode or electrodes which are earth grounded. The GEC connects the GSC, EGC's, and BJ's to the grounding electrode system (GES) which creates a single point ground. **(See Examples 5-1 and 5-2 for a step by step procedure for sizing properly GEC's based upon DC or AC grounded electrical power sources).**

6 Calculating Conductor Ampacities

Operating temperature is the temperature of conductors when they are carrying current. The allowable ampacities listed in Tables 310-16 through 310-19 may be used where there are three or less current-carrying conductors in a raceway or cable, or direct burial in the earth. Four or more current-carrying conductors routed through a raceway or cable or subjected to ambient temperatures above 86°F can increase the operating temperature. Therefore, their allowable ampacities per Tables 310-16 through 310-19 have to be derated to allow for the higher operating temperature of the conductors per correction factors to Tables 310-16 through 310-19 and adjustment factors per Table 310-15(b)(2)(a) to Tables 0-2,000 volts.

Quick Reference

Ampacities Per Table 310-16
310-10; Table 310-16; Correction Factors; Table 310-15(b)(2)(a)
To Tables 0-2000 Volts

Where there is not more than three current-carrying conductors in a raceway or cable, the allowable ampacities listed in Table 310-16 may be used. If four or more current-carrying conductors or surrounding temperature exceeding 86°F is present, derating factors must be applied, based upon their conditions of use.

Example 6-1(a). What is the allowable ampacity (based on Table current) of the conductors in Illustration 6-1(a)?

Sizing ampacity

Step 1: Calculating ampacity
Table 310-16; 310-15(b)(2)(a)
3 #12 cu. conductors allow 25 A

Solution: The allowable ampacity is 25 amps.

ADDITIONAL HELP 39:

Calculating the AC resistance of the conductors in Ill. 6-1(a) based on Given Facts.

- Calc. AC R
  Table 9, Ch. 9

$$R = \frac{150'}{1{,}000} \times 2.0$$

R = 0.3

Solution: The AC resistance is 0.3 ohms.

Example 6-1(b). What is the allowable ampacity of the current-carrying conductors in Illustration 6-1(b)?

Sizing ampacity

Step 1: Calculating ampacity
Table 310-16
#12 THHN cu. = 30 A

Step 2: Applying derating factors
Table 310-15(b)(2)(a); 310-10(4)
30 A x 80% = 24 A

Solution: The allowable ampacity is 24 amps.

ADDITIONAL HELP 40:

Calculating the resistance of the conductors in Ill. 6-1(b) based on Given Facts.

- Calc. Mills of wire
  Table 8, Ch. 9
  Mills = .092 dia. x 1,000
  Mills = 92
- Calc. R of conductors

$$R = \frac{12R \times 250' \ (3{,}000)}{\text{mills } 92^2 \ (8{,}464)}$$

R = .3544

Solution: The resistance of the conductors is .3544.

Example 6-1(c). What is the current-carrying capacity of the conductors in Illustration 6-1(c)?

Sizing ampacity

Step 1: Calculating ampacity
Table 310-16
#12 THHN cu. = 30 A

Step 2: Applying derating factors
Table 310-15(b)(2)(a); 310-10(4)
30 A x 70% = 21 A

Solution: The allowable ampacity is 21 amps.

ADDITIONAL HELP 41:

Calculating the resistance of the conductors in Ill. 6-1(c) based on Given Facts.

- Calc. R at 50°C

$$R = 1.98 \times \frac{\text{Zero R} + °C}{\text{Zero R} + °C}$$

$$R = 1.98 \times \frac{234.5 + 50 \ (284.5)}{234.5 + 25 \ (259.5)} \ (1.09634)$$

R = 2.171

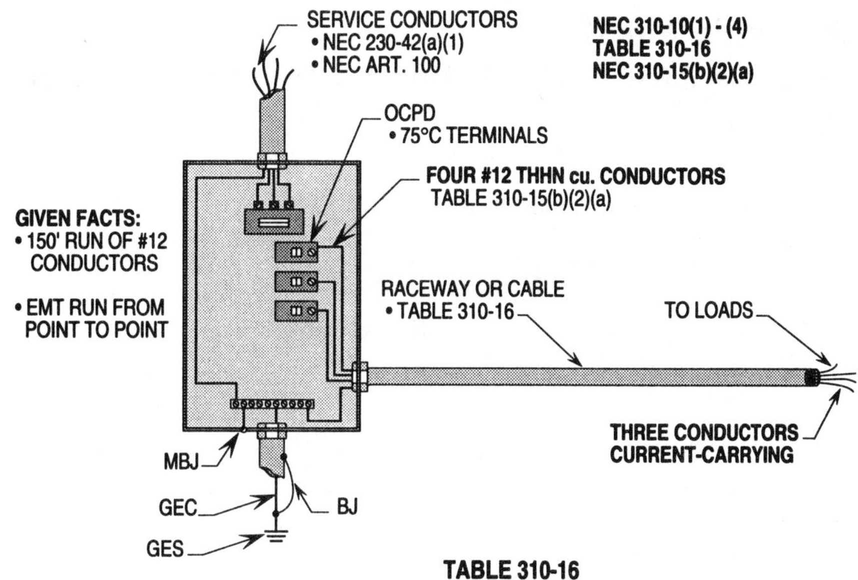

Ill. 6-1(a)

TABLE 310-16
310-15(b)(2)(a); TABLE 310-15(b)(2)(a)

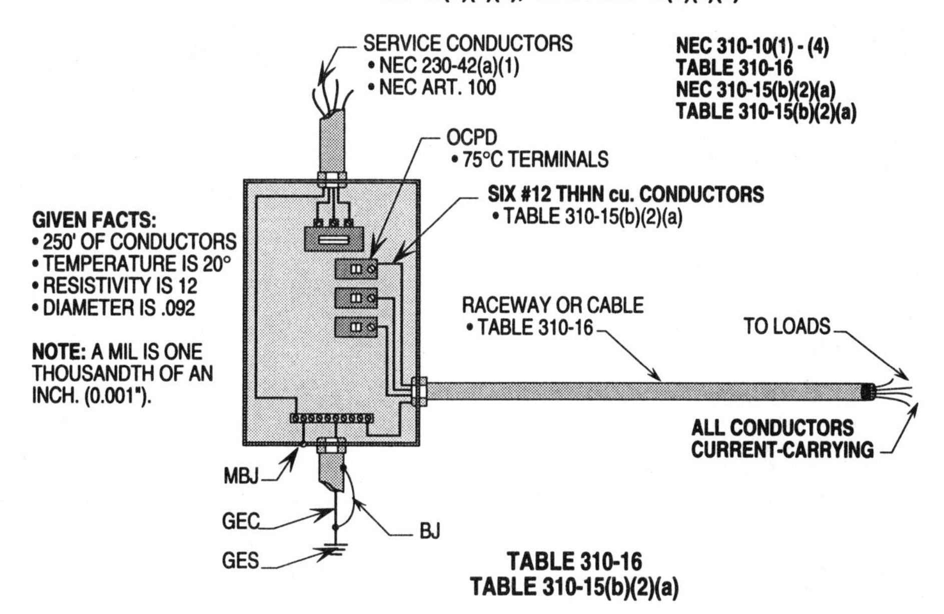

Ill. 6-1(b)

TABLE 310-16
TABLE 310-15(b)(2)(a)

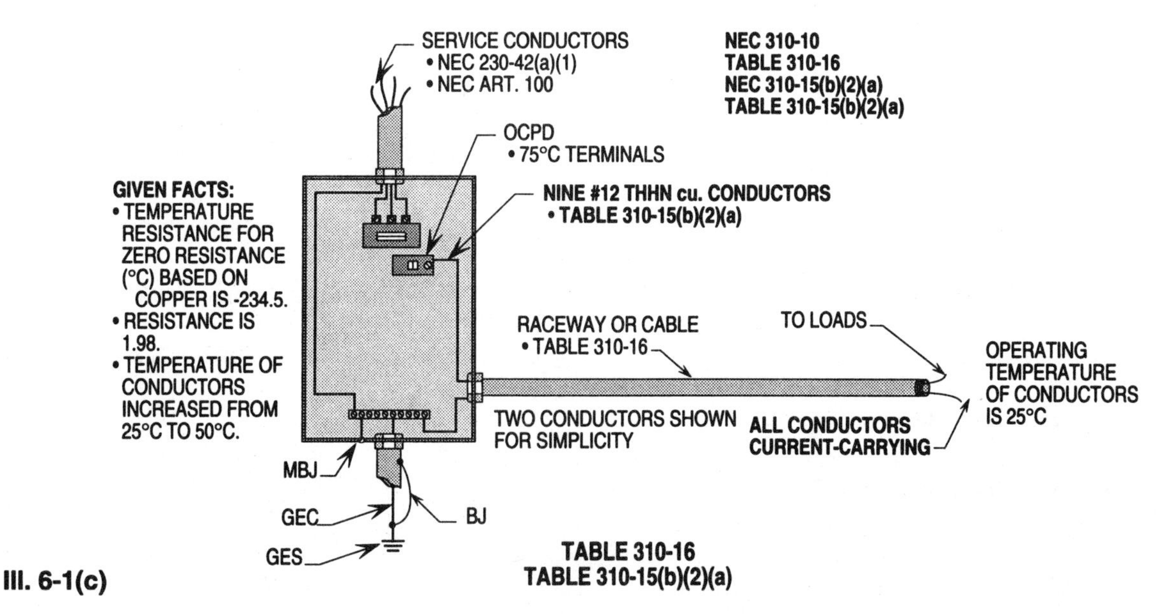

Ill. 6-1(c)

TABLE 310-16
TABLE 310-15(b)(2)(a)

Ampacities Per Table 310-16
310-10(1) - (4); Table 310-16; Correction Factors; Adjustment Factors To Table 310-15(b)(2)(a)

Four or more current-carrying conductors in a conduit or raceway which are exposed to surrounding ambient temperatures above 86° F are required to be derated.

Example 6-1(d). What is the allowable ampacity for the conductors in Illustration 6-1(d)?

Sizing ampacity

Step 1: Calculating ampacity
Table 310-16
#12 THHN cu. = 30 A

Step 2: Applying derating factors
Correction factors to Table 310-16
30 A x 91% = 27.3 A

Solution: The allowable ampacity is 27.3 amps.

ADDITIONAL HELP 42:

Calculating the square mil area of the busbar in Given Facts of Ill. 6-1(d).

- Calc. Mills of busbar
  1/4" x 1,000 = 250 mils
  3' x 1,000 = 3,000 mils
- Calc. Mils of sq. mils
  sq. Mils = 250 x 3,000 x 1.273
  sq. Mils =954,750

Solution: The square mils area is 954,750.

Example 6-1(e). What is the allowable ampacity of the conductors in Illustration 6-1(e)?

Sizing ampacity

Step 1: Calculating ampacity
Table 310-16
#12 THHN cu. = 30 A

Step 2: Applying derating factors
Adjustment factors to Table 310-15(b)(2)(a)
30 A x 80% = 24 A

Step 3: Applying derating factors
Correction factors to 310-16
24 A x 91% = 21.8 A

Solution: The allowable ampacity is 21.8 amps.

ADDITIONAL HELP 43:

Calculating the size THWN cu. conductors to supply a load of 16 amps at an ambient temperature of 102°F.

- Calc. THWN cu. conductor
  A = 16 A load x 1.10
  A =17.6
- Selecting size conductor
  17.6 A requires #12 cu.

Solution: The size THWN copper conductors are No. 12.

Note: To find the multiplier above, divide 1 by .91 (1÷ .91 = 1.10) per Table 310-16 based on a correc-

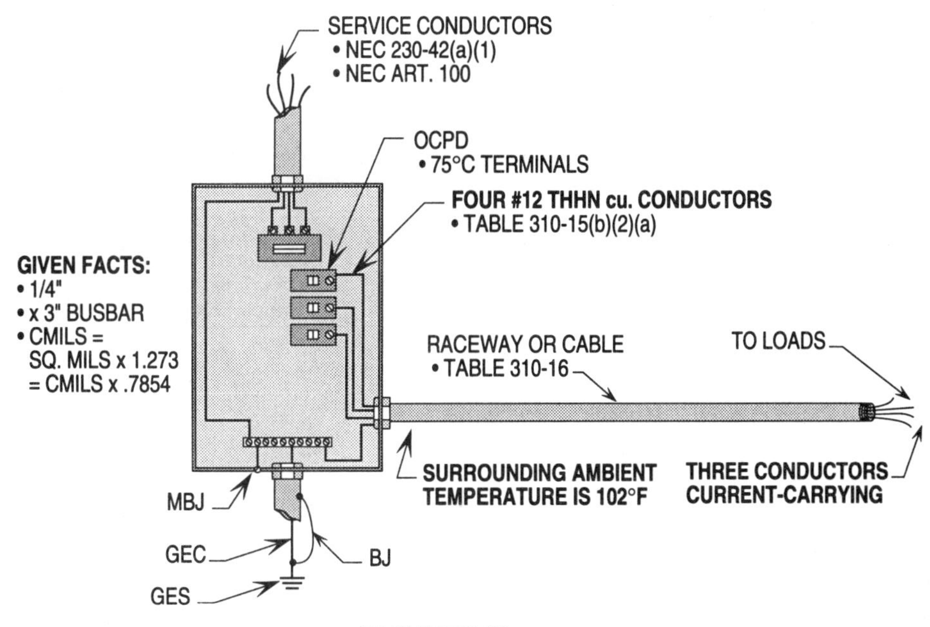

TABLE 310-16
CORRECTION FACTORS TO TABLE 310-16

Ill. 6-1(d)

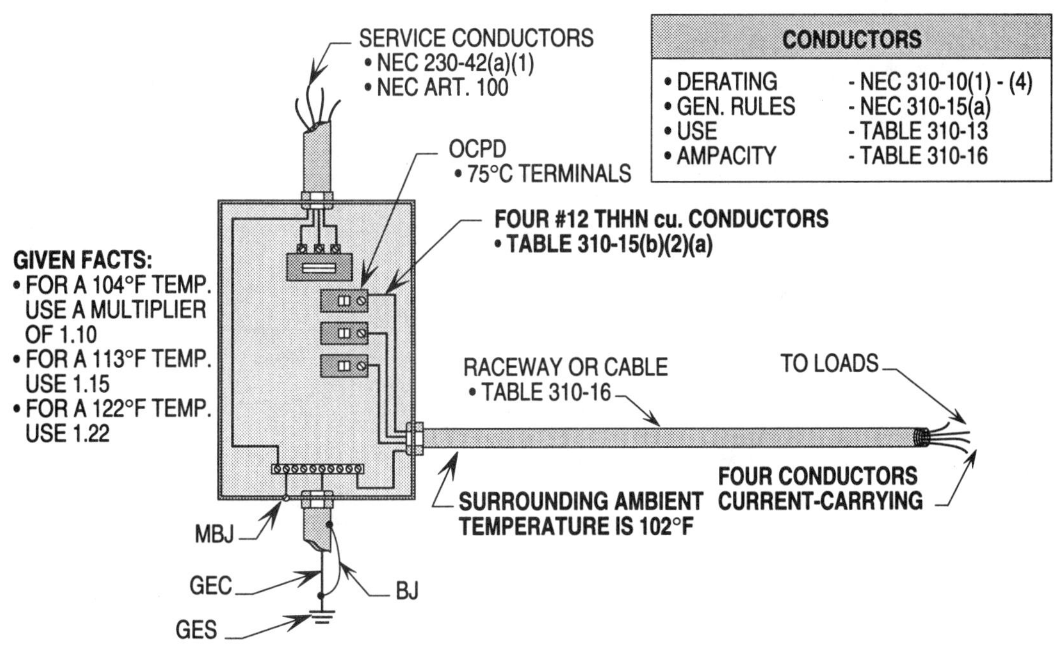

TABLE 310-16
ADJUSTMENT FACTORS TO TABLE 310-15(b)(2)(a)

Ill. 6-1(e)

Ampacities Per Table 310-16
310-10(1) - (4); Table 310-16; Correction Factors; Adjustment Factors To Table 310-15(b)(2)(a)

Larger size conductors are sometimes required, due to derating ampacity of conductors. **Note:** It is the rating of the OCPD that determines the size of the branch-circuit.

Example 6-1(f). What is the allowable ampacity and size OCPD for the conductors in Illustration 6-1(f)?

Size
- **NEC 210-3**

Conductors
- **NEC 210-19(a)**

OCPD's
- **NEC 210-20(a)**

Sizing ampacity

Step 1: Calculating ampacity
Table 310-16
#10 THWN cu. = 40 A

Step 2: Applying adjustment factors
Table 310-15(b)(2)(a)
40 A x 70% = 28 A

Step 3: Applying correction factors
Table 310-16
28 A x 82% = 22.96 A

Solution: The allowable ampacity is 22.96 amps.

Sizing OCPD

Step 1: Finding OCPD
Table 310-16; correction factors
Table 310-15(b)(2)(a) adjustment factors
22.96 A requires 20 A OCPD

Solution: The size OCPD is 20 amps.

ADDITIONAL HELP 44:

Calculating the allowable ampacity for loading half of the conductors in installation 2.

- Finding A
  Table B-310-11; Table 310-16
  30 A x 70% = 21 A

Solution: The allowable ampacity is 21 A.

ADDITIONAL HELP 45:

Calculating the allowable ampacity for loading 18 of the conductors in installation 2.

- Finding A
  Table B-310-11, Ex. 2

$$A_2 = \sqrt{\frac{0.5 \times N}{E}} \times A_1 \times \%$$

$$A_2 = \sqrt{\frac{0.5 \times 24}{18}} \; (.816) \times 30 \text{ A} \times 70\%$$

$$A_2 = 17 \text{ amps}$$

Solution: The allowable ampacity is 17 amps.

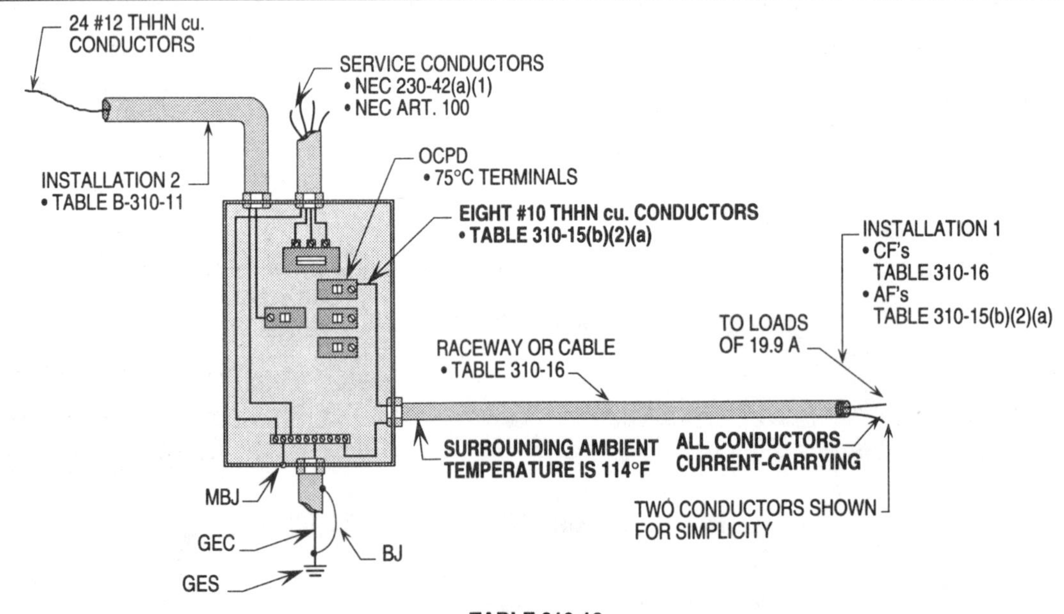

TABLE 310-16
ADJUSTMENT FACTORS TO TABLE 310-15(b)(2)(a)
CORRECTION FACTORS TO TABLE 310-16

Ill. 6-1(f)

Calculating Ampacity Per Table 310-17
Table 310-17

Table 310-17 deals with copper, copper-clad, and aluminum conductors up to 2,000 volts where they are used as single conductors in free air.

Example 6-2. What is the allowable ampacity of the conductors in Illustration 6-2?

Sizing ampacity

Step 1: Calculating allowable ampacity
Table 310-17
#8 cu. = 70 A

Solution: The allowable ampacity is 70 amps.

ADDITIONAL HELP 46:

Calculating the square mils when circular mils are known.

- Calculating Sq. mils of No. 8 cu.
  Sq. mils = CM x .7854
  Sq. mils = 16,510 x .7854
  Sq. mils = 12,967

Solution: The square mils of a No. 8 copper conductor is 12,967.

Note: When using the above formula, .7854 is the constant to be used everytime when finding the square mills.

NEC 310-10(1) - (4); NEC 225-6(a)(1); (2); NEC 225-10; TABLE 310-17

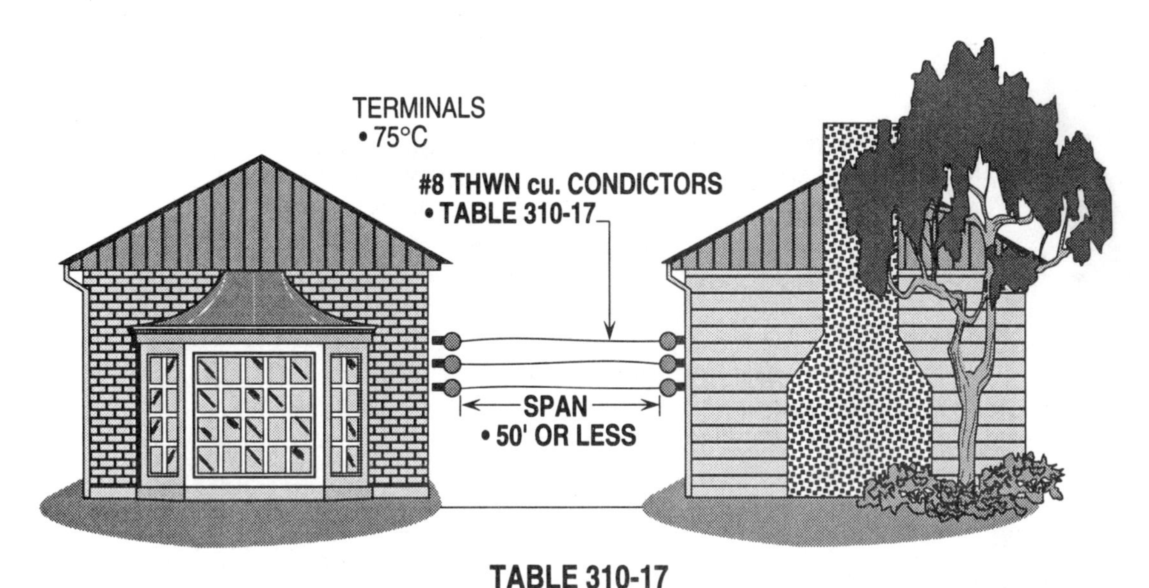

Ill. 6-2

Calculating Ampacity Per Table 310-18
Table 310-18

Table 310-18 contains conductors with higher temperature ratings which range from 302°F, 392°F, and 482°F respectively. These conductors which have higher temperature ratings may be utilized in locations involved with high ambient temperatures. Conductors are routed in raceway or cable. The allowable ampacity must be adjusted according to terminal ratings, number of current-carrying conductors, and ambient temperatures.

Example 6-3. What is the allowable ampacity of the conductor in Illustration 6-3?

Sizing ampacity

Step 1: Calculating allowable ampacity
Table 310-18
#10 FEP cu. = 60 A

Solution: The allowable ampacity is 60 amps.

ADDITIONAL HELP 47:

Calculating the CM area of 1/4" round conductor.

- Calc. mils of 1/4" wire
  Mils = 1/4" x 1,000
  Mils = 250

- Calc. CM area
  CM = Diameter in mils^2
  CM = 250 x 250
  CM = 62,500

Solution: The CM area of a 1/4 in. round conductor is 62,500.

Note: 1 in. is equal to 1,000 mils.

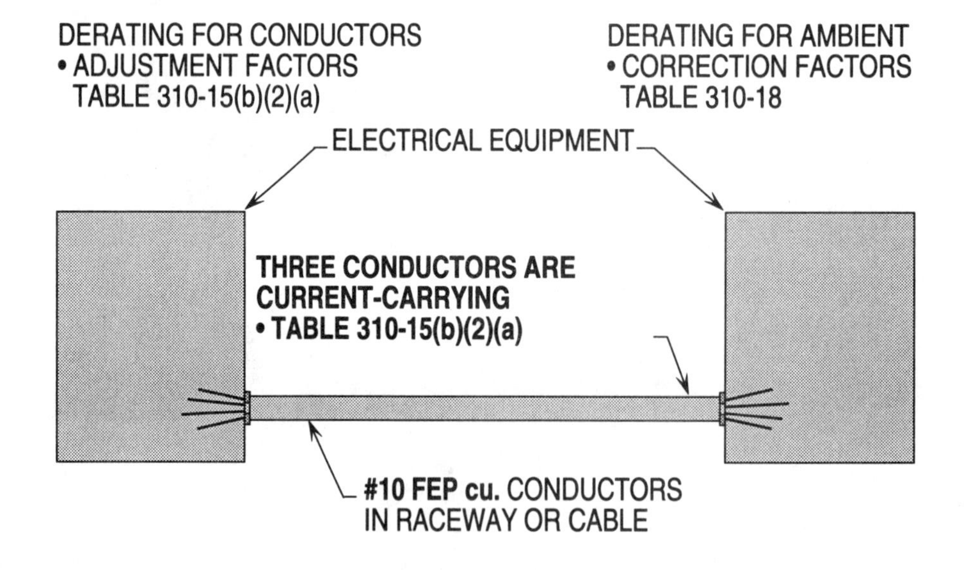

TABLE 310-18

Ill. 6-3

Calculating Ampacity Per Table 310-19
Table 310-19

Table 310-19 contains conductors with higher temperature ratings which range from 302°F to 482°F. Conductors must be routed in free air to use these ratings.

The allowable ampacity must be adjusted according to terminal ratings, number of current-carrying conductors, and ambient temperatures.

Example 6-4. What is the allowable ampacity for the conductors in Illustration 6-4?

Sizing ampacity

Step 1: Calculating allowable ampacities
Table 310-19
#8 FEP cu. = 124 A

Solution: The allowable ampacity is 124 amps.

ADDITIONAL HELP 48:

Calculating the CM of a copper conductor that is 500 ft. long having a resistance of 4 ohms. (k = 12 for copper)

- Calc. CM

$$CM = \frac{L \times K}{R}$$

$$CM = \frac{500' \times 12}{4R}$$

$$CM = 1{,}500$$

Solution: The CM area is 1,500.

DERATING
• AMBIENT - CORRECTION FACTORS TO TABLE 310-19
• CONDUCTORS - ADJUSTMENT FACTORS TO TABLE 310-15(b)(2)(a)

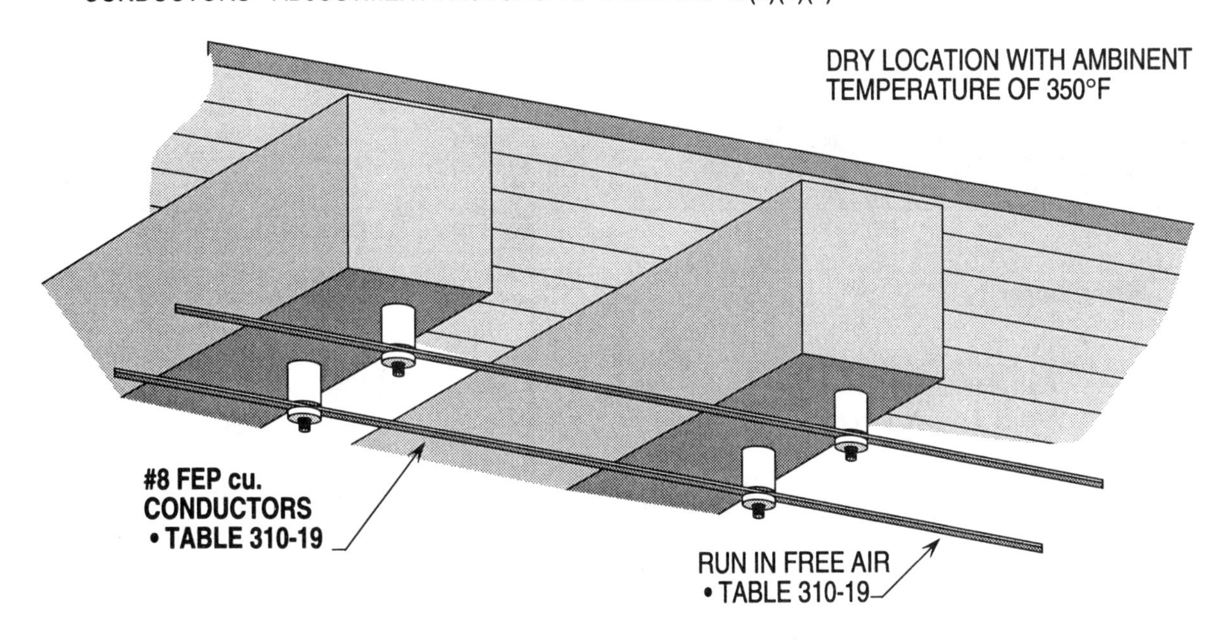

TABLE 310-19

Ill. 6-4

Calculating Ampacity Per Tables 310-69 Through 310-86 Tables 310-67 Through 310-86

Tables 310-69 through 310-86 list maximum continuous ampacities for copper and aluminum solid dielectric insulated conductors rated over 2,000 to 35,000 volts.

Example 6-5(a). What is the ampacities for the conductors in Illustration 6-5(a)?

Sizing ampacity

Step 1: Calculating ampacity
Figure 310-60; Table 310-77
#3/0 cu. = 250 A

Solution: The ampacity is 250 amps.

ADDITIONAL HELP 49:

Calculating the diameter of a No. 3/0 copper conductor when CM is known.

- Calc. diameter in mils
  Table 8, Ch. 9

$$\text{Dia. in mils} = \sqrt{\text{CM area}}$$

$$\text{Dia. in mils} = \sqrt{167{,}800}$$

$$\text{Dia. in mils} = 409.6$$

Solution: The diameter in mils is 409.6.

Example 6-5(b). What is the ampacities for the conductors in Illustration 6-5(b)?

Sizing ampacity

Step 1: Calculating ampacity
Figure 310-60; Table 310-77
#3/0 cu. = 210 A

Solution: The ampacity is 210 amps

ADDITIONAL HELP 50:

Calculating the length of a coil of cu. wire having a resistance of .10 ohms and a CM area of 167,800.

- Calc. length

$$L = \frac{R \times CM}{K}$$

$$L = \frac{.10R \times 167{,}800\ CM}{12}$$

$$L = 1{,}398\ ft.$$

Solution: The length of the wire is about 1,398 ft.

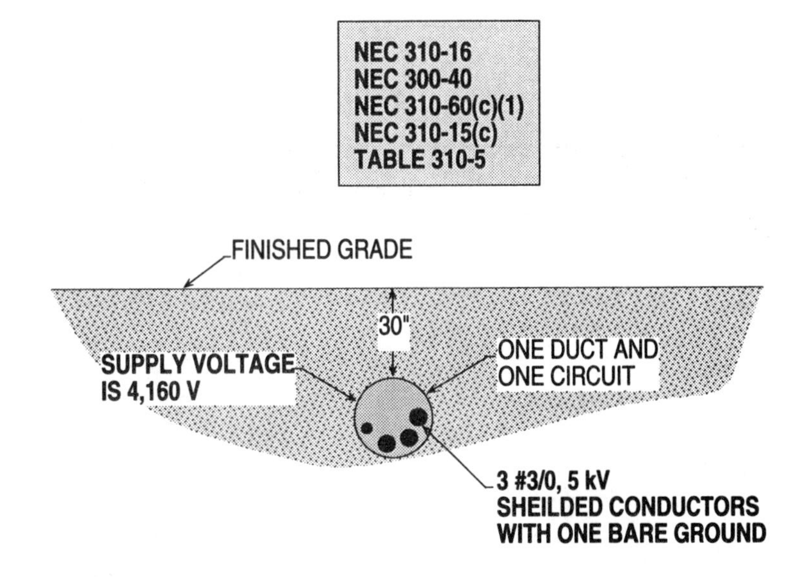

FIGURE 310-60 TO TABLES 310-67 TO 86; TABLE 310-77

Ill. 6-5(a)

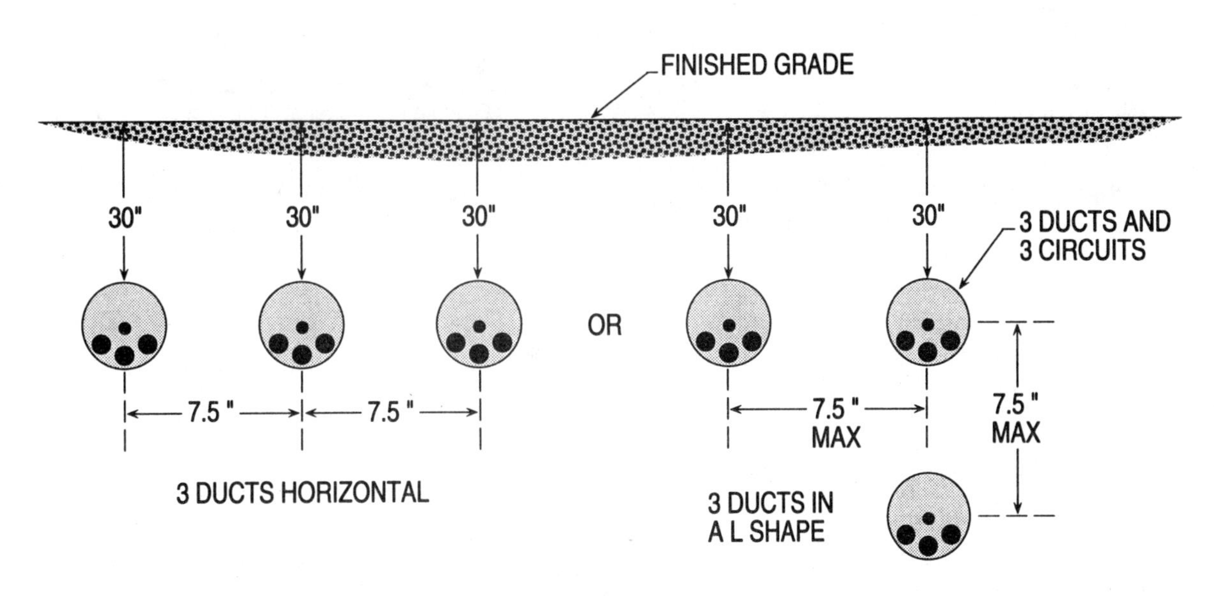

ALL DUCTS CONTAINS 3 #3/0, 5 kV SHIELDED CONDUCTORS AND SUPPLIED BY 4,160 VOLTS

FIGURE 310-60 TO TABLES 310-67 TO 86; TABLE 310-77

Ill. 6-5(b)

Calculating Ampacity Per Tables 310-69 Through 310-86
Tables 310-67 Through 310-86

Tables 310-69 through 310-86 list maximum continuous ampacities for copper and aluminum solid dielectric insulated conductors rated over 2,000 to 35,000 volts.

Example 6-5(c). What is the ampacities for the conductors in Illustration 6-5(c)?

Sizing ampacity

Step 1: Calculating ampacity
Figure 310-60; Table 310-77
#3/0 cu. = 260 A

Solution: The ampacity is 260 amps.

ADDITIONAL HELP 51:

Calculating the temperature in Fahrenheit when the temperature in centigrade is 35°.

- Calc. temperature °F

Temp. °F = (9/5 x temp. C) plus 32

Temp. °F = (9/5 x 35) plus 32

Temp. °F = 95°

Solution: The temperature in °F is 95°.

Note: See Appendix F in this book.

Example 6-5(d). What is the ampacities for the conductors in Illustration 6-5(d)?

Sizing ampacity

Step 1: Calculating ampacity
Figure 310-60; Table 310-77
#3/0 cu. = 210 A

Solution : The ampacity is 210 amps

ADDITIONAL HELP 52:

Calculating the temperature in centigrade when the temperature in Fahrenheit is 95°.

- Calc. Temp. °C

Temp. °C = 5/9 x (temp. F - 32)

Temp. °C = 5/9 x (95 - 32)

Temp. °C = 35°

Solution: The temperature in °C is 35°.

Note: See Appendix F in this book.

NEC 310-16
NEC 300-40
NEC 310-60(c)(1)
NEC 310-15(c)
TABLE 310-5

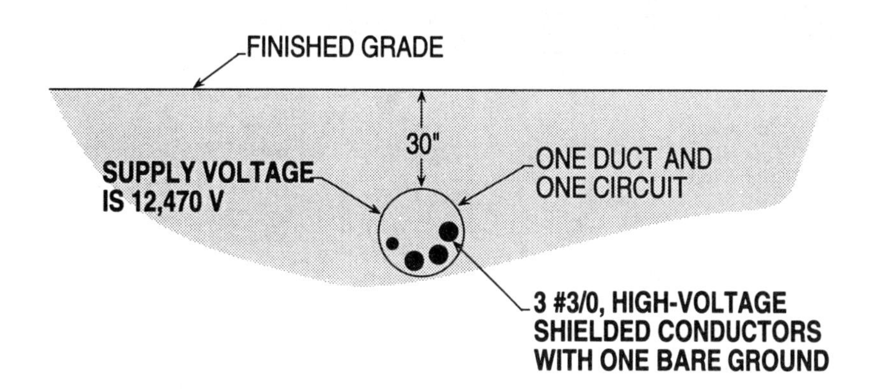

FIGURE 310-60 TO TABLES 310-67 TO 86; TABLE 310-77

Ill. 6-5(c)

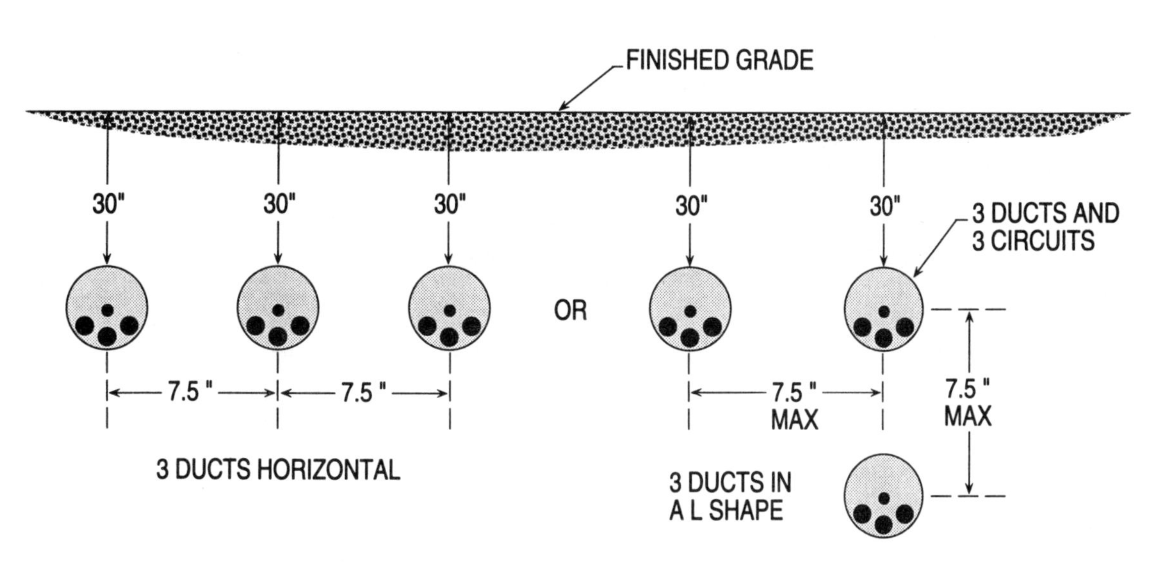

ALL DUCTS CONTAINS 3 #3/0, HIGH-VOLTAGE SHIELDED CONDUCTORS AND SUPPLIED BY 12,470 VOLTS

FIGURE 310-60 TO TABLES 310-67 TO 86; TABLE 310-77

Ill. 6-5(d)

Calculating Ampacity Per Tables 310-69 Through 310-86
Tables 310-69 Through 310-86

Conduits and conductors having a covering more than 30 inches below finished grade shall have the ampacity of the conductors derated based upon the additional depth.

Example 6-5(e). What is the ampacity for the conductors in Illustration 6-5(e)?

Sizing ampacity

Step 1: Calculating length
310-60(c)(2)(a);
Table 310-77
300' run x 25% = 75'

Step 2: Comparing length
310-60(c)(2)(a);
Table 310-77
50' is less than 75'

Step 3: Calculating ampacity
Table 310-77
#3/0 cu. = 250 A

Solution: The ampacity is 250 amps.

Example 6-5(f). What is the ampacity for the conductors in Illustration 6-5(f)?

Sizing ampacity

Step 1: Calculating length
310-15(c)(2)(b);
Note 1 to Figure 310-60
24" below 30" requires 12% derate

Step 2: Applying derating factors
310-15(c)(2)(b);
Table 310-77
250 A (#3/0) x 12% = 30 A
250 A - 30 A = 220 A

Solution: The ampacity is 220 amps.

ADDITIONAL HELP 53:

Calculating ampacity using another method.

- Calc. ampacity
  310-60(c)(2)(b)
  250 A x 88% = 220 A

Solution: The ampacity is 220 amps.

Note: The value 88 percent is found by subtracting 12 percent from 100 percent (100% - 12% = 88%).

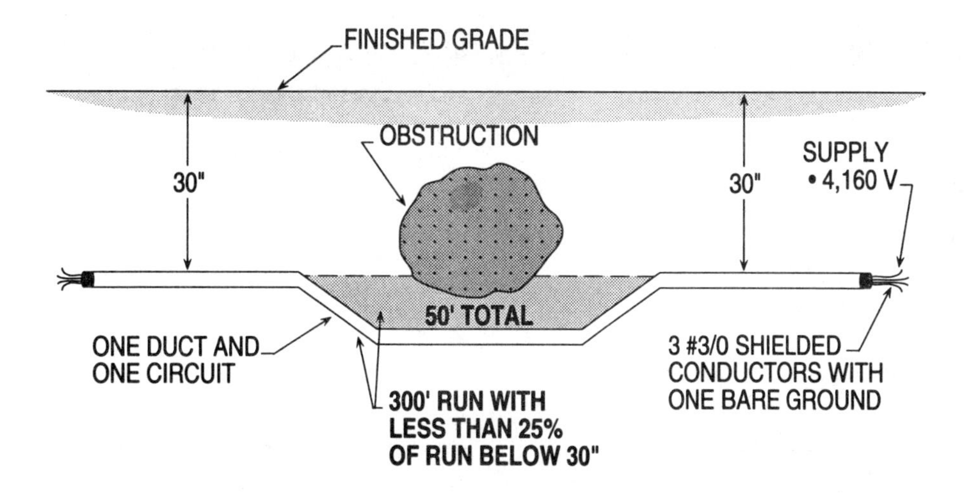

FIGURE 310-60 TO TABLES 310-67 TO 86
310-60(c)(2)(b); TABLE 310-77

Ill. 6-5(e)

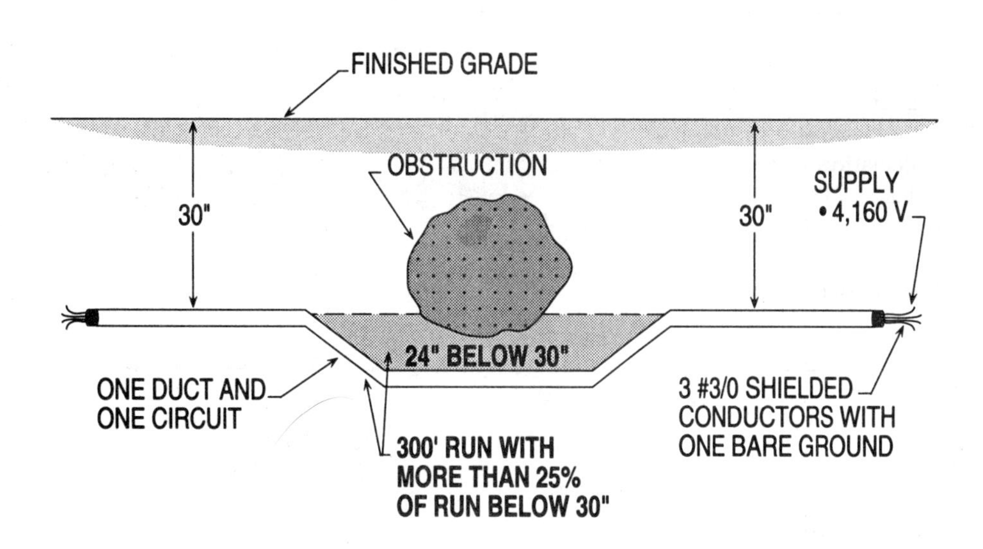

FIGURE 310-60 TO TABLES 310-67 TO 86
310-60(c)(2)(b); NOTE 1 TO FIGURE 310-60; TABLE 310-77

Ill. 6-5(f)

Calculating Ampacity Per Tables 310-69 Through 310-86

Tables 310-69 through 310-86 list maximum continuous ampacities for copper and aluminum solid dielectric insulated conductors rated over 2,000 to 35,000 volts.

Example 6-5(g). What is the ampacities for the conductors in Illustration 6-5(g)?

Sizing ampacity

Step 1: Calculating ampacity
Table 310-85
#3/0 cu. = 315 A

Solution: The ampacity is 315 amps.

Example 6-5(h). What is the ampacities for the conductors in Illustration 6-5(h)?

Sizing ampacity

Step 1: Calculating ampacity
Table 310-86
#3/0 cu. = 245 A

Solution: The ampacity is 245 amps

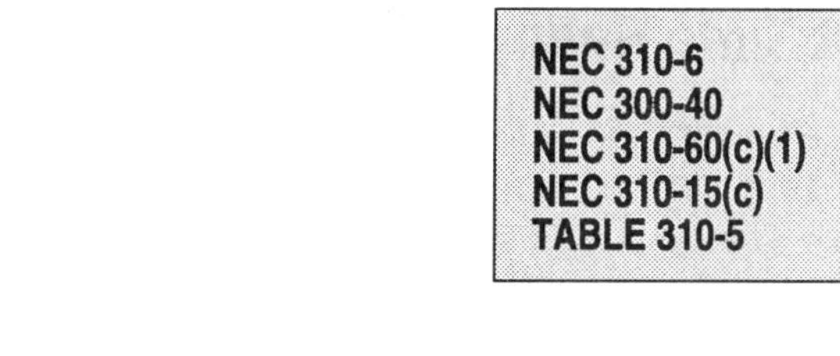

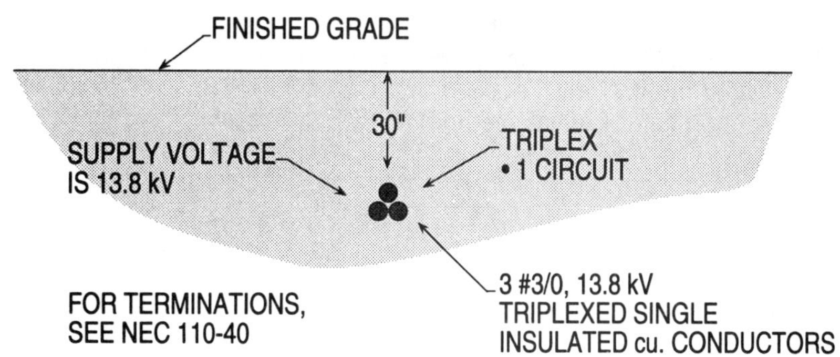

TABLE 310-85

Ill. 6-5(g)

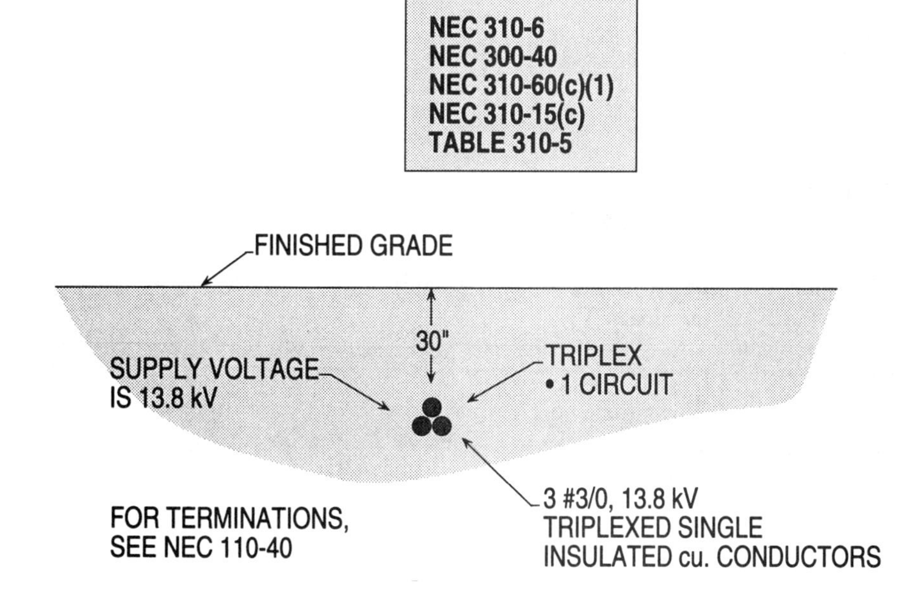

TABLE 310-86

Ill. 6-5(h)

Calculating Long And Short Time Ratings Of Conductors 310-10

The long time current rating of conductors are determined by selecting the ampacity of the conductor based upon its size, material, and insulation.

The short time current rating of copper conductors are determined by dividing the CM rating of the conductor by 42.3 where the conductors are in the same raceway or cable assembly. If a single conductor is installed in a raceway or routed by itself, the CM rating is divided by 30. **Note:** The short time current rating in amps for copper conductors is based for a duration of five seconds or less. (Rule of thumb method)

Example 6-6(a). Referring to Illustration 6-6(a), what is the long time ampacity rating of one of three No. 10 THHN copper conductors in a metal raceway system used at 60°?

Step 1: Find the rating
Table 310-16
#10 THHN cu. = 30 A

Solution: **The No. 10 THHN copper conductors have an ampacity rating of 30 amps at 60°C terminals per NEC 240-3(d).**

Example 6-6(b). Referring to Illustration 6-6(b), what is the short time current rating of a No. 10 THHN copper conductor in a conduit system with other No. 10 conductors?

Step 1: Finding CM
Table 8, Ch. 9
#10 = 10,380 CM

Step 2: Find the rating
I = CM ÷ 42.3
I = 10,380 CM ÷ 42.3 = 245 A

Solution: **The short time current rating for five seconds is 245 amps for each No. 10 in the conduit at 60°C terminals.**

Example 6-6(c). Referring to Illustration 6-6(c), what is the short time current rating of an insulated No. 6 THHN copper grounding electrode that is run in conduit to a metal water pipe?

Step 1: Finding CM
Table 8, Ch. 9
#6 = 26,240 CM

Step 2: Find the rating
I = CM ÷ 30
I = 26,240 ÷ 30 = 875 A

Solution: **The short time current rating of the No. 6 is 875 amps.**

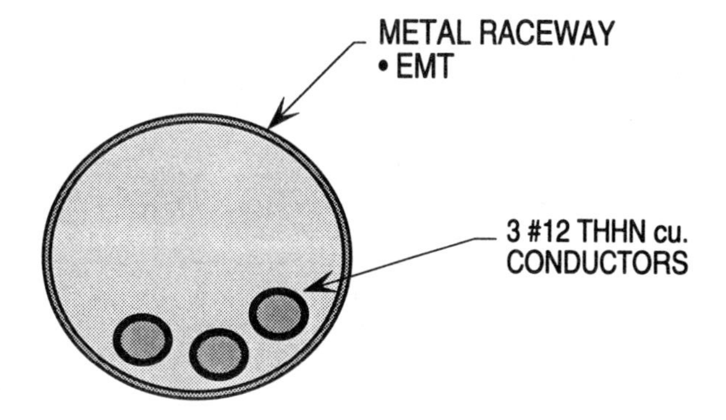

Ill. 6-6(a)

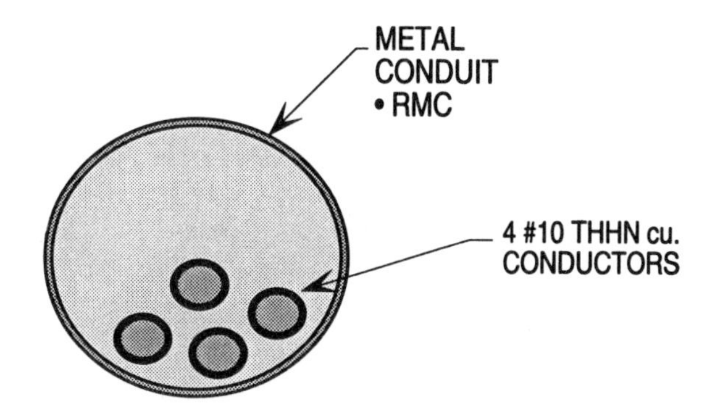

Ill. 6-6(b)

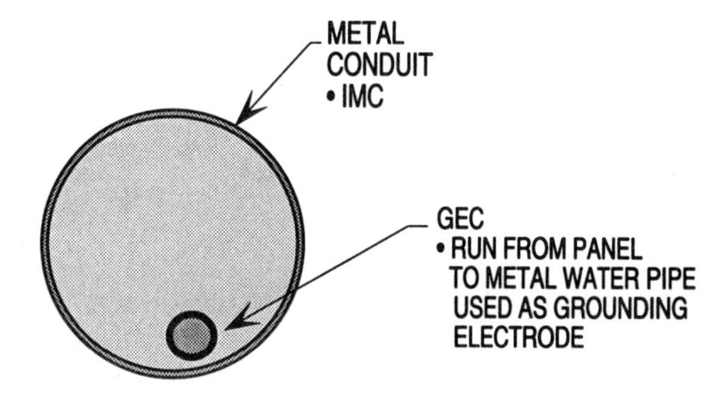

Ill. 6-6(c)

7

Raceway and Cable Trays

The number of conductors that are allowed in raceway systems is covered in Appendix C of the National Electrical Code® (NEC®) in Tables C1 through C12(A) for conductors all of the same size with the same type of insulation. Tables 4 and 5 are utilized for conductors of different sizes with the square inch rating of each conductor being multiplied by the number to derive the total square inch area. The total square inch rating is used to select the raceway from Table 4 to Chapter 9. Where the conductors are all the same size with the same type of insulation, the total number is used to select the size raceway from Tables C1 through C12(A).

Cable tray systems are used to support conductors and cables or combinations of such. Cable trays are open, raceway-like support assemblies which are sized based upon the diameters or square-inch area of the conductors or cables in which they are designed and selected to support.

Note: The size of the conduit selected depends upon the particular type of raceway system that is used and the type of insulation on the conductors.

Quick Reference

Sizing Conduits
Notes to Tables to Chapter 9

Conduits with conductors that have the same type of insulation and are the same size may be selected by using Tables C1 through C12(A) of the NEC.

Conduits enclosing different size conductors are sized by selecting the square inch area of each conductor and multiplying by the number. This total is used to select the size conduit per Table 4 in Chapter 9.

Example 7-1(a). Determine the size rigid metal conduit to enclose the copper conductors in Illustration 7-1(a).

Sizing RMC

Step 1: Finding size RMC
Table C8, Appendix C
4 #6 requires 3/4"

Solution: A 3/4 in. rigid metal conduit is required.

Example 7-1(b). Determine the size rigid metal conduit to enclose the copper conductors in Illustration 7-1(b).

Sizing RMC

Step 1: Finding sq. in. area
Table 5, Ch. 9
#14 THWN = .0097 sq. in. area
#12 THWN = .0133 sq. in. area
#10 THWN = .0211 sq. in. area

Step 2: Calculating sq. in. area
Table 5, Ch. 9
.0097 sq. in. x 12 = 0.1164 sq. in. area
.0133 sq. in. x 4 = 0.0532 sq. in. area
.0211 sq. in. x 4 = 0.0844 sq. in. area
Total = .254 sq. in. area

Step 3: Selecting size conduit
Table 4, Ch. 9
.254 sq. in. area requires 0.355 sq. in.

Solution: A 1 in. rigid metal conduit is required.

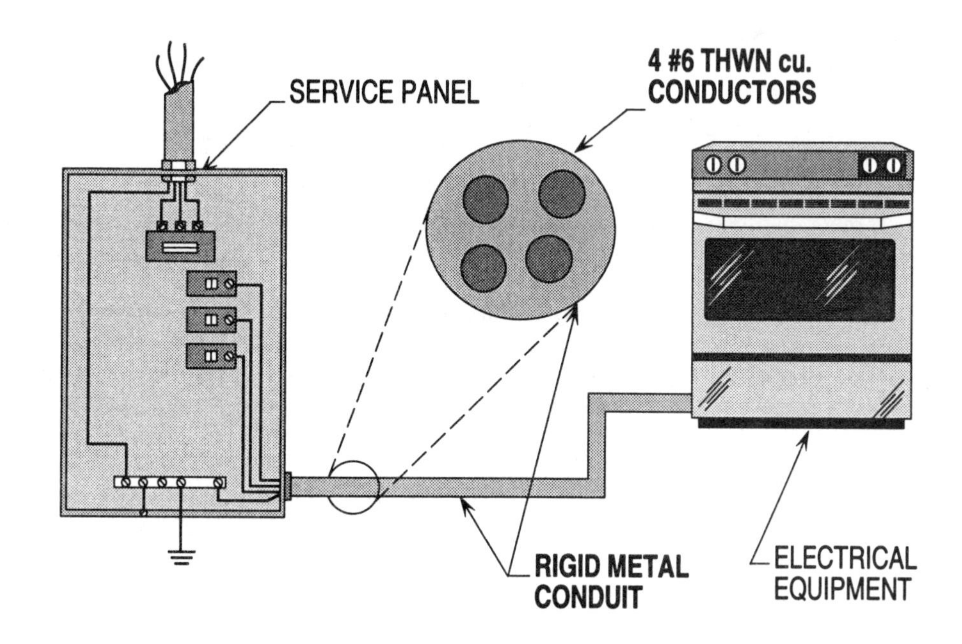

Ill. 7-1(a)

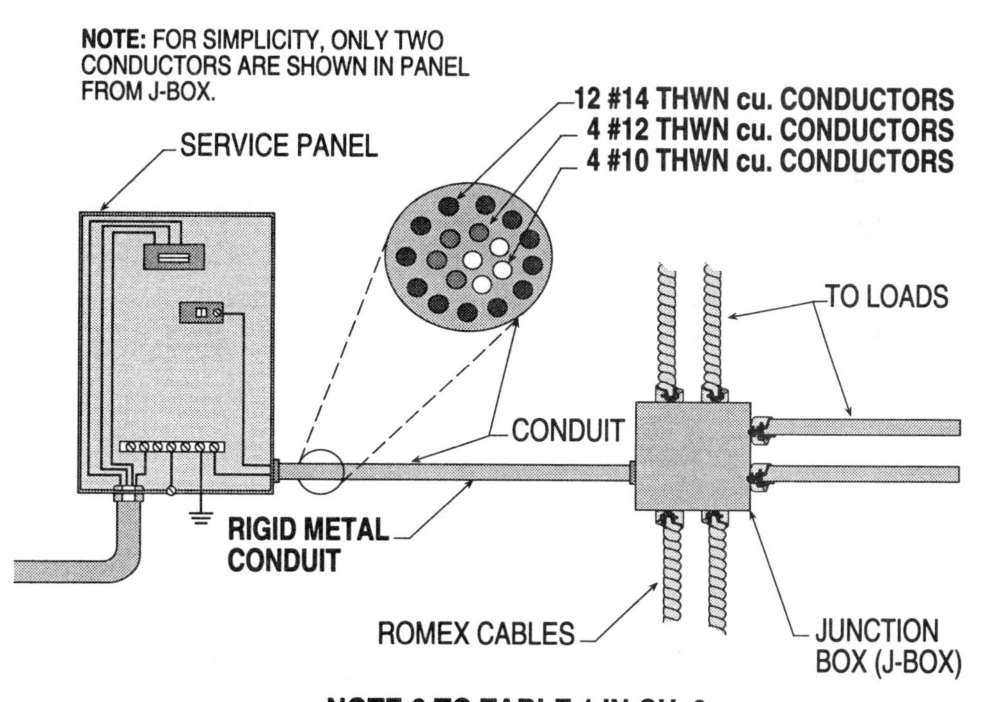

Ill. 7-1(b)

Sizing Nipples
Note 4 to Table 1 in Ch. 9

The difference in a nipple and a conduit is that a nipple is 24 in. or less in length. A conduit system is any run over 24 in. long in length.

A conduit system with more than two conductors is allowed 40 percent fill per Table 1 and 4 in Ch. 9, while a nipple is permitted to have 60 percent fill per Note 3 to Chapter 9.

Nipples are sized by finding the square inch area of each conductor and multiplying by the number. The total 100 percent fill in square inches in Table 4 in Chapter 9 is selected based upon the size conduit and this value is multiplied by 60 percent. If the total square inch area of all the conductors is less or equal to the total of the square inch area produced by applying the 60 percent factor, the size of the nipple is selected based upon that conduit used in the calculation.

Example 7-2. Determine the size EMT nipple required to enclose the copper conductors in Illustration 7-2.

Sizing nipple

Step 1: Finding sq. in. area
Table 5, Ch. 9
#12 THHN = .0133 sq. in. area
#14 THHN = .0097 sq. in. area
#10 THHN = .0211 sq. in. area
8 THHN = .0366 sq. in. area

Step 2: Calculating sq. in. area
Table 5, Ch. 9
.0133 sq. in. x 9 = .1197 sq. in.
.0097 sq. in. x 9 = .0873 sq. in.
.0211 sq. in. x 12 = .2532 sq. in.
.0366 sq. in. x 14 = .5124 sq. in.
Total = .9726 sq. in.

Step 3: Finding EMT at 100% total fill
Table 4, Ch. 9
1 1/2" EMT has a 100% total of 2.036 sq. in.

Step 4: Applying 60% fill
Note 4 to Table 1 in Ch. 9
sq. in. area x 60% = fill area
2.036 x 60% = 1.2216 sq. in.

Step 5: Selecting the nipple
Table 4, Ch. 9
1.2216 sq. in. is greater than .9726 sq. in.

Solution: A 1 1/2 in. EMT nipple is required.

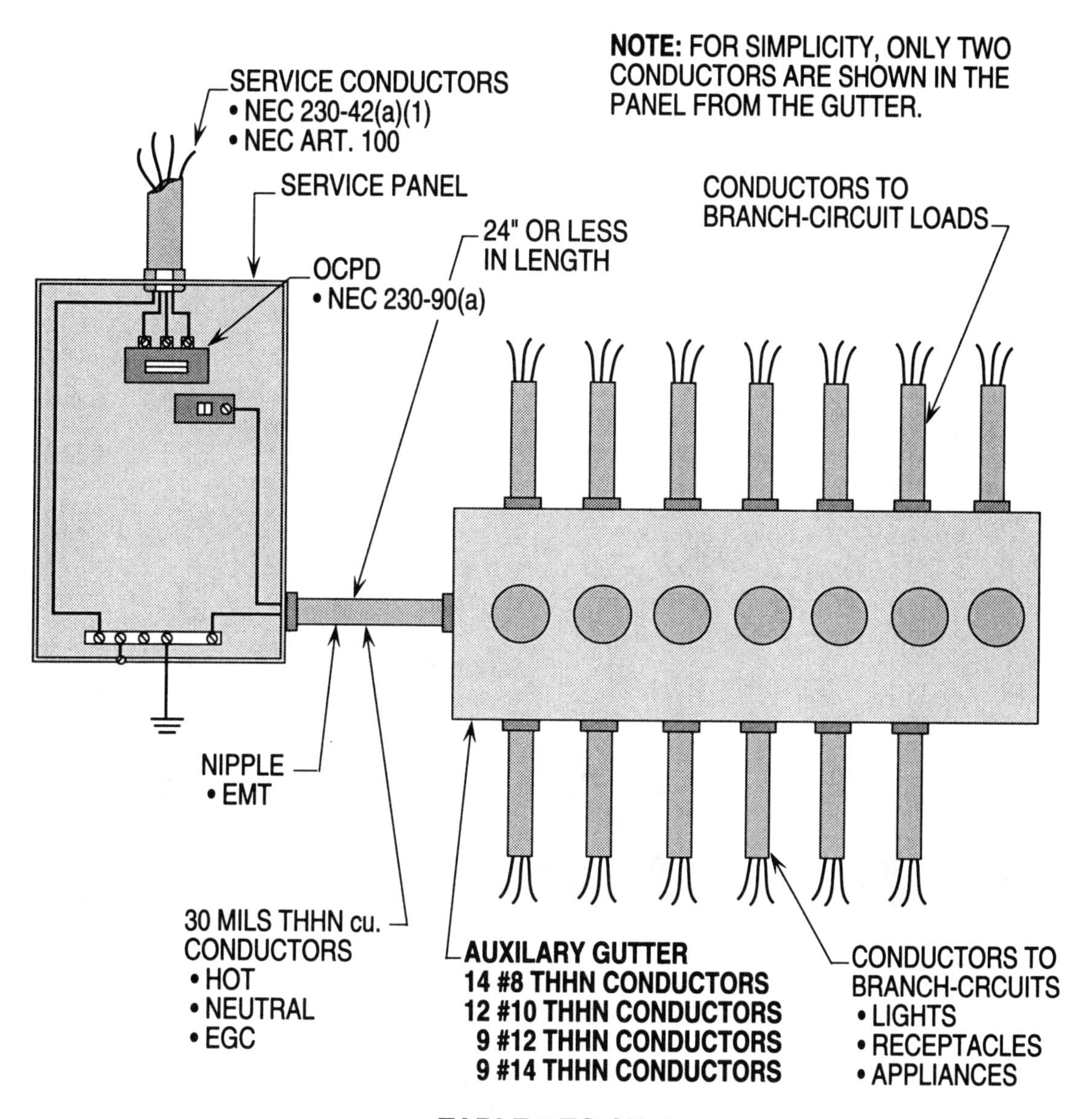

TABLE 5 TO CH. 9
TABLE 4 TO CH. 9
NOTE 4 TO TABLE 1 IN CH. 9
Ex. 3 TO TABLE 310-15(b)(2)(a)

Ill. 7-2

Sizing Raceways Using Bare Conductors Notes (3); (8) to Table 1 in Ch. 9

Notes (3) and (8) cover the dimensionms to be used based upon the conductors being insulated or uninsulated (bare) when they are pulled through raceway systems.

Example 7-3. Determine the square inch area for the insulated and noninsulated conductors in Illustration 7-3?

Insulated - Table 5	Bare - Table 8
#14 THWN = .0097 sq. in.	#14 Bare = .004 sq. in.
#12 THWN = .0133 sq. in.	#12 Bare = .006 sq. in.
#10 THWN = .0211 sq. in.	#10 Bare = .011 sq. in.

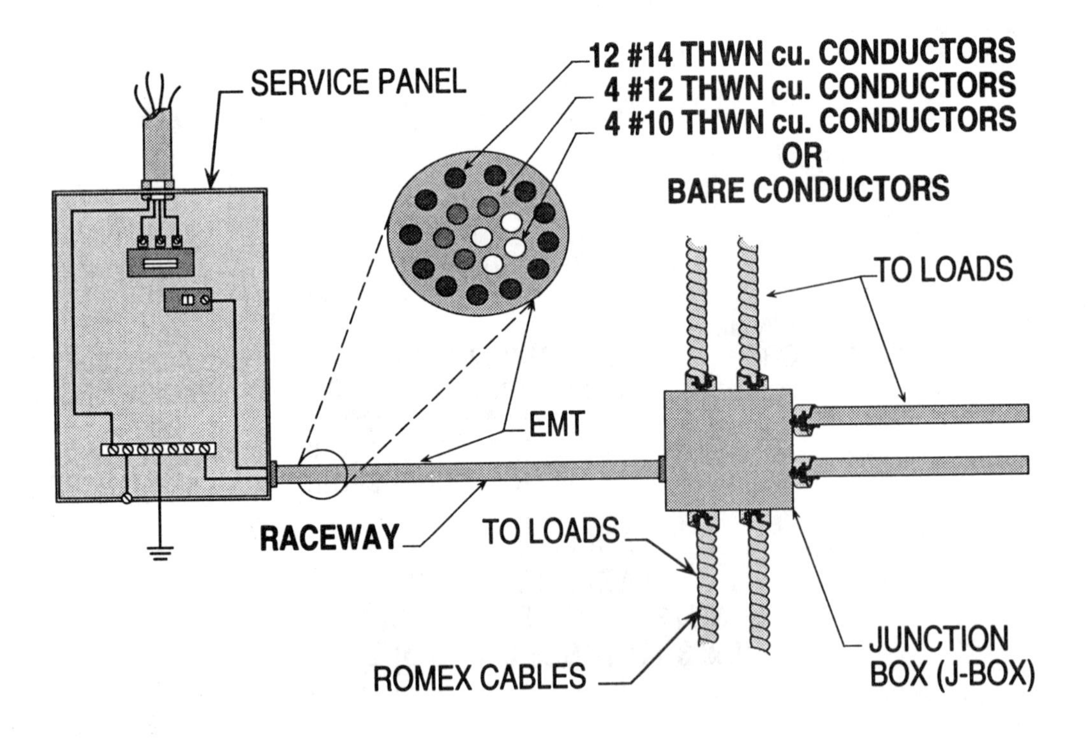

Ill. 7-3

Determining Number of Conductors in Raceway Note (7) to Table 1 in Ch. 9

Note (7) to Table 1 in Ch. 9 clarifies the rules for calculating the maximum number of conductors permitted in a conduit or tubing where they are all of the same size. Note that the next higher whole number can be used to determine the maximum number of conductors permitted when the calculation results in a decimal of 0.8 or larger.

Example 7-4. Can 63 No. 14 THWN cu. conductors be pulled through the 1 1/4 in. RMC in Illustration 7-4?

Step 1: Finding fill area
Table 5, Ch. 9
1 1/4" RMC = .610 sq. in.

Step 2: Finding sq. in. of #14 cu.
Table 5, Ch. 9
#14 THWN cu. = .0097 sq. in.

Step 3: Performing calculation

$$\frac{1\ 1/4" = .610 \text{ sq. in.}}{\#14 = .0097 \text{ sq. in.}} = 62.89$$

Solution: **Yes, 63 No. 14 cu. conductors can be pulled in 1 1/4 in. RMC.**

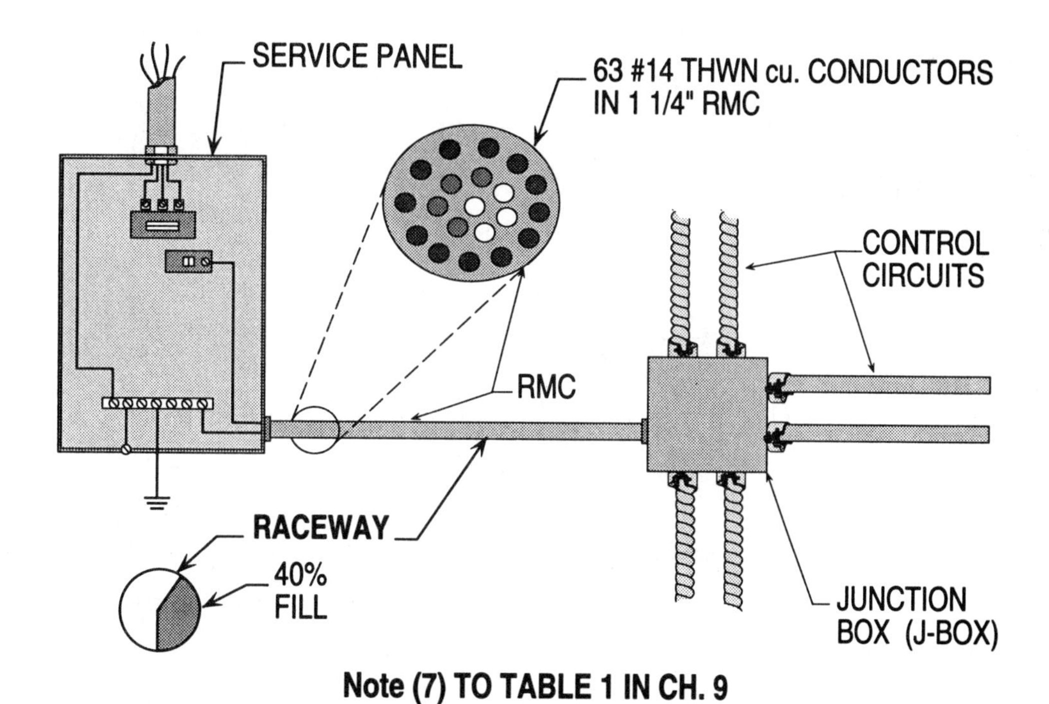

Note (7) TO TABLE 1 IN CH. 9

Ill. 7-4

Sizing a Sleeve Enclosing A Multiconductor Note (9) to Table 1 in Ch. 9

Note (9) to Table 1 in Ch. 9 clarifies the rules for calculating the percentage of conduit fill area for multiconductor cables with two or more conductors which are treated as a single conductor. For cables that have elliptical cross sections, the cross-sectional area calculation must be based on using the major diameter of the ellipse as a circle diameter.

Example 7-5. What size RMC is required to enclose the 1.625 in. multiconductor cables in Illustration 7-5?

Step 1: Calculate sq. in.
Note 9 to Table 1 in Ch. 9
Dia^2 x .7854
1.625 x 1.625 x .7854 = 2.074 sq. in.

Step 2: Selecting RMC
Table 4, Ch. 9
2.074 sq. in. = 3" RMC

Solution: **A 3 in. RMC is required.**

Note 1: The 40 percent fill column of Table 4 in Ch. 9 was used to ensure proper size raceway.

Note 2: When using the above formula, .7854 is the constant to be used everytime when finding the square mils.

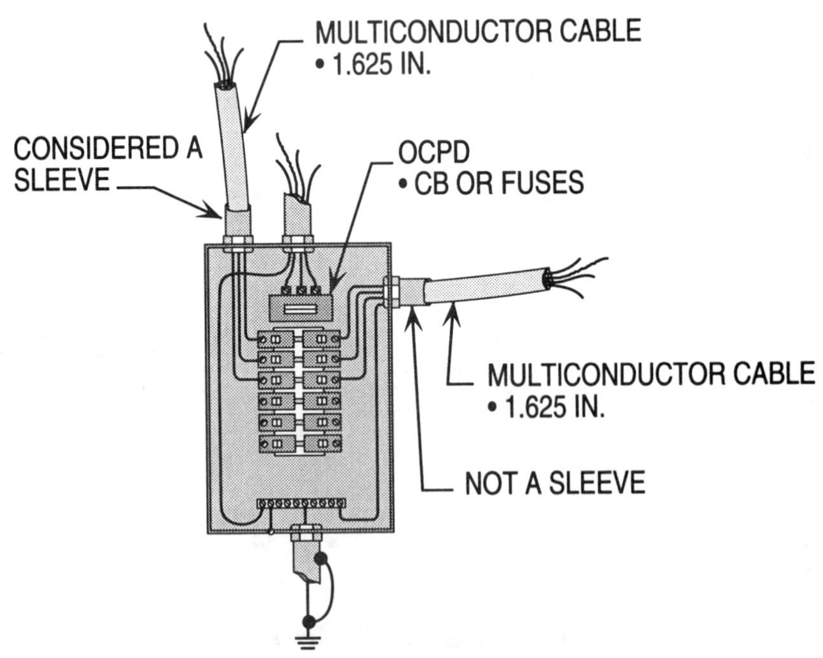

NOTE (9) TO TABLE 1 IN CH. 9

Ill. 7-5

Sizing Flexible Metal Conduit
Chapter 9, Table 1; 350-12

When sizing the number of conductors permitted for flexible metal conduit, the conduit shall not exceed the percentages of fill specified in Chapter 9, Table 1.

Example 7-6. Determine the size flexible metal conduit to enclose the copper conductors in Illustration 7-6?

Sizing flexible metal conduit

Step 1: Finding size conduit
Table C3, Appendix C
4 #6 requires 3/4"

Solution: **A 3/4 in. flexible metal conduit is required.**

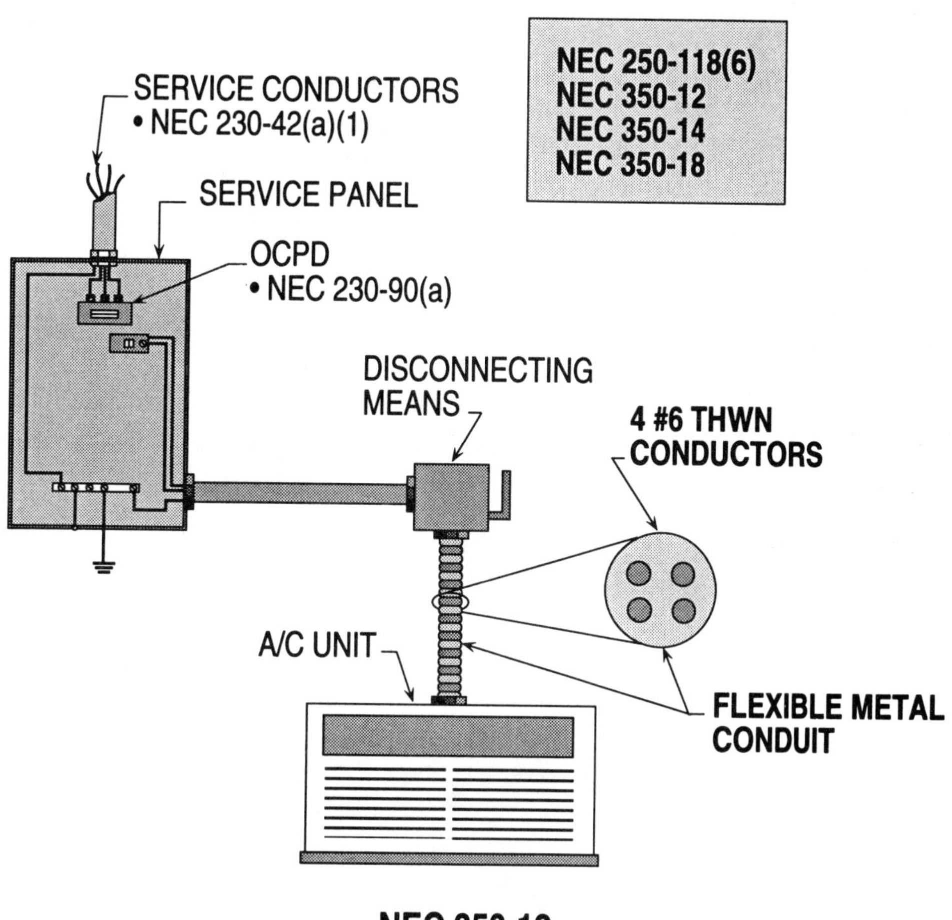

Ill. 7-6

Sizing Liquidtight Flexible Metal Conduit
Chapter 9, Table 1; 351-6

When sizing the number of conductors permitted for liquidtight flexible metal conduit, the conduit shall not exceed the percentages of fill specified in Chapter 9, Table 1.

Example 7-7. Determine the size liquidtight flexible metal conduit to enclose the copper conductors in Illustration 7-7?

Sizing liquidtight flexible metal conduit

Step 1: Finding size conduit
Table C7, Appendix C
4 #8 requires 3/4"

Solution: **A 3/4 in. liquidtight flexible metal conduit is required.**

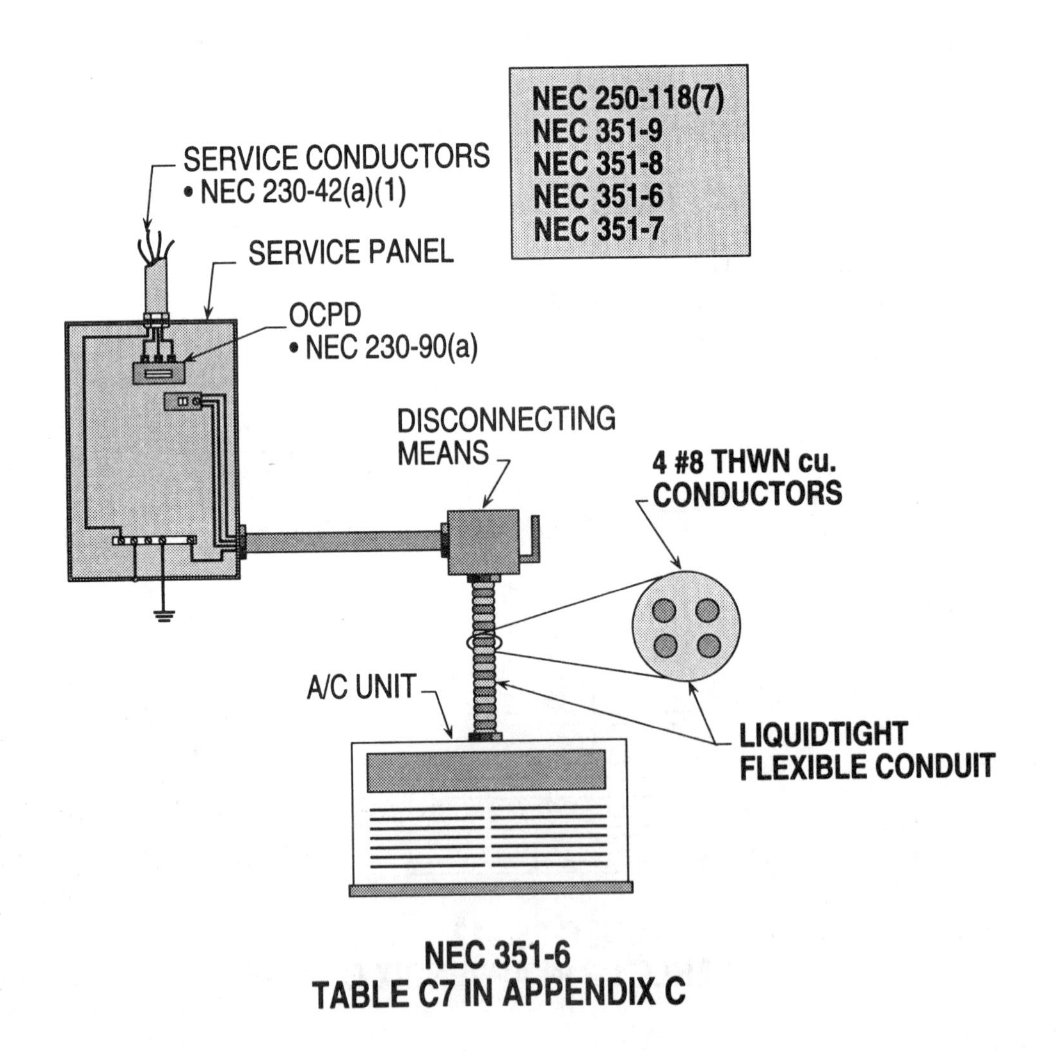

Ill. 7-7

Sizing Ventilated Cable Tray - Single Conductor 318-10; Table 318-10

When sizing single conductor cables for cable trays, cables smaller than 1,000 KCMIL are based on Table 318-10, Column 1. A combination of cables rated, 1,000 KCMIL and larger are based on Table 318-10, Column 2. The total diameter of all cables 1,000 KCMIL and larger must not exceed the width for ladder or ventilated-trough trays. The total sum of all cables must not exceed the inside width for ventilated channel-type trays.

Example 7-8. Determine the size cable tray required for the copper conductors in Illustration 7-8?

Sizing ventilated cable tray

Step 1: Finding sq. in. area
Table 5, Ch. 9
#300 KCMIL THHN = .4608 sq. in.
#1,000 KCMIL THHN = 1.3478 sq. in.

Step 2: Calculating sq. in. area
318-10(a)(2); Table 318-10, Col. 1
.4608 sq. in. x 14 = 6.4512 sq. in.
1.3478 sq. in. x 10 = 13.478 sq. in.
Total area = 19.9292 sq. in.

Step 3: Selecting size cable tray
Table 318-10, Col. 1
19.9292 sq. in. requires 24 in width.
24 in. wide tray requires 26 sq. in.

Solution: A 24 in. wide cable tray has 26 sq. inches per Table 318-10 is required.

SINGLE CONDUCTOR CABLES

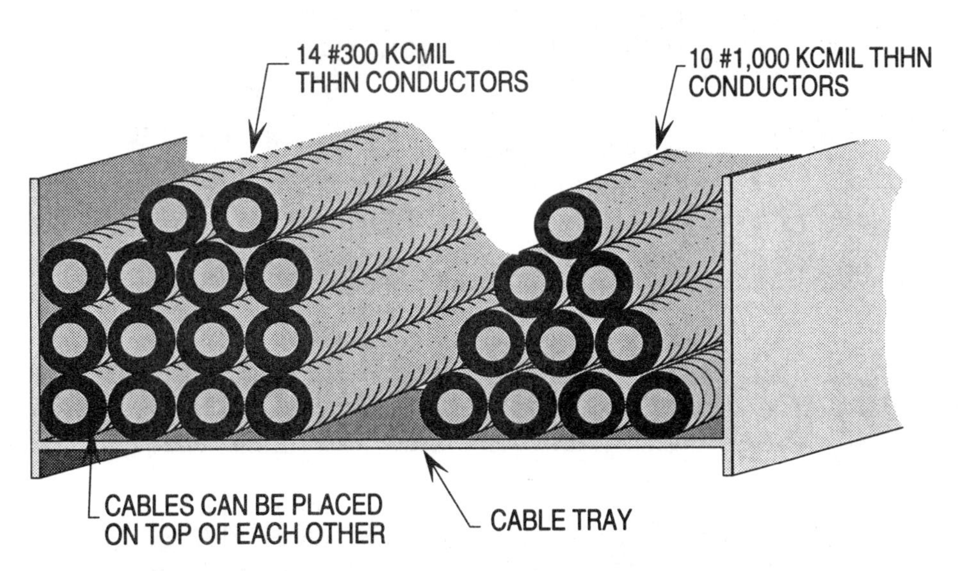

TABLE 5, CH. 9
NEC 318-10(a)(2)
TABLE 318-10, COL.'s 1; 2

Ill. 7-8

Sizing Ventilated Cable Tray - Multiconductor 318-9; Table 318-9

When sizing multiconductor cables for cable trays, cables smaller than No. 4/0 are based on Table 318-9, Column 1. A combination of cables larger than No. 4/0 are based on Table 318-9, Column 2. The diameter of all cables No. 4/0 and larger must be added and the total must not exceed the width of ladder or ventilated-trough trays.

Example 7-9. Determine the size cable tray required for the copper conductors in Illustration 7-6?

Sizing ventilated cable tray

Step 1: Finding sq. in. area
318-9(a)(3)
#4/0 and smaller = 22 sq. in.
#4/0 and larger = 12.6 in. diameter

Note: The sq. in. area of No. 4/0 and smaller has already been calculated, including the diameter of No. 4/0 and larger.

Step 2: Calculating sq. in. area
318-9(a)(3)
#4/0 and smaller = 22.00 sq. in.
#4/0 and larger (12.6" x 1.2) = 15.12 sq. in.
Total = 37.12 sq. in.

Step 3: Selecting size cable tray
Table 318-9, Col. 2
37.12 sq. in. requires 36 in.
36 in. wide tray = 42 in.

Solution: A 36 in. wide cable tray is required.

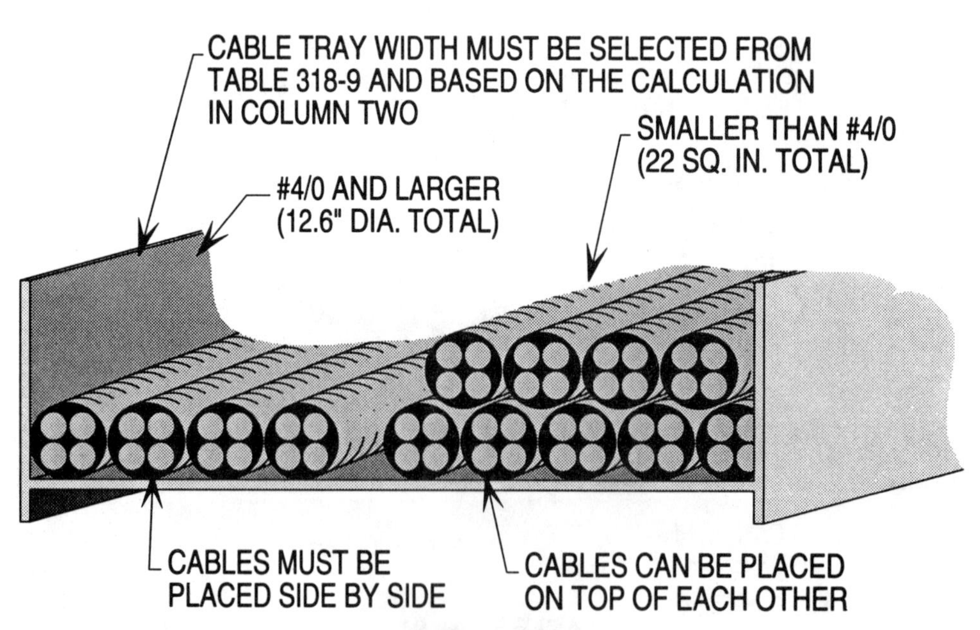

Ill. 7-9

Sizing Cable Tray As EGC
318-7; Table 318-7(b)(2)

Cable trays can be used as equipment grounding conductors based upon their size per Table 318-7(b)(2). Table 250-122 is used to select the size overcurrent protection device (OCPD) based upon the thickness of the metal of the cable tray used.

Example 7-10. What size bonding jumper (BJ) and OCPD will the cable tray in Illustration 7-10 clear safely?

Sizing EG of Alu. cable tray

Step 1: Calculating OCPD for alu. tray
Table 318-7(b)(2)
1.50 sq. in. tray clears 1,600 A

Solution: A 1.50 sq. in. alu. cable tray is capable of clearing a 1,600 amp OCPD.

Sizing EG of steel cable tray

Step 1: Calculating OCPD for steel tray
Table 318-7(b)(2)
1.50 sq. in. tray clears 600 A

Solution: A 1.50 sq. in. steel cable tray is capable of clearing a 600 amp OCPD.

Sizing bonding jumper for Alu. and metal cable tray

Step 1: Calculating OCPD
Table 318-7(b)(2)
1.50 sq. in. alu. tray clears 1,600 A
1.50 sq. in. metal tray clears 600 A

Step 2: Calculating BJ
250-102(d); Table 250-122
1,600 A requires #350 KCMIL alu.
600 A requires #1 cu.

Solution: The size bonding jumper required is No. 350 KCMIL alu. for the alu. tray and No. 1 cu. for the metal tray.

NEC 318-5(e); NEC 318-6(a); NEC 318-6(j); NEC 318-7; Table 318-7(b)(2); NEC 250-102(d); Table 250-122

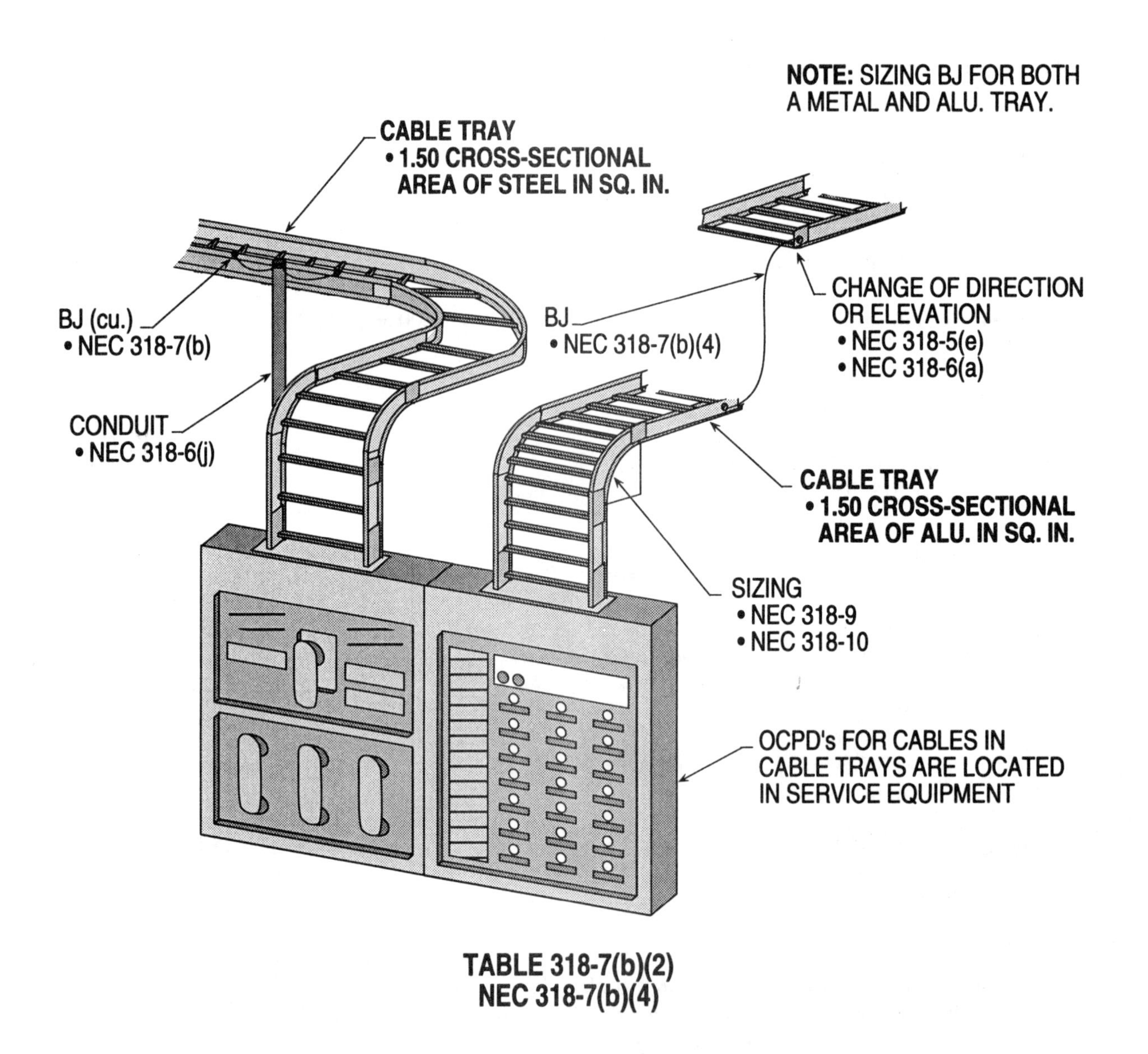
NOTE: SIZING BJ FOR BOTH A METAL AND ALU. TRAY.
CABLE TRAY
• 1.50 CROSS-SECTIONAL AREA OF STEEL IN SQ. IN.
BJ (cu.)
• NEC 318-7(b)
BJ
• NEC 318-7(b)(4)
CHANGE OF DIRECTION OR ELEVATION
• NEC 318-5(e)
• NEC 318-6(a)
CONDUIT
• NEC 318-6(j)
CABLE TRAY
• 1.50 CROSS-SECTIONAL AREA OF ALU. IN SQ. IN.
SIZING
• NEC 318-9
• NEC 318-10
OCPD's FOR CABLES IN CABLE TRAYS ARE LOCATED IN SERVICE EQUIPMENT
TABLE 318-7(b)(2)
NEC 318-7(b)(4)

Ill. 7-10

Sizing Ampacity Of Control And Signaling Cables In Cable Trays 318-9(b); (d)

Ladder or ventilated trough trays may be filled with control or signaling cables or both up to 50 percent of the cross-sectional area of their trays up to 6 in. deep. Solid bottom trays are computed using the same procedure except 40 percent is utilized as the allowable multiplier.

Example 7-11. What size cable tray is required for each cable tray in Illustration 7-11?

Sizing case 1

Step 1: Calculating cable tray for case 1
318-9(b)
18 x .35 sq. in. = 6.3 sq. in.
35 x .15 sq. in. = 5.25 sq. in.
Total = 11.55 sq. in.

Step 2: Applying multiplier
318-9(b)
11.55 sq. in. / 50% = 23.1 sq. in.

Step 3: Applying depth of tray
318-9(b); Table 318-9
23.1 sq. in. / 4" = 5.775 in.

Solution: A cable tray with a 6 in. width is required.

Sizing case 2

Note: In the case of a solid bottom tray the multiplier is 40 percent and the results would be 7.2 in. (11.55 ÷ 40% = 28.88 ÷ 4 = 7.2") which a 12 in. cable tray is required per Table 318-9.

Step 1: Calculating solid bottom tray
318-9(d)
16 x .170 = 2.72 sq. in.

Step 2: Selecting cable tray
Table 318-9
2.72 sq. in. requires 6"

Solution: The size cable tray required is 6 in. per the NEC.

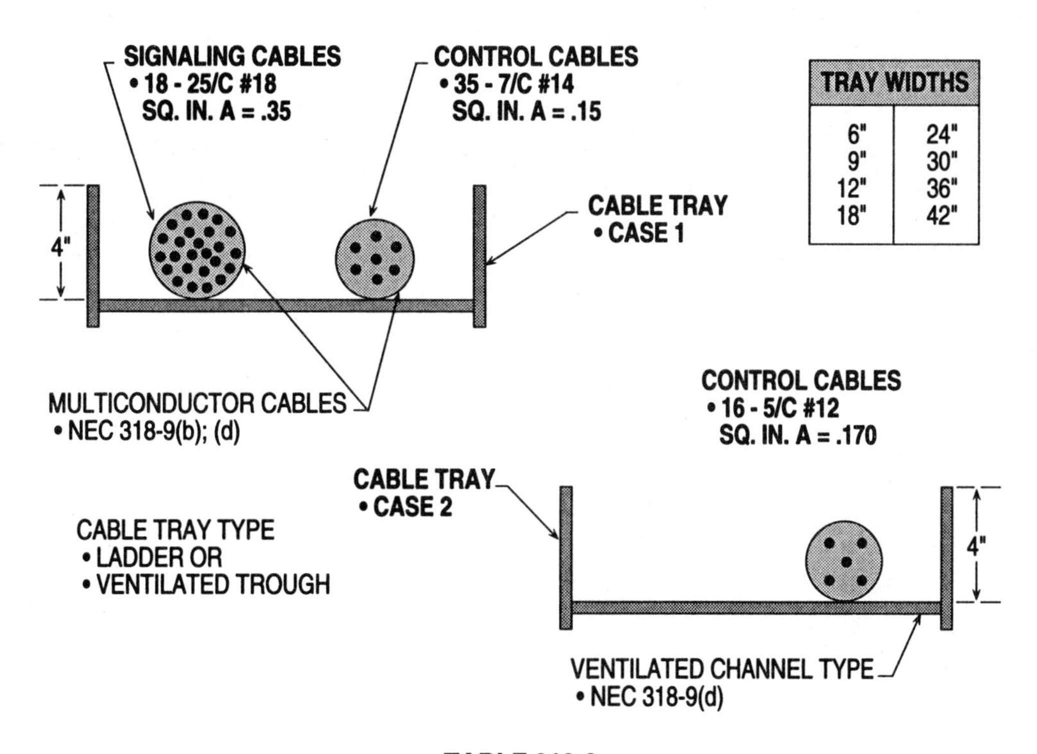

TRAY WIDTHS	
6"	24"
9"	30"
12"	36"
18"	42"

TABLE 318-9
NEC 318-9(b); (d)

Ill. 7-11

Sizing Underfloor Raceways
354-5

When sizing the number of conductors permitted in underfloor raceways, the combined cross-sectional area of all conductors or cables shall not exceed 40 percent of the interior cross-sectional area of the raceway.

Example 7-12. Determine the size underfloor raceway to enclose the copper conductors in Illustration 7-12?

Sizing underfloor raceway

Step 1: Finding sq. in. area
Table 5, Ch. 9
#2/0 THWN = .2223 sq. in.
#4/0 THWN = .3237 sq. in.
#300 KCMIL = .4608 sq. in.

Step 2: Calculating sq. in. area
Table 5, Ch. 9
.2223 sq. in. x 3 = .6669 sq. in.
.3237 sq. in. x 3 = .9711 sq. in.
.4608 sq. in. x 3 = 1.3824 sq. in.
Total = 3.0204 sq. in.

Step 3: Finding raceway
354-5
sq. in. area ÷ 40% = total fill
3.0204 ÷ 40% = 7.551 sq. in.

Step 4: Selecting raceway
Chart
7.551 sq. in. requires 16 sq. in.
4" x 4" = 16"

Solution: **A 4 in. x 4 in. underfloor raceway is required.**

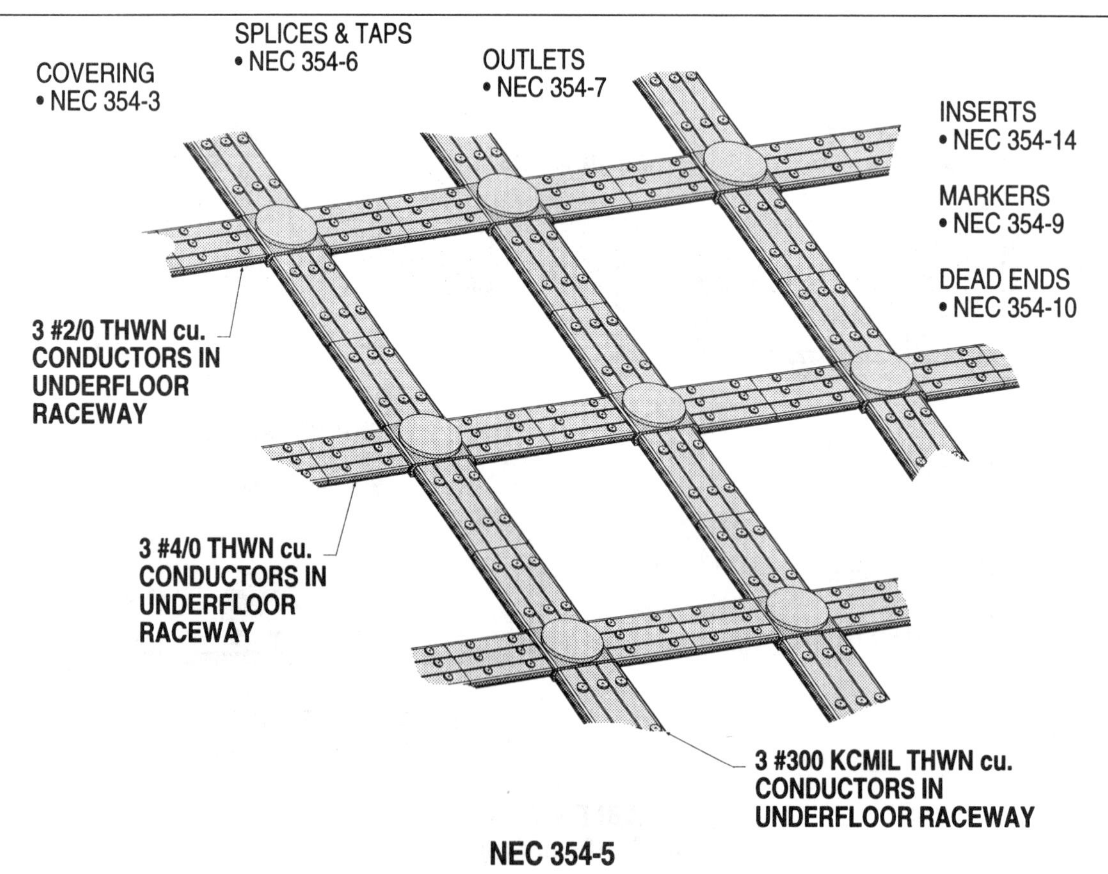

Ill. 7-12

Sizing Wireways
362-5

When sizing the number of conductors permitted in wireways, the sum of cross-sectional areas of all contained conductors at any cross section of the wireway shall not exceed 20 percent of the interior cross-sectional area of the wireway.

Example 7-13. Determine the size wireway to enclose the copper conductors in Illustration 7-13?

Sizing wireway

Step 1: Finding sq. in. area
Table 5, Ch. 9

#12 THHN	=	.0133 sq. in.
#10 THHN	=	.0211 sq. in.
# 8 THWN	=	.0366 sq. in.
#1/0 THW	=	.2223 sq. in.

Step 2: Calculating sq. in. area
Table 5, Ch. 9

.0133 sq. in. x 20	=	.266 sq. in.
.0211 sq. in. x 4	=	.0844 sq. in.
.0366 sq. in. x 4	=	.1464 sq. in.
.2223 sq. in. x 6	=	1.3338 sq. in.
Total	=	1.8306 sq. in.

Step 3: Finding wireway
362-5
sq. in. area ÷ 20% = total fill
1.8306 ÷ 20% = 9.153 sq. in.

Step 4: Selecting wireway
Chart
9.153 sq. in. requires 16 sq. in.
4" x 4" = 16"

Solution: A 4 in. x 4 in. wireway is required.

Note: For splices and taps, only 12 sq. in. area is allowed. (4" x 4" x 75% = 12 sq. in.)

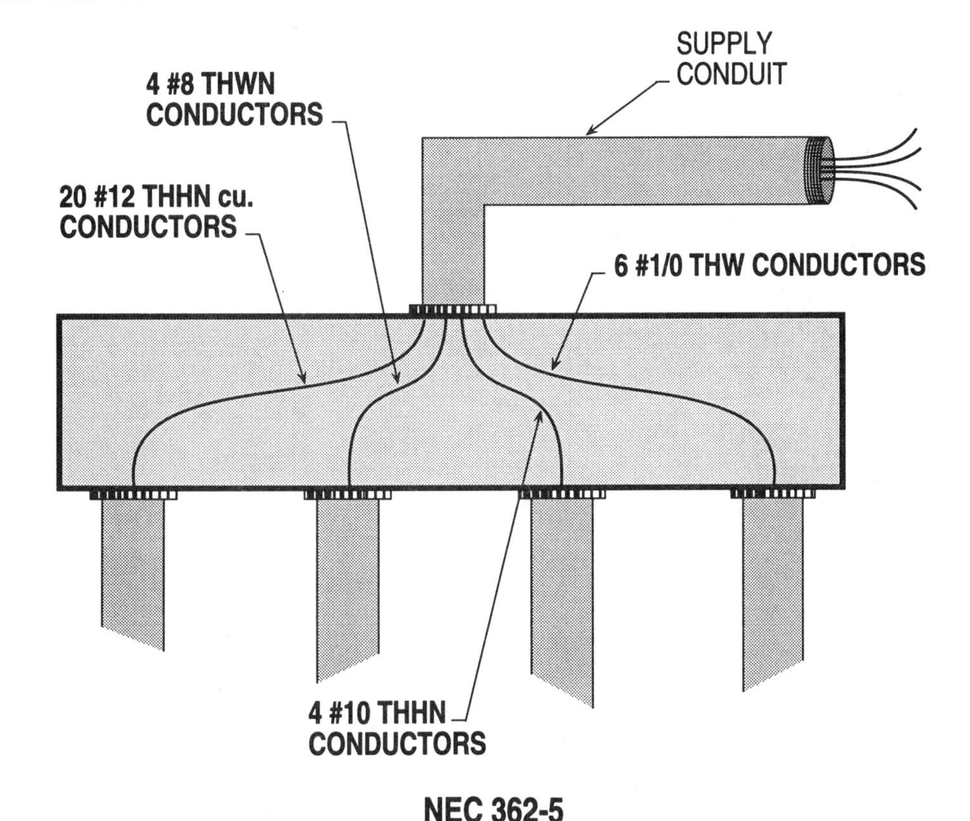

NEC 362-5

Ill. 7-13

Sizing Gutters 374-5

The number of current-carrying conductors permitted in an auxiliary gutter without derating per Table 310-15(b)(2)(a) to Tables 310-16 through 310-19 is 30 or less per NEC 374-6.

The size of an auxiliary gutter is determined by dividing the total square inch area of the conductors by 20 percent fill area for installing conductors. The total square inch area of the conductors is found by multiplying the square inch area per Table 5, Ch. 9 of each conductor which is based on the size and insulation of each conductor placed in the auxiliary gutter.

Example 7-14. Determine the size auxiliary gutter required to house the copper conductors in Illustration 7-14?

Sizing auxiliary gutter

Step 1: Finding sq. in. area
Table 5, Ch. 9
#250 KCMIL = .397
#4/0 THWN = .3237
#1/0 THWN = .1855

Step 2: Calculating sq. in. area
Table 5, Ch. 9
.397 sq. in. x 3 = 1.191 sq. in.
.3237 sq. in. x 3 = .9711 sq. in.
.1855 sq. in. x 3 = .5565 sq. in.
Total = 2.7186 sq. in.

Step 3: Finding gutter
374-5
sq. in. area ÷ 20% = total fill
2.7186 sq. in. ÷ 20% = 13.593 sq. in.

Step 4: Selecting gutter
Chart
13.593 sq. in. requires 16 sq. in.
4" x 4" = 16"

Solution: A 4 in. x 4 in. auxiliary gutter is required.

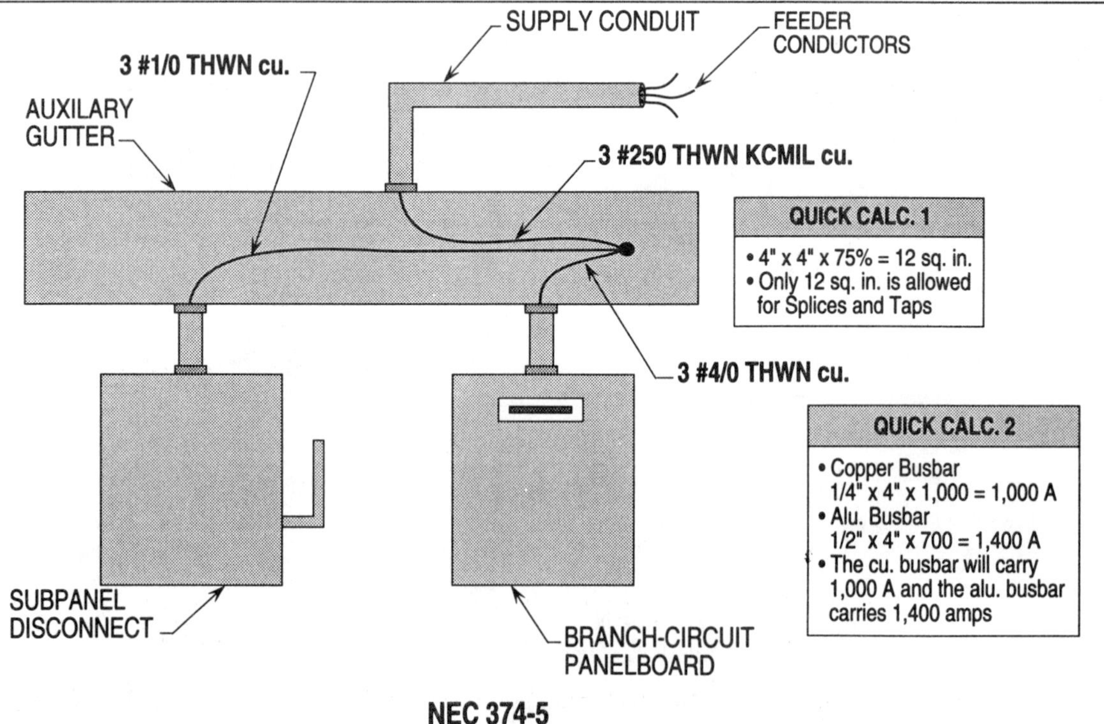

Ill. 7-14

When Conductors Can Jam In A Conduit Pull
FPN 2 to Table 1 in Ch. 9

When an electrician is pulling three conductors or cables through a raceway, if the ratio of such raceway (inside diameter) to the conductor or cable (outside diameter) falls between 2.8 and 3.2, jamming can occur and the next larger size raceway is recommended to be used.

Example 7-15. Determine if jamming of conductor could occur in the EMT in Illustration 7-15?

Step 1: Finding OD (outside diameter) in.
FPN 2 to Table 1 in Ch. 9
#250 KCMIL = .711 OD in.

Step 2: Finding ID (inside diameter) in.
Table 4; Ch 9
2" EMT = 2.067 ID in.

Step 3: Finding if jamming can occur.
FPN 2 to Table 1 in Ch. 9
2.067 ID ÷ .71100 = 2.91

Step 4: Verification of jamming
FPN 2 to Table 1 in Ch. 9
Jamming 2.8 thru 3.2 (2.91)

Solution: **Jamming could occur so a 2 1/2 in. EMT should be used.**

Note that 2.91 falls between 2.8 thru 3.2, therefore, jamming could occur.

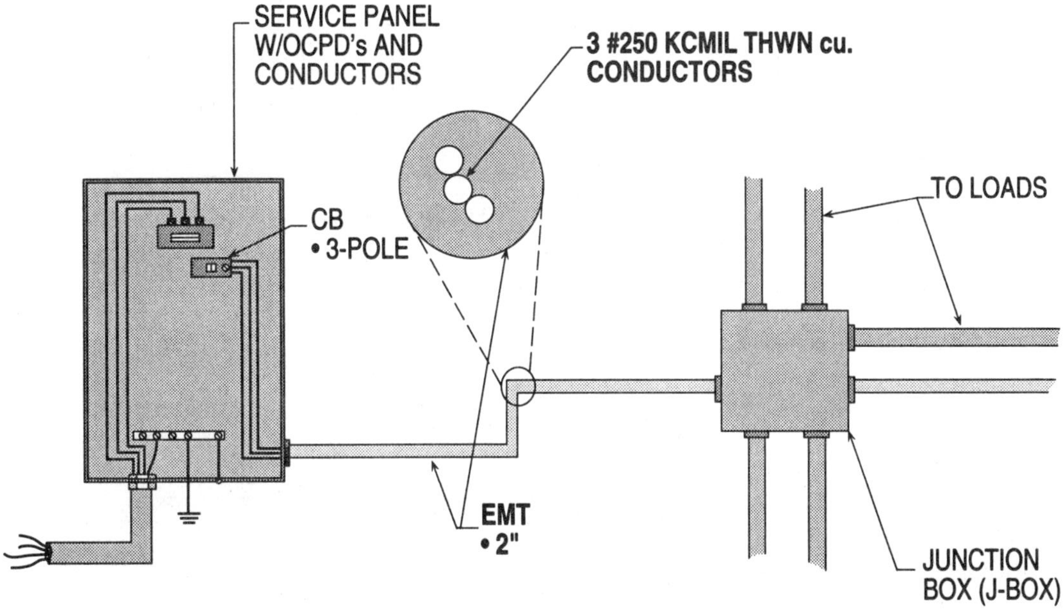

FPN 2 TO TABLE 1 IN CH. 9

Ill. 7-15

Determining If Expansion Joints Are Needed In A PVC Run Exposed To High Temperatures
NEC 347-9; Table 347-9(A)

Expansion fittings for rigid nonmetallic conduit must be provided to compensate for thermal expansion and contraction where the length change in accordance with NEC 347-9 and Table 347-9(A), is expected to be 0.25 in. or greater in a straight run between securely mounted items such as boxes, cabinets, elbows, or other conduit terminations.

Example 7-16. How many expansion joints are needed for a 200 ft. run of RNMC between two J-boxes if each has an expansion capacity of 6 in.? The average temperature for the year is 28°F to 100°F.

Step 1: Finding inch change
347-9; Table 347-9(A)
100°F has a 4.1 in. change per 100 ft.

Step 2: Finding total inch for run
Table 347-9(A)
200 ft. run = 4.1 x 2 = 8.2

Step 3: Finding number needed
Table 347-9(A)
= 8.2 / 6 in. expansion capacity
= 1.36

Step 4: Verifying number
Table 347-9(A)
8.2" is greater than 6"
1.36 rounds up to 2

Step 5: Locating expansions joints
200' run / 2 = 100'

Solution: Two expansion joints are needed at approximately every 100 ft. intervals

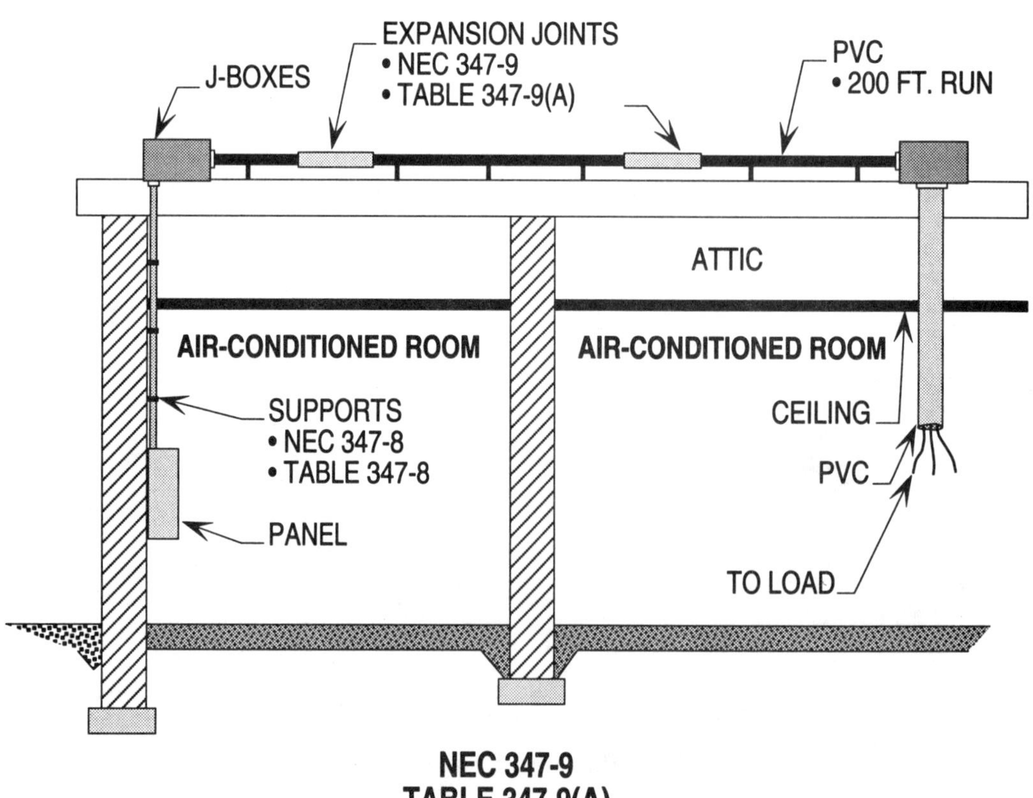

Ill. 7-16

8

Boxes and Fittings

Article 370 of the National Electrical Code® (NEC®) regulates the use of all types of boxes and fittings. When they are utilized for splicing, tapping, or pulling conductors.

Table 370-16(a) is used to select boxes when all conductors are the same size. The total counted conductors based upon the items taking up space in the box which are listed in NEC 370-16(b)(1) through (b)(5) are used to size the box per Table 370-16(a). Boxes housing conductors of different sizes are selected from Table 370-16(a) based upon the total cubic-inch rating of the conductors.

The cubic-inch rating of conductors are found in Table 370-16(b) which are determined by the size of the conductors. Tables 370-16(a) and (b) are used where the conductors are No. 18 through No. 6 respectively. NEC 370-28(a)(1) and (a)(2) are utilized to size junction boxes for conductors that are No. 4 and larger. Multipliers are used based upon straight, angle, or U-pulls to compute the values to size the junction box.

Quick Reference

Number Of Conductors In Octagonal Box
Table 370-16(a)

Boxes containing conductors that are the same size shall have their total fill space determined by adding the number of conductors plus additional conductors for each fitting or device. The cubic inch value listed in Table 370-16(b) for the largest conductor in the box is used to determine the fill capacity for the fittings. For devices, use the size conductor connecting to its terminals.

Boxes containing conductors that are not the same size are computed on the cubic inch rating of each size conductor listed in Table 370-16(b). The cubic inch rating of each conductor is added together plus the number of fittings or devices. The cubic inch rating used for each fitting is based on the largest conductor entering the box. The cubic inch rating for yokes or straps are based on the size conductor that is connected to the switch or receptacle.

Example 8-1(a). Determine the size octagon box required to support the lighting fixture in Illustration 8-1(a)?

Sizing octagon box

Step 1: Counting conductors
370-16(b)(1) thru (b)(5)

2 #12 hots	= 2
2 #12 neutrals	= 2
2 #12 EGC's	= 1
1 fixture stud	= 1
1 hickey	= 1
3 #18 fixture wires	= 0
plus EGC	= 0
Total	= 7

Step 2: Selecting box
Table 370-16(a)
7 #12 conductors requires a 4" x 2 1/8" box

Solution: A 4 in. x 2 1/8 in. octagon box is required.

Example 8-1(b). Determine the size octagon box required to support the lighting fixture in Illustration 8-1(b)?

Sizing octagon box

Step 1: Counting conductors
370-16(b)(1) thru (b)(5); Table 370-16(b)

2 #12 hot	= 2.25 cu. in. x 2	= 4.5	cu. in.
2 #12 neutral	= 2.25 cu. in. x 2	= 4.5	cu. in.
2 #12 EGC	= 2.25 cu. in. x 1	= 2.25	cu. in.
1 #16 EGC		= 0	cu. in.
2 #16 fix. wires (1.75 cu. in. x 2)		= 3.5	cu. in.
Total		= 14.75	cu. in.

Step 2: Selecting box
Table 370-16(a)
14.75 cu. in. requires
4" x 1 1/2" box

Solution: A 4 in. x 1 1/2 in. octagon box is required.

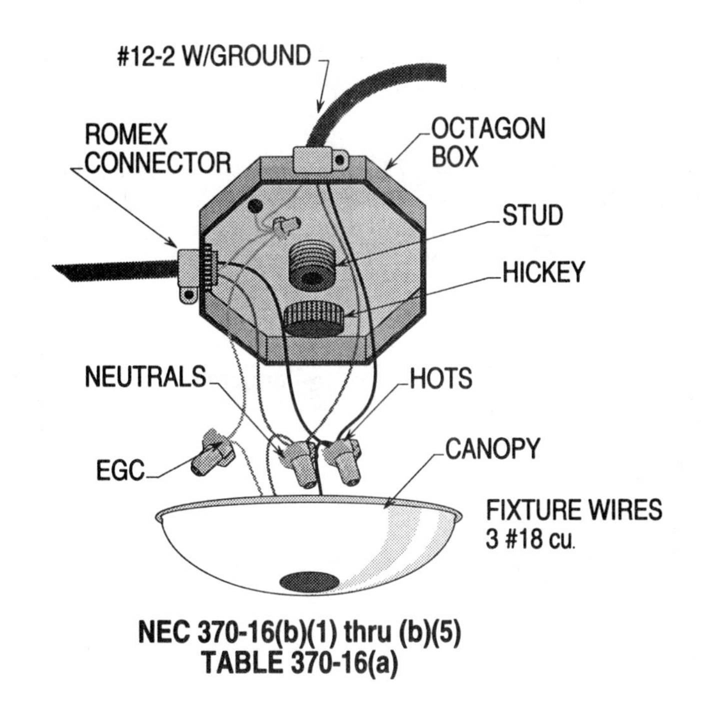

NEC 370-16(b)(1) thru (b)(5)
TABLE 370-16(a)

Ill. 8-1(a)

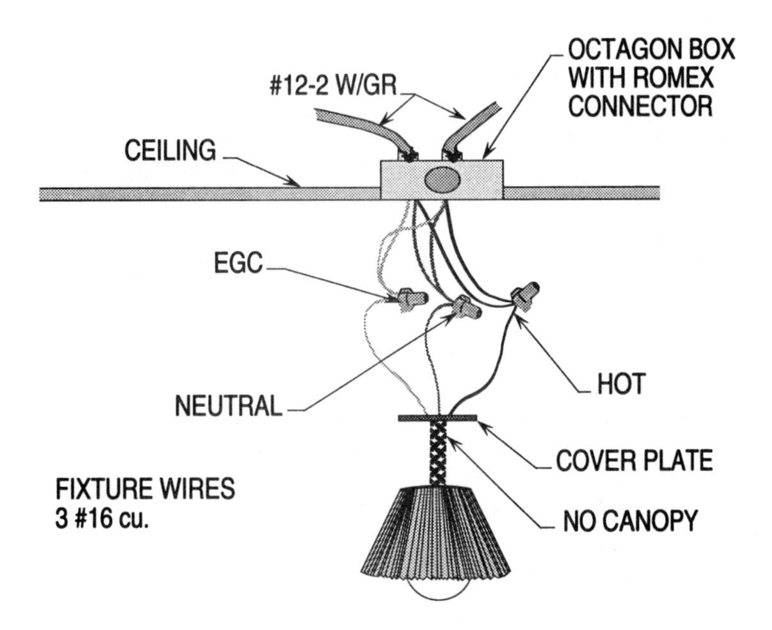

NEC 370-16(b)(1) thru (b)(5)
TABLE 370-16(b)
TABLE 370-16(a)

Ill. 8-1(b)

Number Of Conductors In Square Box 370-16(b)

Section 370-16(a) lists the items in or out of the box and the number of conductors that shall be added for each item based on the condition of use. Where all conductors are the same size, each item in or out of the box are counted based on these same size conductors.

Sections 370-16(b)(1) through (b)(5) refers to Table 370-16(b) which requires the cubic inch rating of each conductor in the box to be used for determining the fill space in cubic inches for selecting the box size. NEC 370-16(a) refers to Table 370-16(a) for selecting the cubic inch rating of each fitting, cable clamps, and etc. in the box based on the larger conductor entering the box. The cubic inch rating for each yoke is based on the size conductor connecting to the device terminal.

Example 8-2(a). Determine the size square box required in Illustration 8-2(a)?

Sizing square box

Step 1: Counting conductors
370-16(b)(1) thru (b)(5)

4 #12 hot	=	4
4 #12 neutrals	=	4
4 #12 EGC's	=	1
2 cable clamps	=	1
Total	=	10

Step 2: Selecting box
Table 370-16(a)
10 #12 conductors requires
4 11/16" x 1 1/4"

Solution: A 4 11/16 in. x 1 1/4 in. square box is required.

Example 8-2(b). Determine the size square box required in Illustration 8-2(b)?

Sizing square box

Step 1: Counting conductors
370-16(b)(1) thru (b)(5); Table 370-16(b)

2 #12 hots	2.25 cu. in. x 2	=	4.5	cu. in.
2 #12 neutrals	2.25 cu. in. x 2	=	4.5	cu. in.
2 #14 hots	2 cu. in. x 2	=	4	cu. in.
2 #14 neutrals	2 cu. in. x 2	=	4	cu. in.
2 #12 EGC's	2.25 cu. in. x 1	=	2.25	cu. in.
2 #14 EGC's	2 cu. in. x 0	=	0	cu. in.
2 cable clamps	2.25 cu. in. x 1	=	2.25	cu. in.
Total		=	21.5	cu. in.

Step 2: Selecting box
Table 370-16(a)
21.5 cu. in. requires
4 11/16" x 1 1/4"

Solution: A 4 11/16 in. x 1 1/4 in. square box is required.

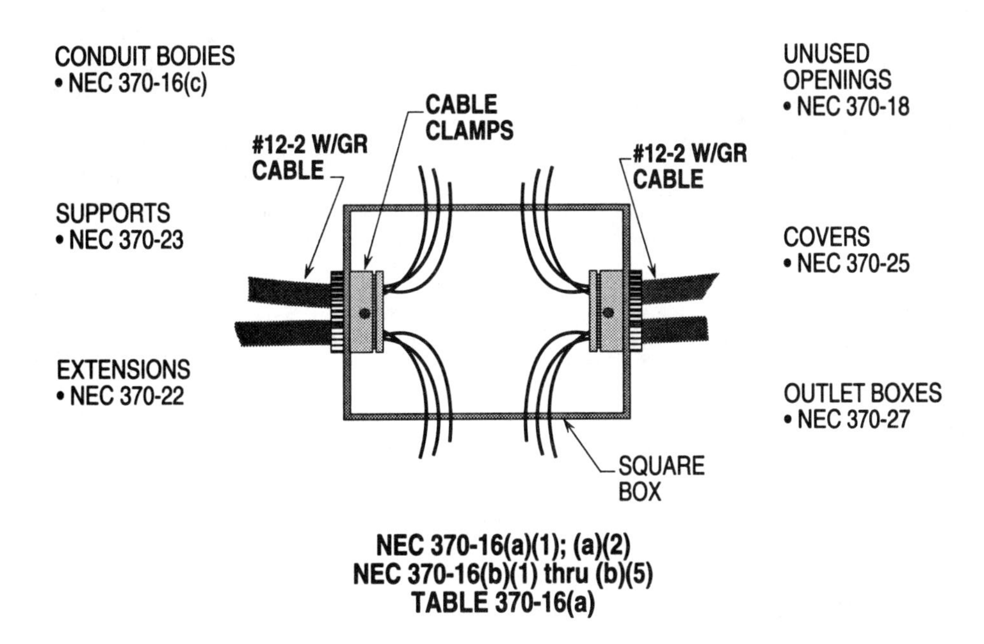

Ill. 8-2(a)

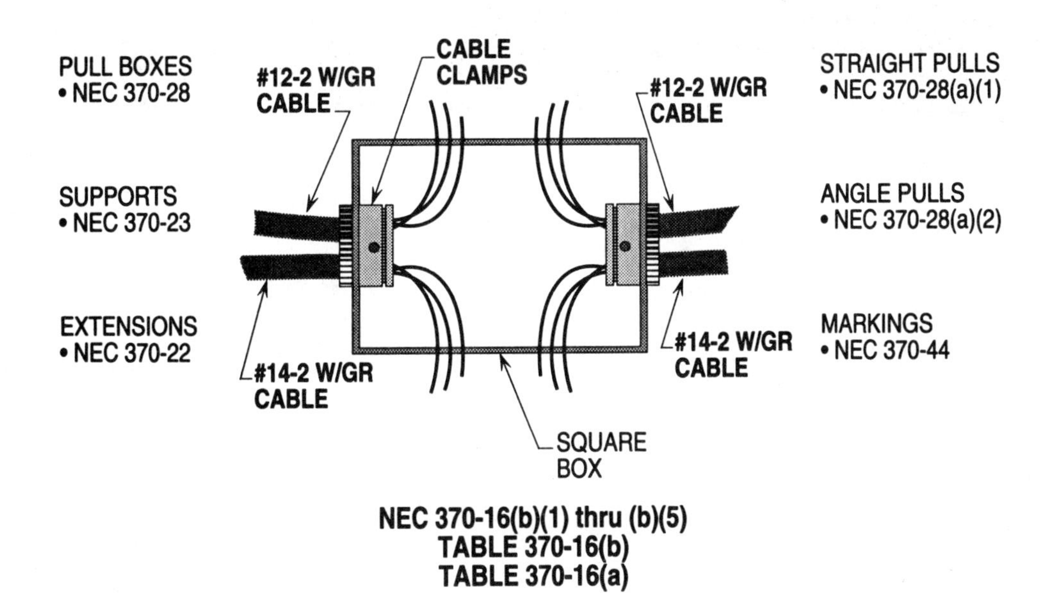

Ill. 8-2(b)

Number Of Conductors In Device Box
Table 370-16(a)

Sections 370-16(b)(1) through (b)(5) contains specific rules that shall be applied where the device box has the same size conductors with the various types of fittings, devices, etc. all present in the box. Devices are required to be counted as two conductors to allow more fill (space) to accommodate the different size devices that are available in the industry today. The count of two conductors is based upon the size conductors connecting to the terminals of the device mounting on the yoke.

Section 370-16(b) requires the cubic inch rating of each conductor from Table 370-16(b) to be utilized where conductors are not the same size in the box. The cubic inch rating of each conductor is multiplied by the number in the box and the total cubic inch value is used to determine the size device box in Table 370-16(a).

Example 8-3(a). Determine the size device box required in Illustration 8-3(a)?

Sizing device box

Step 1: Counting conductors
370-16(b)(1) thru (b)(5)

1 #14 straight thru	= 2
2 #14 hots	= 2
2 #14 neutrals	= 2
2 #14 EGC's	= 1
1 Receptacle	= 2
2 #14 pigtails	= 0
1 #14 bonding jumper	= 0
Total	= 9

Step 2: Selecting box
Table 370-16(a)
9 #14 conductors requires
3" x 2" x 3 1/2"

Solution: A 3 in. x 2 in. x 3 1/2 in. device box is required.

Example 8-3(b). Determine the size device box required in Illustration 8-3(b)?

Sizing device box

Step 1: Counting conductors
370-16(b)(1) thru (b)(5); Table 370-16(b)

2 #12 hots	2.25 cu. in. x 2	= 4.5 cu. in.
2 #12 neutrals	2.25 cu. in. x 2	= 4.5 cu. in.
1 receptacle	2.25 cu. in. x 2	= 4.5 cu. in.
2 pigtails		= .0 cu. in.
1 bonding jumper		= .0 cu. in.
Total		= 13.5 cu. in.

Step 2: Selecting box
Table 370-16(a)
13.5 cu. in. requires
3" x 2" x 2 3/4"

Solution: A 3 in. x 2 in. x 2 3/4 in. device box is required.

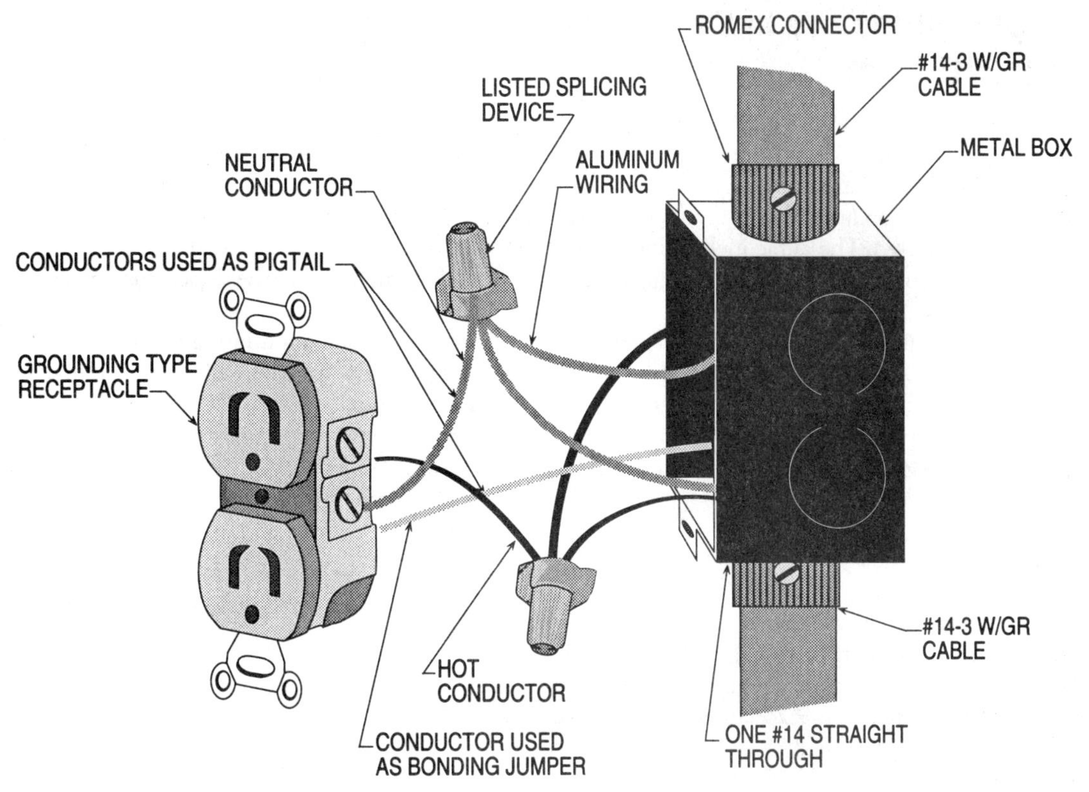

Ill. 8-3(a)

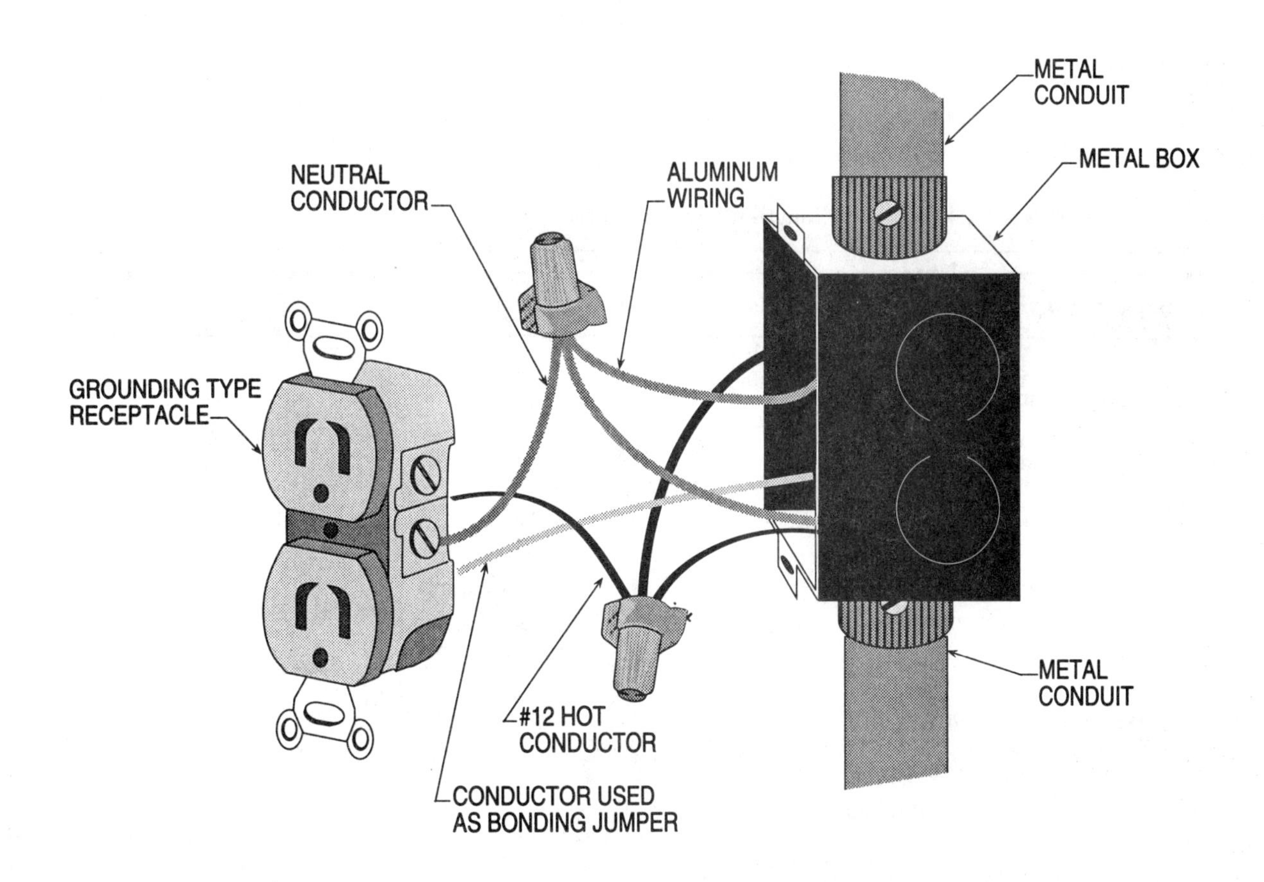

Ill. 8-3(b)

Number Of Conductors In Other Boxes 370-16(a); 370-16(b)

Because there are not any Tables available with the number of conductors permitted in junction boxes for conductors that are the same size, a calculation applying the cubic inch value of each conductor must be done. By multiplying the cubic inch rating of each conductor by the number in the box and using this total cubic inch rating, the proper size junction box may be selected. The cubic inch rating of each conductor is found in Table 370-16(b) for conductors No. 18 through No. 6. The cubic inch fill space of a junction box is found by multiplying the dimensions of the box.

To select the proper size junction box housing different size conductors, the cubic inch rating of each conductor from Table 370-16(b) must be selected and multiplied by the number of conductors for each cubic inch rating. The total computation of cubic inch ratings is used to select the proper size junction box.

Example 8-4(a). Determine the size junction box required in Illustration 8-4(a)?

Sizing junction box

Step 1: Counting cu. in.
370-16(b)(1) thru (b)(5); Table 370-16(b)
8 #12 conductors 2.25 cu. in. x 8 = 18 cu. in.
8 #12 conductors 2.25 cu. in. x 8 = 18 cu. in.
8 #12 conductors 2.25 cu. in. x 8 = 18 cu. in.
8 #12 conductors 2.25 cu. in. x 8 = 18 cu. in.
4 #12 conductors 2.25 cu. in. x 4 = 9 cu. in.
4 #12 conductors 2.25 cu. in. x 4 = 9 cu. in.
Total = 90 cu. in.

Step 2: Selecting box
Box chart
6" x 4" x 4" box = 96 cu. in.
96 cu. in. will contain 90 cu. in.

Solution: A 6 in. x 4 in. x 4 in. junction box is required.

Example 8-4(b). Determine the size junction box required in Illustration 8-4(b)?

Sizing junction box

Step 1: Counting conductors
370-16(b)(1) thru (b)(5); Table 370-16(b)
16 #10 conductors 2.5 cu. in. x 16 = 40 cu. in.
20 #14 conductors 2 cu. in. x 20 = 40 cu. in.
24 #12 conductors 2.25 cu. in. x 24 = 54 cu. in.
12 # 6 conductors 5 cu. in. x 12 = 60 cu. in.
Total = 194 cu. in.

Step 2: Selecting box
Box chart
8" x 6" x 6" = 288 cu. in.
192 cu. in. will contain
194 cu. in.

Solution: A 8 in. x 6 in. x 6 in. junction box is required.

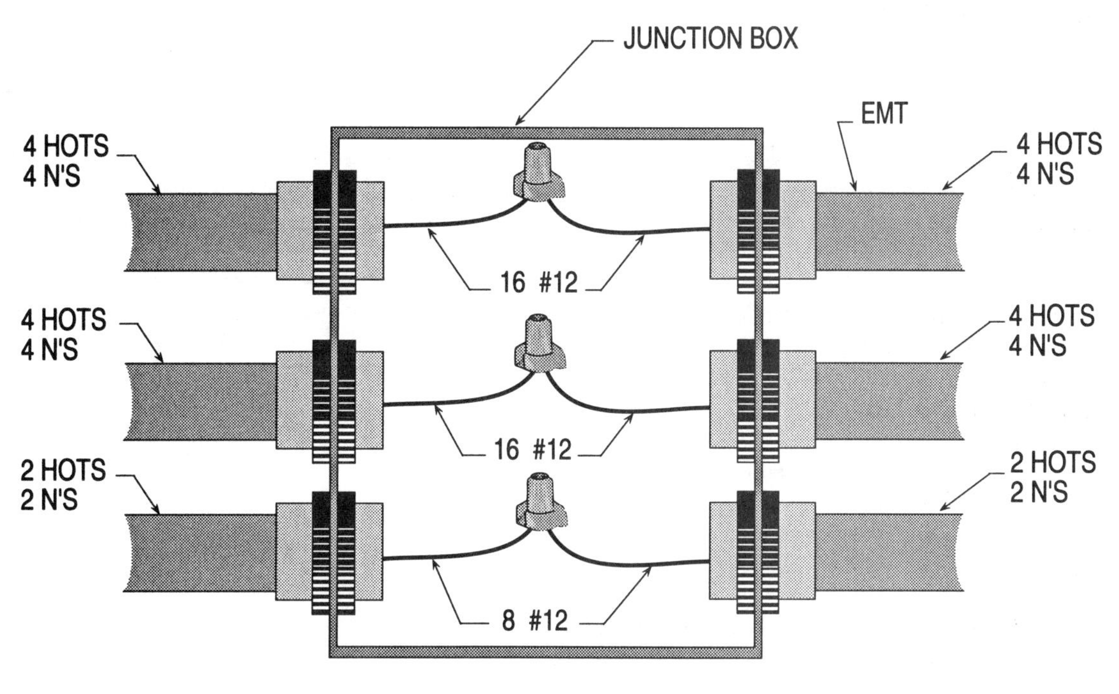

Ill. 8-4(a)

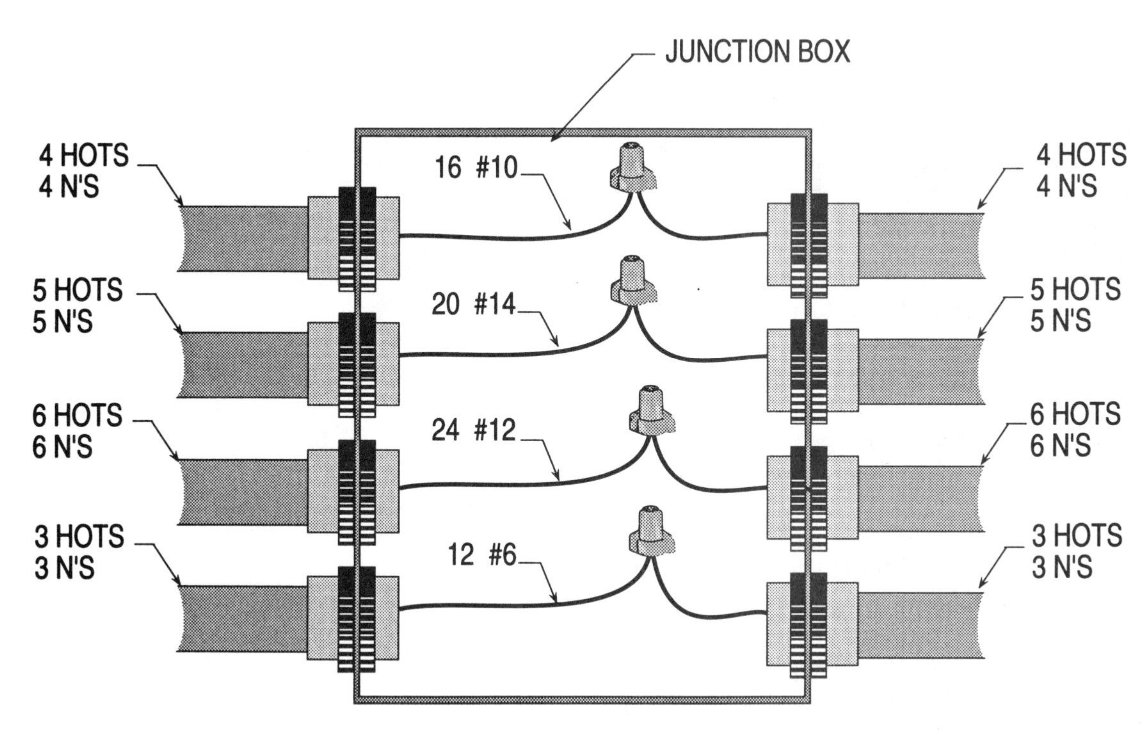

Ill. 8-4(b)

Sizing 600 Volt Or Less - Pull Boxes 370-28(a)(1); (2)

Junction boxes used for a straight pull are sized at 8 times the largest conduit entering the box.

Junction boxes used for a angle or U pull are sized at 6 times the largest conduit plus the remaining conduits sharing the same wall. Where there are more than one row of conduits per wall, the row that provides the maximum distance must be used for the calculation.

Example 8-5(a). What is the minimum size junction box for the straight pulls in Illustration 8-5(a)?

Sizing junction box

Step 1: Calculating straight pull with one conduit
370-28(a)(1)
Min. size = largest conduit x 8
Min. size = 2 1/2" x 8 = 20"

Solution: The minimum length of the junction box is 20 in.

Step 1: Calculating straight with more than one conduit
370-28(a)(1)
Min. size = largest conduit x 8
Min. size = 3" x 8 = 24"

Solution: The minimum length of the junction box is 24 in.

Example 8-5(b). What is the minimum size junction box for the angle pulls in Illustration 8-5(b)?

Sizing junction box

Step 1: Calculating angle pull with one conduit
370-28(a)(2)
Min. size = largest conduit x 6
Min. size = 3" x 6 = 18"

Solution: The minimum size junction box is 18 in. x 18 in.

Step 1: Calculating angle pull with more than one conduit
370-28(a)(2)
Min. size = largest conduit x 6 (plus remaining)
Min. size = 3 1/2" x 6 + 2 1/2" + 1 1/2"
Min. size = 25"

Solution: The minimum size junction box is 25 in. x 25 in.

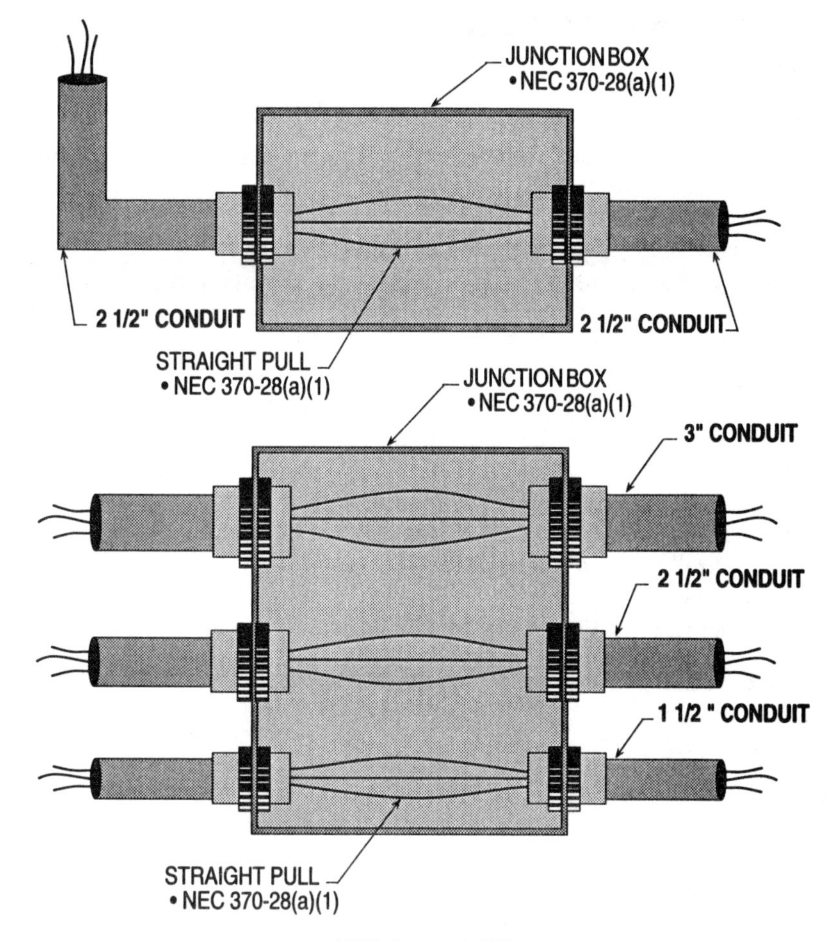

NEC 370-28(a)(1)

Ill. 8-5(a)

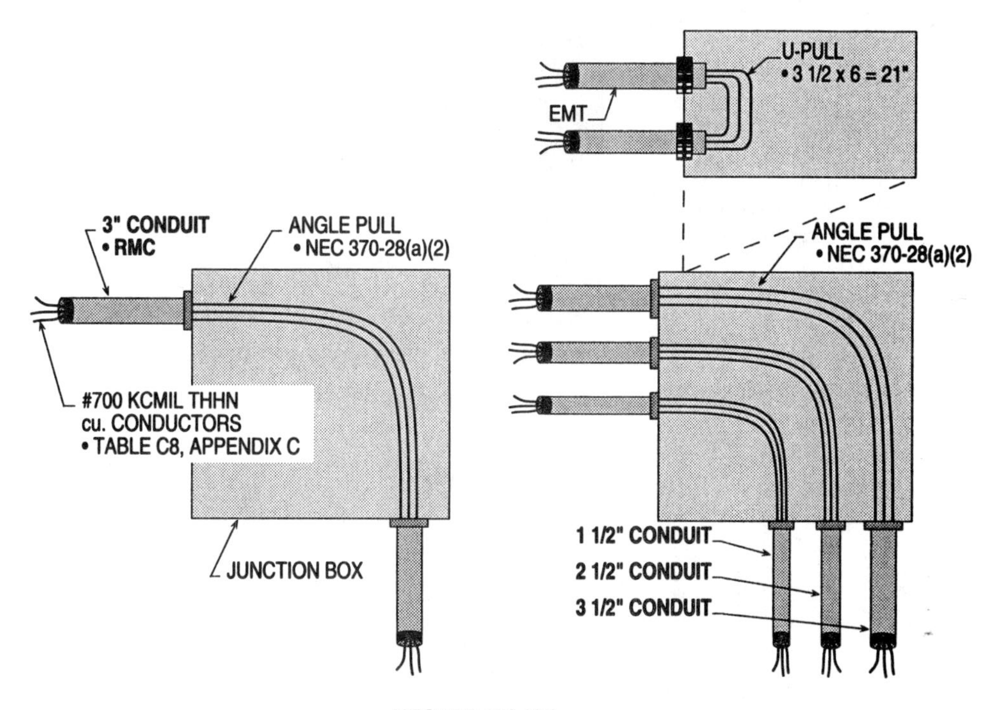

NEC 370-28(a)(2)

Ill. 8-5(b)

Sizing 600 Volt Or Less - Pull Boxes 370-28(a)(2)

Junction boxes with multiple rolls of conduits shall be sized based upon the row which produces the greater distance after each roll of conduits have been calculated.

Example 8-5(c). What is the minimum size junction box for the angle pull in Illustration 8-5(c)?

Sizing junction box

Step 1: Calculating angle pull with rolls of conduits
370-28(a)(2)
Min. size 1st roll = largest conduit x 6 (plus remaining)
Min. size 1st roll = 2 1/2" x 6 + 2 1/2" + 2 1/2" + 2 1/2" = 22 1/2"
Min. size 2nd roll = 3" x 6 + 3" + 3" = 24 "

Solution: The minimum size junction box is 24 in x 24 in.

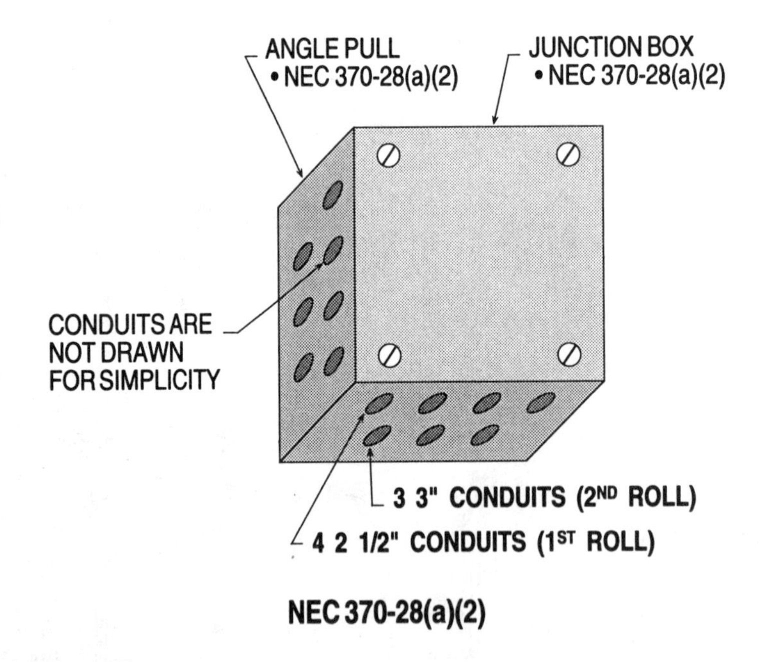

Ill. 8-5(c)

Sizing Over 600 Volts - Pull Boxes 370-71(a); (b)

When sizing junction boxes, the length of the box must be 48 times the outside diameter of the largest cable or conductor entering the box. For angle or U pulls, the minimum width and length are at least 36 times the outside diameter of the largest cable or conductor.

Example 8-6. What is the minimum size junction box for the straight and angle pull in Illustration 8-6?

Sizing straight pull

Step 1: Calculating straight pull
370-71(a)
Min. size = dia. x 48
Min. size = .71 x 48
Min. size = 34"

Solution: The minimum size length of the junction box is 34 in.

Note: Use 32 x the diameter for nonshielded conductors.

Sizing angle pull

Step 1: Calculating angle pull
370-71(b)
Min. size = dia. x 36
Min. size = .646 x 36
Min. size = 23.3"

Solution: The minimum size junction box is 23.3 in. x 23.3 in.

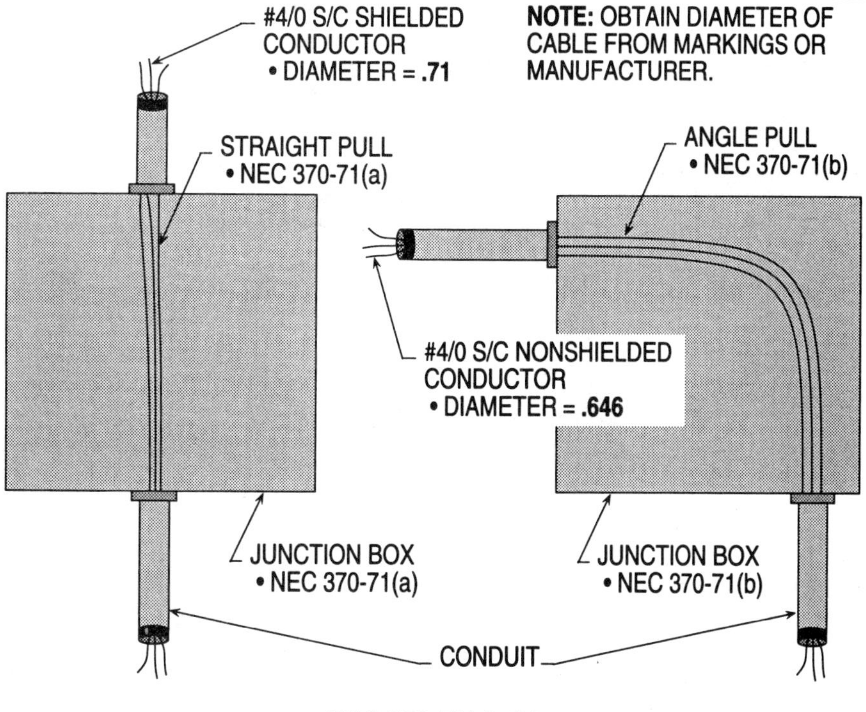

NEC 370-71(a); (b)

Ill. 8-6

Sizing Bending Space For Cables in Manholes 370-54, Ex. ; 370-71(b)

Bending space for cables in manholes is calculated based on straight or angle pulls through the manholes. The multiplier times cable(s) is selected based on 600 volts or less or over 600 volt cables.

Example 8-7. What is the minimum bending space for the cables in Illustration 8-7 where a straight and angle pull is used?

Sizing straight pull for long dimension (six ducts)

Step 1: Calculating bending space
370-54, Ex.; 370-71(a)
Bending space = 48 x 1/6' (8.02) + 23 x 1/6' (3.84)
Bending space = 8.02 + 3.84
Bending space = 12' (rounded up)

Solution: **The minimum bending space is 12 ft. for a straight pull.**

Note: Use 32 x the diameter for nonshielded cables.

Sizing angle pull (long dimension)

Step 1: Calculating bending space
370-54, Ex.; 370-71(b)(1)
Bending space = 36 x 1/6' (6.01) + 23 x 1/6' (3.84)
Bending space = 6.01 + 3.84
Bending space = 10' (rounded up)

Sizing angle pull (short-dimension)

Step 1: Calculating bending space
370-54, Ex.; 370-71(b)(1)
Bending space = 36 x 1/6' (6.01) + 16 x 1/6' (2.67)
Bending space = 6.01 + 2.67
Bending space = 9' (rounded up)

Solution: **The minimum bending space is 10 ft. for long dimension and 9 ft. for short dimension for the angle pull.**

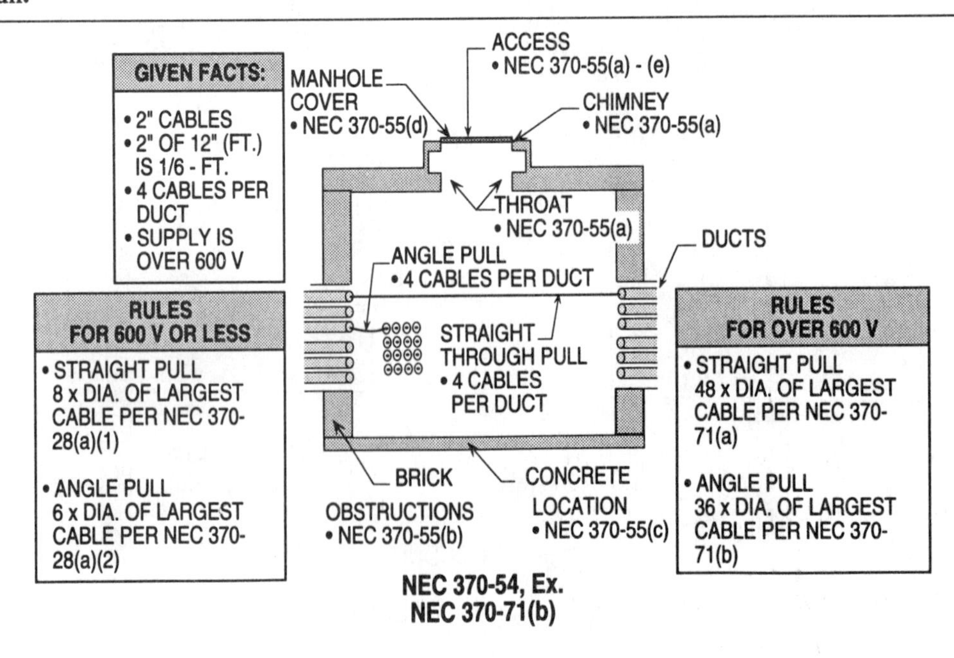

Ill. 8-7

Cabinets, Cutout Boxes, and Meter Socket Enclosures

The space inside enclosures is required to have sufficient space to terminate conductors to lugs of overcurrent protection devices or busbars. This space is required to protect the conductors and equipment from physical damage when they are being terminated.

Table 373-6(a) is utilized to provide the proper space to terminate conductors formed into a L-bend configuration.

Table 373-6(b) is selected for conductors that are formed into a S or Z-bend configuration.

The minimum wire-bending space at terminals and correct width of wiring gutters in inches is determined by the number of conductors connected to each terminal.

Quick Reference

Sizing For L-Bends
384-3(g); Table 373-6(a)

Panelboards must comply with the provisions of NEC 384-3(g) which requires the minimum gutter space in the panel to meet the clearances in Table 373-6(a) for L-bends.

If the lugs in the panelboard are removable, the clearance can be reduced from the dimension in Table 373-6(b) based on the number and size of conductors connected to each lug.

Example 9-1. Determine the minimum clearances between the terminating lugs and the side of a panelboard employing a L-bend in Illustration 9-1?

Sizing L-bend

Step 1: Finding the minimum clearance
Table 373-6(a)
1 #250 KCMIL per lug = 4 1/2"

Solution: A 4 1/2 in. minimum clearance is required.

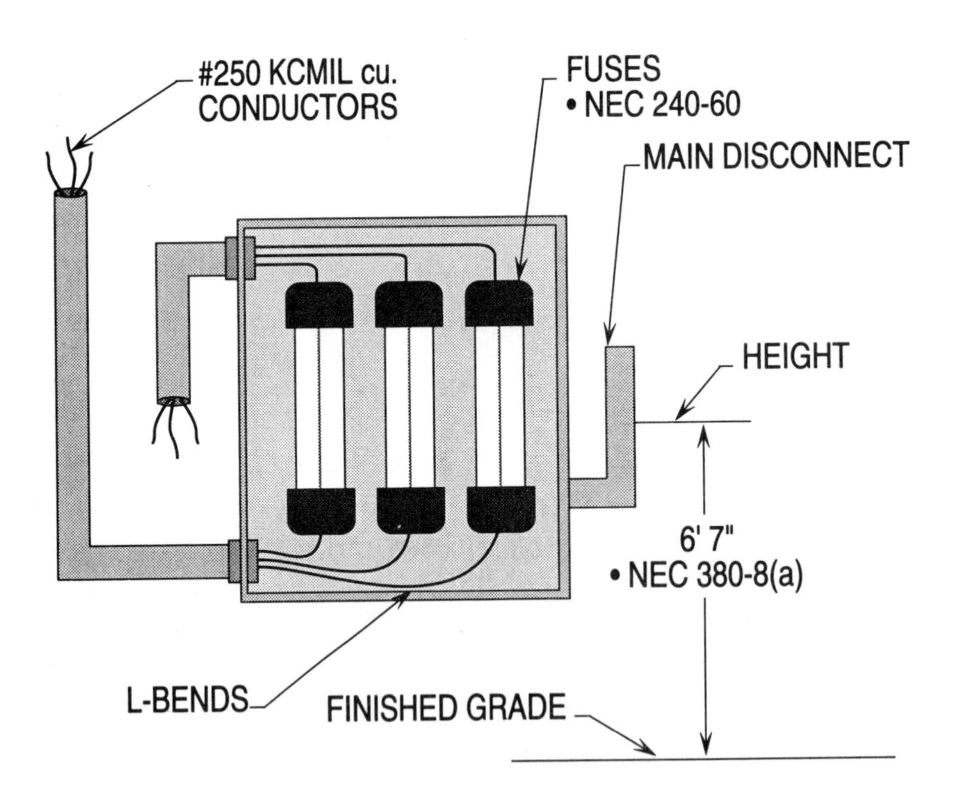

Ill. 9-1

Sizing For S Or Z-Bends
384-3(g); Table 373-6(b)

Panelboards must comply with the provisions of NEC 384-3(g) which requires the minimum gutter space in the panel to meet the clearances in Table 373-6(b) for S or Z-bends.

If the lugs in the panelboard are removable, the clearance can be reduced from the dimension in Table 373-6(b) based on the number and size of conductors connected to each lug.

Example 9-2. Determine the minimum clearance for the copper conductors terminating to the lugs in a panelboard entering from the bottom wall in Illustration 9-2?

Sizing S or Z-bend

Step 1: Finding the minimum clearance
Table 373-6(b)
1 #2/0 per lug = 6"

Solution: The minimum clearance required is 6 in..

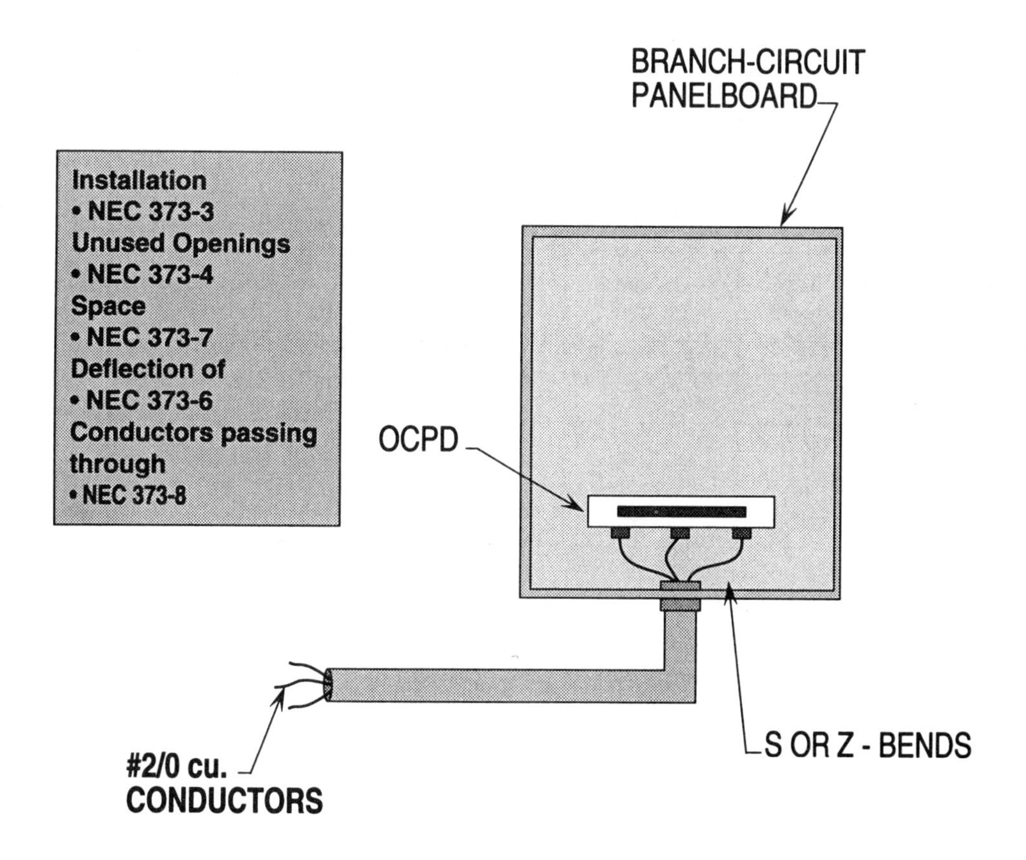

Ill. 9-2

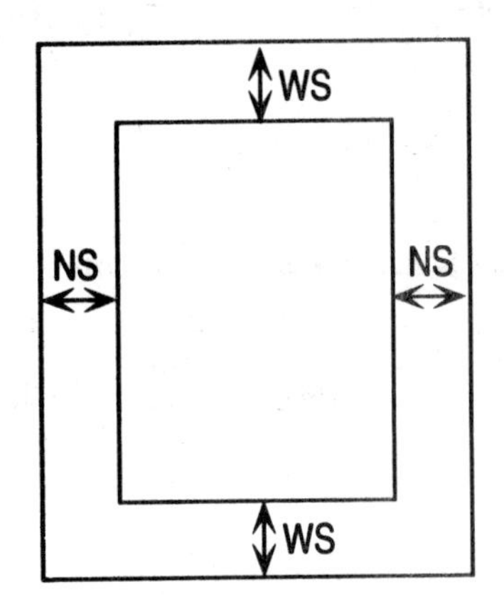

NS = NARROW SPACE PER TABLE 373-6(a)

WS = WIDE SPACE PER TABLE 373-6(b)

384-35 AND Ex.'s 1 AND 2
(NOTE: ALSO SEE Ex.'s 3 AND 4)

General rule	Lighting Panelboard not rated over 225 amps	Rules for any Panlboard
Top and Bottom space must comply with Table 373-6(b) and sides must comply with Table 373-6(a)	Top or Bottom space must comply with Table 373-6(b) for space opposite lugs	One of the two required bending spaces can be at the side of the panel-board if necessary

General Rule for Panelboards			
Top is WS	Bottom is WS	Right side is NS	Left side is NS
Panelboard 225 amps or less			
Top is WS	Bottom is NS	Right side is NS	Left side is NS
Any Panelboard			
Top is NS	Bottom is WS	Right side is NS	Left side is WS

Ill. 9-3 shows the narrow and wide space requirements for housing conductors terminating to OCPD's and Lugs in lighting and appliance branch-circuits panelboards or power panels.

10

Flexible Cord/Cable and Fixture Wire

Flexible cords are not to be used as a permanent wiring method. However, they may be used as wiring methods for construction sites and other similar types of use per NEC 305-4. The ampacity is determined based upon the number of current-carrying conductors used in each flexible cord per Column A or B to Table 400-5(A).

Fixture wire of smaller sizes than the branch-circuit conductors may be used between the junction box and fixture sockets, ballasts, transformers, etc. The maximum size OCPD's for branch-circuits and fixture wiring is found in NEC 240-4(b)(2).

Quick Reference

Ampacity Of Flexible Cords And Cables 400-5

The ampacity of cords depends upon the number of conductors used as current-carrying conductors, which is based on the type.

Example 10-1. What are the ampacities of the conductors in each flexible cord in Illustration 10-1?

Sizing flexible cord for case 1

Step 1: Calculating A for case 1
Table 400-5(A), Col. A; 400-5
#2-3/c cord = 80 A

Solution: The allowable ampacity is 80 amps for the three conductor cord.

Sizing flexible cord for case 2

Step 1: Calculating A for case 2
Table 400-5(A), Col. A; 400-5
#4-3/c cord = 60 A

Solution: The allowable ampacity is 60 amps for the three conductor cord.

Sizing flexible cord for case 3

Step 1: Calculating A for case 3
Table 400-5(A), Col. A; 400-5
#6-4/c cord = 45 A x 80% = 36 A

Solution: The allowable ampacity is 36 amps for the four conductor cord

Sizing flexible cord for case 4

Step 1: Calculating A for case 4
Table 400-5(A), Col. A; 400-5
#12-7/c cord = 20 A x 70% = 14 A

Solution: The allowable ampacity is 14 amps for the seven conductor cord.

Sizing flexible cord for case 5

Step 1: Calculating A for case 5
Table 400-5(A), Col. A; 400-5
#10-10/c cord = 25 A x 50% = 12.5 A

Solution: The allowable ampacity is 12.5 amps for the ten conductor cord.

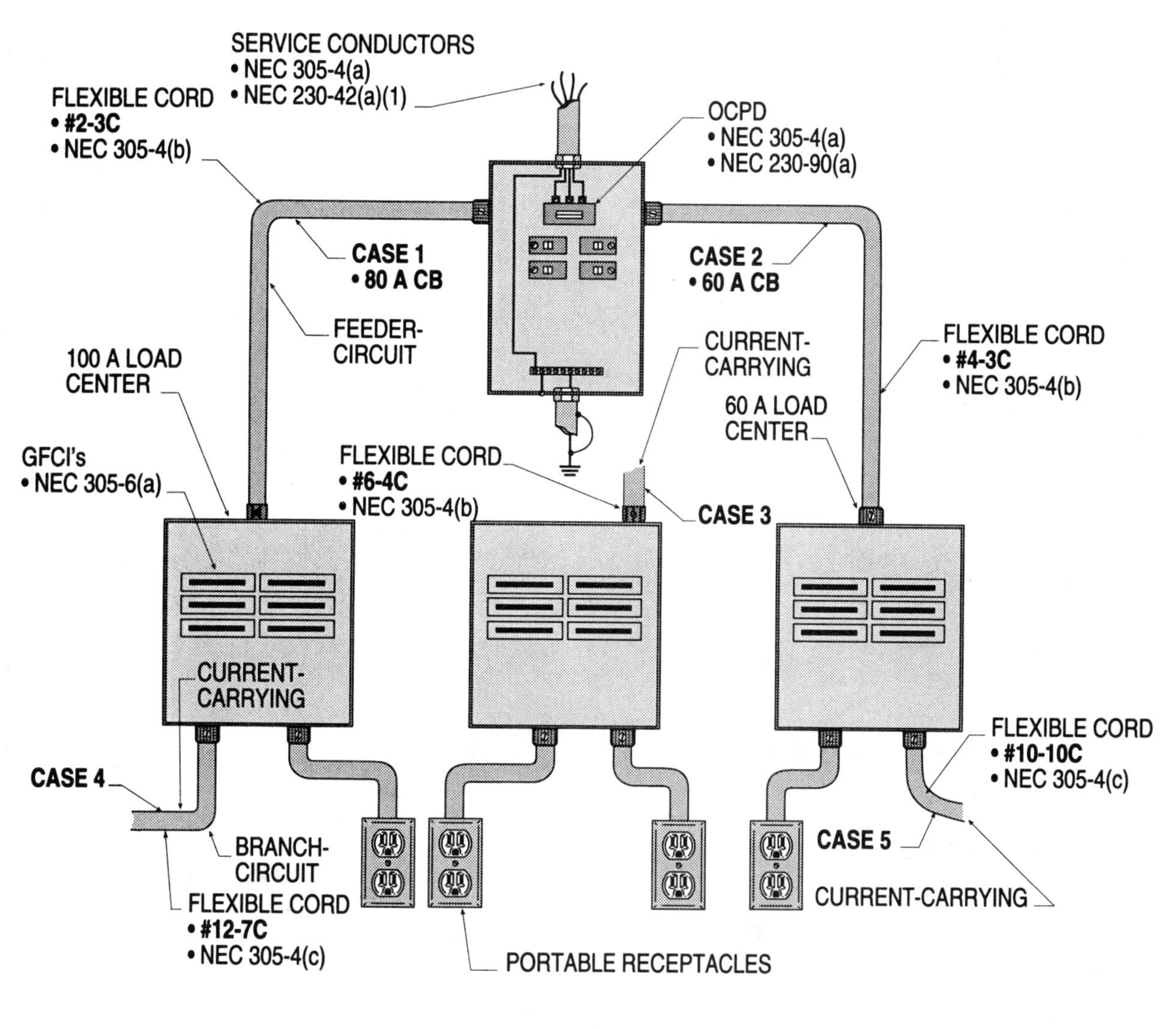

NEC 400-7(a)(11)
TABLE 400-5(A)
NEC 400-5

Ill. 10-1

Number Of Fixture Wire Conductors In Raceway Table 402-5

The allowable ampacity of fixture wire is found in Table 402-5 and their conditions of use is listed in Table 402-3. The number allowed in conduit or tubing is determined by Note 2 to Chapter 9 Tables.

Example 10-2. What is the allowable ampacity for the fixture wires and size rigid metal conduit required for the listed wiring method and assembly in Illustration 10-2?

Sizing fixture wire

Step 1: Calculating allowable ampacity
Table 402-5; Table 402-5
#16 fixture wire = 8 A

Step 2: Applying derating factors
402-5, FPN; Table 310-15(b)(2)(a)
8 A x 80% = 6.4 A

Solution: The allowable ampacity of the No. 16 fixture wire is 6.4 amps.

Sizing conduit

Step 1: Calculating conduit size
Table C8, Appendix C
4 #16 TFN conductors requires 1/2"

Solution: The size rigid metal conduit required is 1/2 in..

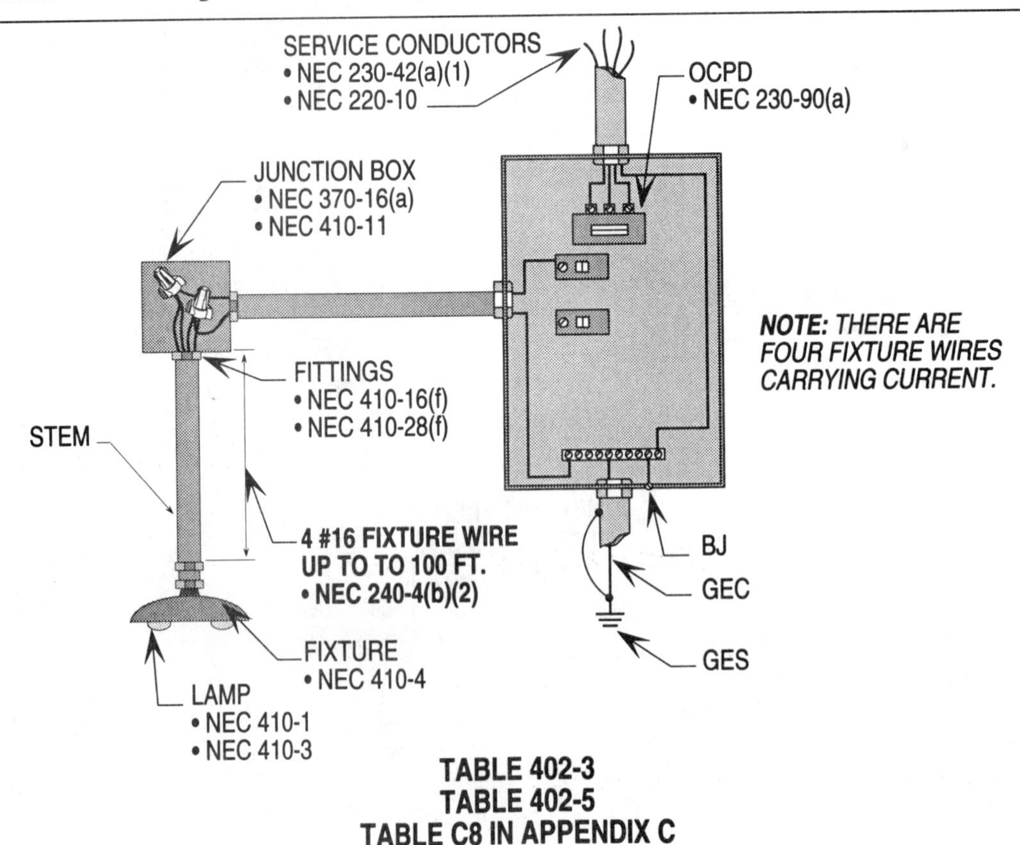

Ill. 10-2

Low-Voltage Lighting Systems
Art. 411

The procedure for calculating the load for low-voltage lighting systems is to compute the primary amps or VA by 125 percent for continuous operation or 100 percent for noncontinuous operation.

Example 10-3. What is the VA rating for the low-voltage lighting system in Illustration 10-3?

Finding VA

Step 1: 411-2
Pri. VA = 3,600 VA

Solution: **The VA rating for the low-voltage lighting system is 3,600 VA.**

Calculating VA for continuous rating

Step 1: 220-10; 230-42(a)(1)
3,600 VA x 125% = 4,500 VA

Solution: **The VA rating for continuous operation is 4,500 VA.**

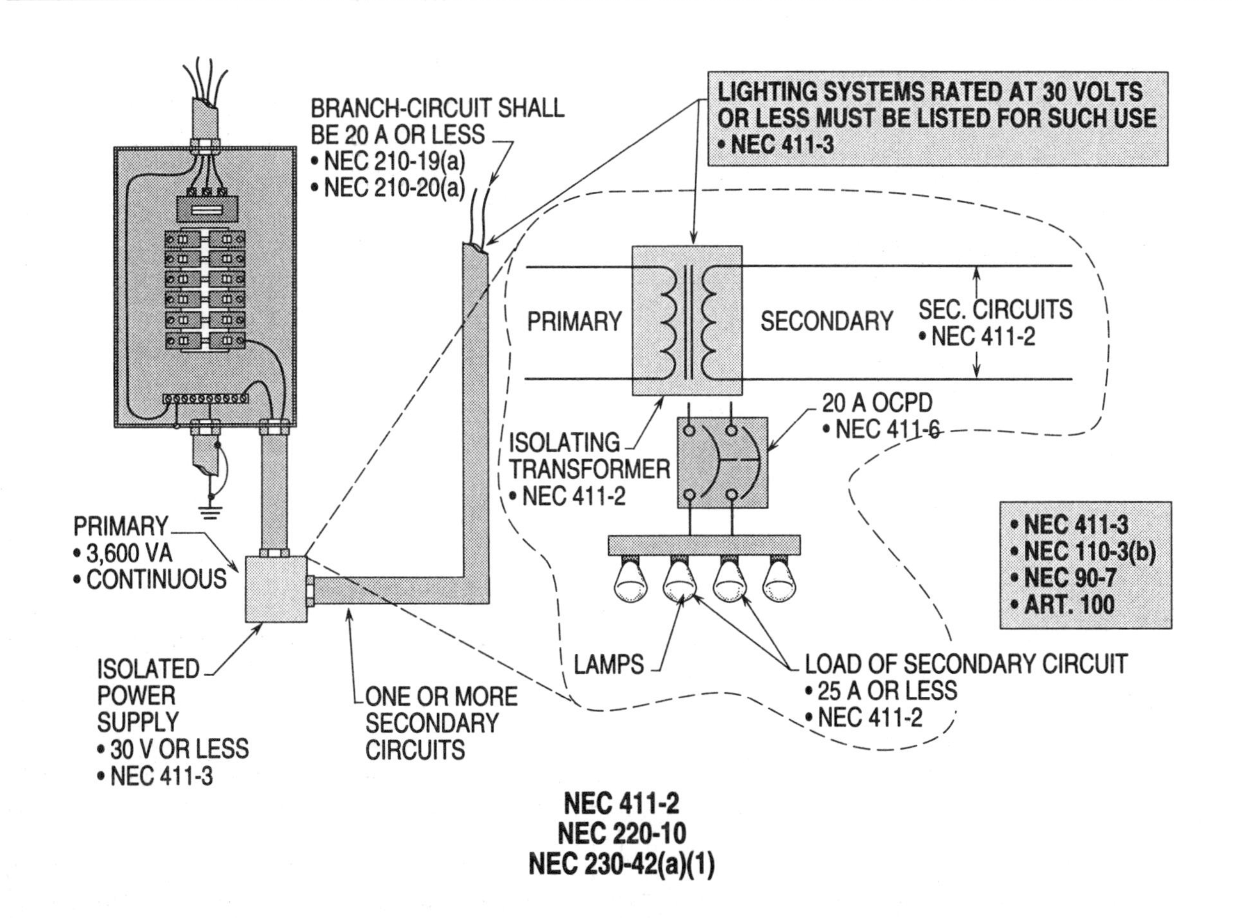

Ill. 10-3

11 Appliances

Appliances are defined in Article 100 of the National Electrical Code® (NEC®) as utilization equipment that generally are not considered the industrial type. They are normally built in standardized sizes or types and installed as units to perform one or more functions such as clothes washing, air-conditioning, food mixing, deep frying, etc.

The overcurrent protection devices (OCPD's) and conductors along with other elements are computed at 100 percent for noncontinuous operation and 125 percent for continuous operation.

Overcurrent protection devices must be sized and selected to allow appliances to start and run properly.

Appliances are either cord-and-plug connected or wired in permanently with an approved wiring method.

Quick Reference

Sizing Branch-circuits 422-10(a)

The elements for circuits supplying power to appliances are required to be sized based upon continuous (load x 125%) or noncontinuous operation (load x 100%).

Example 11-1. What size OCPD and conductors are required to protect and supply power to the appliances in Illustration 11-1?

Sizing OCPD

Step 1: Calculating load
422-10(a); 210-20(a)
15 A x 125% = 18.75 A

Step 2: Selecting OCPD
422-11; 240-3(b); 240-6(a)
18.75 A requires 20 A

Solution: The size OCPD is 20 amps

Sizing conductors

Step 1: Selecting conductors
Table 310-16; 210-19(a); 240-3(d)
(See asterisk)
18.75 A requires #12 cu.

Solution: The size conductors are No. 12 THWN copper conductors.

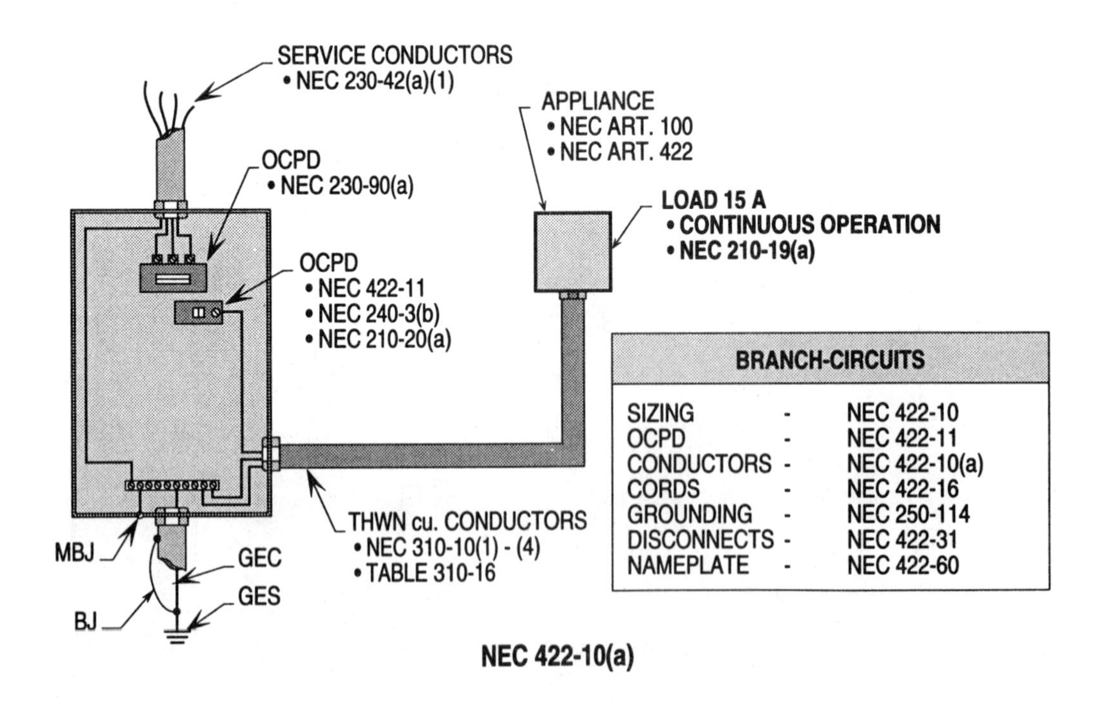

Ill. 11-1

Branch-circuit For Storage-type Water Heater 422-13

When sizing the branch-circuit for storage-type water heaters the overcurrent protection devices and conductors must be sized not less than 125 percent of the heating load per NEC 422-13 to prevent tripping open the OCPD and disconnecting all the elements connected into the circuit.

Example 11-2. Determine the size conductors and OCPD required for the water heater in Illustration 11-2?

Sizing conductors

Step 1: Finding amperage
I = 5,500 VA ÷ 240 V = 22.9 A

Step 2: Calculating the A
422-13; 422-10(a)
22.9 A x 125% = 28.6 A

Step 3: Selecting the conductors
336-26; Table 310-16; 240-3(d)
28.6 A load requires #10 cu.

Sizing OCPD

Step 1: Calculating load for OCPD
422-13
22.9 A x 125% = 28.6 A

Step 2: Selecting the OCPD
422-13; 240-3(b); 240-6(a)
28.6 A load requires a 30 A OCPD

Solution: No. 10-2 w/ground nonmetallic sheathed cable with a 30 amp OCPD is required.

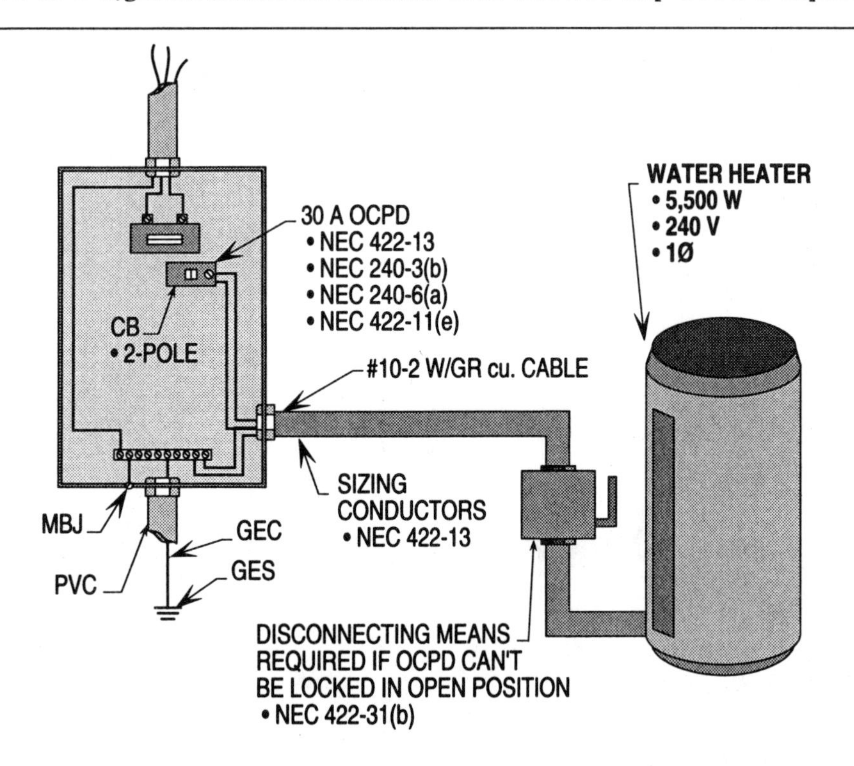

Ill. 11-2

Sizing OCPD For Surface Heating Elements 422-11(b)

Surface mounted heating elements with a demand load of more than 60 amps per Table 220-19 shall have its supply power subdivided into two or more circuits and protected at 50 amps or less.

Example 11-3. What size OCPD's and conductors are required to supply power to the appliance in Illustration 11-3?

Sizing OCPD

Step 1: Calculating OCPD's
422-11(b)
A = 24 kW x 1,000 / 240 V
A = 100 A

Step 2: Subdividing load
422-11(b)
100 A / 2 = 50 A

Step 3: Selecting OCPD
422-11(b); 422-11(a); 240-3(b)
two circuits at 50 A each

Solution: **There are two circuits at 50 amps each required to protect the appliance and circuits.**

Sizing conductors

Step 1: Calculating conductors
422-10(a); Table 310-16
50 A requires #8 cu.

Solution: **The size conductors for each circuit is No. 8 THWN copper.**

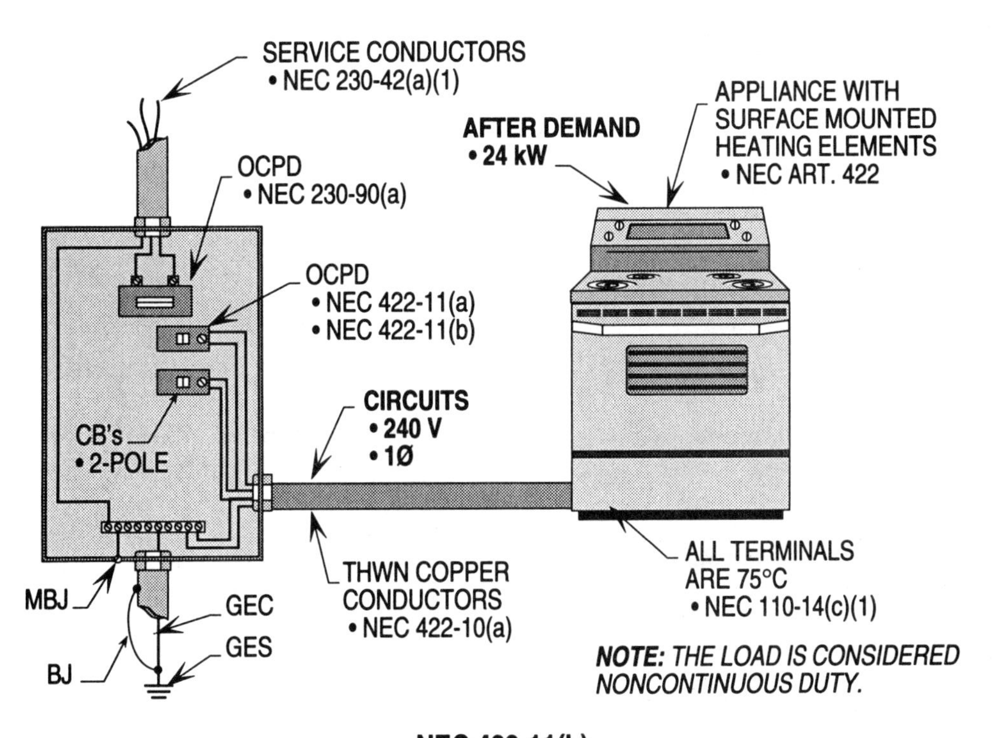

NEC 422-11(b)

Ill. 11-3

Sizing OCPD For Single Nonmotor Appliance 422-11(e)(2)

When individual circuits supply only one "nonmotor" appliance rated at 13.3 amps or more, the OCPD is limited to 150 percent of the FLC rating of the appliance. If the OCPD trips and is increased up to 150 percent of the appliance rating and does not correspond to a standard OCPD, the next size may be used.

Example 11-4(a). What size OCPD's and conductors are required to supply power to each appliance in Illustration 11-4(a)?

Sizing OCPD for case 1

Step 1: Calculating OCPD for cir. 1
422-11(e)(2); 422-10(a)
31 A x 125% = 38.8 A

Step 2: Selecting OCPD for cir. 1
240-3(b); 240-6(a)
38.8 A requires 40 A

Solution: The size OCPD for circuit 1 is 40 amps.

Sizing conductors for circuit 1

Step 1: Selecting conductors for cir. 1
Table 310-16; step 1 above
38.8 A requires #8 cu.

Solution: The size conductors for circuit 1 is No. 8 THWN copper.

Sizing OCPD for case 2

Step 1: Calculating OCPD for cir. 2
422-11(e)(2); 422-10(a)
12.5 A x 125% = 15.6 A

Step 2: Selecting OCPD
240-3(b); 240-6(a)
15.6 A requires 20 A

Solution: The size OCPD for circuit 2 is 20 amps.

Sizing conductors for circuit 2

Step 1: Selecting conductors for cir. 2
Table 310-16; 240-3(d); step 1 above
15.6 A requires #12 cu.

Solution: The size conductors for circuit 2 is No. 12 copper.

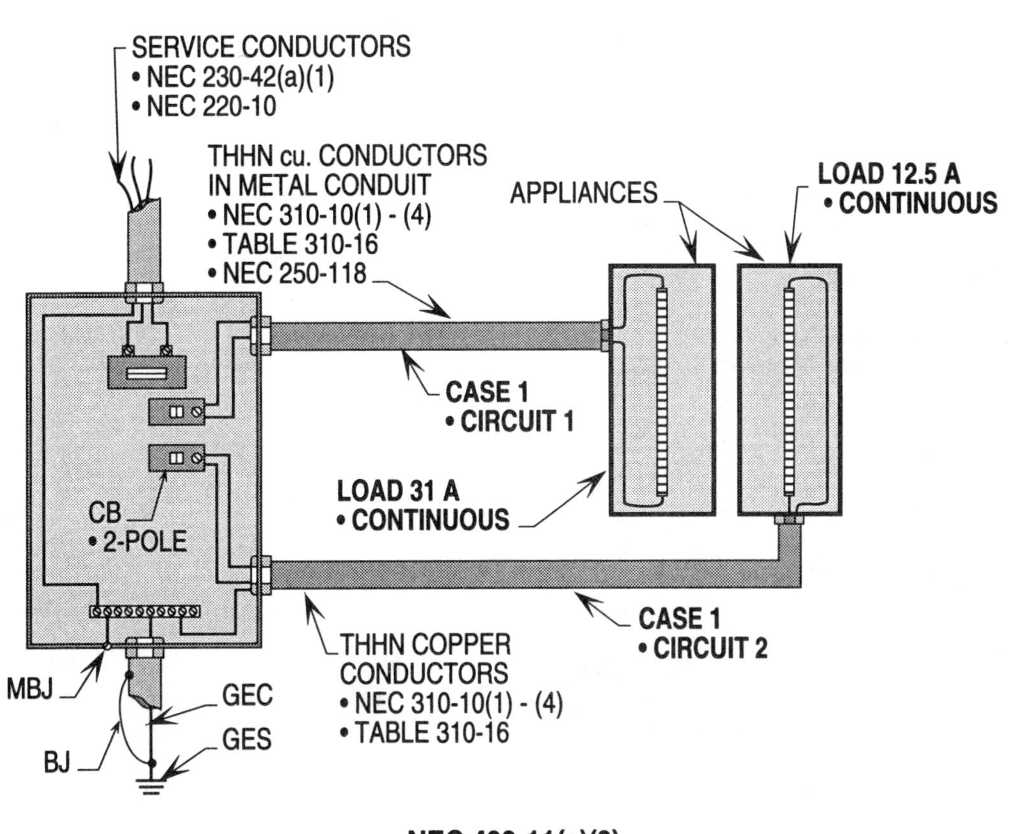

NEC 422-11(e)(2)

Ill. 11-4(a)

Sizing OCPD For Single Nonmotor Appliance 422-11(e)(3)

The OCPD protecting only one "nonmotor" appliance rated at 16.7 amps or more (13.3 x 125% = 16.7 A), which is supplied by an individual circuit, may be sized at 125 percent or 150 percent of the FLA of such appliance.

Example 11-4(b). What is the minimum and maximum size OCPD allowed for the appliance in Illustration 11-4(b)?

Sizing min. OCPD

Step 1: Calculating min. OCPD
422-11(e)(2); 422-10(a)
15.5 A x 125% = 19.4 A

Step 2: Selecting min. OCPD
240-3(b); 240-6(a)
19.4 A requires 20 A

Solution: The minimum size OCPD is 20 amps.

Sizing max. OCPD

Step 1: Calculating max. OCPD
422-11(e)(3); 240-3(b)
15.5 A x 150% = 23.3 A

Step 2: Selecting max. OCPD
422-11(e)(3); 240-6(a)
23.3 A allows 25 A OCPD

Solution: The maximum size OCPD is 25 amps.

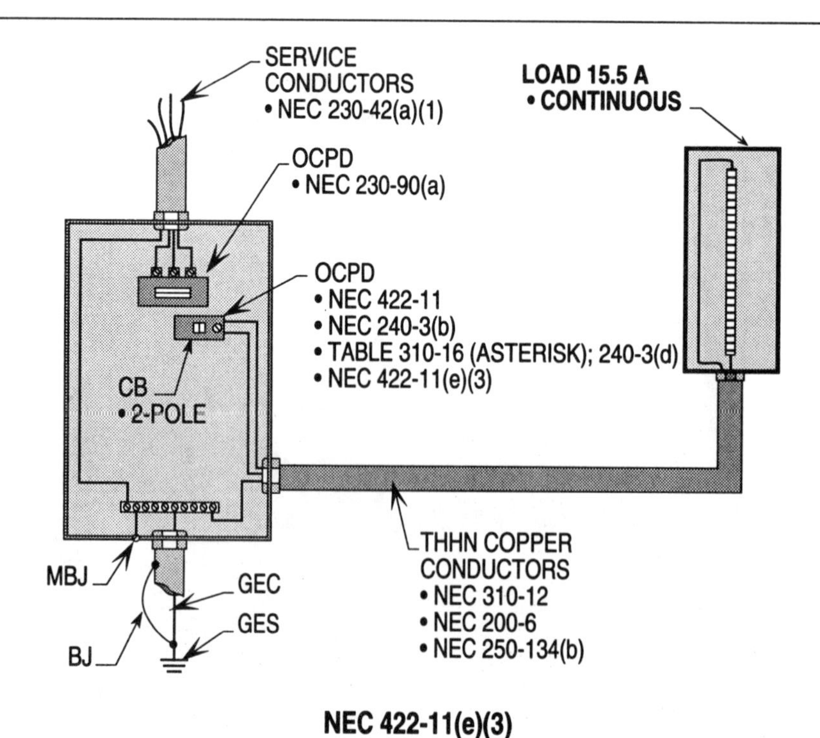

Ill. 11-4(b)

12

Heating and Deicing Equipment

Heating equipment loads are usually calculated at 125 percent of the heating elements plus the other loads. There are cases when the heating elements and other loads are computed at not less than 125 percent of the combined load. This rule is intended to provide safe and reliable circuits which will not create a fire hazard.

Heating equipment may include such equipment as heating cables, unit heaters, boilers, central systems, pipeline and vessel systems, etc.

The conductors utilized to supply heating equipment loads are sized per Table 310-16 and selected for use per Table 310-13. The allowable ampacities of conductors are determined by the requirements listed in NEC 310-10(1) through (4). OCPD's are sized per NEC 240-3(a) through (g) and selected for ratings per NEC 240-6(a). Their condition of use are found in NEC 240-60(c) and NEC 240-83(c) and (e) and NEC 240-85.

Quick References

Individual Branch-circuits 424-3(a)

Individual circuits supplying fixed electric space heating equipment can be any size. Multioutlets serving such equipment is allowed to be 15, 20, or 30 amp OCPD's respectively.

Example 12-1. What size OCPD and THWN cu. conductors are required to supply power to the baseboard heaters in Illustration 12-1?

Sizing OCPD for case 1

Step 1: Calculating load for case 1
424-3(a); 424-3(b)
24 A (12 A + 12 A) x 125% = 30 A

Step 2: Selecting OCPD for case 1
240-3(b); 240-6(a)
30 A requires 30 A OCPD

Solution: The size OCPD required is 30 amps.

Sizing conductors for case 1

Step 1: Selecting conductors for case 1
Table 310-16; 424-3(a); step 1 above
30 A requires #10 cu.

Solution: The size conductors required are No. 10 THWN copper.

Sizing OCPD for case 2

Step 1: Calculating load for case 2
424-3(a); 424-3(b)
40 A x 125% = 50 A

Step 2: Selecting OCPD for case 2
240-3(b); 240-6(a)
50 A requires 50 A OCPD

Solution: The size OCPD required is 50 amps.

Sizing conductors for case 2

Step 1: Selecting conductors for case 2
Table 310-16; 424-3(a); step 1 above
50 A requires #8 cu.

Solution: The size conductors are No. 8 THWN copper.

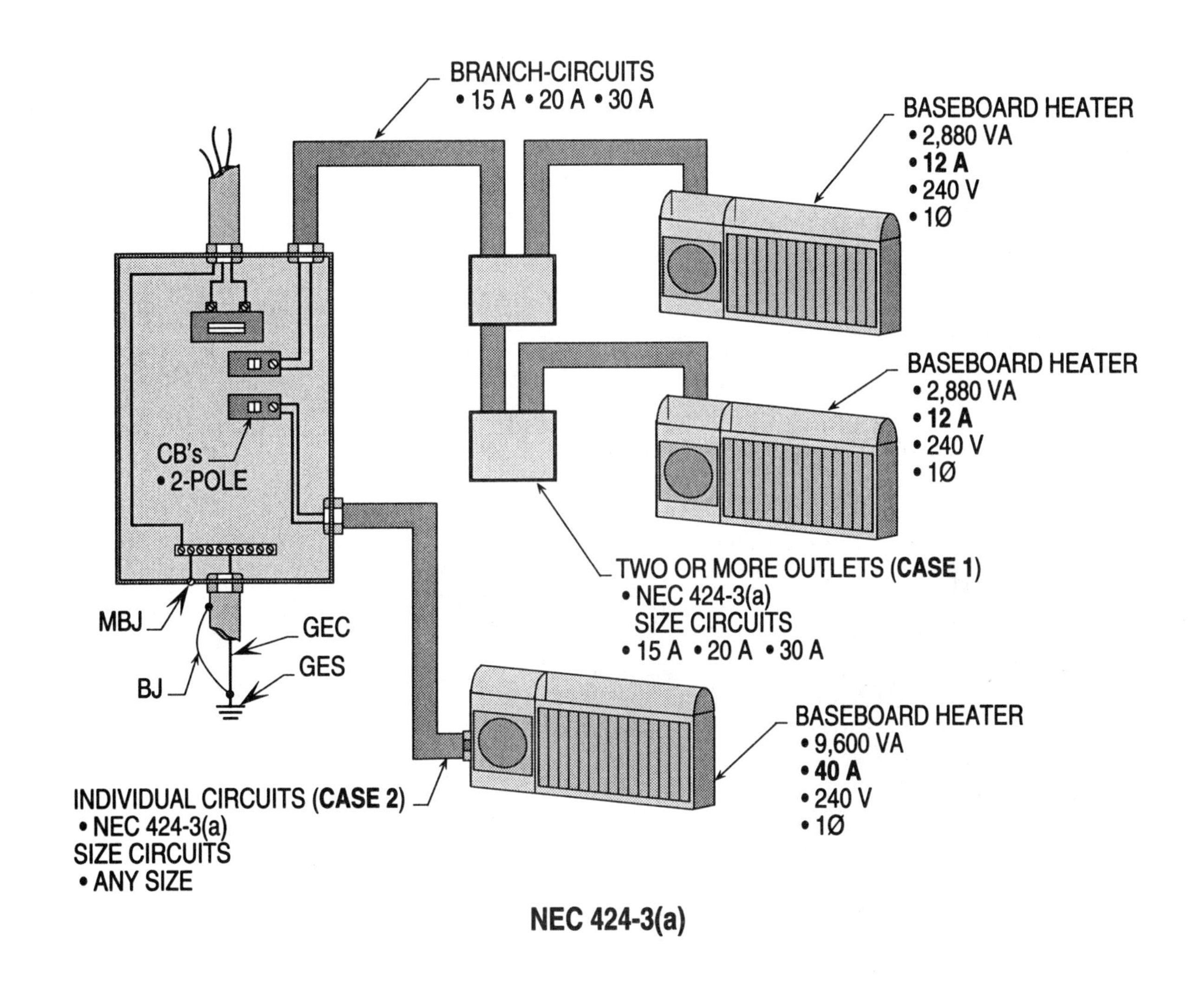

NEC 424-3(a)

Ill. 12-1

Sizing Space Heating Circuits 424-3(b)

The branch-circuit elements for circuits supplying electric space heating units are computed at 125 percent of the heating elements FLA plus the FLA of the blower motor.

Example 12-2. What size OCPD and conductors are required to supply power to the heating unit in Illustration 12-2?

Sizing conductors

Step 1: Calculating A
424-3(b); 220-2(a)
I = 20 kVA x 1,000 ÷ 240 V = 83 A

Step 2: Calculating load
424-3(b)
(83 A + 3 A) x 125% = 107.5 A

Step 3: Selecting conductors
Table 310-16
107.5 A load requires #2 THWN cu.

Solution: No. 2 THWN copper conductors are required.

Sizing OCPD

Step 1: Calculating load
424-3(b)
83 A + 3 A x 125% = 107.5 A

Step 2: Selecting OCPD based on load
240-6(a); 240-3(b)
110 A is the next higher standard size

Solution: The size CB is 110 amps.

Note 1: OCPD can be sized up to 125 amp based upon No. 2 THWN cu. conductor per NEC 240-3(b).

Note 2: See NEC 430-24, Ex. 2. on page 13-33.

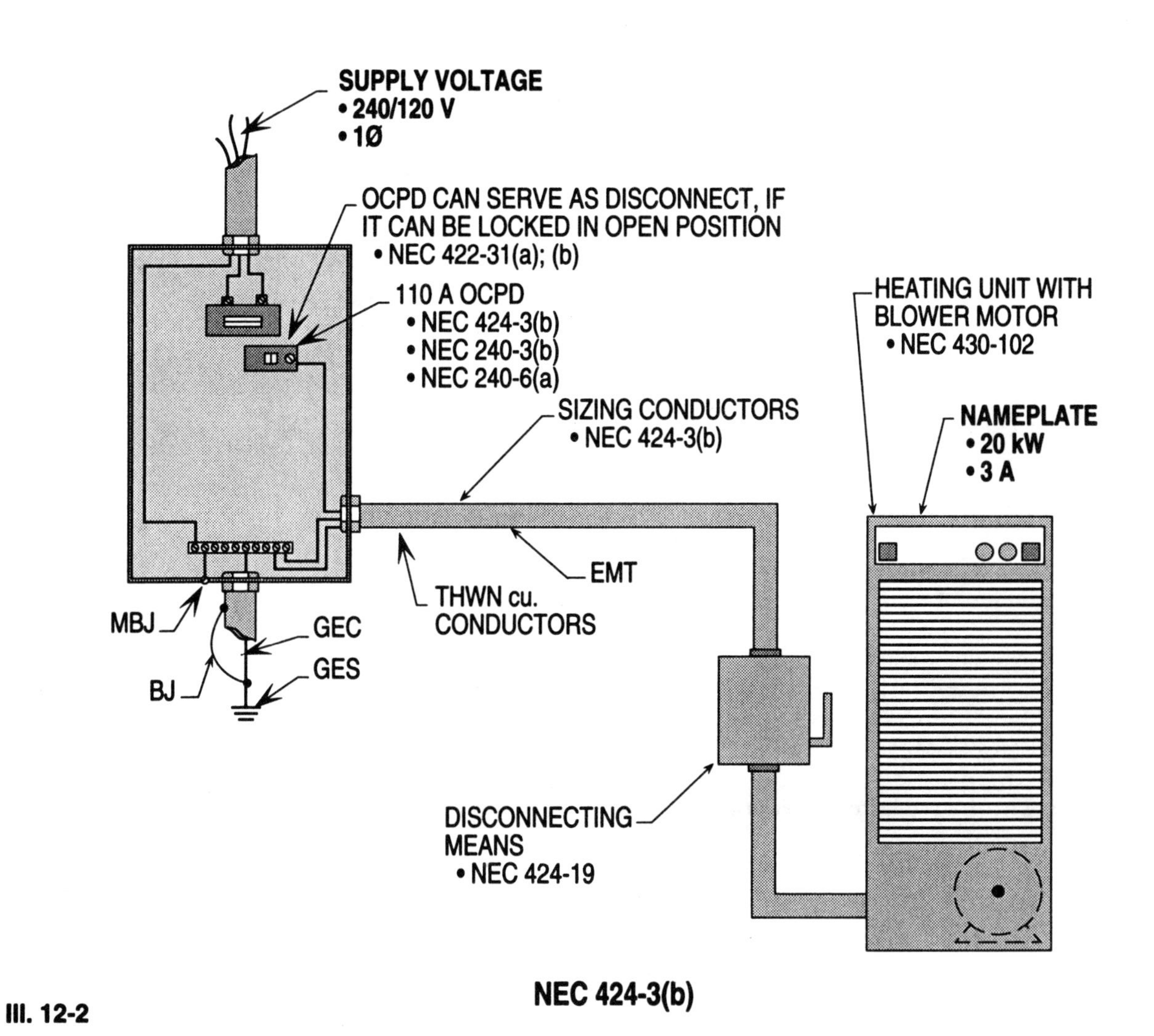

NEC 424-3(b)

Ill. 12-2

Sizing Subdivided Element Loads 424-22(b)

The elements in resistance-type space heating units rated over 48 amps must be subdivided by the manufacturer into loads of not over 48 amps. If the OCPD of such elements is separate, the conductors between OCPD's and the elements shall be computed at 125 percent of the load.

Example 12-3. What size OCPD's and conductors are required between the disconnect and heating unit?

Sizing OCPD

Step 1: Calculating load for elements
424-22(b)
Load = 5 kW x 6 x 1,000 / 240 V
Load = 125 A

Step 2: Calculating subdivided load
424-24(b)
Load = 125 A / 3
Load = 42 A

Step 3: Calculating motor load
Table 430-148
1/3 HP = 3.6 A

Step 4: Calculating total load
424-3(b); 424-22(e)
Run #1: 42 A + 3.6 A x 125% = 57 A
Run #2; 3: 42 A x 125% = 52.5 A

Step 5: Selecting OCPD
424-22(c); 424-3(b); 240-3(b); 240-6(a)
Run #1: 57 A requires 60 A OCPD
Run #2; 3: 52.5 A requires 60 A OCPD

Solution: **The size OCPD is 60 amp for each run.**

Sizing conductors

Step 1: Selecting conductors
424-22(e); Table 310-16; Step 4 above
Run #1: 57 A requires #6 cu.
Run #2 & 3: 52.5 A requires #6 cu.

Solution: **The size conductors in each are required to be No. 6 THWN copper.**

Determining kW when rated (240 V, 1Ø) voltage is greater than the supply voltage (208 V, 1Ø)

Step 1: Finding Multiplier
Multiplier = app. V ÷ htg. unit V = X^2

Step 2: Calculating Multiplier
Multiplier = 208 V ÷ 240 V =
8.66666667 x 8.66666667
Multiplier = .751111112

Step 3: Applying Multiplier
kW = kW of heating unit x multiplier
kW = 30 x .751111112
kW = 22.53

Solution: **kW = 22.53**

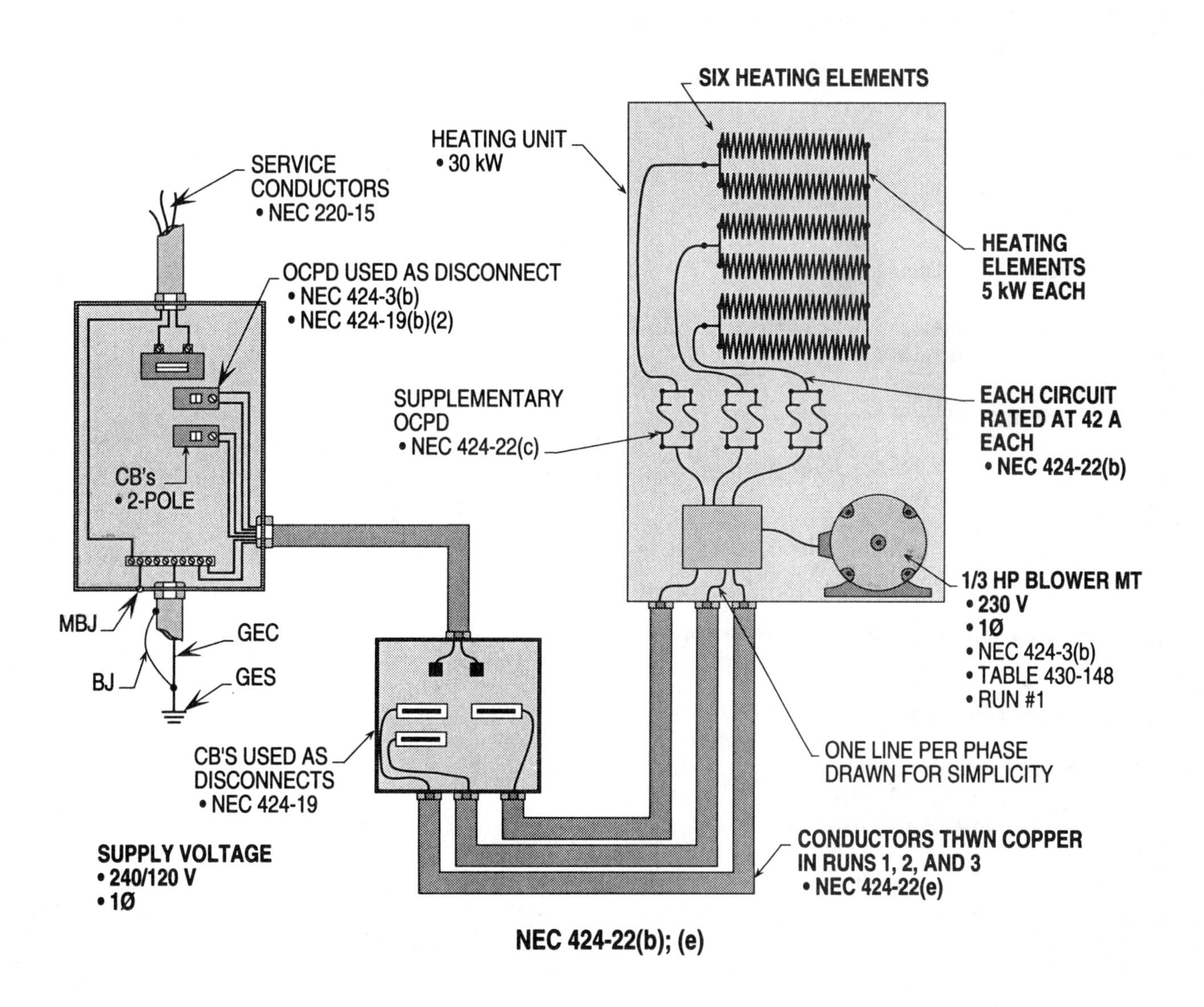
SIX HEATING ELEMENTS
HEATING UNIT
• 30 kW
SERVICE CONDUCTORS
• NEC 220-15
OCPD USED AS DISCONNECT
• NEC 424-3(b)
• NEC 424-19(b)(2)
HEATING ELEMENTS 5 kW EACH
SUPPLEMENTARY OCPD
• NEC 424-22(c)
EACH CIRCUIT RATED AT 42 A EACH
• NEC 424-22(b)
CB's
• 2-POLE
1/3 HP BLOWER MT
• 230 V
• 1Ø
• NEC 424-3(b)
• TABLE 430-148
• RUN #1
MBJ
GEC
BJ
GES
CB'S USED AS DISCONNECTS
• NEC 424-19
ONE LINE PER PHASE DRAWN FOR SIMPLICITY
SUPPLY VOLTAGE
• 240/120 V
• 1Ø
CONDUCTORS THWN COPPER IN RUNS 1, 2, AND 3
• NEC 424-22(e)
NEC 424-22(b); (e)

Ill. 12-3

Sizing Snow And Deicing Circuits 426-4

The OCPD and conductors for deicing and snow melting equipment must be sized at 125 percent of the current in amps of the load supplied.

Example 12-4. What size OCPD and conductors are required to supply power to the heating cable in Illustration 12-4? (using a raceway to enclose conductors)

What size OCPD and No. 12-2 w/ground NM cable is required to supply power to the receptacle outlet and heating cable in Illustration 12-4?

Sizing OCPD

Step 1: Calculating load
426-4
Load = 1,800 VA / 120 V
Load = 15 A

Step 2: Calculating OCPD
426-4
15 A x 125% = 18.75 A

Step 3: Selecting OCPD
426-4; 240-3(b); 240-6(a)
18.75 A requires 20 A OCPD

Solution: The size of the OCPD is 20 amps.

Sizing conductors

Step 1: Selecting conductors
336-26; Table 310-16
18.75 A requires #12 cu.

Solution: The size conductors in the NM cable or raceway must be No. 12 copper.

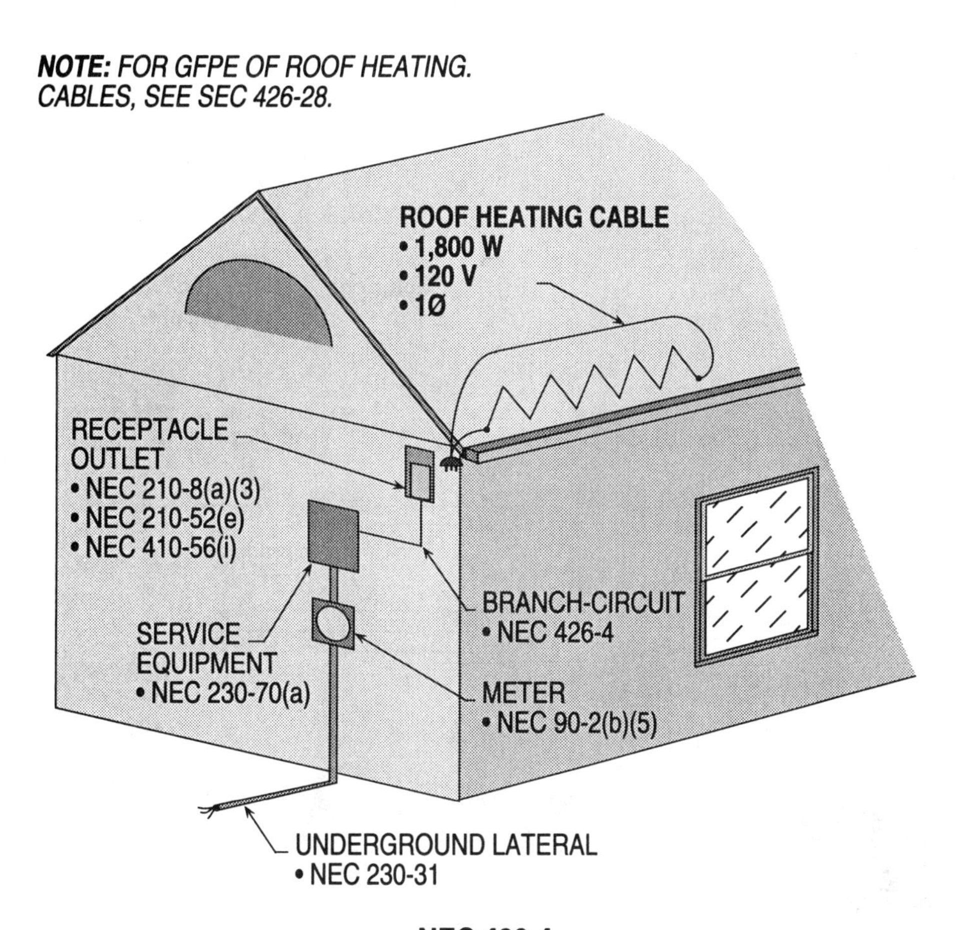

NEC 426-4

Ill. 12-4

Sizing Fixed Heating For Pipelines
427-4

The OCPD and conductors for pipeline systems or vessels must be sized at 125 percent of the current in amps of the load serviced.

Example 12-5. What size OCPD and conductors are required to supply power to the heating cable in Illustration 12-5?

Sizing OCPD

Step 1: Calculating load
427-4
22 A x 125% = 27.5 A

Step 2: Selecting OCPD
427-4; 240-3(b); 240-6(a)
27.5 A requires 30 A OCPD

Solution: **The size of the OCPD is 30 amps.**

Sizing conductors

Step 1: Selecting conductors
427-4: Table 310-16; 240-3(d)
27.5 A requires #10 cu.

Solution: **The size conductors are No. 10 THWN copper.**

Note: For heating cable protection, Review NEC 427-22 very carefully.

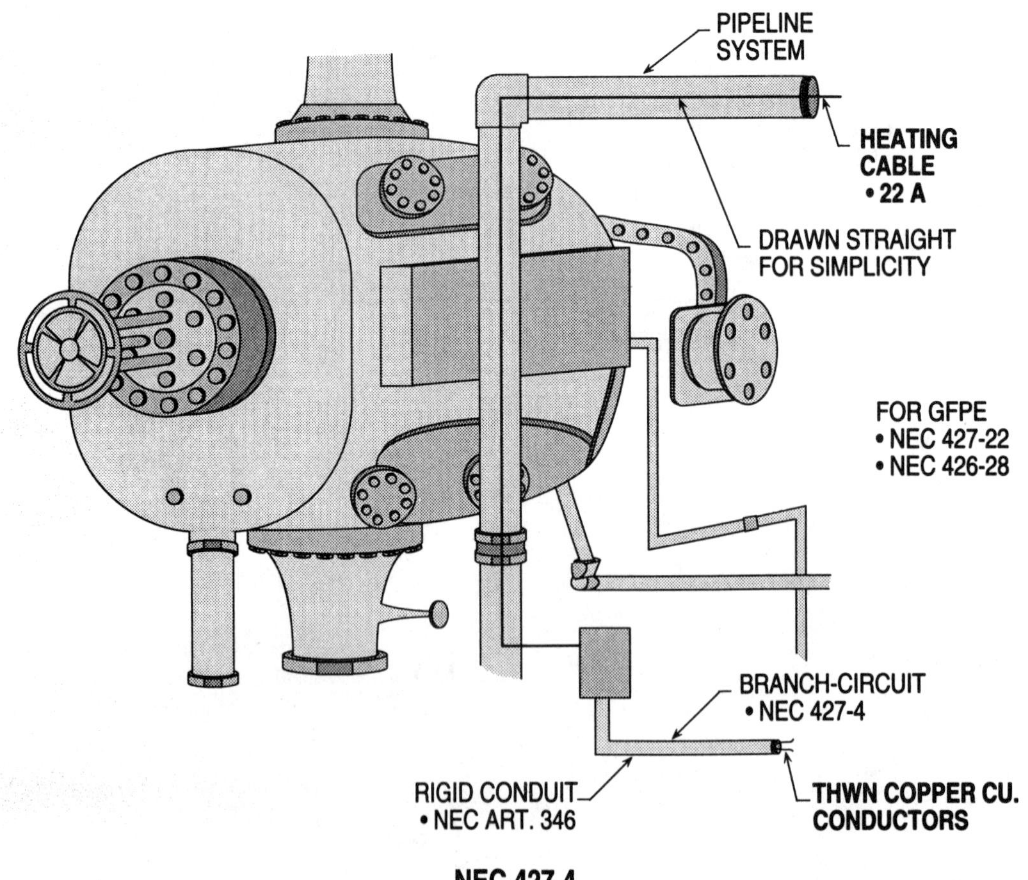

Ill. 12-5

Motors, Motor Circuits and Controllers

There are three currents that have to be found before designing and selecting the elements to make up circuits supplying power to motors. The first current to be determined is the nameplate amps on the motor. This current rating in amps is used to size the overloads (OL's) to protect the motor windings and conductors. The second current to be found is the full-load amps (FLA) from Table 430-148 for single-phase and Table 430-150 for three-phase. This current rating in amps is used to size all the elements of the circuit except the overload (OL) protection. The third current that must be found is the locked-rotor current (LRC) in amps from Table 430-7(b) and Tables 430-151(A) or (B). Note that Table 430-7(b) is used for motors with code letters and Tables 430-151(A) and (B) are used for motors with design letters. The OCPD must be sized large enough to hold this current rating (LRC) in amps and permit the motor to start and run.

Quick Reference

Adjustable Speed Drive Systems
430-2

If power-conversion equipment is part of an adjustable speed drive system, the circuit conductors, OCPD's and other related elements are sized on a basis of rated input current of the unit.

Example 13-1. What size OCPD and conductors are required to supply power to the packaged variable-speed drive systems in Illustration 13-1?

Sizing OCPD

Step 1: Finding load
430-2
Rated input = 150 A

Step 2: Selecting OCPD on rated input
430-52(b); 240-3(b); 240-6(a)
150 A requires 150 A OCPD

Solution: The size OCPD is 150 amps.

Note 1: See NEC 430-2 for overload requirements.

Note 2: The AHJ may allow a 200 OCPD based on ampacity of conductors.

Sizing conductors

Step 1: Sizing conductors
430-2; 430-22(a), Ex. 2
150 A x 125% = 187.5 A

Step 2: Selecting conductors
Table 310-16
187.5 A requires #3/0 cu.

Solution: The size conductors required are No. 3/0 THWN cu.

ADDITIONAL HELP 54:

If the motor in Ill. 13-1 has a supply of 68.5 kW and it's output is 85 HP, what would be the efficiency of the motor?

- Setting up formula to find watts
  Watts = 85 HP x 746 W
  Watts = 63,410
- Setting up formula to find efficiency

$$\text{Eff.} = \frac{\text{output}}{\text{input}}$$

$$\text{Eff.} = \frac{63{,}410\text{ W}}{68{,}500\text{ W}}$$

Eff. = 92%

Note: The constant 746 W is used for 1 HP.

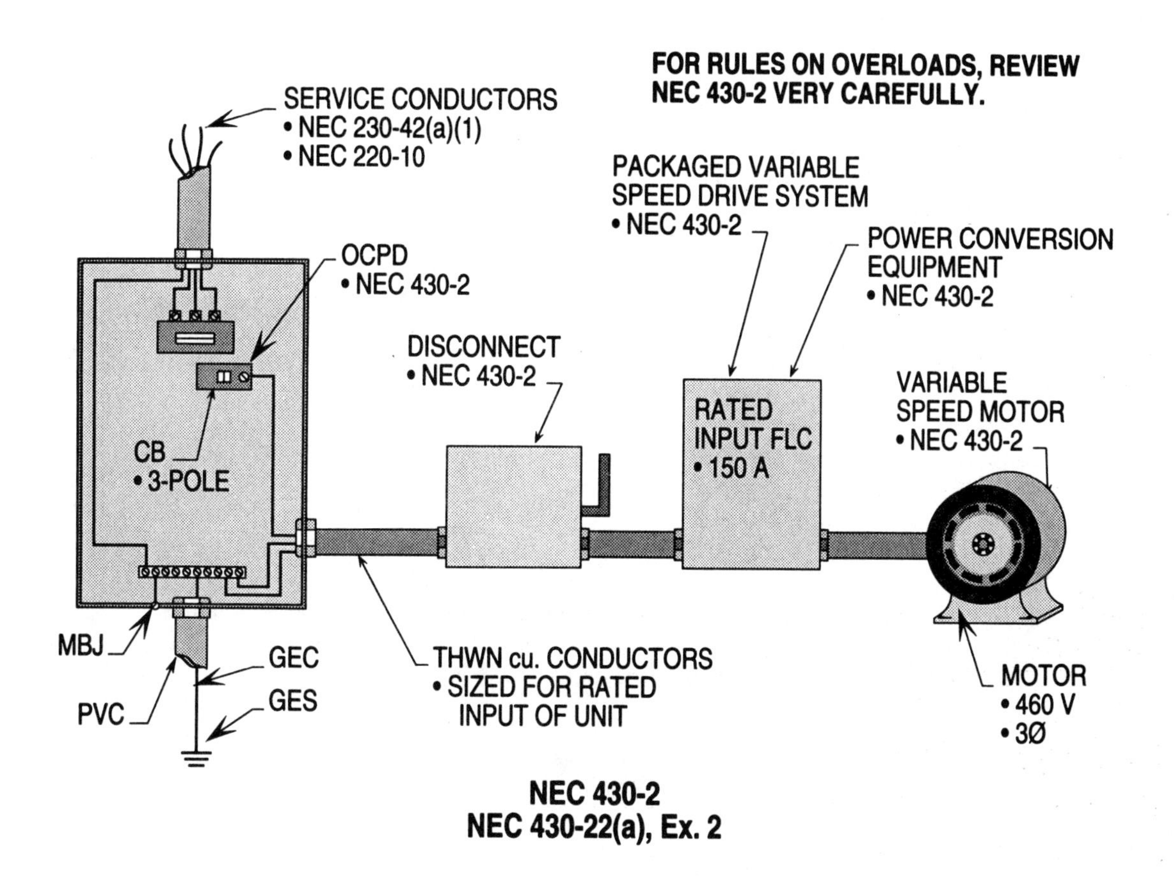

NEC 430-2
NEC 430-22(a), Ex. 2

Ill. 13-1

Part-winding Motors
430-3

When sizing the overcurrent protection device for a part-winding motor, the requirements are listed in NEC 430-3. When selecting the protective device per NEC 430-52(c)(1), the percentages to be applied are found in Table 430-152. Only half the percentages are applied in Table 430-152 for selecting the overcurrent protection device, since only half of the motors horsepower is used for starting.

Example 13-2. What is the size time delay fuse required for the part-winding motor in Illustration 13-2?

Sizing TDF

Step 1: Calculating motor FLC
430-6(a)(1); Table 430-150
40 HP = 104 A

Step 2: Calculating applied percentage
430-52(c)(1); Table 430-152
TDF = 175%

Step 3: Calculating amperage
430-3
104 A x 175% = 182 A
one-half
182 ÷ 1/2 = 91 A

Step 4: Selecting TDF
430-52(c)(1); 240-3(g); 240-6(a)
91 A = 90 A TDF's

Solution: The size fuses required for the starting winding and second winding is 90 amps.

Note: OCPD could be sized at 100 amps, per NEC 430-52(c)(1), Ex. 1.

ADDITIONAL HELP 55:

Calculating the amount of amps that the motor in Ill. 13-2 takes from line.

- Setting up formula to find Watts
  Watts = 40 HP x 746 W
  Watts = 29,840
- Setting up formula to find Volts

  Volts = 230 V x 1.732 x eff. x PF
  Volts = 304.7454
- Setting up formula to find Amps

  $$\text{Amps} = \frac{29{,}840\text{ W}}{304.7454\text{ W}}$$

  Amps = 97.9

Solution: The angle pull from the line is 97.9 amps.

Note: The constant 746 W is used for 1 HP.

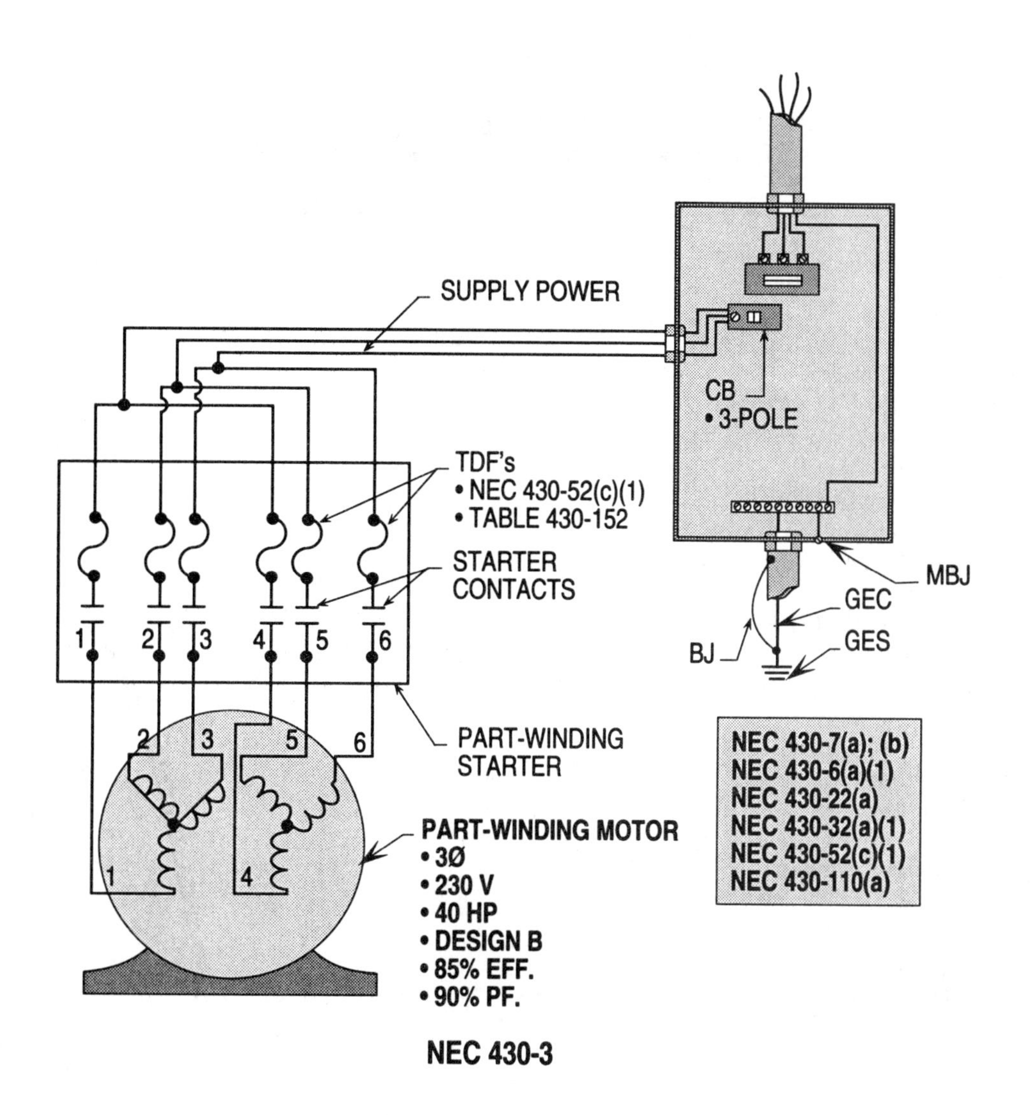
SUPPLY POWER
CB
• 3-POLE
TDF's
• NEC 430-52(c)(1)
• TABLE 430-152
STARTER
CONTACTS
MBJ
GEC
GES
BJ
1
2
3
4
5
6
PART-WINDING
STARTER
PART-WINDING MOTOR
• 3Ø
• 230 V
• 40 HP
• DESIGN B
• 85% EFF.
• 90% PF.
NEC 430-7(a); (b)
NEC 430-6(a)(1)
NEC 430-22(a)
NEC 430-32(a)(1)
NEC 430-52(c)(1)
NEC 430-110(a)
NEC 430-3

Ill. 13-2

Torque Motors
430-6(b)

Torque motors are built to operate in the LRC stalled position, in this way exerting a torque without rotating. The locked rotor current (LRC) marked on the nameplate of the motor is used for all determinations.

Example 13-3. What size OCPD and conductors are required to supply power to the torque motor in Illustration 13-3?

Sizing OCPD

Step 1: Finding amps
430-6(b)
LRC = 120 A

Step 2: Selecting OCPD
430-52(b); 240-3(b); 240-6(a)
120 A requires 125 A OCPD

Solution: **The size OCPD is 125 amps.**

Sizing conductors

Step 1: Selecting conductors
430-6(b); Table 310-16
120 A requires #1 cu.

Solution: **The size conductors are No. 1 THWN cu. conductors.**

Locked Rotor Indicating Code Letters
430-7(b); Table 430-7(b)

Since code letters are not listed in Tables 430-151(A) and (B) and Table 430-152, their LRC in amps must be computed per NEC 430-7(b) and Table 430-7(b). For calculating procedure to derive LRC in amps, based on Code Letter, see Example 13-7 and Illustration 13-7 on page 13-14.

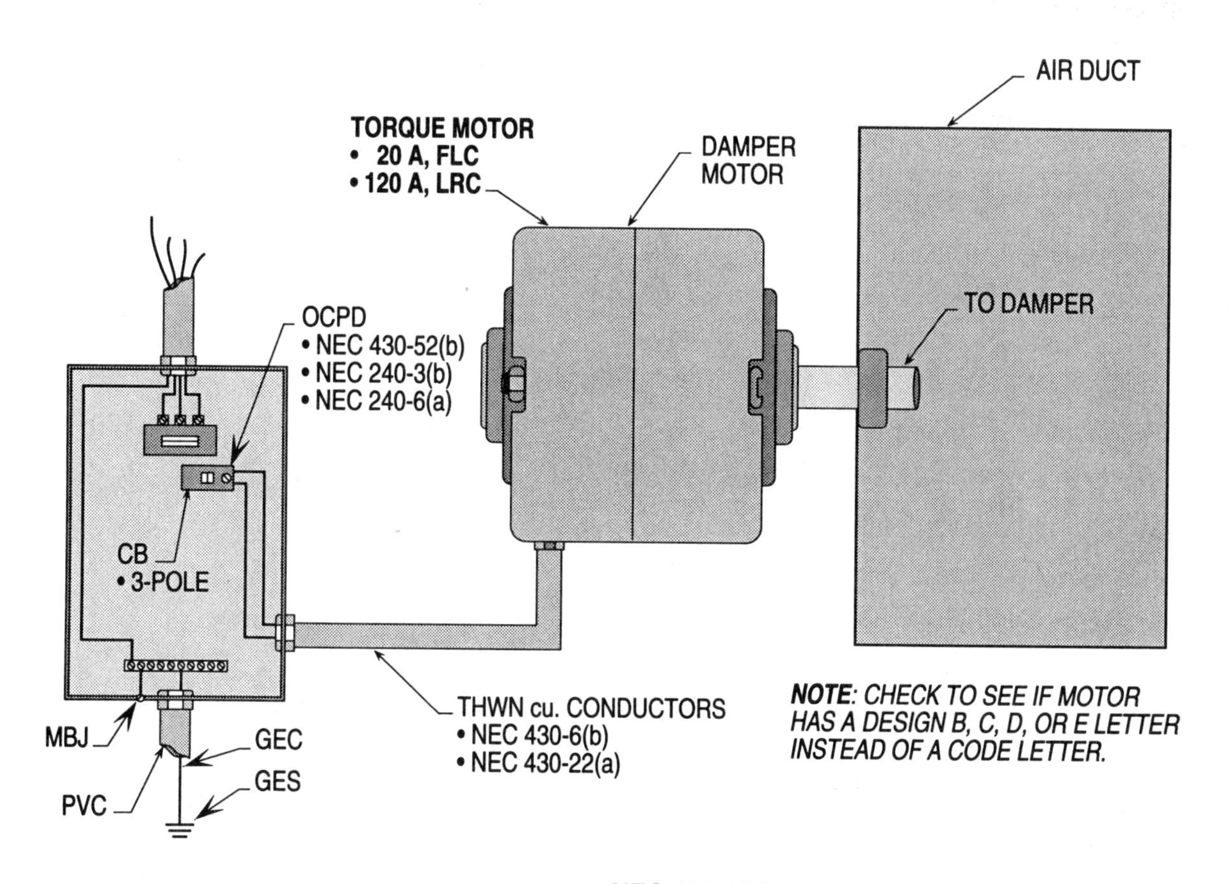

NEC 430-6(b)

Ill. 13-3

Wound-rotor Motors
430-23(a); 430-23(b); 430-23(c)

When sizing for continuous duty of a wound-rotor motor, the secondary conductors connecting between the motor and the controller must be sized at not less than 125 percent of the full-load secondary current of the motor in amps.

When sizing for other than continuous duty of a wound-rotor motor, and the motor is used for periodic duty, the secondary conductors are sized at not less than 125 percent of the secondary current. When the conductors are sized by the classification of service, for the correct percentages to apply, refer to Table 430-22(b).

When sizing a resistor that is separate from the controller of a wound-rotor motor, Table 430-23(c) requires the loads between the resistor bank and controller be sized according to the resistor duty classification.

Example 13-4. Using the following items size the THWN copper conductors required for the motor and resistor bank in Illustration 13-4?
1. The motor's power conductors
2. The motor's secondary conductors
3. The resistor bank's conductors

Sizing motor's power conductors

Step 1: Finding amperage
430-6(a)(1); Table 430-150
60 HP = 77 A

Step 2: Calculating amperage
430-23(a)
77 A x 125% = 96.25 A

Step 3: Selecting conductors
240-3(g); Table 310-16
96.25 A = #3 THWN cu. conductors

Solution: The size THWN cu. copper conductors are required to be No. 3.

Sizing motor's secondary conductors

Step 1: Finding amperage
430-6(a)(1); Table 430-150
FLC = 38.5 A

Step 2: Calculating amperage
430-23(a)
38.5 A x 125% = 48.13 A

Step 3: Selecting conductors
310-10(2); Table 310-16
48.13 A = #8 THWN cu. conductors

Solution: The size THWN cu. copper cond uctors are required to be No. 8.

Sizing resistor bank conductors

Step 1: Finding amperage
430-6(a)(1); Table 430-150
FLC = 38.5 A

Step 2: Calculating amperage
430-23(c)
38.5 A x 110% = 42.35 A

Step 3: Selecting conductors
310-10(2); Table 310-16
42.35 A = #8 THWN cu. conductors

Solution: The size THWN cu. copper conductors are required to be No. 8.

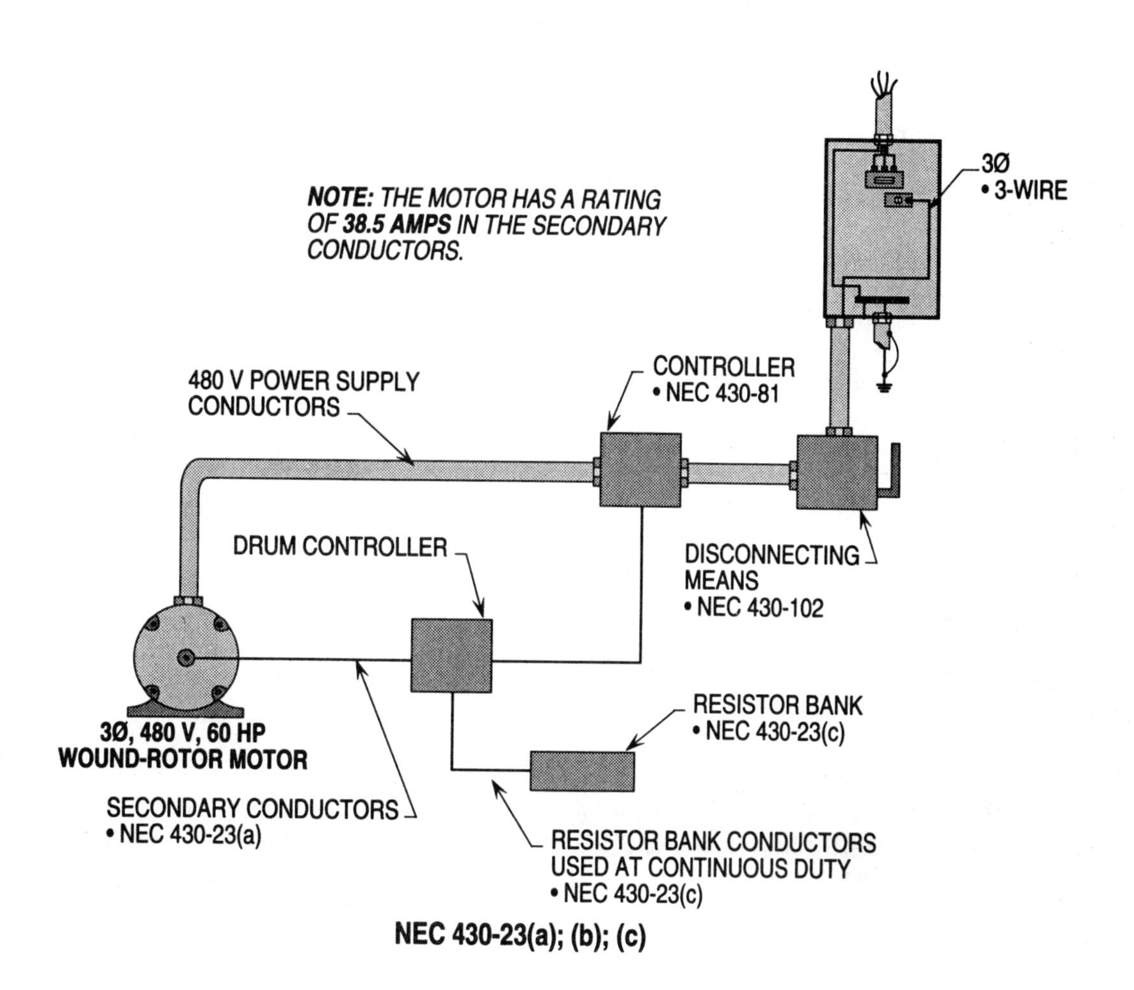

NEC 430-23(a); (b); (c)

Ill. 13-4

DC Motors
430-29

Conductors connecting a motor controller to power accelerating and dynamic braking resistors in the armature circuit of DC motors are required to be sized by the percentages listed in Table 430-29.

Example 13-5(a). What size conductors are required for the dynamic breaker resistor bank mounted separately from the controller and DC motor in Illustration 13-5(a)?

Step 1: Calculating load of power resistor
430-29; Table 430-29
55 A x 55% = 30.25 A

Step 2: Selecting conductors
430-29; Table 310-16
30.25 A requires #10 cu.

Solution: The size conductors are No. 10 THWN copper.

Example 13-5(b). What size conductors are required for the dynamic breaker resistor bank mounted separately from the controller and DC motor in Illustration 13-5(b)?

Step 1: Calculating load of power resistor
430-29; Table 430-29
55 A x 75% = 41.25 A

Step 2: Selecting conductors
430-29; Table 310-16
41.25 A requires #8 cu.

Solution: The size conductors are No. 8 THWN copper.

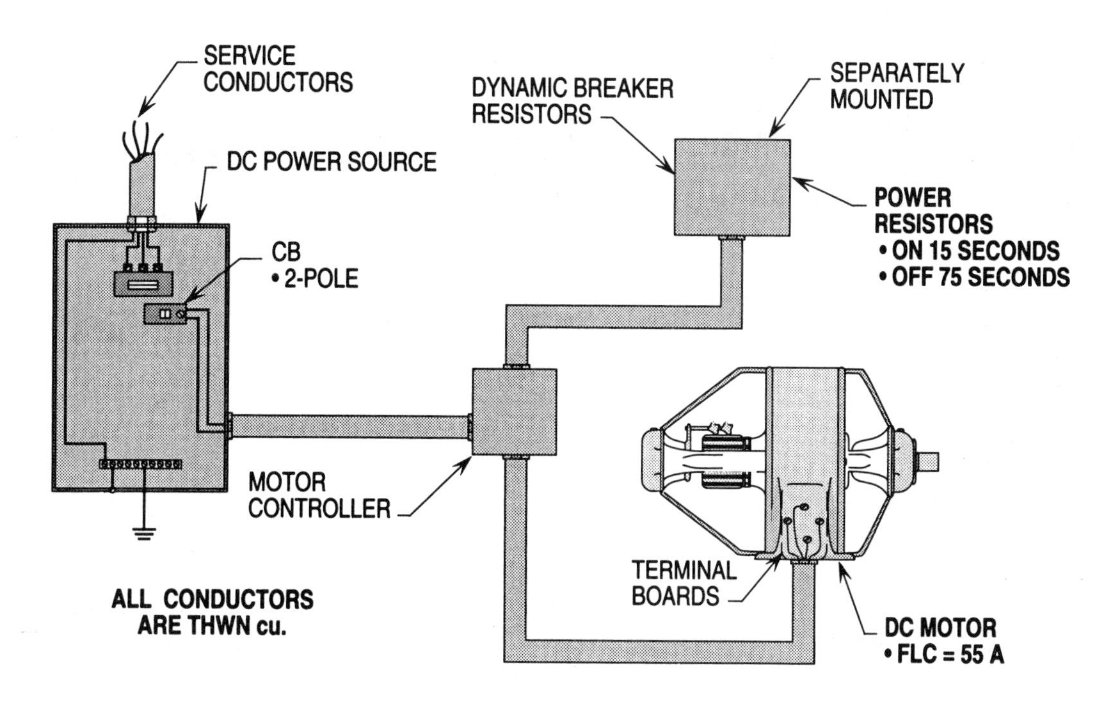

Ill. 13-5(a)

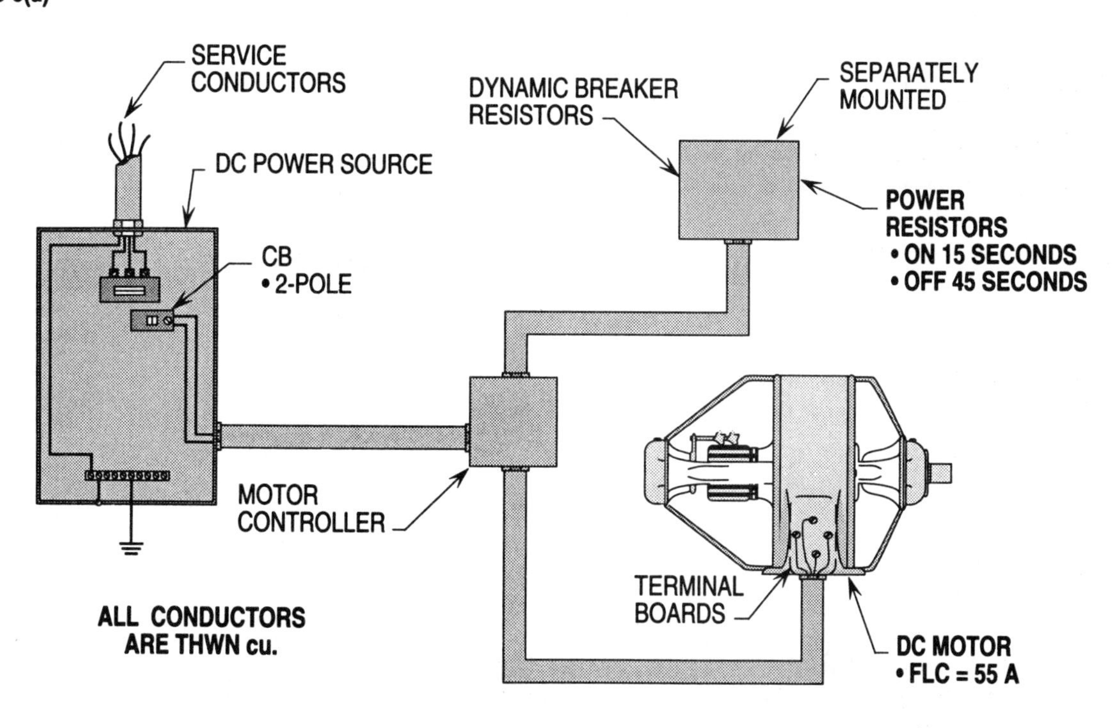

Ill. 13-5(b)

DC Motors
430-29; 430-22(a)

The conductors supplying power to a DC motor must be sized at 125 percent of FLC of the motor. OCPD's shall be sized to carry the starting current of the motor.

Example 13-6. What size OCPD and conductors are required for the DC motor in Illustration 13-6?

Sizing OCPD

Step 1: Calculating A to size OCPD
430-52(c)(1)
LRC based on arm. = 240 V / .5 (1/2 ohm)
LRC based on arm. = 480 A

Step 2: Calculating OCPD
Table 430-147; 430-52(c)(1);
Table 430-152
55 A x 150% = 82.5 A

Step 3: Selecting OCPD
240-3(g); 240-6(a);
430-52(c)(1), Ex. 1
82.5 A requires 90 A OCPD

Step 4: Verifying starting of motor
- CB must hold 480 A starting current of motor ÷ 3 (CB holds about 3 times its rating) = 160 A
- 160 A requires 175 A CB
- 175 A CB holds 525 A (175 A x 3 = 525 A)
- 525 A holds 480 A of LRC

Step 5: Applying max. OCPD
430-52(c)(1), Ex. 2(c)
Max. = 55 A x 400% = 220 A

Solution: A 200 amp OCPD may be used. However, a 175 amp OCPD will allow the motor to start and run per NEC 430-52(c)(1), Ex. 2(c).

Sizing conductors

Step 1: Finding FLA
430-6(a)(1); Table 430-147
15 HP = 55 A

Step 2: Calculating load
430-22(a)
55 A x 125% = 68.75 A

Step 3: Selecting conductors
310-10(2); Table 310-16
68.75 A requires #4 cu.

Solution: The size THWN copper conductors are No. 4 cu.

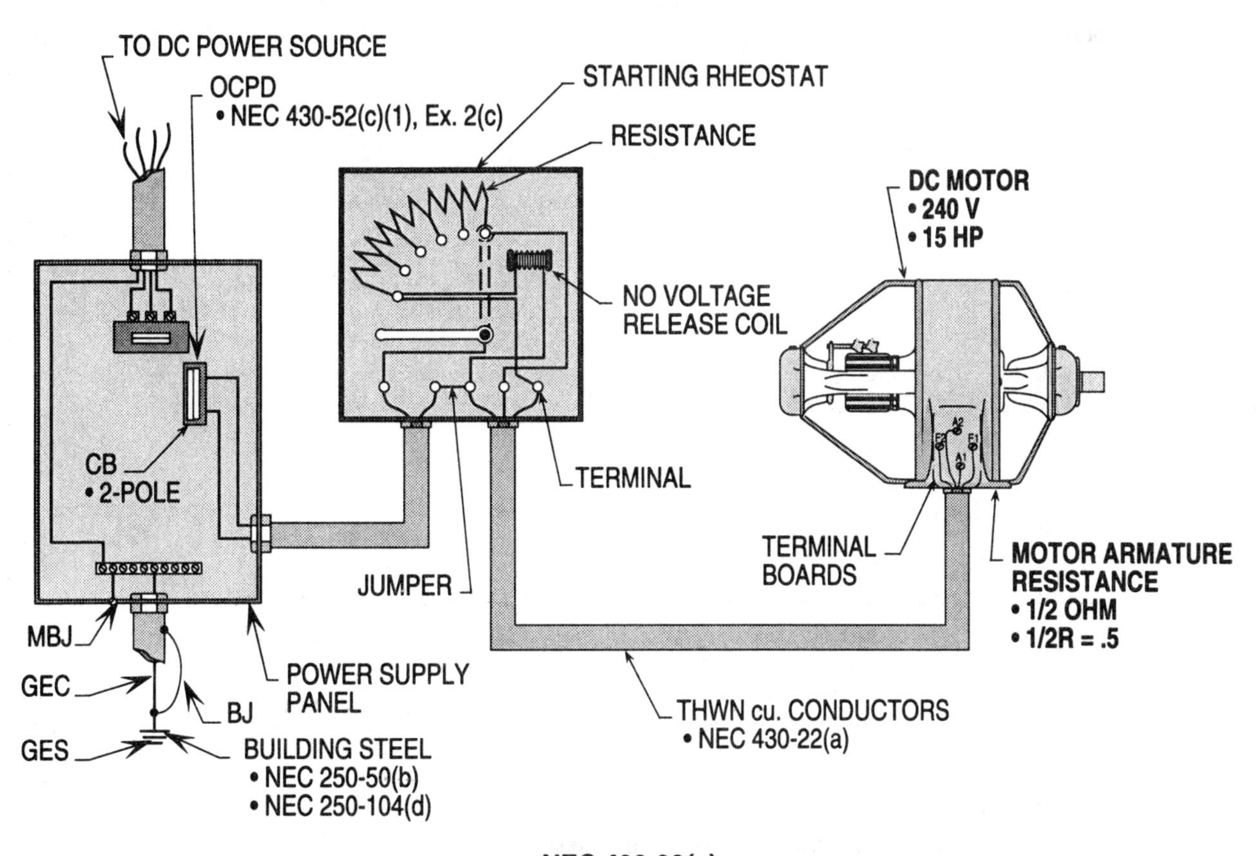

NEC 430-22(a)
TABLE 430-147

Ill. 13-6

Locked-rotor Indicating Code Letters Table 430-7(b); Table 430-152

For motors with code letters, the locked rotor current in amps must be found by using the kVA per the HP values found in Table 430-7(b), based upon the code letter on the motor's nameplate. Note that the code letters are no longer listed in the Table 430-152. Such code letters can be used per NEC 430-7(b) to determine the actual starting current of a motor. Code letters A through H (min.) are used for motors starting and running under normal conditions. Locked-rotor current of motors listed in Tables 430-151(A) or (B) are calculated at approximately six times the FLC of the motor, per Tables 430-148 and 430-150 respectively for most motors.

Example 13-7. What is the locked-rotor current for the motor in Illustration 13-7?

Finding locked-rotor current

Step 1: Finding kVA multiplier
430-7(b); Table 430-7(b)
Code letter G = 6.29 kVA

Step 2: Applying formula

$$LRC = \frac{\text{kVA per HP} \times \text{HP} \times 1{,}000}{V \times 1.732}$$

$$LRC = \frac{6.29 \times 50 \times 1{,}000}{240 \times 1.732}$$

LRC = 756 A

Solution: The locked-rotor current is 756 amps.

Note 1: Code letters are used to find locked rotor currents only if they are listed on the motor's nameplate. Table 430-152 of the 1996 NEC no longer lists code letters. The motor's design is used,instead of the code letter. See NEC 430-7(a)(9) and 430-7(b) for more details concerning these requirements.

Note 2: The LRC in amps for motors with code letters cannot be found in Tables 430-151(A) and (B). The LRC in amps for these motors must be calculated per Table 430-7(b) as shown in Example 13-7.

ADDITIONAL HELP 56:

Calculating the LRC in amps if the motor in Illustration 13-7 had a code letter B.

- Setting up formula to find LRC

$$LRC = \frac{\text{kVA per HP} \times \text{HP} \times 1{,}000}{V \times 1.732}$$

$$LRC = \frac{3.54 \text{ kVA} \times 50 \text{ HP} \times 1{,}000}{240 \text{ V} \times 1.732}$$

LRC = 425.5 A

Solution: The LRC in amps is 425.5.

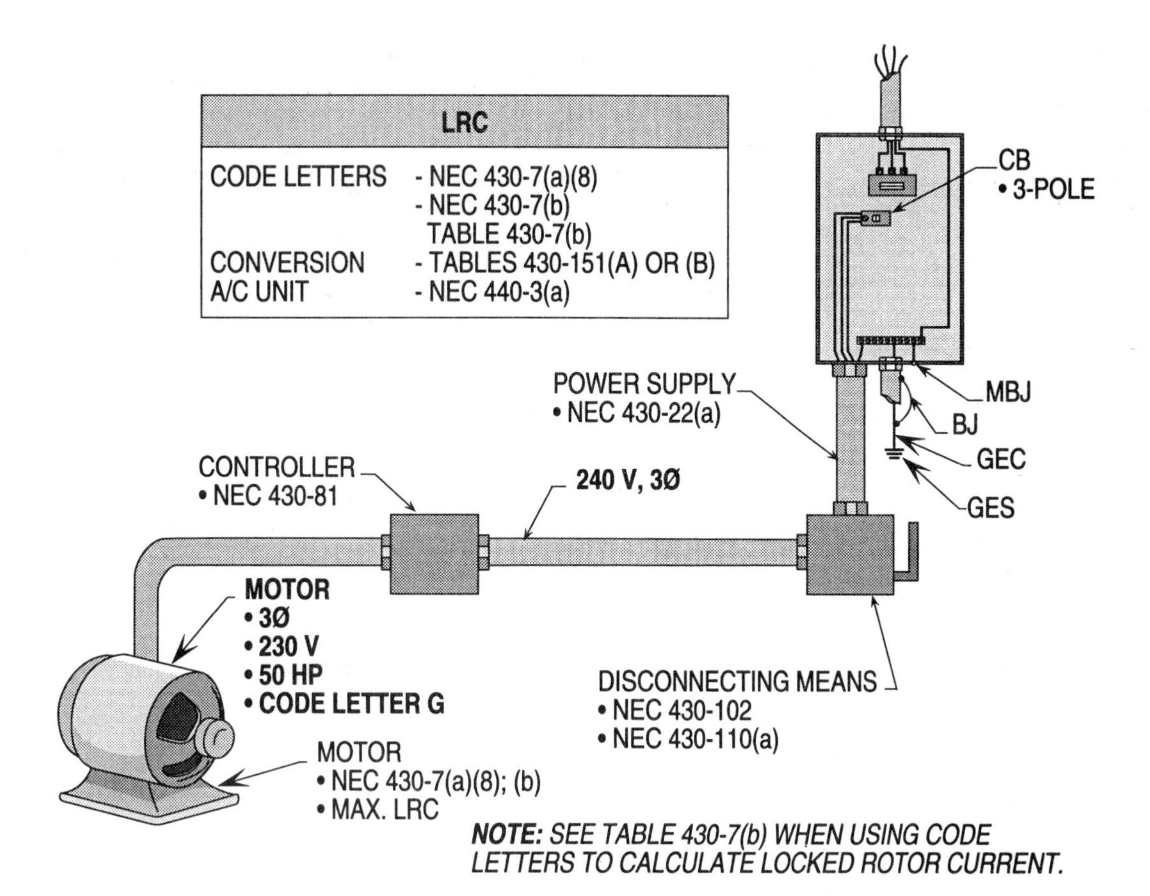

Ill. 13-7

ADDITIONAL HELP 57:

Calculating the LRC in amps for the above motor using the rule of thumb method instead of performing the actual calculation procedure based on code letter in the Example Problem 13-7.

- Applying rule of thumb method
  Code letters A through G
  Table 430-150
  230 V, 3Ø, 50 HP = 130 A

- Calculating LRC in A
  130 A x 6 = 780 A

Solution: The LRC in amps using Rule of Thumb method which is 6 times FLA of Table 430-148 for single-phase motors and Table 430-150 for three-phase motors.

Duty Cycle
Table 430-22(b)

When sizing the classification of service for a motor, the conductors are sized from the percentage of nameplate current rating in amps listed in Table 430-22(b). Conductors are sized for supplying individual motors used for short-time, intermittent, periodic, or varying duty per Table 430-22(b). See Article 100 for definitions of the different duties.

Example 13-8. What is the size THWN copper conductors required to supply the intermittent-duty motor in Illustration 13-8?

Sizing conductors

Step 1: Finding amperage
430-6(a)(1); Table 430-150
100 HP motor = 124 A

Step 2: Finding percentage
430-22(b); Table 430-22(b)
30 minute rated = 90%

Step 3: Calculating amperage
Table 430-150; Table 430-22(b)
124 A x 90% = 112 A

Step 4: Selecting conductors
310-10(2); Table 310-16
112 A = #2 THWN cu. conductors

Solution: No. 2 THWN cu. conductors are required.

Note: For power conversion equipment, see page 13-2 of this chapter.

ADDITIONAL HELP 57:

Calculating the load in amps for the intermittent-duty motor used as a continuous rated motor.

- Selecting percentage
  430-22(b); Table 430-22(b)
  percentage = 140%

- Calculating amp
  430-22(b); Table 430-22(b)
  124 A x 140% = 173.6 A

Solution: The load is 173.6 amps.

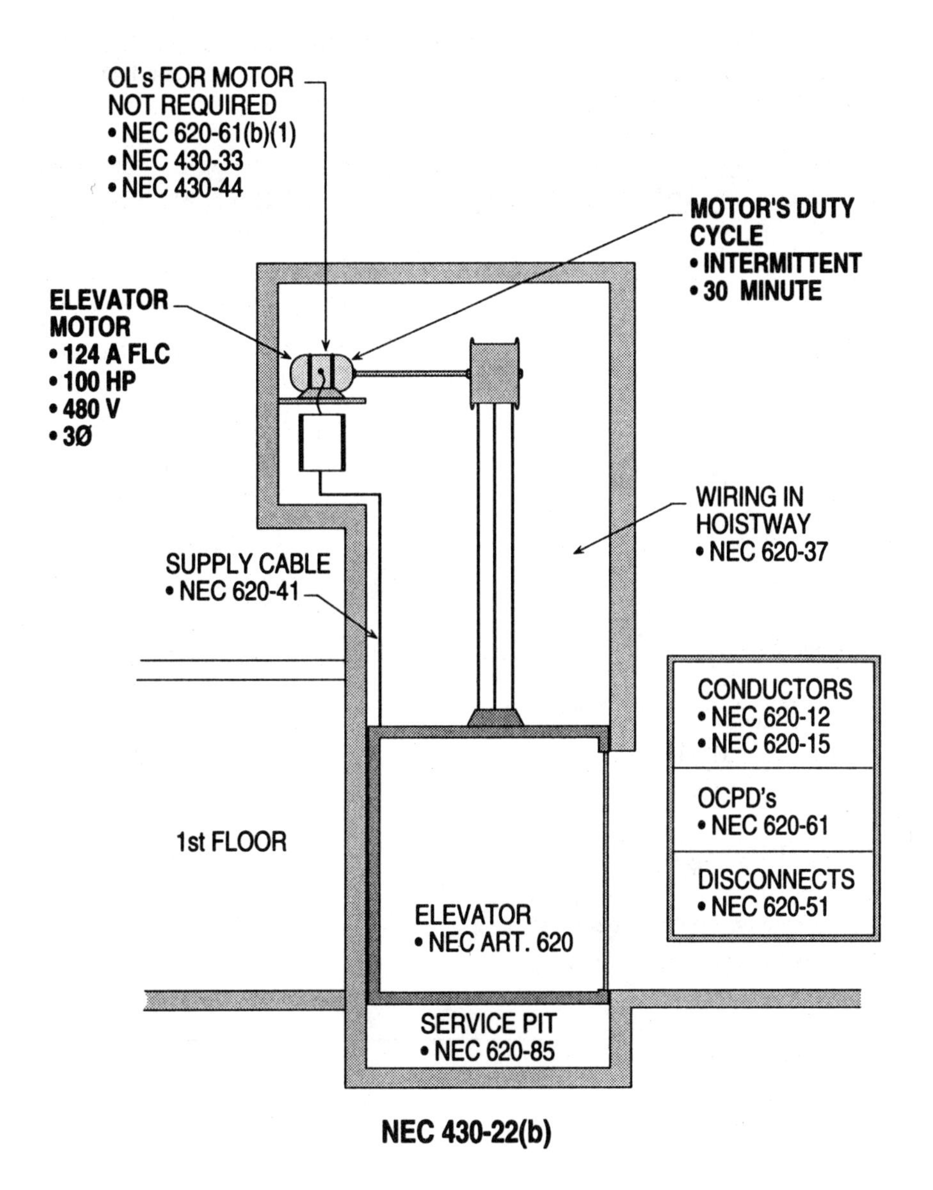

NEC 430-22(b)

Ill. 13-8

Duty-cycle Service 430-22(b)

Motors that operate on short-time duty cycles, with periods of rest in between may have their conductors calculated at a percent less than 125 percent per Table 430-22(b). This permissive rule is due to the conductors having a chance to cool during periods of driving the load.

Example 13-9. What are the size THHN copper conductors required to supply power to the motor in Illustration 13-9?

Sizing conductors

Step 1: Finding FLA
430-22(b); Table 430-150; 430-6(a)(1)
100 HP = 124 A

Step 2: Calculating load
430-22(b); Table 430-22(b)
124 A x 85% = 105.4 A

Step 3: Selecting conductors
310-10(2); Table 310-16
105.4 A requires #2 cu.

Solution: The size THHN copper conductors are No. 2.

Note: For power conversion equipment, see page 13-2 of this chapter.

ADDITIONAL HELP 58:

Calculating the amps for the motor in Illustration 13-9 if it was used as a continuous rated motor on an industrial machine at varying duty operation.

- Selecting percentage
  430-22(b); Table 430-22(b)
  percentage = 200%

- Calculating amps
  430-22(b); Table 430-22(b); Step 1 above
  124 A x 200% = 248

Solution: The load is 248 amps.

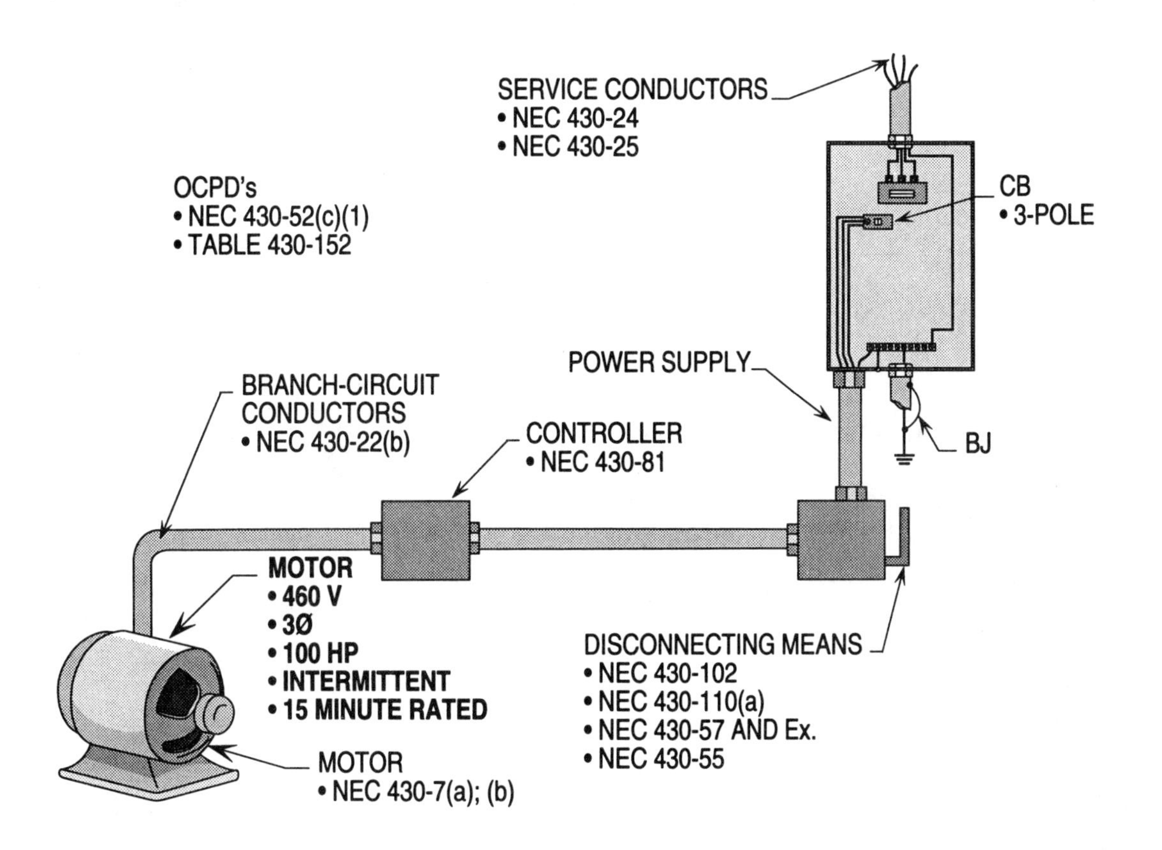
SERVICE CONDUCTORS
• NEC 430-24
• NEC 430-25
OCPD's
• NEC 430-52(c)(1)
• TABLE 430-152
CB
• 3-POLE
POWER SUPPLY
BRANCH-CIRCUIT CONDUCTORS
• NEC 430-22(b)
CONTROLLER
• NEC 430-81
BJ
MOTOR
• 460 V
• 3Ø
• 100 HP
• INTERMITTENT
• 15 MINUTE RATED
DISCONNECTING MEANS
• NEC 430-102
• NEC 430-110(a)
• NEC 430-57 AND Ex.
• NEC 430-55
MOTOR
• NEC 430-7(a); (b)

Ill. 13-9

Single Motor Load
430-22(a)

The conductors supplying power to a single motor must be calculated at 125 percent of the FLA of the motor per Table 430-148 for single-phase and Table 430-150 for three-phase. The ampacity of the conductors increased by 125 percent protects the insulation of the conductors from overloads.

Example 13-10. What size conductors are required to supply power to the motors in Illustration 13-10?

Single-phase motors

Step 1: Finding FLA
430-6(a)(1); Table 430-148
2 HP = 13.2 A

Step 2: Calculating load
430-22(a)
13.2 A x 125% = 16.5 A

Step 3: Selecting conductors
310-10(2); Table 310-16
16.5 A requires #14 cu.

Solution: **The size THWN copper conductors are No. 14.**

Three-phase motors

Step 1: Finding FLA
430-6(a)(1); Table 430-150
25 HP = 74.8 A

Step 2: Calculating load
430-22(a)
74.8 A x 125% = 93.5 A

Step 3: Selecting conductors
310-10(2); Table 310-16
93.5 A requires #3 cu.

Solution: **The size THWN copper conductors are No. 3.**

Note: For power conversion equipment, see page 13-2 of this chapter.

ADDITIONAL HELP 59:

Calculating amps when HP is known.

- Finding HP
  Ill. 13-10
  HP = 25

- Calculating HP
  Chart

$$A = \frac{HP \times 746\ W}{1.73 \times E \times Eff. \times PF}$$

$$A = \frac{25\ HP \times 746\ W}{1.73 \times 460\ V \times 85\% \times 90\%}$$

$$A = \frac{18{,}650}{608.787}$$

$$A = 30.6$$

Solution: The amps for the 25 HP motor, based on formula, is 30.6 amps.

Note: The constant 746 W is used for 1 HP.

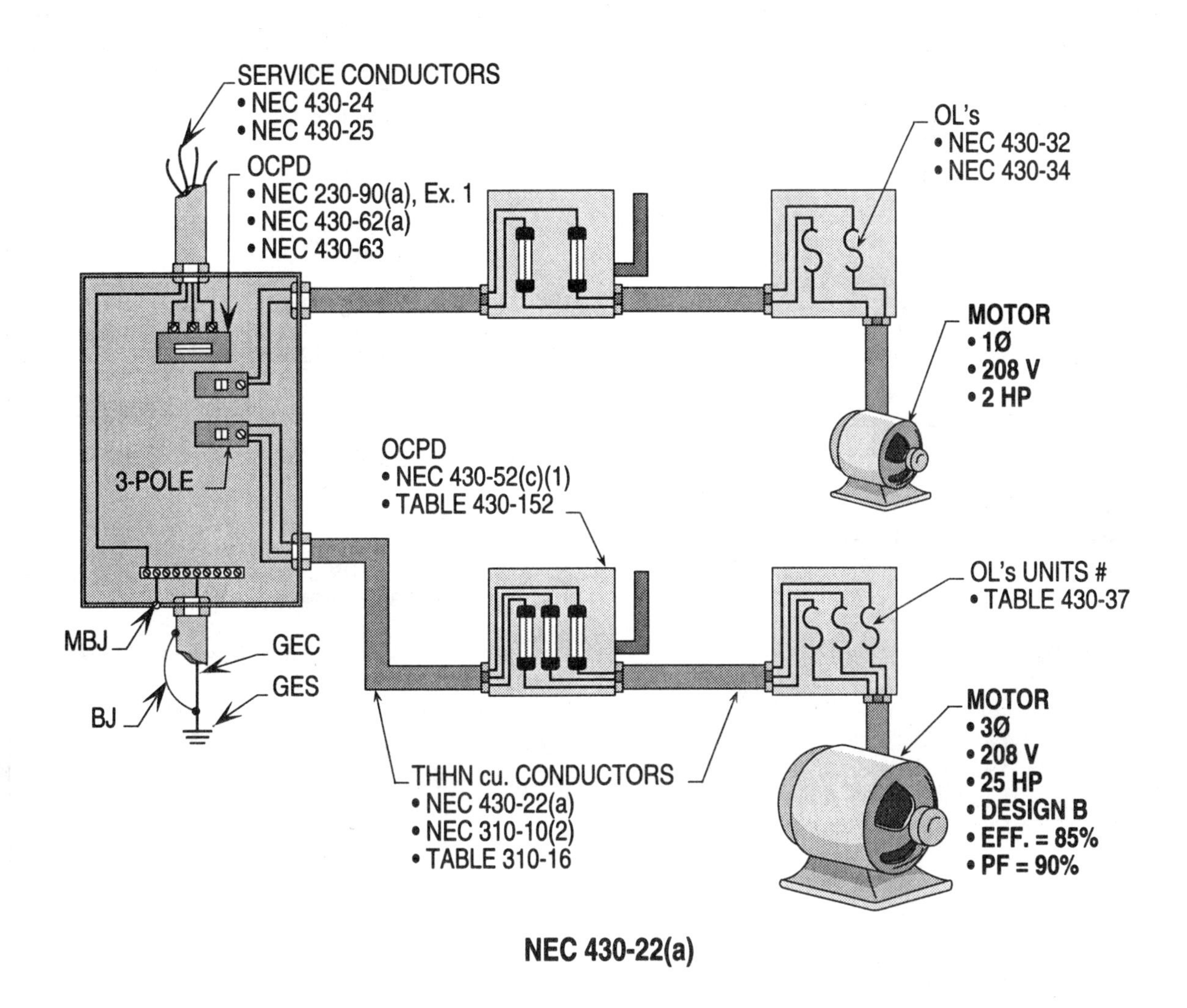

NEC 430-22(a)

Ill. 13-10

Multispeed Motors
430-22(a)

The full-load current (FLC) rating in amps for multispeed motors is taken from the motor's nameplate rather than the Tables. The branch-circuit conductors is based upon the highest current rating listed on the motor's nameplate. The conductors between the controller and motor are sized according to the current rating for each speed.

Example 13-11. What size branch-circuit and multispeed THWN copper conductors are required to supply power to the multispeed motor in Illustration 13-11?

Sizing branch-circuit conductors

Step 1: Finding FLA
430-22(a)
45 A largest amperage

Step 2: Calculating load
430-22(a)
45 A x 125% = 56.25 A

Step 3: Selecting conductors
310-10(2); Table 310-16
56.25 A requires #6 cu.

Solution: The size THWN copper conductors are No. 6.

Sizing multispeed conductors

Step 1: Finding FLA
430-22(a)
3,420 RPM = 27 A
2,280 RPM = 35 A
1,710 RPM = 45 A

Step 2: Calculating load
430-22(a)
27 A x 125% = 33.75 A
35 A x 125% = 43.75 A
45 A x 125% = 56.25 A

Step 3: Selecting conductors
310-10(2); Table 310-16
33.75 A requires #10 cu.
43.75 A requires # 8 cu.
56.25 A requires # 6 cu.

Solution: The size THWN copper conductors are No. 10 cu., No. 8 cu., and No. 6 cu. sized for each speed.

ADDITIONAL HELP 60:

Calculating the HP output for the motor in Ill. 13-11.

- Finding I, E, Eff.
  I = 27 A
  E = 208 V, 3Ø
  EFF. = 85%

- Calculating HP output of motor

$$HP = \frac{I \times 1.73 \times E \times Eff. \times PF}{746\ W}$$

$$HP = \frac{27\ A \times 1.73 \times 208\ V \times 85\% \times 80\%}{746\ W}$$

$$HP = \frac{6{,}606.6624}{746\ W}$$

$$HP = 8.856$$

Solution: The HP output for the 27 amp motor turning at 3,420 RPM's is 8.856.

Note: The constant 746 W is used for 1 HP.

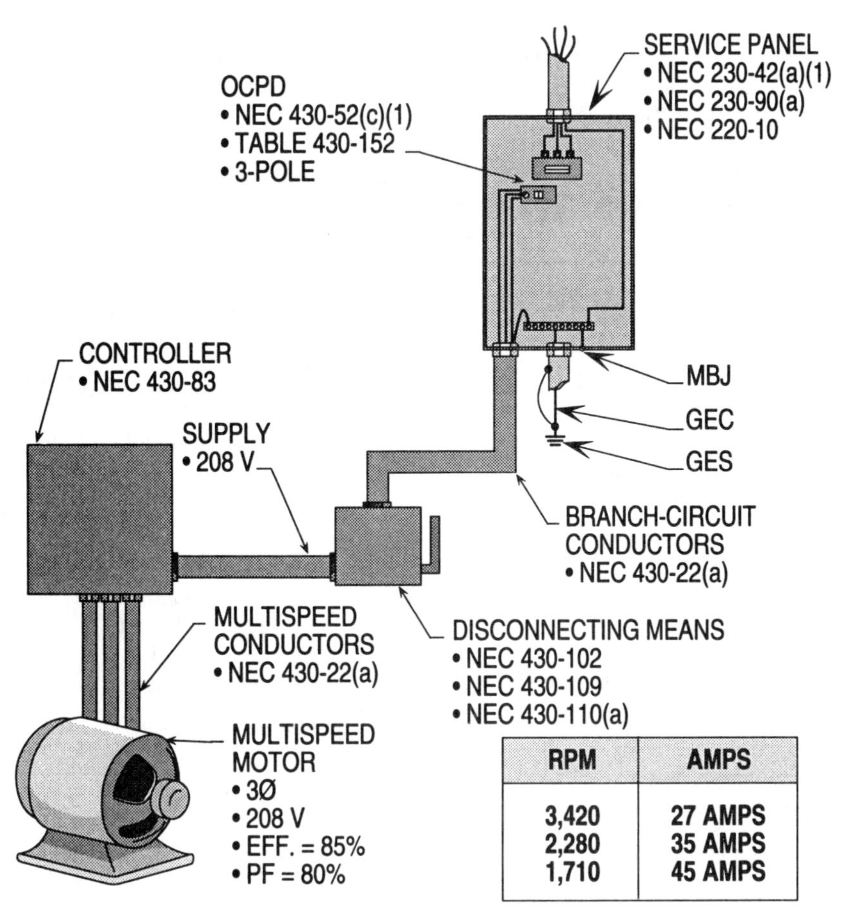

RPM	AMPS
3,420	27 AMPS
2,280	35 AMPS
1,710	45 AMPS

NEC 430-22(a)

Ill. 13-11

ADDITIONAL HELP 61:

Calculating the torque in pound-feet for the motor turning at 3,420 RPM.

- Setting up formulas

$$\text{Torque} = \frac{\text{HP x 5,252}}{\text{RPM}}$$

$$\text{Torque} = \frac{\text{8.856 HP x 5,252}}{\text{3,420 RPM}}$$

$$\text{Torque} = \frac{46{,}511.712}{\text{3,420 RPM}}$$

Torque = 13.59 pound-ft.

Solution: Based on the HP from Additional Help in Example 13-16 the torque in pound-feet is 13.59.

Note: The constant 5,252 is found by dividing 33,000 foot-pounds per minute by 6.2831853, which is found by multiplying π (3.14159265) by 2 (33,000 ÷ 6.2831853 = 5,252).

Multimotor Branch-circuits 430-24

When sizing the conductors for a feeder supplying two or more motors, the full-load current rating in amps of the largest motor is multiplied by 125 percent. The full-load current rating of the remaining motors are added to this total to derive the FLA for sizing the components of the circuit.

If one of the motor's is classified short-time, intermittent, periodic, or varying duty, the motor's full-load current rating in amps can be reduced by the percentages per Table 430-22(b).

Example 13-12(a). What is the size THWN copper feeder required to supply the motors in Illustration 13-12(a)?

Sizing conductors

Step 1: Finding amperage
430-6(a)(1); Table 430-150
40 HP = 52 A
50 HP = 65 A
60 HP = 77 A

Step 2: Calculating amperage
430-24

77 A x 125%	=	96.25 A
	=	65 A
	=	52 A
Total load	=	213.25 A

Step 3: Selecting conductors
310-10(2); Table 310-16
213.25 A = #4/0 THWN cu.

Solution: The size THWN cu. copper conductors are required to be No. 4/0.

ADDITIONAL HELP 62:

Calculating the speed of the 60 HP motor in Ill. 13-12(a).

- Setting up formula

$$\text{RPM} = \frac{\text{F x 120}}{\text{Poles}}$$

$$\text{RPM} = \frac{60 \text{ x } 120}{4}$$

RPM = 1,800

Solution: The RPM of the motor is 1800.

Example 13-12(b). Determine the size THWN copper conductors required to supply the motors in Illustration 13-12(b)?

Sizing conductors

Step 1: Finding amperage
430-6(a)(1); Table 430-150
5 HP = 15.2 A
7 1/2 HP = 22 A
10 HP = 28 A

Step 2: Calculating amperage
430-24; Table 430-22(b)

28 A x 125%	=	35 A
22 A x 85%	=	18.7 A
15.2 A x 85%	=	12.9 A
Total load	=	66.6 A

Step 3: Selecting conductors
Table 310-16
66.6 A = #4 THWN cu.

Solution: The size THWN cu. conductors are required to be No. 4.

ADDITIONAL HELP 63:

Calculating the number of poles for the 60 HP motor in the Additional Help problem above.

- Setting up formula

$$\text{Poles} = \frac{\text{F x 120}}{\text{RPM}}$$

$$\text{Poles} = \frac{60 \text{ x } 120}{1{,}800}$$

Poles = 4

Solution: The number of poles are 4.

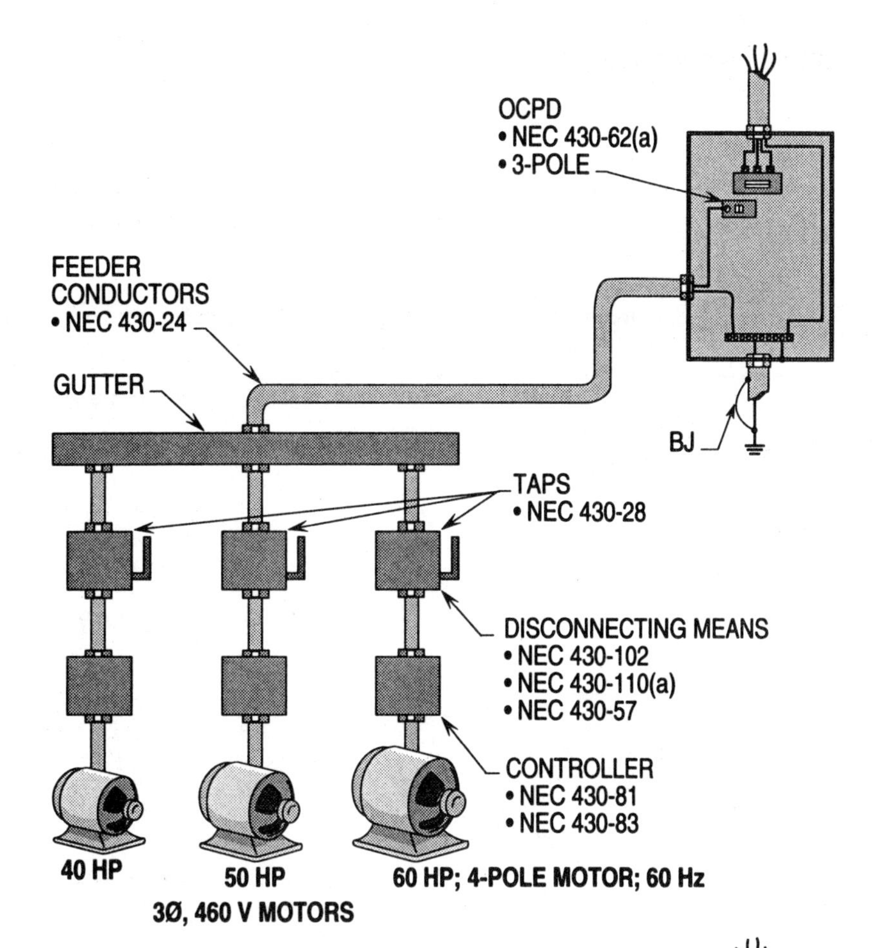

Ill. 13-12(a)

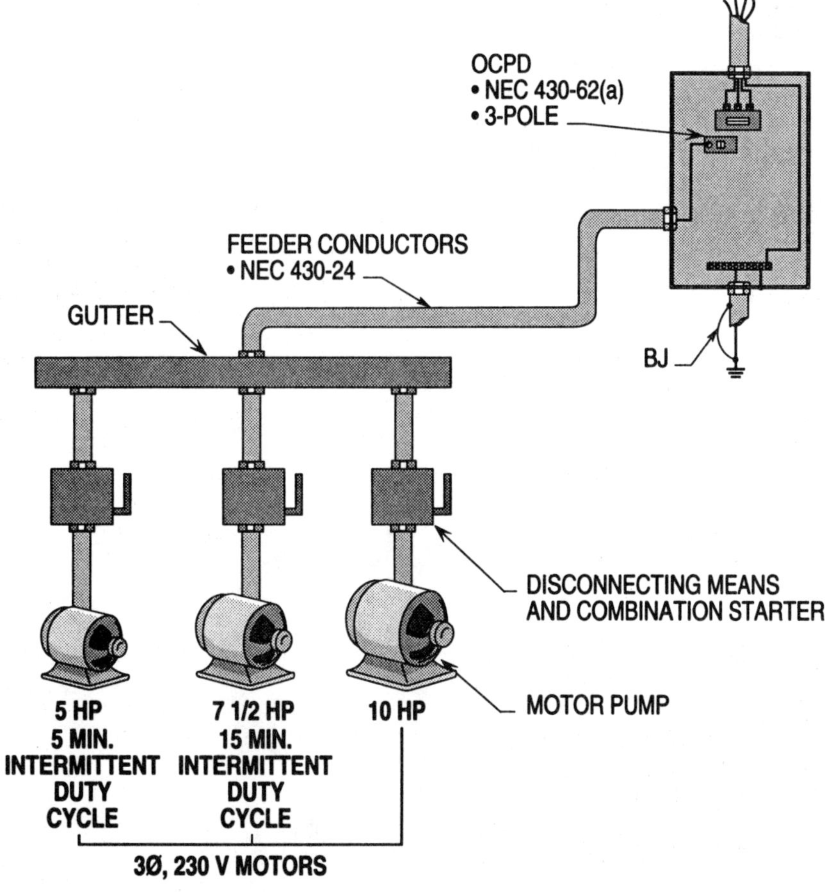

Ill. 13-12(b)

SEVERAL MOTORS ON A FEEDER
430-24

Feeder-circuit conductors supplying several motors must be sized to carry 125 percent of the FLC rating of the highest rated current in amps for any one motor plus the sum of the FLA ratings of all remaining motors on the circuit.

Example 13-13. What size feeder-circuit conductors are required to supply power to the motors in Illustration 13-13?

Step 1: Finding FLA
430-6(a)(1); Table 430-150
10 HP = 14 A
20 HP = 27 A
25 HP = 34 A
30 HP = 40 A

Step 2: Calculating load
430-24
40 A x 125% = 50 A
= 34 A
= 27 A
= 14 A
Total load = 125 A

Step 3: Selecting conductors
310-10(2); Table 310-16
125 A requires #1 cu.

Solution: The size THHN copper conductors are No. 1.

ADDITIONAL HELP 64:

Calculating the efficiency of the 30 HP motor in Ill. 13-13 where the motor draws 24,640 watts of power.

- Setting up formula
  output = HP x 746 W
  output = 30 HP x 746 W
  output = 22,380 W

- Calculating Eff.

$$\text{Eff.} = \frac{\text{Output}}{\text{Input}}$$

$$\text{Eff.} = \frac{22{,}380}{24{,}640}$$

Eff. = 90%

Solution: The efficiency of the 30 HP motor is about 90 percent.

Note: The constant 746 W is used for 1 HP.

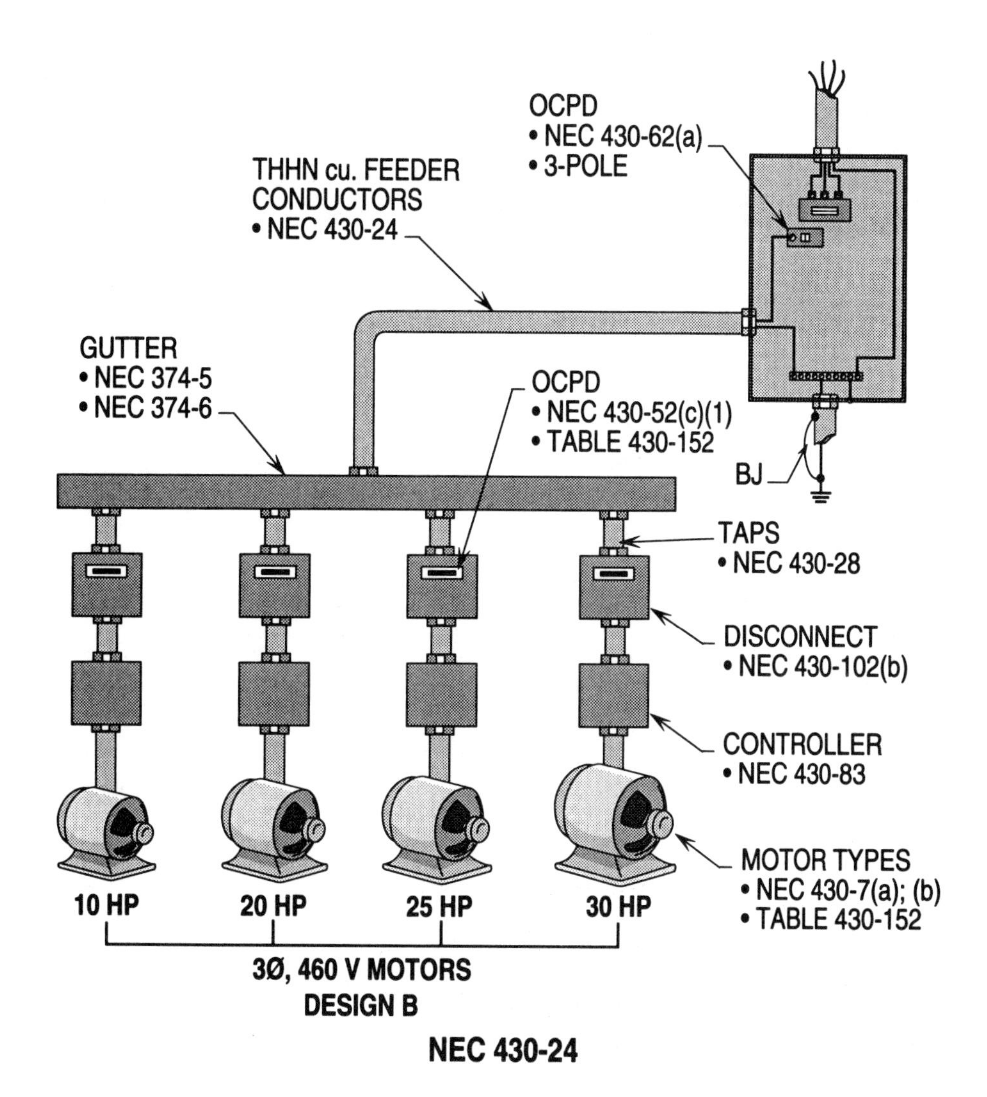
OCPD
• NEC 430-62(a)
• 3-POLE
THHN cu. FEEDER
CONDUCTORS
• NEC 430-24
GUTTER
• NEC 374-5
• NEC 374-6
OCPD
• NEC 430-52(c)(1)
• TABLE 430-152
BJ
TAPS
• NEC 430-28
DISCONNECT
• NEC 430-102(b)
CONTROLLER
• NEC 430-83
MOTOR TYPES
• NEC 430-7(a); (b)
• TABLE 430-152
10 HP
20 HP
25 HP
30 HP
3Ø, 460 V MOTORS
DESIGN B
NEC 430-24

Ill. 13-13

Selecting Largest Motor Based Upon Duty Cycle 430-24, Ex. 1

Feeder-circuit conductors supplying power to two or more motors that are utilized to serve duty cycle loads must have the largest motor selected based upon their conditions of use.

Example 13-14. What is the largest motor load and size THWN copper conductors required to supply power to the motors in Illustration 13-14?

Finding largest motor load

Step 1: Applying demand factors
430-24, Ex. 1; 430-22(b); Table 430-22(b)
100 A x 85% = 85 A
90 A x 110% = 99 A
80 A x 150% = 120 A

Solution: The largest motor load is 120 amps.

Sizing conductors

Step 1: Calculating load
430-24, Ex. 1

Largest motor load	=	120 A
Plus others	=	99 A
	=	85 A
Total load	=	304 A

Step 2: Selecting conductors
310-10(2); Table 310-16
304 A requires #350 KCMIL cu.

Solution: The size THWN copper conductors are No. 350 KCMIL.

ADDITIONAL HELP 65:

If the intermitten motor in Ill. 13-14 pulls 90 amps at 460 volts with an efficiency of 85 percent, how much power in watts is lost?

- Setting up watt formulas

$$\text{Watts} = \frac{90\text{ A} \times 460\text{ V}}{80\%}$$

Watts = 51,750

- Setting up power lost formula
  90 A x 460 V = 41,400 W
  51,750 W - 41,400 W = 10,350 W

Solution: The Watts lost is about 10,350.

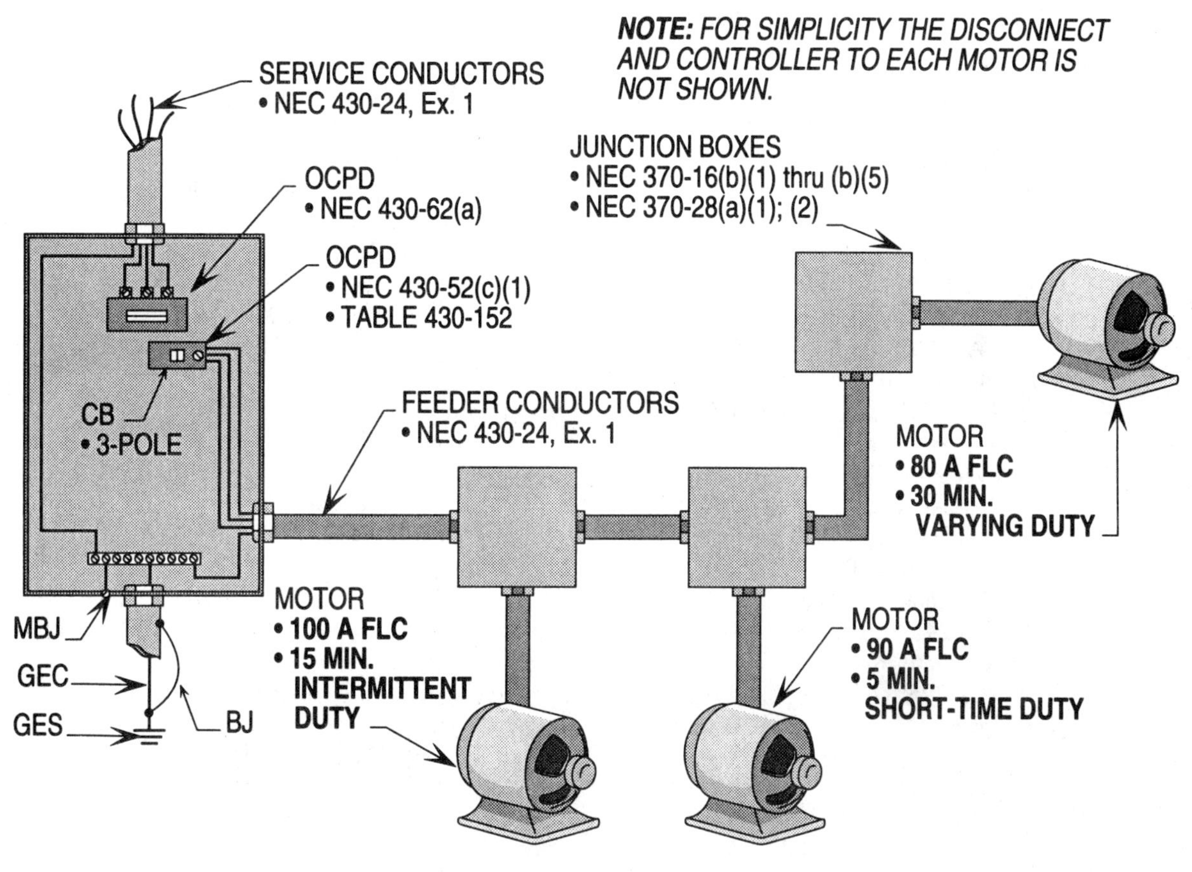

NEC 430-24, Ex. 1

Ill. 13-14

Motor Branch-circuit Conductors - Combination Loads 430-25

When sizing conductors for one or more motor loads with combination loads on the same circuit, the motor load is calculated per NEC 430-22(a) or 430-24. Other loads are calculated per Article 220 and other applicable Articles. Total all loads for the ampacity required for the component in the circuit.

Example 13-15. What is the size OCPD and THWN copper conductors required for the feeder-circuit in Illustration 13-15?

Sizing OCPD based on loads

Step 1: Calculating loads
215-2(a); 215-3; 220-10;
Table 430-148; 220-14

- Lighting load
  40 A x 125% = 50.0 A
- Receptacle load
  37.5 A x 100% = 37.5 A
- Appliance load
  30 A x 125% = 37.5 A
- Heat or A/C load
  10 kVA x 1,000 x 100% / 240 V = 42.0 A
- Motor load
  5 HP = 28 A x 100% = 28.0 A
  1 HP = 16 A x 100% = 16.0 A
  2 HP = 24 A x 100% = 24.0 A
- Largest motor load
  28 A x 25% = 7.0 A

Total load = 242.0 A

Step 2: Selecting OCPD based on load
215-3; 240-3(b); 430-63; 240-6(a)
242 A allows 250 A OCPD

Solution: The size OCPD based upon calculated load is 250 amps.

Sizing OCPD based on motor load

Step 1: Calculating OCPD
430-52(c)(1); 430-62(a); 240-3(g);
Table 430-152
Motor loads

- 5 HP = 28 A x 250% = 70.0 A
- 1 HP = 16 A x 100% = 16.0 A
- 2 HP = 24 A x 100% = 24.0 A

Step 2: Other loads

- Lighting load = 50.0 A
- Receptacle load = 37.5 A
- Appliance load = 37.5 A
- Heating load = 42.0 A

Total load = 277.0 A

Step 3: Selecting OCPD
430-62(a)
250 A OCPD is the next size below 277 A

Solution: The size OCPD based upon motor loads is 250 amps.

Note: In most cases the calculated load produces the largest or same size OCPD unless there is a unusually large motor involved.

Sizing OCPD based on computed load, per NEC 430-63. (See Illustration 13-15 on page 13-32).

Sizing OCPD and conductors

Step 1: Calculating load
215-2(a); 220-10

• Lighting load		
40 A x 125%	=	50.0 A
• Receptacle load		
37.5 A x 100%	=	37.5 A
• Appliance load		
30 A x 125%	=	37.5 A
• Heat load		
42 A x 100%	=	42.0 A
• Motor loads		
5 HP = 28 A x 100%	=	28.0 A
1 HP = 16 A x 100%	=	16.0 A
2 HP = 24 A x 100%	=	24.0 A
• Largest motor load		
28 A x 25%	=	7.0 A
Total load	=	242.0 A

Step 2: Selecting conductors
310-10(2); Table 310-16

- Calculated load at 125% = 242 A
- Sized conductors are #250 KCMIL (255 A)
- 250 A OCPD per 240-3(b) is considered to protect conductors.

Solution: **The No. 250 KCMIL cu. conductors calculated at 125 percent are considered protected by the 250 amp OCPD sized at 125 percent and may be used to supply the feeder-circuit.**

Sizing neutral

Step 1: Calculating load
215-2(a); 220-22; 220-10

• Lighting load		
40 A x 100	=	40.0 A
• Receptacle load		
37.5 A x 100%	=	37.5 A
• Appliance load		
30 A x 100%	=	30.0 A
• Heat		
42 A x 100%	=	42.0 A
• Motor loads		
5 HP = 28 A x 100%	=	28.0 A
1 HP = 16 A x 100%	=	16.0 A
2 HP = 24 A x 100	=	24.0 A
• Largest motor load		
24 A x 25%	=	6.0 A
Total load	=	223.5 A

Step 2: Calculating conductors
Table 310-15(b)(4)(c); 220-22

200 A x 100%	=	200 A
23.5 A x 70%	=	16.45 A
Total	=	216.45 A

Step 3: Selecting conductors
310-10(2); Table 310-16
216.45 A requires #4/0 cu.

Solution: **The size of the conductors based upon 100 percent are No. 4/0 THWN copper.**

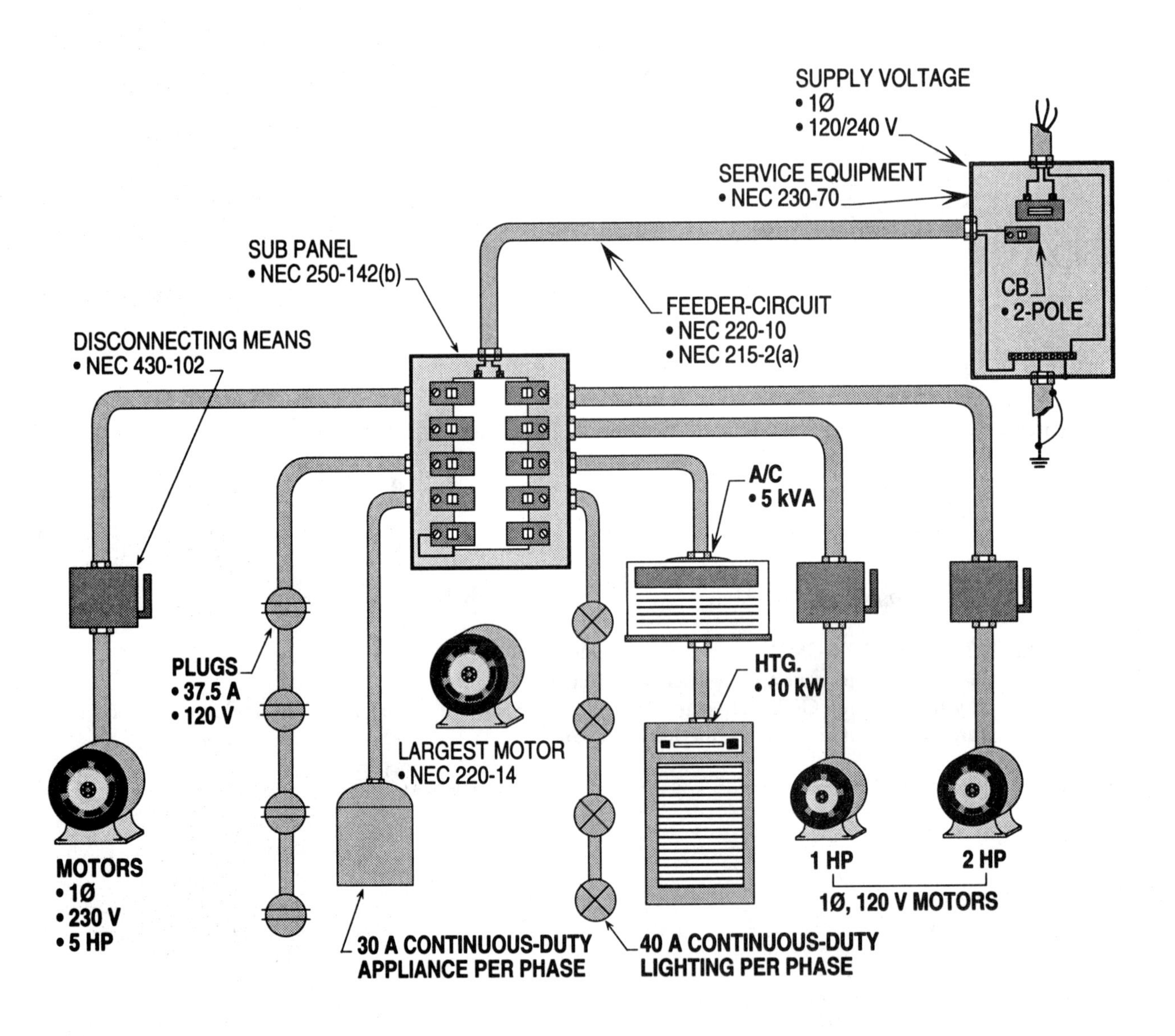
SUPPLY VOLTAGE
• 1Ø
• 120/240 V
SERVICE EQUIPMENT
• NEC 230-70
SUB PANEL
• NEC 250-142(b)
FEEDER-CIRCUIT
• NEC 220-10
• NEC 215-2(a)
CB
• 2-POLE
DISCONNECTING MEANS
• NEC 430-102
A/C
• 5 kVA
HTG.
• 10 kW
PLUGS
• 37.5 A
• 120 V
LARGEST MOTOR
• NEC 220-14
MOTORS
• 1Ø
• 230 V
• 5 HP
1 HP
2 HP
1Ø, 120 V MOTORS
30 A CONTINUOUS-DUTY APPLIANCE PER PHASE
40 A CONTINUOUS-DUTY LIGHTING PER PHASE

Ill. 13-15

Electric Space Heating With Blower Motor 430-24, Ex. 2

Section 424-3(b) must be used to calculate the size conductors to supply power to fixed electric space heating units that are equipped with motor operated equipment.

Example 13-16. What size THWN copper conductors are required to supply power to the heating unit in Illustration 13-16?

Sizing Conductors

Step 1: Finding FLA
430-24, Ex. 2; 424-3(b)

Heating unit = 20,000 VA / 240 V	= 83 A
Blower motor	= 3 A
Total load	= 86 A

Step 2: Calculating load
424-3(b)
86 A x 125% = 107.5 A

Step 3: Selecting conductors
310-10(2); Table 310-16
107.5 A requires #2 cu.

Solution: **The size THWN copper conductors are No. 2.**

Note: This circuit must be computed as a branch-circuit and not a feeder-circuit.

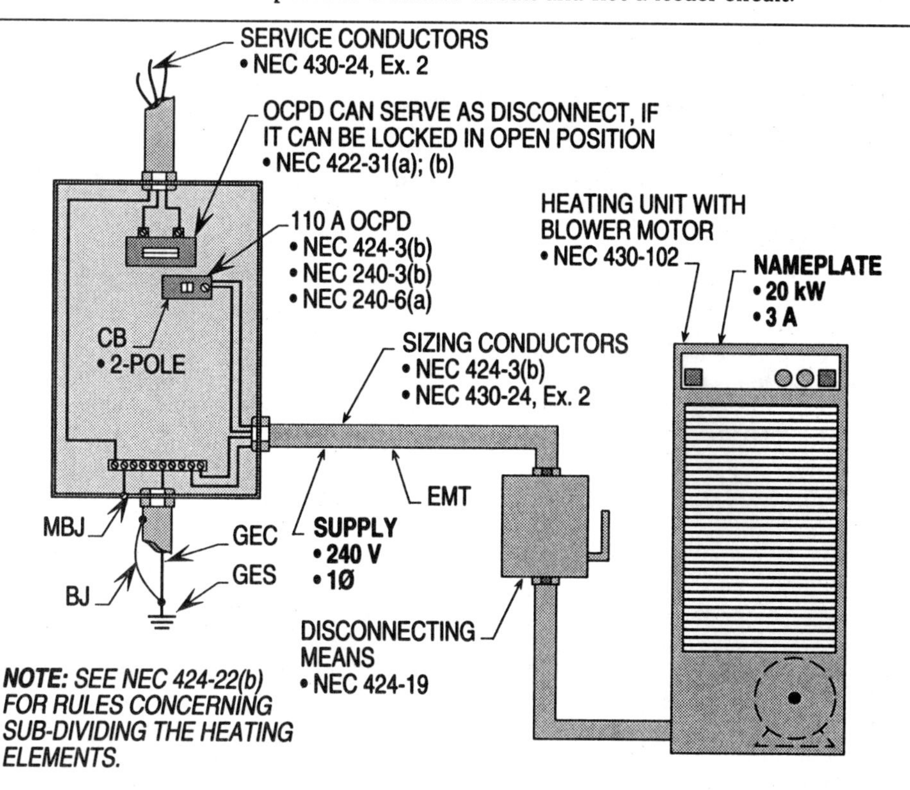

NEC 430-24, Ex. 2

Ill. 13-16

Interlocked Motors
430-24, Ex. 3

Motors that operate with other loads and are interlocked so as not to operate at the same time can have the feeder-circuit conductors based upon the interlocked group producing the greater FLA rating.

Example 13-17. What size THWN conductors are required to supply power to the motors and other loads in Illustration 13-17?

Finding largest load

Step 1: Finding load for motors 1 and 2
430-24, Ex. 3; 220-10
Motor 1 = 24.2 A x 125% = 30.25 A
Motor 2 = 16.7 A x 100% = 16.7 A
Other loads = 40 A x 125% = 50 A
Total load = 96.95 A

Step 2: Finding load for motors 3 and 4
430-24, Ex. 3; 220-10
Motor 3 = 46.2 x 125% = 57.75 A
Motor 4 = 30.8 A x 100% = 30.8 A
Other loads = 40 A x 125% = 50 A
Total load = 138.55 A

Step 3: Selecting largest load
430-24, Ex. 3
138.55 A is greater than 96.95 A

Solution: The greater interlocked load is 138.55 amps.

Sizing conductors

Step 1: Selecting conductors
310-10(2); Table 310-16; Step 2 above
138.55 A requires #1/0 cu.

Solution: The size THWN copper conductors are No. 1/0.

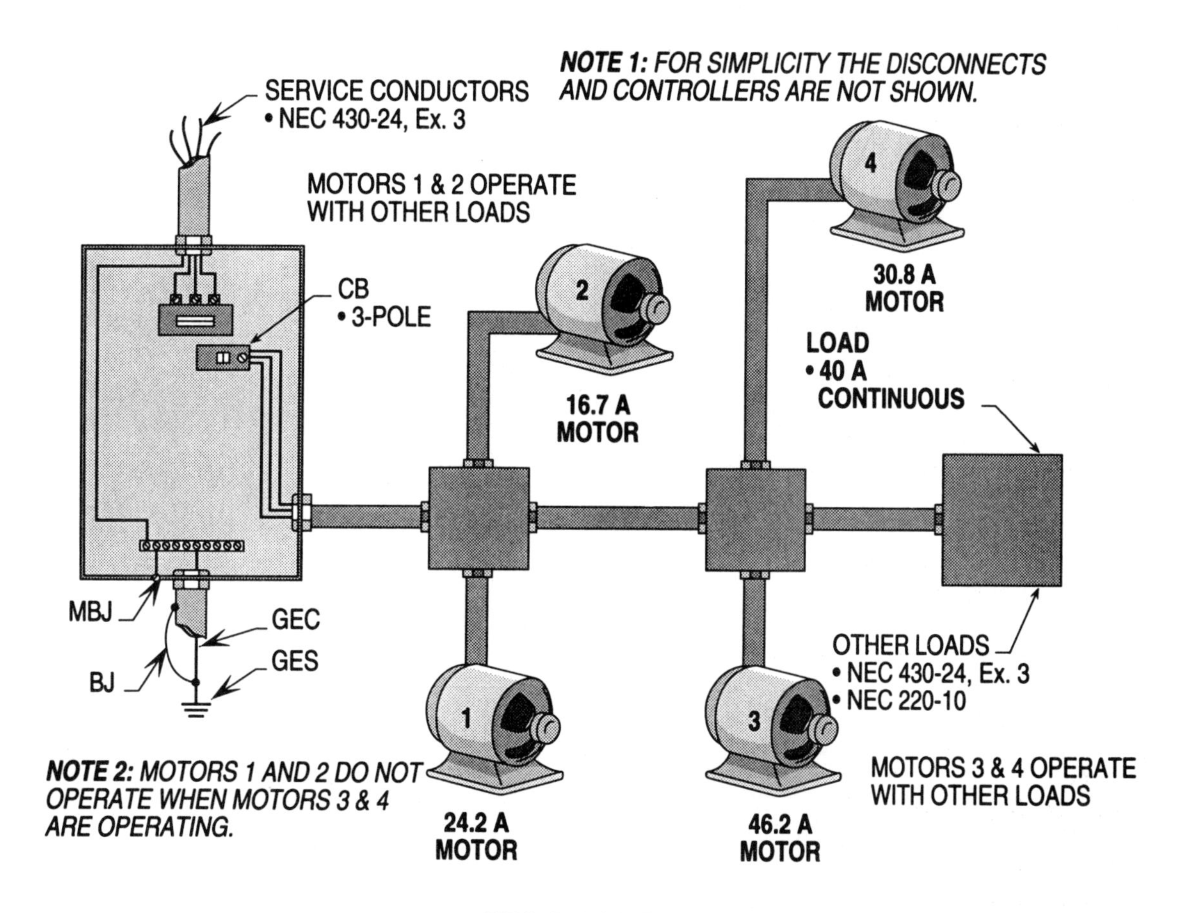

NEC 430-24, Ex. 3

Ill. 13-17

Multimotor And Combination Loads 430-25

The conductors for feeder-circuits supplying power to multimotor and combination equipment loads are required to be sized using the nameplate FLA if listed or by applying the requirements of NEC 430-25 and 430-24.

Example 13-18. What is the load and size conductors for the motor assemblies and other loads in Illustration 13-18?

Sizing load for assembly 1

Step 1: Finding load
430-25; 110-3(b)
Nameplate FLA = 150 A
Other loads FLA (40 A x 125%) = 50 A
Total FLA = 200 A

Solution: The load is 200 amps.

Sizing conductors

Step 1: Selecting conductors
310-10(2); Table 310-16
200 A requires #3/0 cu.

Solution: The size THWN copper conductors are No. 3/0.

Sizing load for assembly 2

Step 1: Calculating load
430-25; 430-24; 220-10
50 HP = 65 A x 125% = 81.25 A
40 HP = 52 A x 100% = 52.0 A
30 HP = 40 A x 100% = 40.0 A
Other loads = 40 A x 125% = 50.0 A
Total load = 223.25 A

Sizing conductors

Step 1: Selecting conductors
310-10(2); Table 310-16
223.25 A requires #4/0 cu.

Solution: The size THWN copper conductors are No. 4/0.

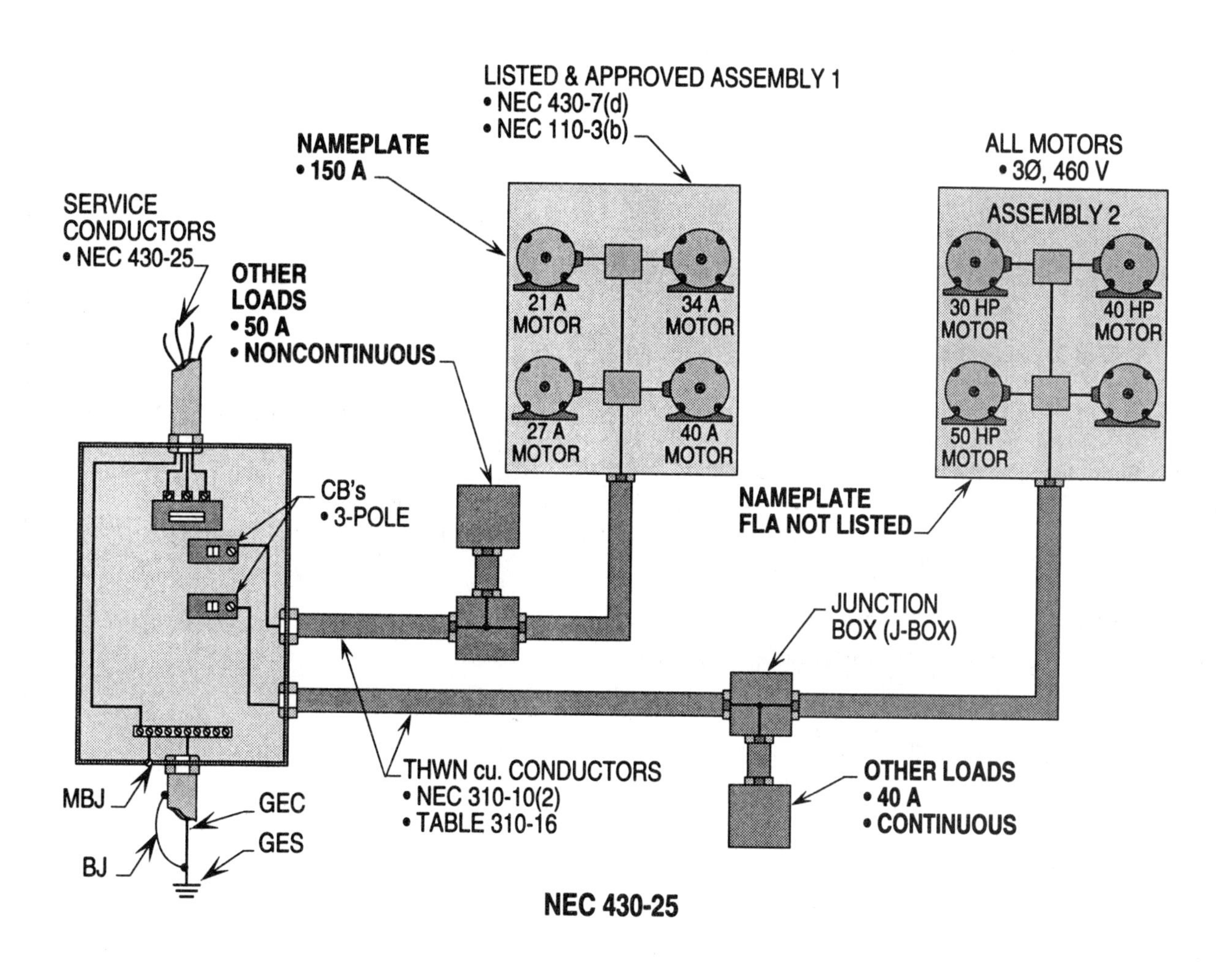

NEC 430-25

Ill. 13-18

Feeder Demand Factor Loads
430-26

There are, in some cases, motor installations where there may be a special situation in which a number of motors are connected to a feeder-circuit. Because of their operations, certain motors do not operate together and the feeder-circuit conductors may be sized based on an historical demand factor.

Example 13-19. What are the size conductors based on the load for the feeder-circuit in Illustration 13-19 when applying the historical documented demand factor?

Sizing largest motor load

Step 1: Calculating amps of motors
Table 430-150

15 HP = 46.2 A x 5	=	231 A
20 HP = 59.4 A x 5	=	297 A
25 HP = 74.8 A x 5	=	374 A
30 HP = 88 A x 5	=	440 A
Total load	=	1,342 A

Step 3: Applying demand factors
430-26
1,342 A x 75% = 1,007 A

Solution: The demand load is 1,007 amps.

Sizing conductors

Step 1: Paralleling 4 times per phase
310-4

$$\text{A per conductor} = \frac{1{,}007 \text{ A}}{4}$$

A per conductor = 252

Step 2: Sizing conductor for feeder
Table 310-16

- 252 A requires #250 THWN KCMIL cu. conductors
- #250 THWN cu. = 255 A
  255 A x 4 = 1,020 A
- 1,020 A supplies 1,007 A

Solution: The size THWN copper conductors are 4 No. 250 KCMIL per phase.

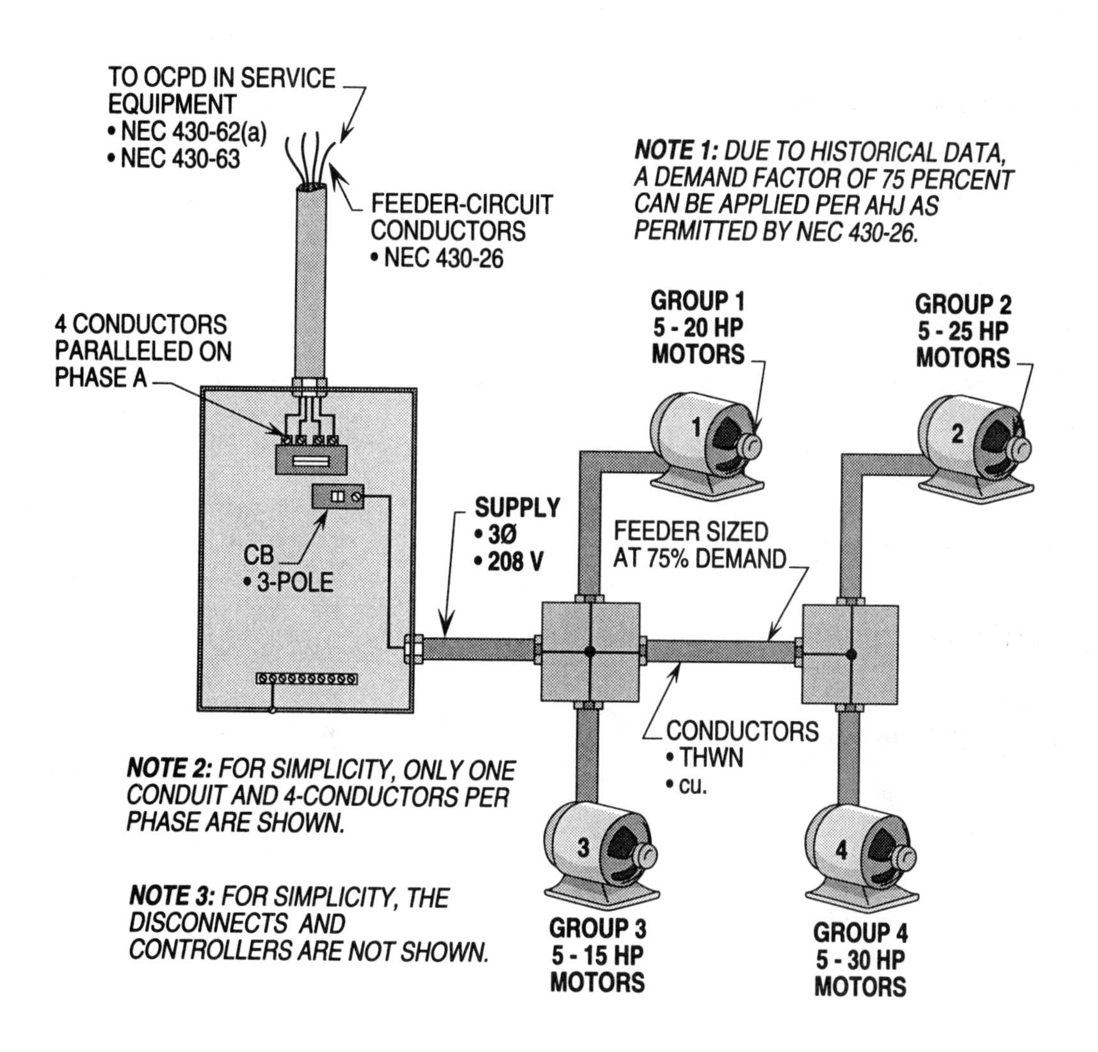
TO OCPD IN SERVICE EQUIPMENT
• NEC 430-62(a)
• NEC 430-63
FEEDER-CIRCUIT CONDUCTORS
• NEC 430-26
NOTE 1: DUE TO HISTORICAL DATA, A DEMAND FACTOR OF 75 PERCENT CAN BE APPLIED PER AHJ AS PERMITTED BY NEC 430-26.
4 CONDUCTORS PARALLELED ON PHASE A
GROUP 1
5 - 20 HP
MOTORS
GROUP 2
5 - 25 HP
MOTORS
1
2
SUPPLY
• 3Ø
• 208 V
FEEDER SIZED AT 75% DEMAND
CB
• 3-POLE
CONDUCTORS
• THWN
• cu.
NOTE 2: FOR SIMPLICITY, ONLY ONE CONDUIT AND 4-CONDUCTORS PER PHASE ARE SHOWN.
3
4
NOTE 3: FOR SIMPLICITY, THE DISCONNECTS AND CONTROLLERS ARE NOT SHOWN.
GROUP 3
5 - 15 HP
MOTORS
GROUP 4
5 - 30 HP
MOTORS

Ill. 13-19

Motor Overload Protection 430-32(a)(1)

The procedure for sizing the overload protection (OL) for motors rated more than one horsepower (HP) is found by multiplying the nameplate FLA of the motor's nameplate by 115 percent or 125 percent. The percentage selected is based upon service factor or temperature rise of the motor.

Example 13-20(a). What size overloads, in amps, are required for the motor in Illustration 13-20(a)?

Sizing OL's using fuses

Step 1: Finding FLA
430-6(a)(2)
Nameplate = 18 A

Step 2: Finding percentage
430-32(a)(1)
SF = 125%
TR = 125%

Step 3: Calculating FLA
430-32(a)(1)
18 A x 125% = 22.5 A

Step 4: Selecting TD fuses
430-32(a)(1); 240-6(a)
22.5 A requires 20 A TDF's

Solution: The size time delay fuses are 20 amps.

Note: 25 TDF's provide backup OL protection.

Sizing OL's in controllers

Step 1: Selecting FLA
430-6(a)(2); 430-32(a)(1)
18 A x 125% = 22.5 A

Solution: The size overloads are selected from a manufacturer's chart based on 18 amps.

Note: The OL units are computed at 22.5 amps when selected from the chart.

Example 13-20(b). What size overloads in amps are required for the motor in Illustration 13-20(b)?

Sizing OL's using fuses

Step 1: Finding FLA
430-6(a)(2)
Nameplate = 18 A

Step 2: Finding percentage
430-32(a)(1)
SF = 115%
TR = 115%

Step 3: Calculating FLA
430-32(a)(1)
18 A x 115% = 20.7 A

Step 4: Selecting TD fuses
430-32(a)(1); 240-6(a)
20.7 A requires 20 A TDF's

Solution: The size time delay fuses are 20 amps.

Sizing OL's in controller

Step 1: Selecting FLA
430-6(a)(2); 430-32(a)(1)
18 A x 115% = 20.7 A

Solution: The size overloads are selected from a manufacturer's chart based upon 18 amps.

Note: The OL units are computed at 20.7 amps when selected from the chart.

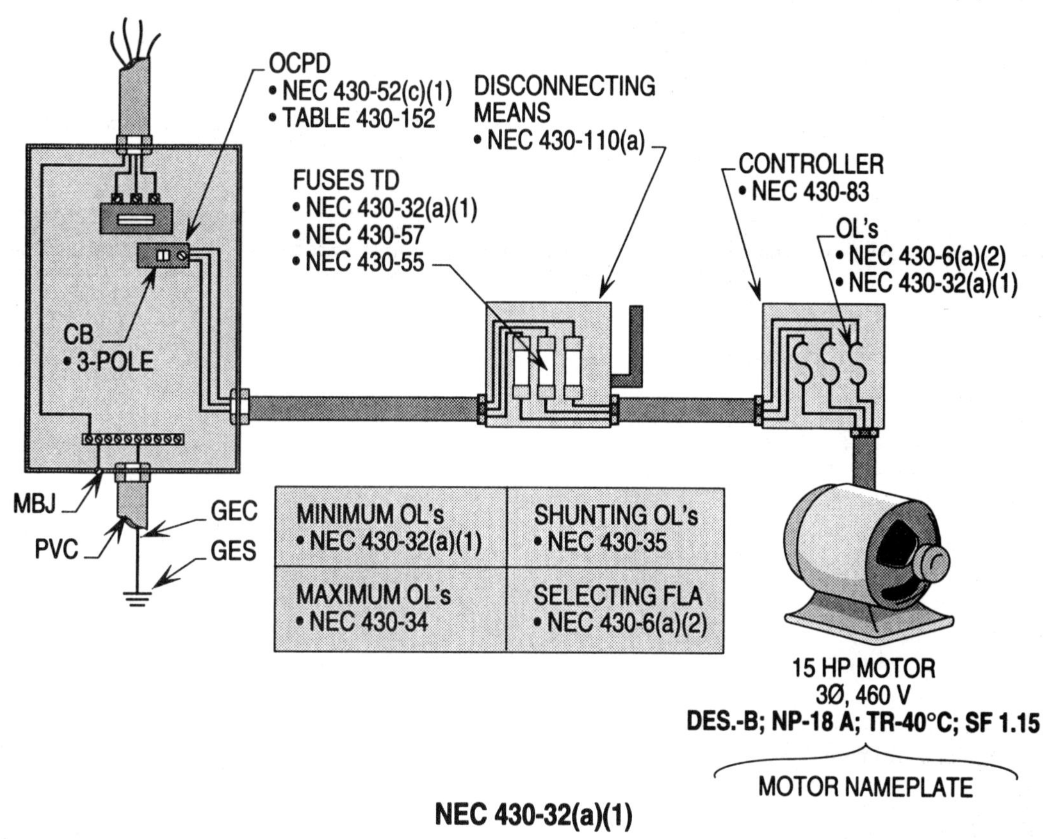

Ill. 13-20(a)

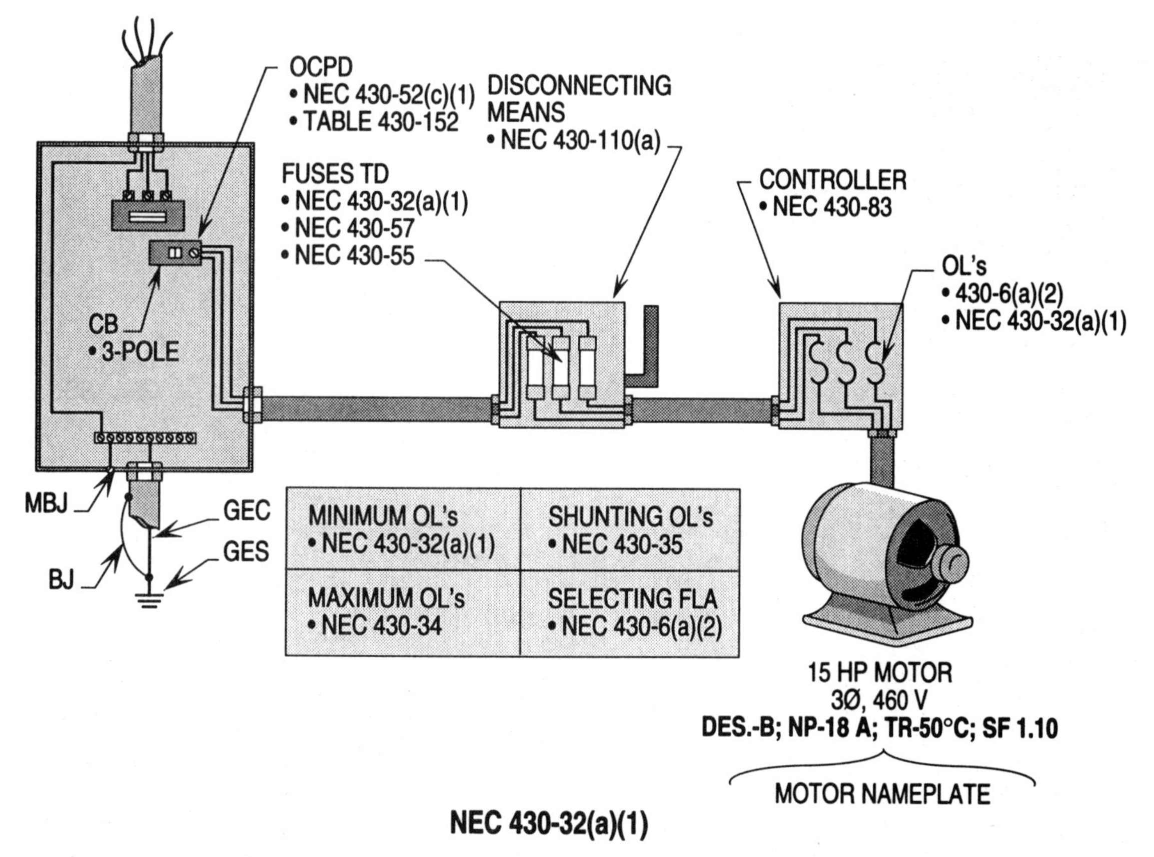

Ill. 13-20(b)

Motor Overload Protection
430-32(a)(2)

A thermal protector built into a motor may be set with ratings greater than permitted for other forms of overload protection, which is designed to protect motor windings.

Example 13-20(c). What size thermal protector (TP) in amps is required for the motor in Illustration 13-20(c)?

Sizing thermal protector

Step 1: Finding FLA
430-32(a)(2); Table 430-148
3/4 HP = 6.9 A

Step 2: Finding percentages
430-32(a)(2)
6.9 A less than 9 A requires 170%

Step 3: Calculating trip current
6.9 A x 170% = 11.73 A

Solution: The ultimate trip current rating for the thermal protector is 11.73 amps.

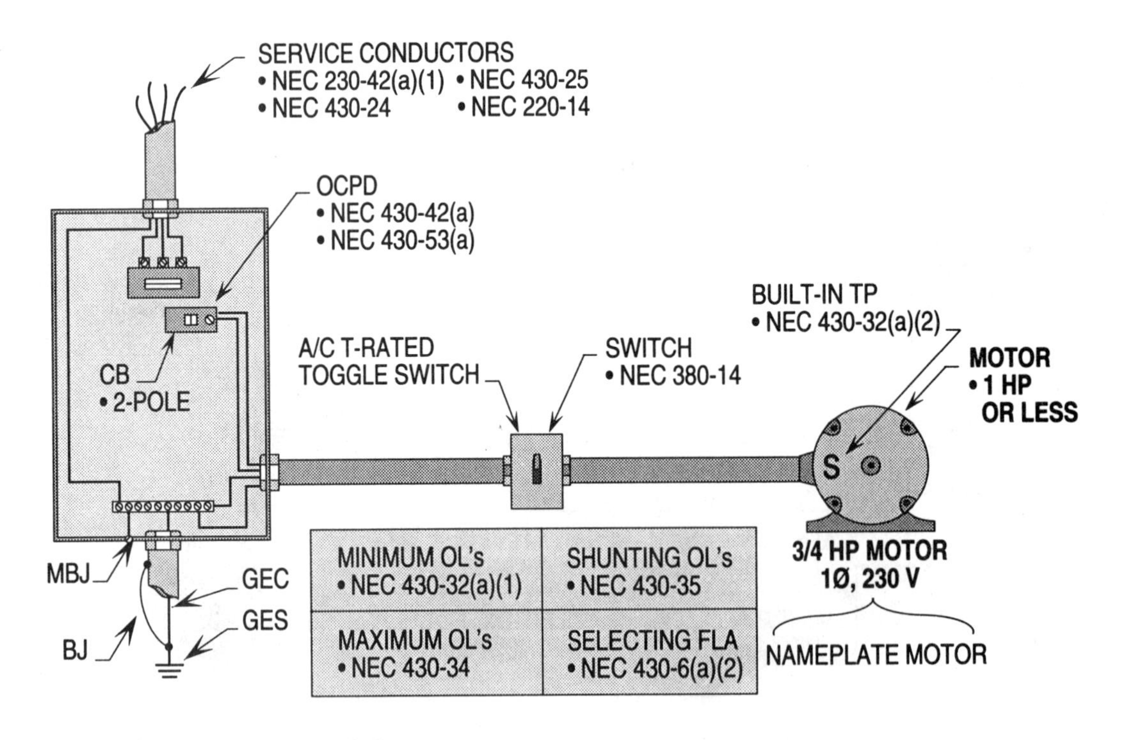

NEC 430-32(a)(2)

Ill. 13-20(c)

Motor Overload Protection
430-34

The procedure for sizing the overload protection (OL) for motors rated more than one horsepower (HP) is found by multiplying the nameplate FLA of the motor's nameplate by 130 percent or 140 percent. The percentage selected is based upon service factor or temperature rise of the motor.

Example 13-21(a). What size overloads (OL's) in amps are required for the motor in Illustration 13-21(a)?

Sizing OL's using fuses

Step 1: Finding FLA
430-6(a)(2)
Nameplate = 18 A

Step 2: Finding percentage
430-34
SF = 140%
TR = 140%

Step 3: Calculating FLA
430-34
18 A x 140% = 25.2 A

Step 4: Selecting TD fuses
430-34
TD fuses are not allowed to be sized at 140%.

Sizing OL's in controller

Step 1: Calculating FLA
430-6(a)(2); 430-34
18 A x 140% = 25.2 A

Solution: **The size overloads are selected based upon 25.2 amps from a manufacturer's chart.**

Note: The OL units are computed at 25.2 amps but can only be increased up to one size in rating.

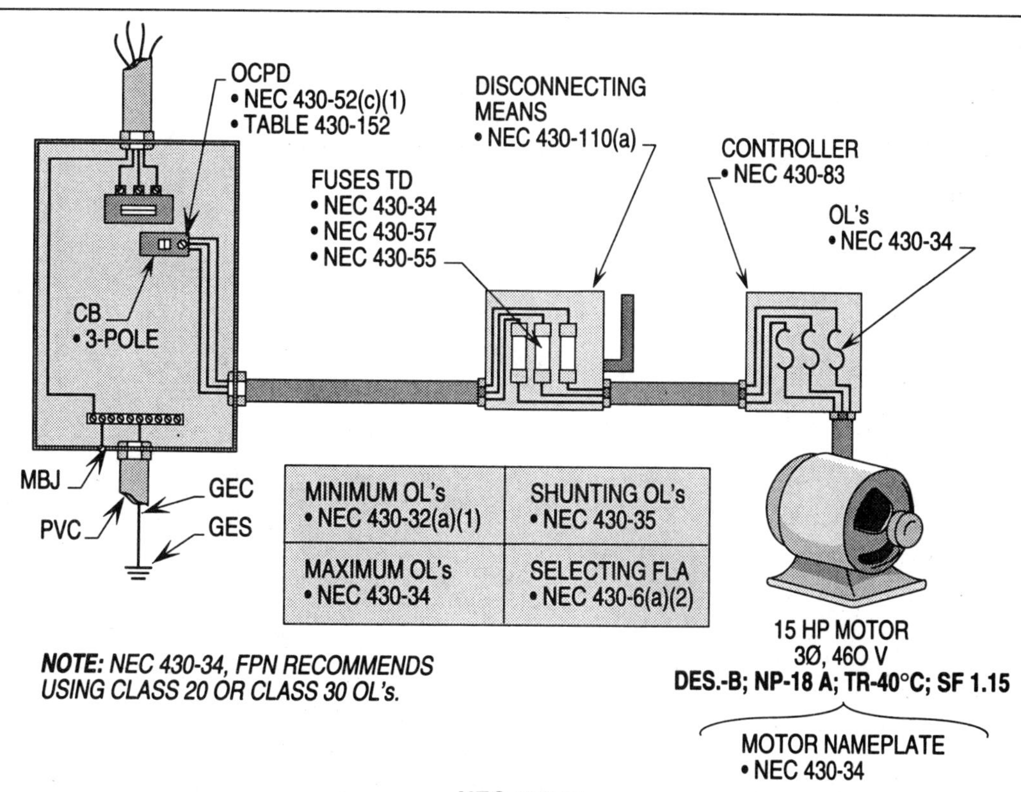

NEC 430-34

Ill. 13-21(a)

Motor Overload Protection
430-34

If an overload relay fails to allow a motor to start and run due to inrush current, the size of the overload relay may be increased to the percentages listed in NEC 430-34.

Example 13-21(b). What size overloads in amps are required for the motor in Illustration 13-21(b)?

Sizing OL's using fuses

Step 1: Finding FLA
430-6(a)(2)
Nameplate = 18 A

Step 2: Finding percentage
430-34
SF = 130%
TR = 130%

Step 3: Calculating FLA
430-34
18 A x 130% = 23.4 A

Step 4: Selecting TD fuses
430-34
TD fuses cannot be used at 130% of the FLA.

Sizing OL's in controller

Step 1: Calculating FLA
430-6(a)(2); 430-34
18 A x 130% = 23.4 A

Solution: **The size overloads are selected from a manufacturer's chart based upon 23.4 amps.**

Note: The OL units are computed at 23.4 amps but can only be increased up to one size rating.

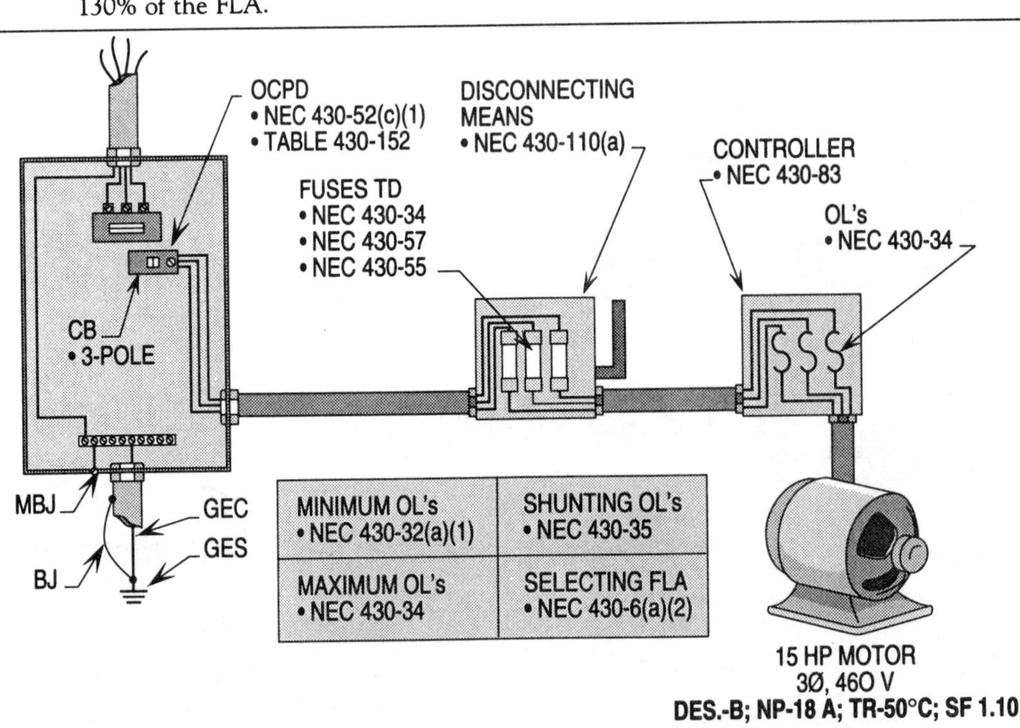

NEC 430-34

Ill. 13-21(b)

OCPD For Single Motor Load
430-52(c)(1)

The OCPD which is used to allow a motor to start and accelerate its driven load must be capable of holding the inrush current of the motor during its starting period. The amount of inrush current is based upon the code letter, or design letter of the motor and the driven load.

Example 13-22(a). What is the minimum size, next size and maximum size nontime delay fuses required for the motor in Illustration 13-22(a)?

Sizing nontime delay fuses

Step 1: Finding FLA
430-6(a)(1); Table 430-150
50 HP = 65 A

Step 2: Finding percentages
430-52(c)(1), Ex.'s 1; 2(a); Table 430-152
Min. size = 300%
Next size = above 300%
Max. size = 400%

Step 3: Calculating size
430-52(c)(1), Ex.'s 1; 2(a)
Min. size = 65 A x 300% = 195 A
Next size = above 300%
Max. size = 65 A x 400% = 260 A

Step 4: Selecting NTDF's
240-3(g); 240-6(a); 430-52(c)(1), Ex.'s 1; 2(a)
Min. size = 175 A NTDF's
Next size = 200 A NTDF's
Max. size = 250 A NTDF's

Solution: **The minimum size is 175 amps, the next size is 200 amps, and the maximum size nontime delay fuse is 250 amps.**

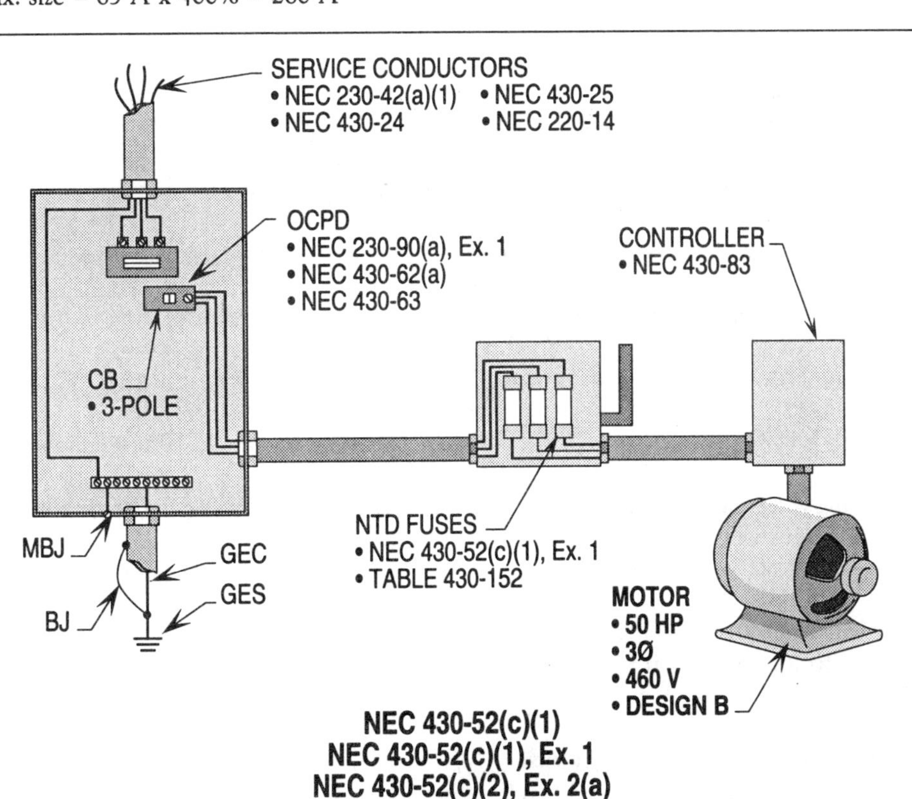

Ill. 13-22(a)

OCPD For Single Motor Load 430-52(c)(1)

Example 13-22(b). What is the minimum size, next size and maximum size time delay fuses required for the motor in Illustration 13-22(b)?

Sizing minimum size TDF's

Step 1: Finding FLA
430-6(a)(1); Table 430-150
50 HP = 130 A

Step 2: Finding percentage
430-52(c)(1); Table 430-152
Min. = 175%

Step 3: Calculating size
430-52(c)(1); Table 430-152
130 A x 175% = 227.5 A

Step 4: Selecting TDF's
240-3(g); 240-6(a)
227.5 A requires 225 A TDF's

Solution: **The minimum size time delay fuses are 225 amps.**

Calculating min. size for OL protection

- 430-57, Ex.; use 125%
- 130 A x 125% = 162.5 A
- 240-6(a); 430-55
- 162.5 A = 150 A TDF's

Sizing next size TDF's

Step 1: Selecting next size
430-52(c)(1), Ex. 1; 240-6(a)
Next size above 227.5 A is 250 A TDF's

Solution: **The next standard size time delay fuses above 227.5 amps is 250 amps.**

Sizing maximum size TDF's

Step 1: Finding percentages
430-52(c)(1), Ex. 2(b); 240-6(a)
Max. = 225%

Step 2: Calculating size
430-52(c)(1), Ex. 2(b); Table 430-152
130 A x 225% = 292.5 A

Step 3: Selecting TDF's
240-3(g); 240-6(a)
292.5 requires 250 A TDF's

Solution: **The next standard size below the 225 percent multiplier is 250 amps.**

Example 13-22(c). What is the minimum and maximum setting of the instantaneous trip circuit breaker in Illustration 13-22(c)?

Sizing minimum setting of inst. CB

Step 1: Finding FLA
430-6(a)(1); Table 430-150
50 HP = 65 A

Step 2: Finding percentages
430-52(c)(3); Table 430-152
Inst. CB = 800%

Step 3: Calculating setting
430-52(c)(3); Table 430-152
65 A x 800% = 520 A

Solution: **The minimum setting is 520 amps.**

Sizing maximum setting of inst. CB

Step 1: Calculating setting
430-52(c)(3), Ex. 1
65 A x 1,300% = 845 A

Solution: **The maximum setting is 845 amps.**

Note: The maximum of 1,300 percent or 1,700 percent is often used to start and run high-efficiency motors.

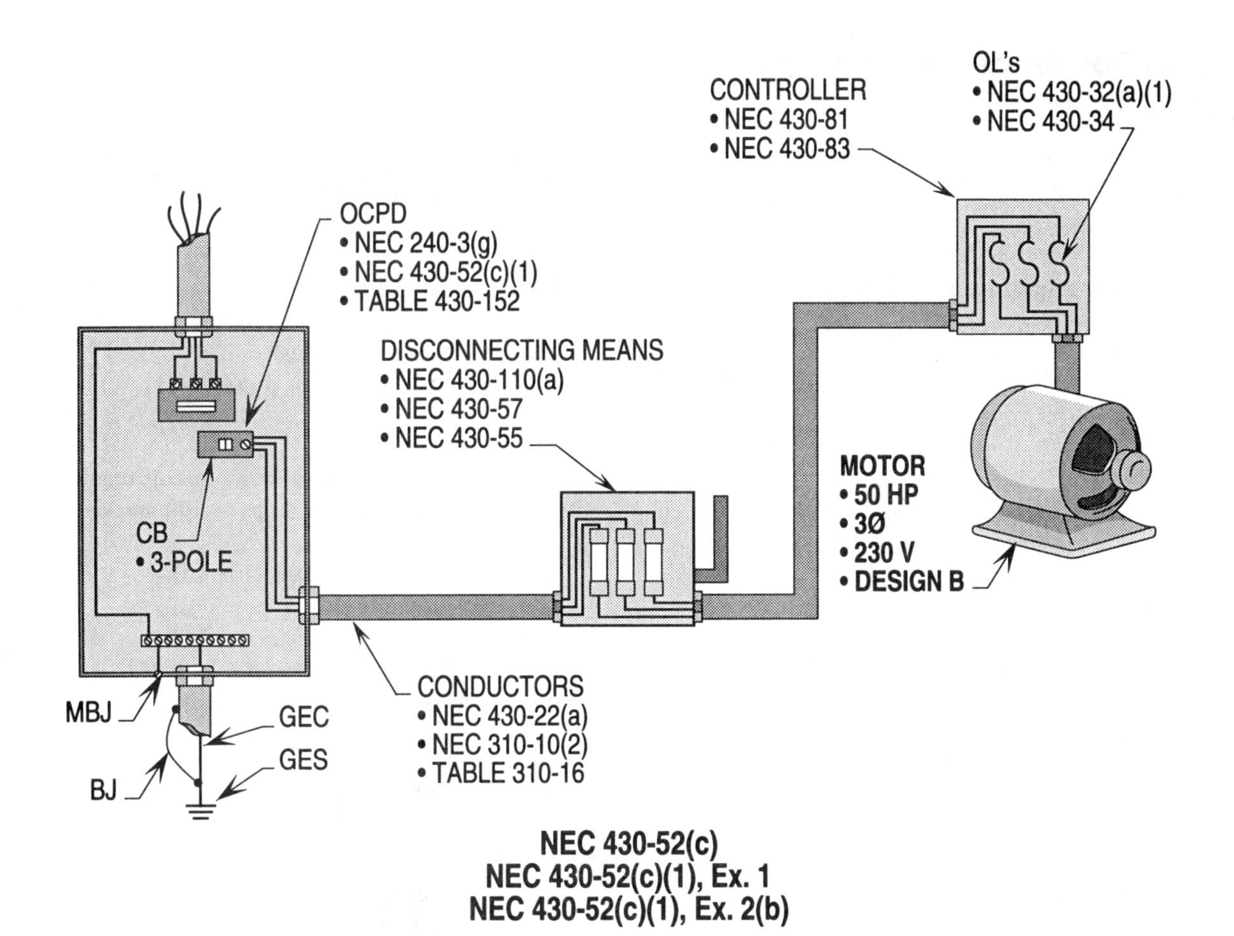

NEC 430-52(c)
NEC 430-52(c)(1), Ex. 1
NEC 430-52(c)(1), Ex. 2(b)

Ill. 13-22(b)

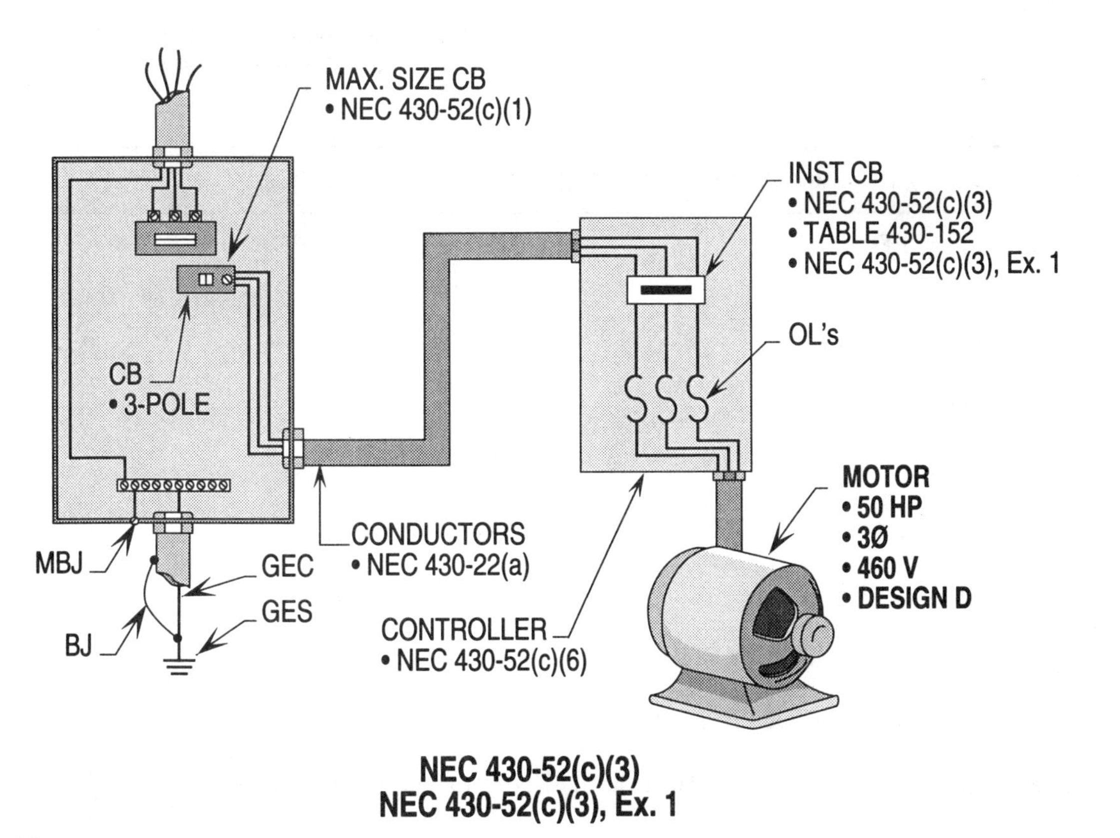

NEC 430-52(c)(3)
NEC 430-52(c)(3), Ex. 1

Ill. 13-22(c)

OCPD For Single Motor Load 430-52(c)(1)

Example 13-22(d). What is the minimum size, next size, and maximum size circuit breaker required for the motor in Illustration 13-22(d)?

Sizing minimum size CB

Step 1: Finding FLA
430-6(a)(1); Table 430-150
50 HP = 143 A

Step 2: Finding percentages
430-52(c)(1); Table 430-152
Min. = 250%

Step 3: Calculating size
430-52(c)(1); Table 430-152
143 A x 250% = 357.5 A

Step 4: Selecting CB
240-3(g); 240-6(a); 430-52(c)(1)
357.5 A requires 350 A INVT CB

Solution: **The minimum size circuit breaker is 350 amps.**

Sizing next size CB

Step 1: Finding percentages
430-52(c)(1), Ex. 1; 240-6(a)
Next size above 357.5 A is 400 A INVT CB

Solution: **The next standard size circuit breaker above 357.5 amps is 400 amps.**

Sizing maximum size CB

Step 1: Finding percentages
430-52(c)(1), Ex. 2(c); 240-6(a)
Max. = 300%

Step 2: Calculating size
430-52(c)(1), Ex. 2(c); Table 430-152
143 A x 300% = 429 A

Step 3: Selecting CB
240-3(g); 240-6(a)
429 A requires 400 A INVT CB

Solution: **The next standard size circuit breaker below 429 amps is 400 amps.**

Calculating OCPD for minimum protection

Sizing min. OCPD

Step 1: Finding LRC
Table 430-151(B)
(802 A x 110% = 882 A)
50 HP = 882 A

Step 2: Calculating Size
$\frac{882\text{ A}}{3}$ (CB holds 3 times its rating) = 294 A

Step 3: Selecting OCPD
240-3(g); 240-6(a)
294 A requiring 300 A OCPD

Solution: **The minimum size OCPD is 300 amps.**

Note: See the Notes to Table 430-150 of the 1987 NEC to obtain the 110 percent rule.

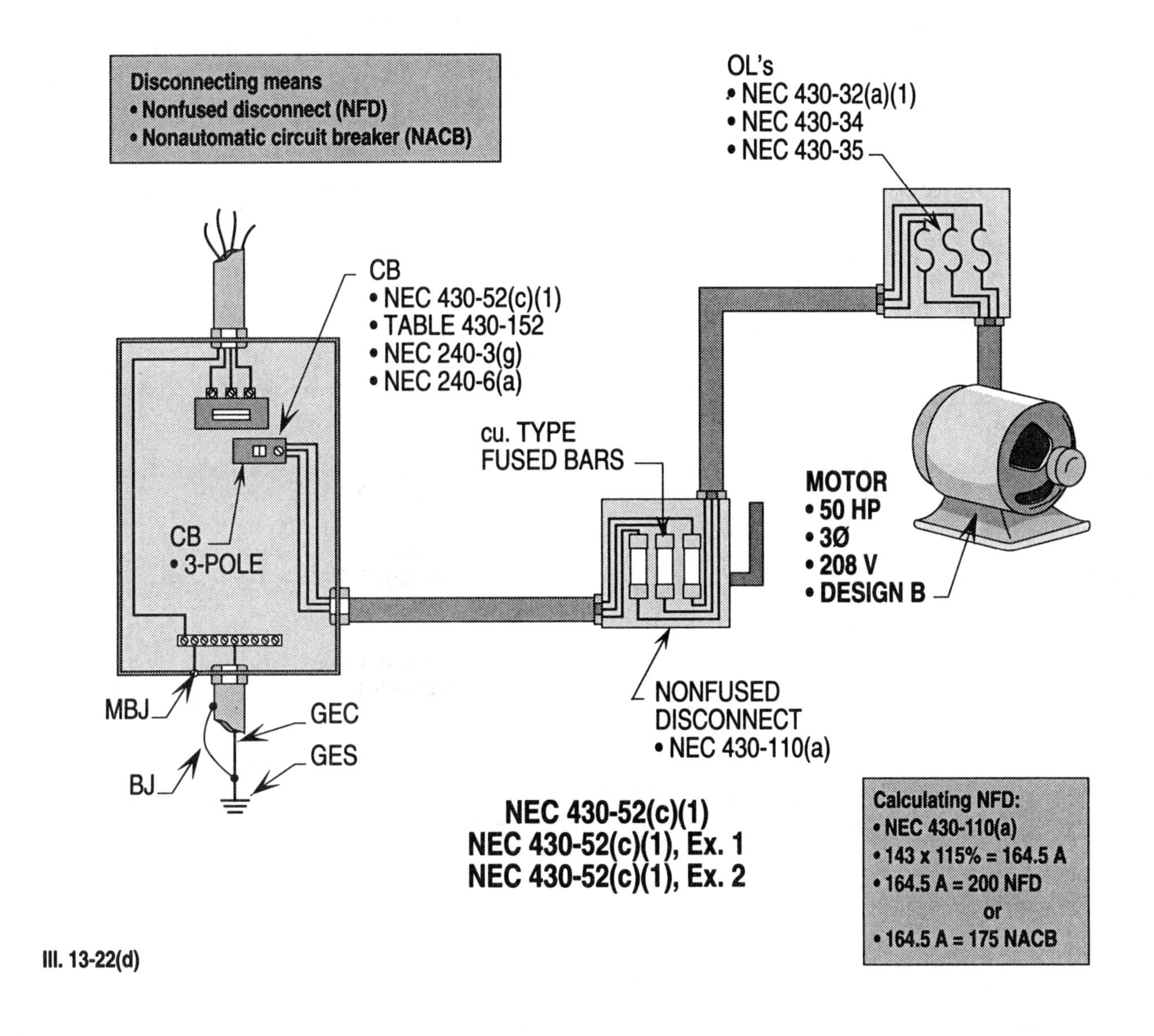
Disconnecting means
• Nonfused disconnect (NFD)
• Nonautomatic circuit breaker (NACB)
OL's
• NEC 430-32(a)(1)
• NEC 430-34
• NEC 430-35
CB
• NEC 430-52(c)(1)
• TABLE 430-152
• NEC 240-3(g)
• NEC 240-6(a)
cu. TYPE
FUSED BARS
MOTOR
• 50 HP
• 3Ø
• 208 V
• DESIGN B
CB
• 3-POLE
MBJ
GEC
GES
BJ
NONFUSED
DISCONNECT
• NEC 430-110(a)
NEC 430-52(c)(1)
NEC 430-52(c)(1), Ex. 1
NEC 430-52(c)(1), Ex. 2
Calculating NFD:
• NEC 430-110(a)
• 143 x 115% = 164.5 A
• 164.5 A = 200 NFD
or
• 164.5 A = 175 NACB

Ill. 13-22(d)

Several Motors On A Branch-circuit 430-53(a)

Two or more motors rated at 1 HP or less and draws 6 amps or less may be connected to a 15 or 20 amp branch-circuit.

Example 13-23(a). Can the motors in Illustration 13-23(a) be connected to the branch-circuit?

Step 1: Finding amps
430-6(a)(1); Table 430-150
1/2 HP = 1.1 A
3/4 HP = 1.6 A
1 HP = 2.1 A

Step 2: Calculating FLA
430-53(a)
1.1 A x 4 = 4.4 A
1.6 A x 3 = 4.8 A
2.1 A x 3 = 6.3 A
Total load = 15.5 A

Step 3: Calculating permitted load
384-16(d); 430-22(a)
20 A CB x 80% = 16 A

Step 4: Verifying loading
430-53(a); 384-16(d)
15.5 A is less than 16 A

Solution: **The ten motors are allowed to be connected to the 20 amp branch-circuit.**

Note 1: Motors are not all started at the same time.

Note 2: Motors rated 480 volts as shown in Ill. 13-23(a) are not allowed to be connected to a 15 amp CB (15 A CB x 80% = 12 A) and branch-circuit per NEC 430-53(a).

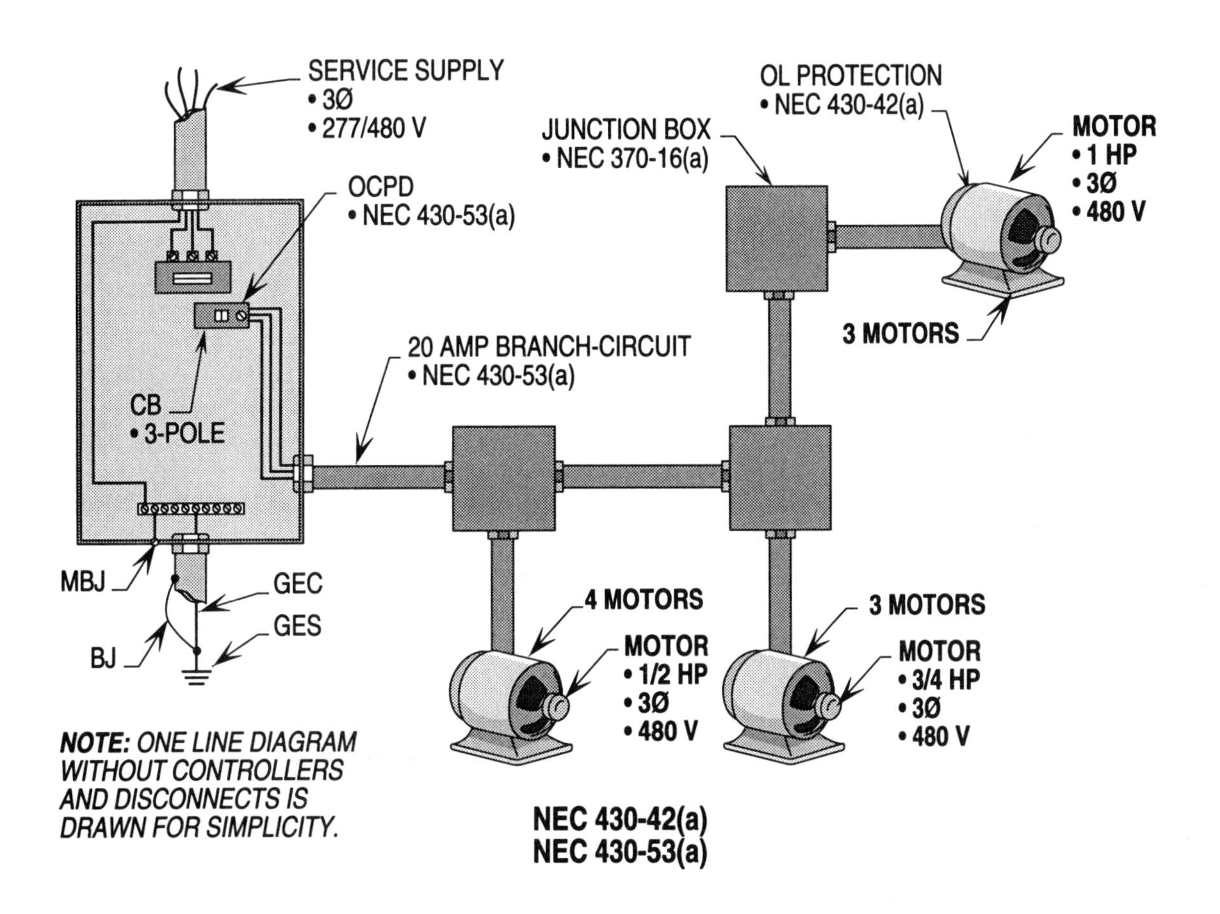
SERVICE SUPPLY
• 3Ø
• 277/480 V
OCPD
• NEC 430-53(a)
CB
• 3-POLE
MBJ
GEC
GES
BJ
JUNCTION BOX
• NEC 370-16(a)
OL PROTECTION
• NEC 430-42(a)
MOTOR
• 1 HP
• 3Ø
• 480 V
3 MOTORS
20 AMP BRANCH-CIRCUIT
• NEC 430-53(a)
4 MOTORS
MOTOR
• 1/2 HP
• 3Ø
• 480 V
3 MOTORS
MOTOR
• 3/4 HP
• 3Ø
• 480 V
NOTE: ONE LINE DIAGRAM WITHOUT CONTROLLERS AND DISCONNECTS IS DRAWN FOR SIMPLICITY.
NEC 430-42(a)
NEC 430-53(a)

Ill. 13-23(a)

Several Motors On A Branch-circuit 430-53(a)

Example 13-23(b). What size circuit breaker is required for the motors in Illustration 13-23(b)?

Step 1: Finding amps
430-6(a)(1); Table 430-148
1/6 HP = 2.2 A
1/4 HP = 2.9 A
1/4 HP = 2.9 A
1/3 HP = 3.6 A

Step 2: Calculating FLA
430-53(b)
2.2 A x 1 = 2.2 A
2.9 A x 1 = 2.9 A
2.9 A x 1 = 2.9 A
3.6 A x 1 = 3.6 A
Total load = 11.6 A

Step 3: Calculating permitted load
384-16(d); Table 210-21(b)(2)
15 A OCPD x 80% = 12 A

Step 4: Verifying load
430-53(b); 384-16(d);
Table 210-21(b)(2)
11.6 A is less than 12 A

Step 5: Protecting smallest motor
430-53(b); Table 430-152
2.2 A x 250% = 5.5 A

Step 6: Selecting OCPD
430-53(b); 240-3(b); 240-6(a)
5.5 A allows 15 A OCPD

Solution: **Section 430-53(b) allows the next circuit breaker which is 15 amps.**

Note: Most inspectors allow this concept due to the next size CB being 15 amps.

Example 13-23(c). What size OCPD and conductors are required for the branch-circuit and each connected motors in Illustration 13-23(c)?

Sizing branch-circuit

Step 1: Finding FLA of motor
430-6(a)(1); Table 430-150
3 HP = 4.8 A
5 HP = 7.6 A
7 1/2 HP = 11 A

Step 2: Calculating branch-circuit FLA
430-22(a); 430-53(b)
4.8 A + 7.6 A + 11 A x 125% = 29.25 A

Step 3: Selecting conductors
430-53(d); Table 310-16
29.25 A requires #10 cu.

Step 4: Selecting OCPD
430-53(d); 240-6(a)
29.25 A requires 30 A OCPD

Solution: **The OCPD for the branch-circuit is 30 amps and the size THHN conductors are No. 10 copper.**

Sizing conductors to each motor

Step 1: Sizing conductors for motor 1
430-53(d); 430-28; 210-3;
240-3(d)
30 A OCPD x 1/3 = 10 A

Step 2: Selecting conductors
430-53(d); Table 310-5;
Table 310-16
10 A requires #14 cu.

Step 3: Sizing conductors for motors 2 and 3
430-53(d)
#10 cu. same as branch-circuit

Solution: **The size THHN conductors to motor 1 is No. 14 copper and the size to motors 2 and 3 are No. 10 copper.**

Note: If terminals were 75°C, check with AHJ to verify 1/3 of OCPD or 1/3 of conductor ampacity for sizing conductors to motor 1.

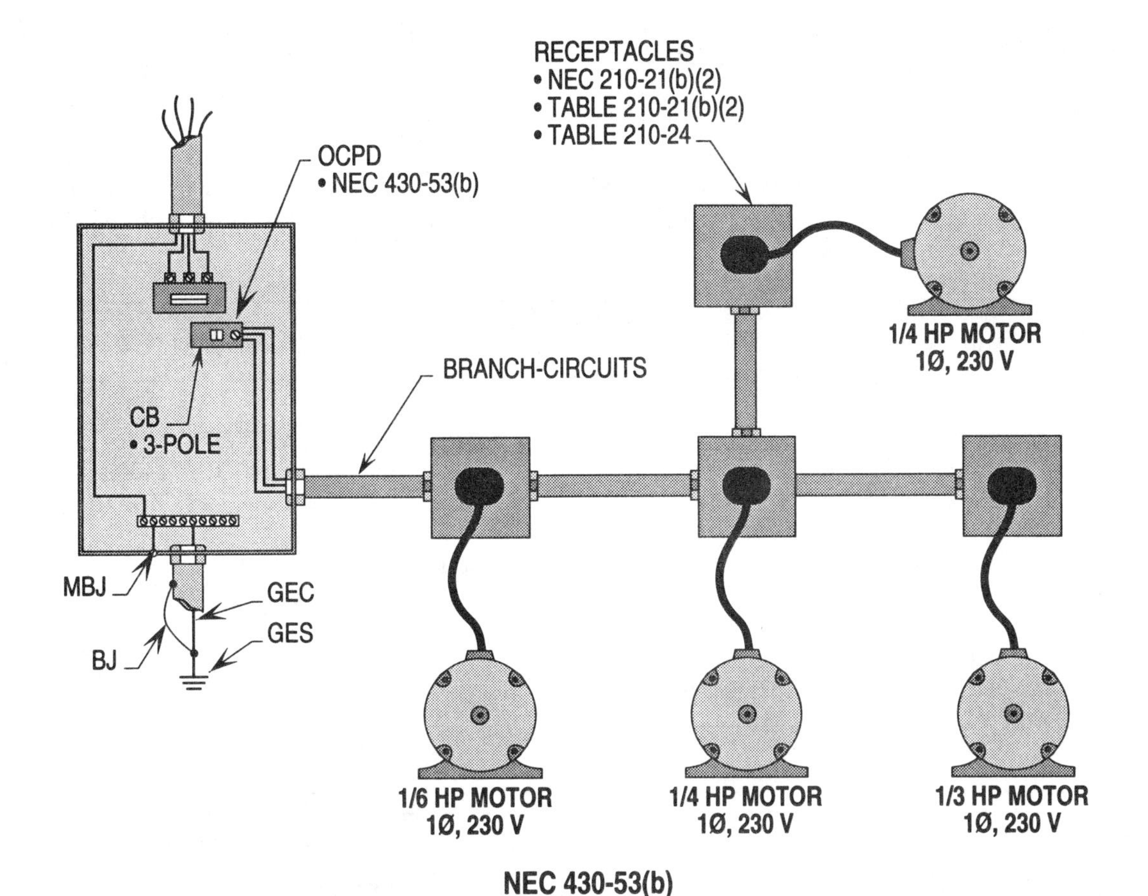

Ill. 13-23(b)

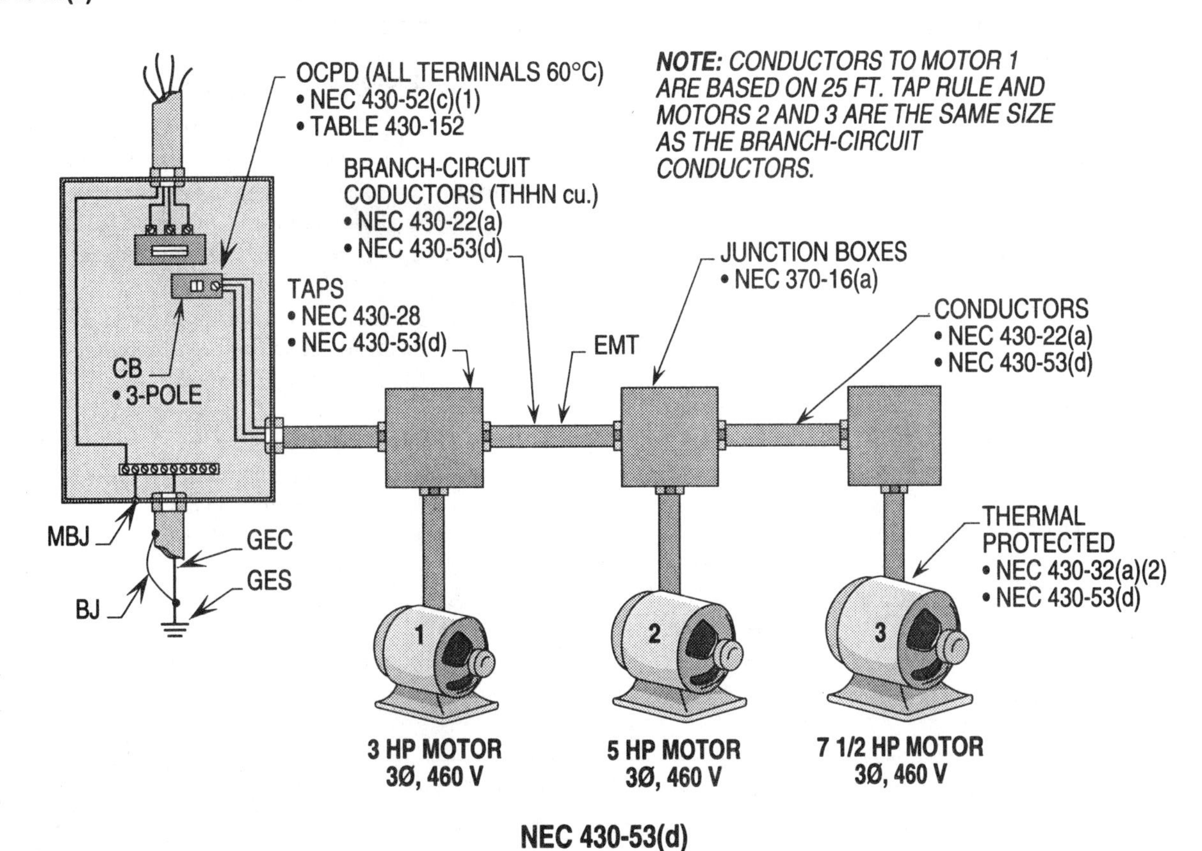

Ill. 13-23(c)

Sizing OCPD For Feeder Loads 430-62(a)

The OCPD for a feeder-circuit supplying two or more motors is based upon the largest OCPD for any motor of the group plus the remaining motors. The procedure requires a selection of OCPD's by rounding down from each calculated value that does not correspond to a standard device.

Example 13-24(a). What size circuit breaker is required for the feeder-circuit in Illustration 13-24(a)?

Step 1: Finding FLA of motors
430-6(a)(1); Table 430-150
10 HP = 14 A
20 HP = 27 A
25 HP = 34 A
30 HP = 40 A

Step 2: Calculating feeder OCPD
430-52(c)(1); Table 430-152; 430-62(a)
40 A x 250% = 100 A + 34 A + 27 A + 14 A = 175 A

Step 3: Selecting OCPD
430-62(a); 240-3(g); 240-6(a)
175 A is a standard OCPD

Solution: The size OCPD required for the feeder-circuit is 175 amp CB.

Example 13-24(b). What size nontime delay fuses are required for the feeder-circuit in Illustration 13-24(b) when applying the rules of Table 430-152?

Step 1: Finding FLA of motors
430-6(a)(1); Table 430-150
10 HP = 14 A
15 HP = 21 A
20 HP = 27 A
25 HP = 34 A

Step 2: Calculating feeder OCPD
430-52(c)(1); Table 430-152;
430-62(a); 240-6(a)
34 A x 300% = 102 A requires 100 A NTDF's
27 A x 300% = 81 A requires 80 A NTDF's
21 A x 300% = 63 A requires 60 A NTDF's
14 A x 300% = 42 A requires 40 A NTDF's

Step 3: Calculating OCPD
430-62(a); 430-52(c)

Largest OCPD	=	100 A
Plus remaining motors	=	27 A
	=	21 A
	=	14 A
Total amps	=	162 A

Step 4: Selecting OCPD
430-62(a); 240-3(g); 240-6(a)
162 A requires 150 A NTDF's
(162 A is not a standard OCPD)

Solution: The size nontime delay fuses for the feeder-circuit are 150 amps based upon the motor's Design letters.

Note 1: There is no exception in NEC 430-62(a) to allow the next higher size OCPD above 162 amps.

Note 2: The AHJ may allow Ex. 1 to NEC 430-52(c)(1) to be applied, which would require 110 amp NTDF's and not 100 amp NTDF's as in Step 2.

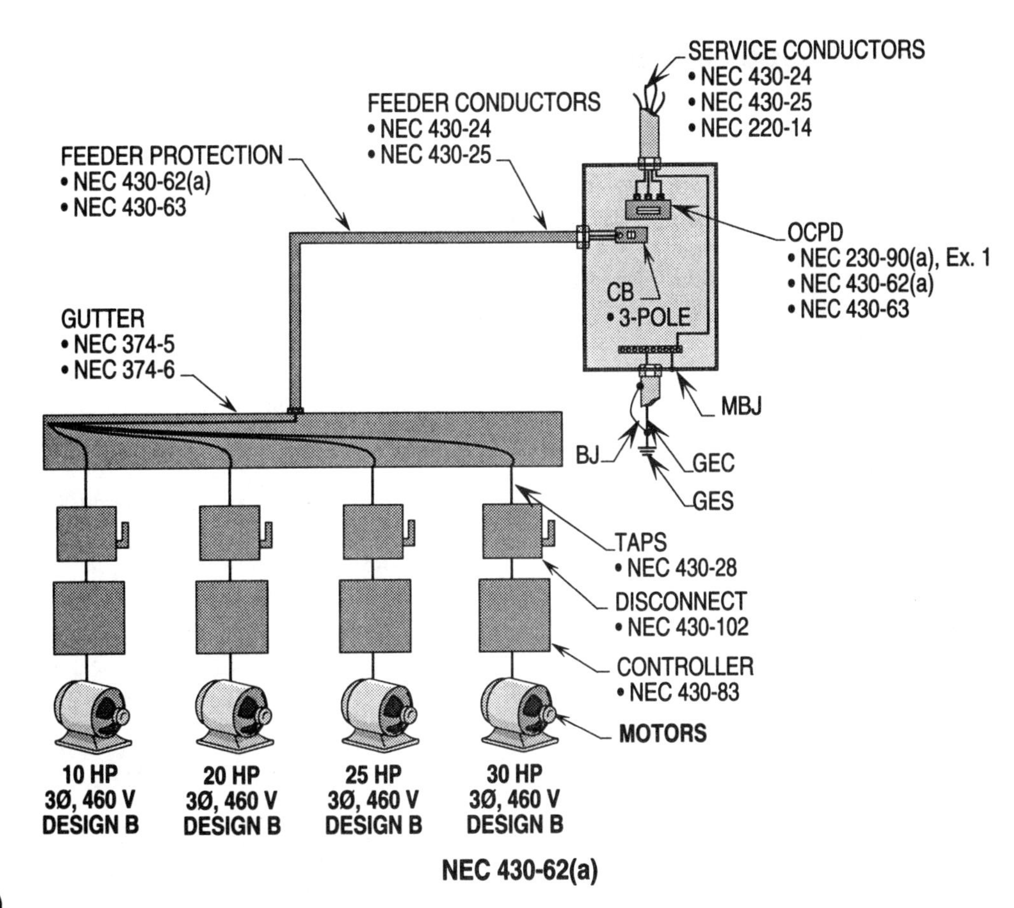

Ill. 13-24(a)

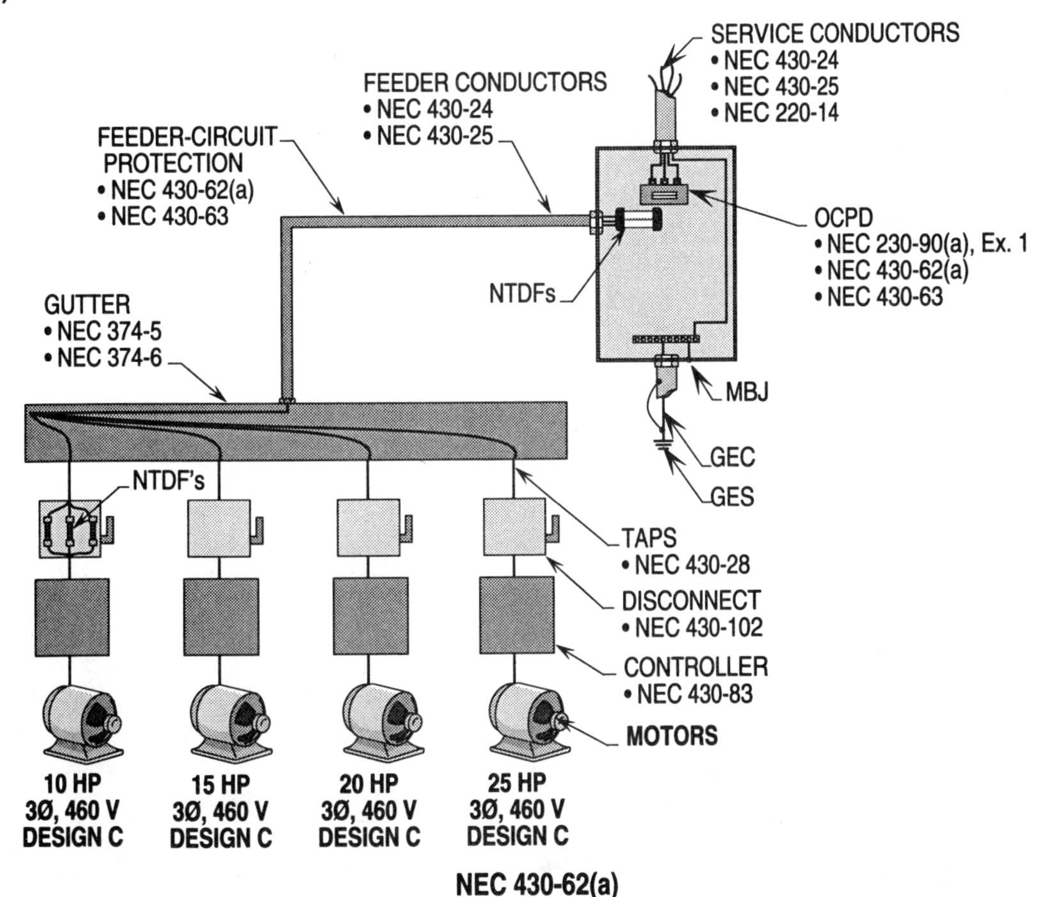

Ill. 13-24(b)

Sizing OCPD For Feeder Loads 430-62(a)

The size OCPD for a feeder-circuit supplying two or more motors is determined by selecting the next lower rating in each calculation based upon the percentages per Table 430-152 times the FLA per Table 430-150.

Example 13-24(c). What size time delay fuses are required for the feeder-circuit in Illustration 13-24(c) when applying the rules of Table 430-152?

Step 1: Finding FLA of motors
430-6(a)(1); Table 430-150
10 HP = 14 A
15 HP = 21 A
20 HP = 27 A
40 HP = 52 A

Step 2: Calculating feeder OCPD
430-52(c)(1); Table 430-152;
430-62(a); 240-6(a)
52 A x 175% = 91 A
requires 91 A OCPD = 90 A
Plus other motors
in group = 27 A
= 21 A
= 14 A
Total amps = 152 A

Step 3: Selecting OCPD
430-62(a); 240-3(g); 240-6(a)
152 A requires 150 A OCPD
(152 A not a standard OCPD)

Solution: The size OCPD required for the feeder-circuit is 150 amps.

Matching Values using 6 x FLC
Table 430-151(B); use 6 x FLC's
- **14 A + 21 A + 27 A + 52 A = 114 A**
- **114 A LRC x 6 = 684 A**
- **684 A LRC**
  150 A TDF x 5 = 750 A
  (150 A x 5 = 750 A)
- **750 A allows all 4 motors to start**

Example 13-24(d). What size instantaneous trip CB is required for the feeder-circuit in Illustration 13-24(d)?

Step 1: Finding FLA of motor
430-6(a)(1); Table 430-150
10 HP = 14 A
15 HP = 21 A
20 HP = 27 A
25 HP = 34 A

Step 2: Calculating feeder OCPD
430-52(c)(3); Table 430-152;
430-62(a), Ex.; 240-6(a)
34 A x 800% = 272 A
requires a 272 A Setting
Plus other motors = 272 A
in group = 27 A
= 21 A
= 14 A
Total amps = 334 A

Step 3: Selecting OCPD
430-62(a); 240-3(g); 240-6(a)
334 A is the minimum setting

Matching Values using 6 x FLC
Table 430-151(B); use 6 x FLC's
- **14 A + 21 A + 27 A + 34 A = 96 A**
- **96 A LRC x 6 = 576 A**
- **576 A LRC**
  334 A setting = 334 A
- **334 A will not allow all 4 motors to start at same time.**

Solution: The minimum setting for the INST. CB for the feeder-circuit is 334 amps.

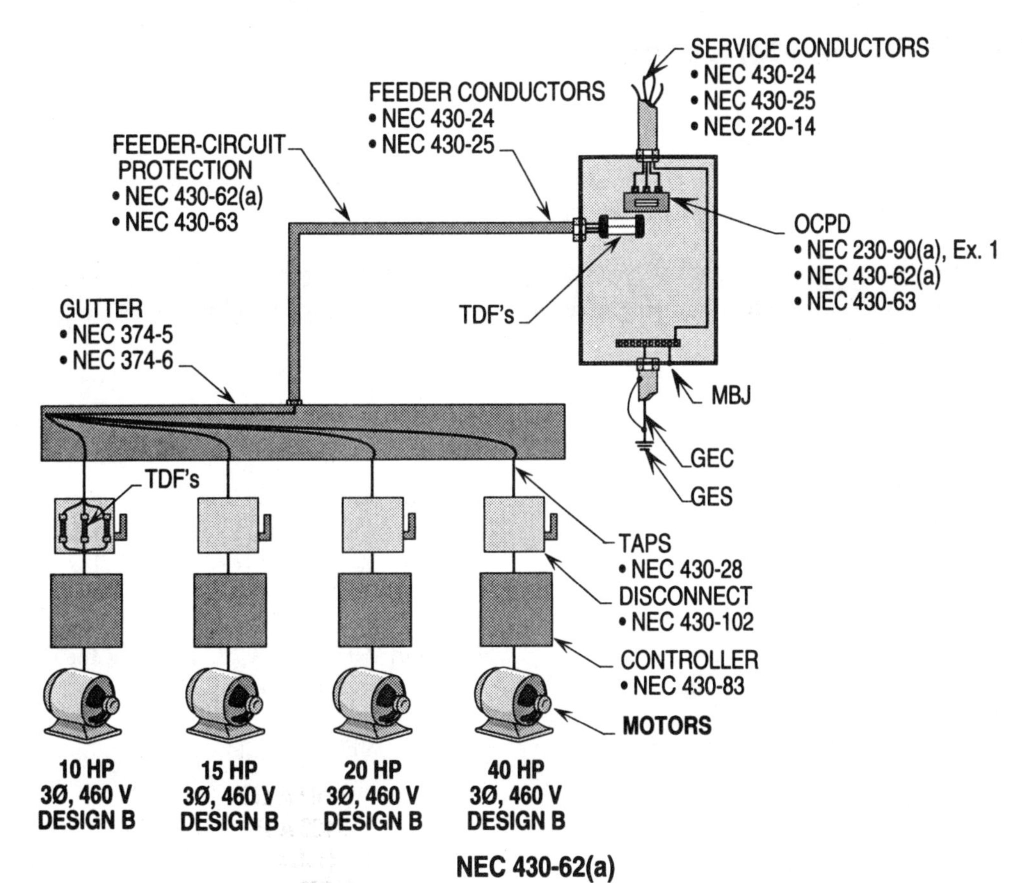

Ill. 13-24(c)

SERVICE CONDUCTORS
• NEC 430-24
• NEC 430-25
• NEC 220-14

FEEDER CONDUCTORS
• NEC 430-24
• NEC 430-25

FEEDER-CIRCUIT PROTECTION
• NEC 430-62(a)
• NEC 430-63

OCPD
• NEC 230-90(a), Ex. 1
• NEC 430-62(a)
• NEC 430-63

INST CB IN MOTOR CONTROL CENTER

CB
• 3-POLE

GUTTER
• NEC 374-5
• NEC 374-6

MBJ

GEC

GES

INST. CB

TAPS
• NEC 430-28

DISCONNECT
• NEC 430-102

CONTROLLER WITH INST. CB
• NEC 430-83
• NEC 430-52(c)(6)

MOTORS

10 HP 3Ø, 460 V DESIGN C
15 HP 3Ø, 460 V DESIGN C
20 HP 3Ø, 460 V DESIGN C
25 HP 3Ø, 460 V DESIGN C

NEC 430-62(a)

Ill. 13-24(d)

Sizing OCPD For Feeder Loads
430-62(a)

The OCPD for a feeder-circuit supplying two or more motors may be selected based upon the ampacity of the feeder conductors, if this procedure provides the greater size device.

Example 13-24(e). What size circuit breaker is required for the feeder-circuit in Illustration 13-24(e) when applying the Ex. 1 to 430-52(c)(1)?

Sizing CB based on motor load

Step 1: Finding FLA of motor
430-6(a)(1); Table 430-150
10 HP = 14 A
15 HP = 21 A
20 HP = 27 A
25 HP = 34 A

Step 2: Calculating feeder OCPD
430-52(c)(1), Ex. 1; Table 430-152;
430-62(a); 240-6(a)
34 A x 250% = 85 A

requires 90 A OCPD	= 90 A
Plus other motors in group	= 27 A
	= 21 A
	= 14 A
Total amps	=152 A

Step 3: Selecting OCPD
430-62(a); 240-3(g); 240-6(a)
152 A requires 150 A OCPD
(152 A not a standard OCPD)

Solution: **The size OCPD required for the feeder-circuit is 150 amps.**

Note 1: NEC 430-62(a) does not allow the next size OCPD above 152 amps per steps 2 and 3 above.

Note 2: The AHJ may require an 80 A OCPD instead of a 90 A OCPD as in Step 2. This means that the applications of Ex. 1 to NEC 430-52(c)(1) would not be allowed.

Example 13-24(f). What size OCPD is required for the feeder-circuit in Illustration 13-24(f) when applying the Ex. 1 to 430-52(c)(1)?

Sizing OCPD based on motor load

Step 1: Calculating OCPD
430-52(c)(1), Ex. 1; 430-62(a);
240-3(g); Table 430-152; Table 430-148
Motor loads

• 10 HP = 50	A x 250% = 125 A	= 125 A
• 1 HP = 16	A x 100%	= 16 A
• 2 HP = 24	A x 100%	= 24 A
• 5 HP = 28	A x 100%	= 28 A
• 5 HP = 28	A x 100%	= 28 A
Total load		= 221 A

Step 2: Other loads

• Lighting loads	= 40 A x 125%	= 50 A
• Receptacle loads	= 37.5 A x 100%	= 37.5 A
• Appliance loads	= 30 A x 125%	= 37.5 A
• Heat load		= 42 A
Total load		= 167 A

Step 3: All loads
430-62(a)

• Motor Loads	= 221 A
• Other loads	= 167 A
Total loads	= 388 A

Step 4: Selecting OCPD
430-62(a)
350 A OCPD is the next size below 388 A

Solution: **The size OCPD based upon motor loads is 350 amps.**

Note: In most cases the calculated load produces the largest or same size OCPD unless there is an unusually large motor involved.

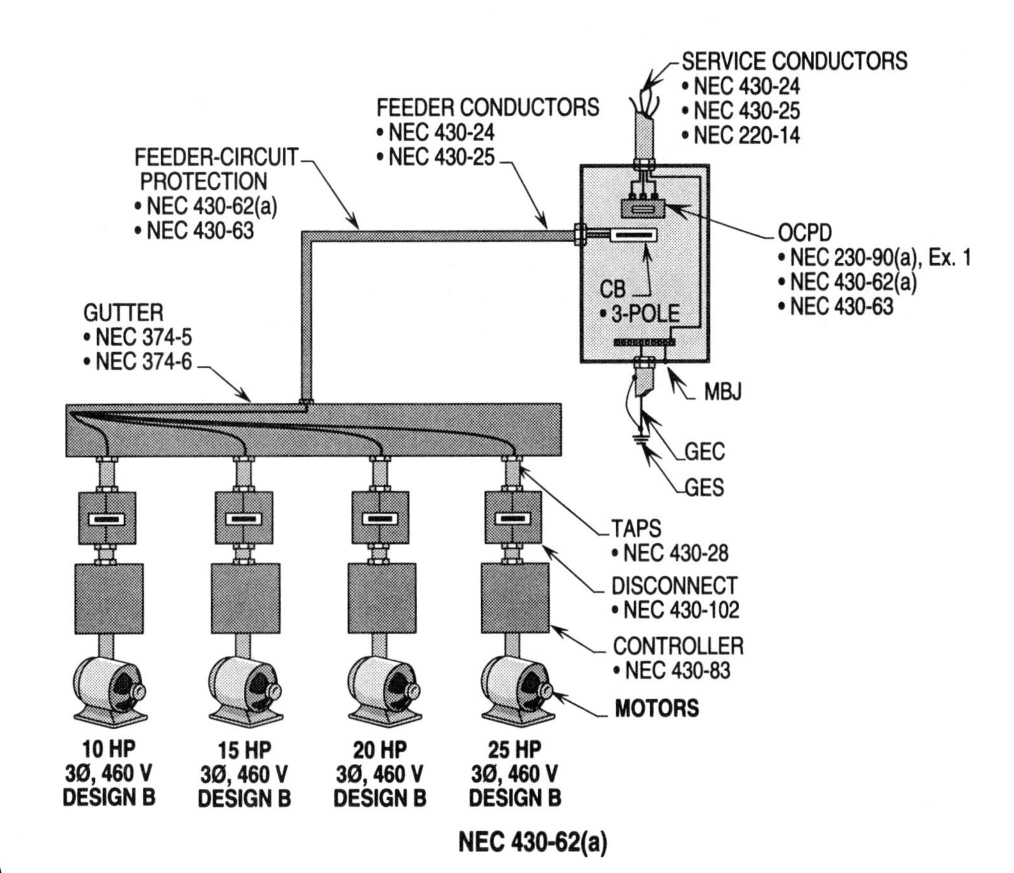

Ill. 13-24(e)

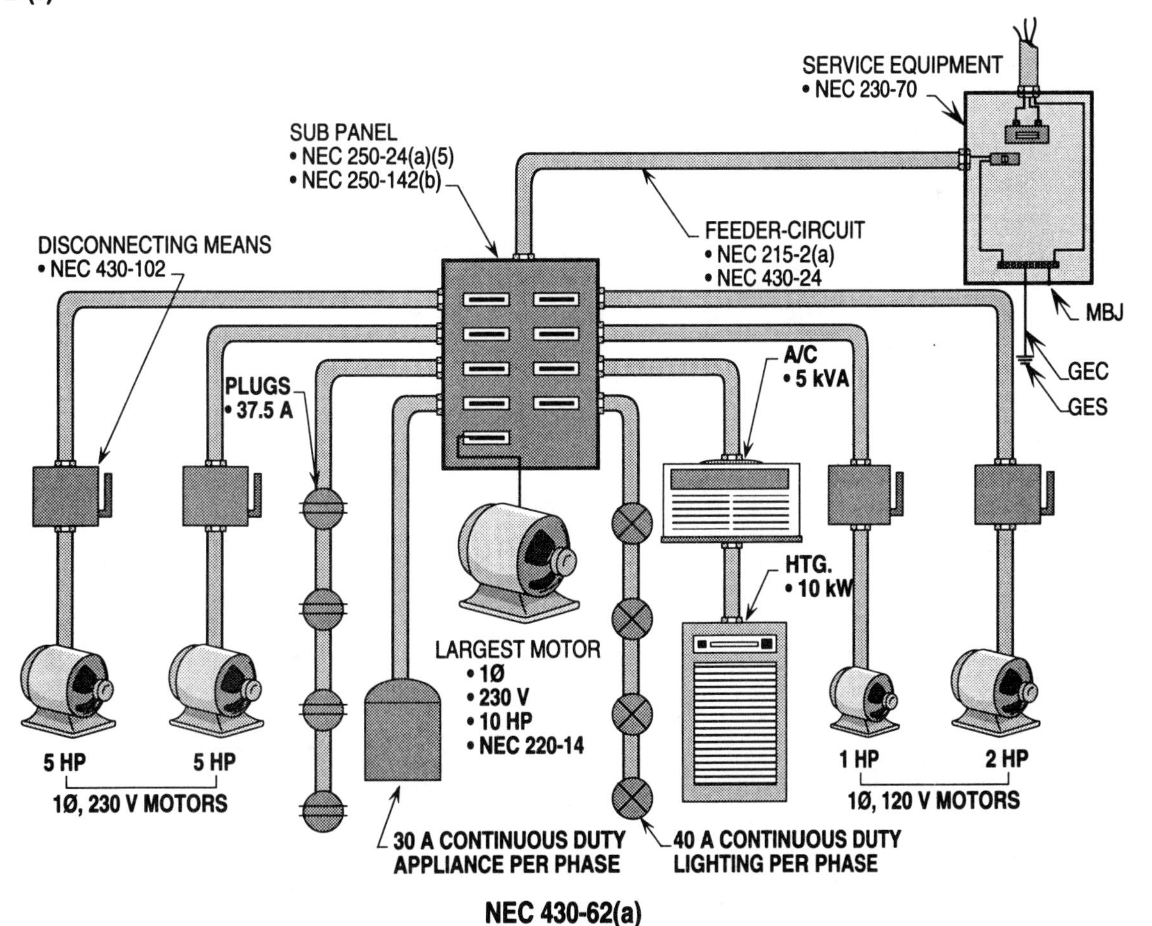

Ill. 13-24(f)

Sizing OCPD For Feeder Loads 430-63

Feeder-circuits may be utilized in supplying one or more motors plus other loads. The size OCPD is computed per Articles 430, 220 and 215 of the NEC.

Example 13-24(g). What size OCPD is required for the feeder-circuit in Illustration 13-24(g)?

Sizing OCPD based on conductors

Step 1: Calculating loads
215-3; 215-2(a); Table 430-148;
220-14; 430-63

- Lighting load
  40 A x 125% = 50.0 A
- Receptacle load
  37.5 A x 100% = 37.5 A
- Appliance load
  30 A x 125% = 37.5 A
- Heat or A/C load
  10 kW x 1000 x 100% / 240 V = 42.0 A
- Motor load
  10 HP = 50 A x 100% = 50.0 A
  5 HP = 28 A x 100% = 28.0 A
  5 HP = 28 A x 100% = 28.0 A
  1 HP = 16 A x 100% = 16.0 A
  2 HP = 24 A x 100% = 24.0 A
- Largest motor load
  50 A x 25% = 12.5 A

Total load = 325.5 A

Step 2: Selecting conductors
310-10(2); Table 310-16
325.5 A requires #400 KCMIL THWN cu.

Step 3: Selecting OCPD based on conductors
215-3; 240-3(b); 430-63
400 KCMIL THWN cu. = 335 A
335 A requires 350 A OCPD

Solution: The size OCPD is allowed to be a 350 amp CB based upon amps of conductors.

Note 1: The calculated load allows a 350 amp circuit breaker per Example 13-24(f).

Note 2: If the largest OCPD for any motor of the group is calculated and added to the remaining loads, the same OCPD is usually produced as in Step 3 above.

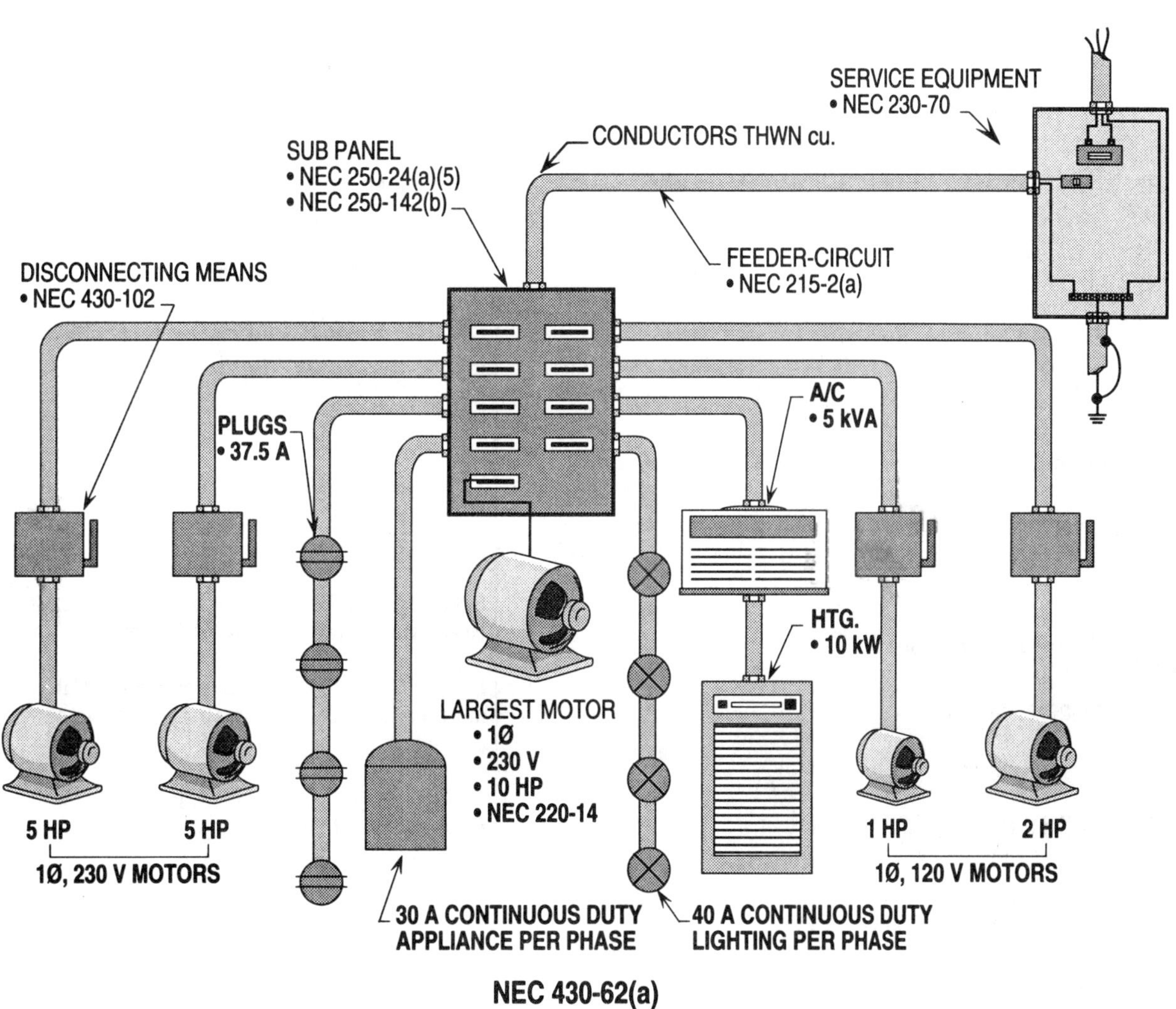

SERVICE EQUIPMENT
• NEC 230-70
CONDUCTORS THWN cu.
SUB PANEL
• NEC 250-24(a)(5)
• NEC 250-142(b)
FEEDER-CIRCUIT
• NEC 215-2(a)
DISCONNECTING MEANS
• NEC 430-102
PLUGS
• 37.5 A
A/C
• 5 kVA
HTG.
• 10 kW
LARGEST MOTOR
• 1Ø
• 230 V
• 10 HP
• NEC 220-14
5 HP
5 HP
1Ø, 230 V MOTORS
1 HP
2 HP
1Ø, 120 V MOTORS
30 A CONTINUOUS DUTY APPLIANCE PER PHASE
40 A CONTINUOUS DUTY LIGHTING PER PHASE
NEC 430-62(a)

Ill. 13-24(g)

Sizing Disconnecting Means 430-110(a); 430-57

The disconnecting means to deenergize a motor can be sized by multiplying the full load current of the motor by 115 percent. It can also be selected by the size fuse or circuit breaker required to allow the motor to start and run.

Example 13-25. What size disconnecting means is required for the motor in Illustration 13-24(h), if a nonfused or fused disconnect is utilized?

Sizing Nonfused Disconnect

Step 1: Finding FLA
430-6(a)(1); Table 430-150
50 HP = 65 A

Step 2: Calculating size
430-110(a)
65A x 115% = 74.75 A

Step 3: Selecting size
Chart
74.75 A requires 100 A

Solution: The size disconnect required is 100 amps.

Note: The size disconnect using a nonautomatic CB is 80 amps.

Sizing Fused Disconnect

Step 1: Finding FLA
430-6(a)(1); Table 430-150
50 HP = 65 A

Step 2: Calculating size TDF's
430-6(a)(1); Table 430-152
Design B allows 175%
65 A x 175% = 113.8 A

Step 3: Selecting size
Chart
113.8 A requires 200 A

Solution: The size disconnect required is 200 amps.

Note: The size disconnect using an automatic CB is 150 amps.
(65 A x 250% = 162.5 A =150 A per Table 430-152 and NEC 430-52(c)(1))

ADDITIONAL HELP 66:

Calculating the HP rating for a 20 HP, 460 volt, three-phase motor having a nameplate current of 31 amps.

- Finding A of motor
  Table 430-150
  20 HP = 27 A

- Calculating HP rating

$$HP = \frac{31\text{ A} - 27\text{ A}}{34\text{ A} - 27\text{ A}} = \frac{4}{7}$$

$$HP = \frac{4 \times (25 - 20\text{ HP})}{7} = \frac{20}{7}$$

$$HP = \frac{20}{7}$$

HP = 2.857
HP = 2.857 + 20
HP = 22.9

Solution: The HP rating of the motor is 22.9 due to the nameplate current of 31 amps.

Note 1: The nameplate current of the 20 HP motor is 31 amps which is greater than the Table 430-150 current of 27 amps.

Note 2: The value 34 amps is based on the nameplate current of the next largest motor.

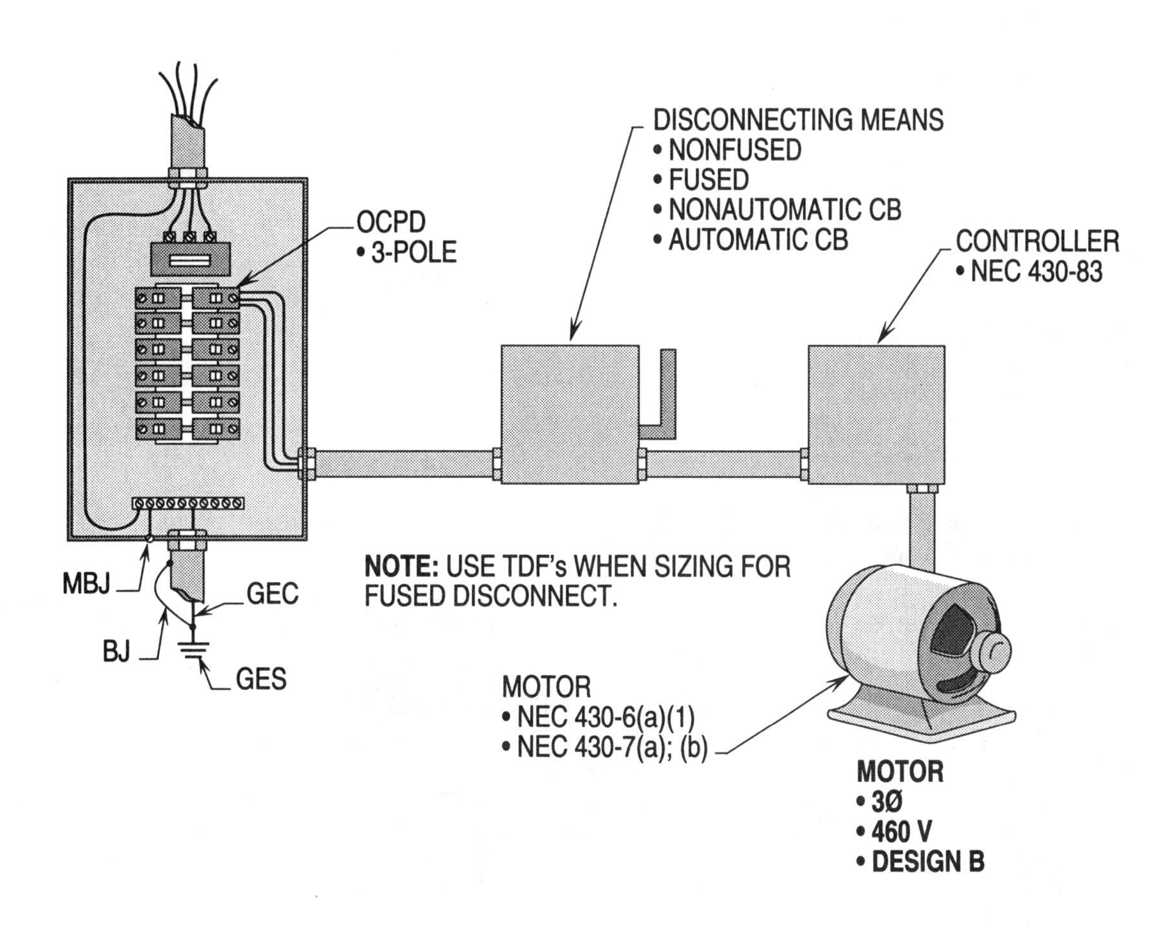

NEC 430-57
NEC 430-110(a)

Ill. 13-24(h)

Sizing and protecting control circuit conductors 430-72(b)(2); Table 430-72(b)

Control circuit conductors that do not extend beyond the control enclosure are selected per Column B of Table 430-72(b). Remote control circuit conductors are selected from Column C of Table 430-72(b) based upon the upstream OCPD.

Example 13-25. What size OCPD is required for the control-circuit conductors supplying the start and stop station in Illustration 13-25?

Sizing OCPD for remote control circuits

Step 1: Finding amperage
Table 310-16
No. 14 = 20 A

Step 2: Applying percentage based on OCPD
430-72(b)(2); Table 430-72(b), Column C
15 A x 300% = 45 A OCPD

Solution: The overcurrent protection device is required to be 45 amps.

Sizing OCPD when controls are on cover

Step 1: Finding amperage
Table 310-17
No. 14 = 25 A

Step 2: Applying percentage
430-72(b)(2); Table 430-72(b), Column B
25 A x 400% = 100 A OCPD

Solution: The overcurrent protection device is required to be 100 amps.

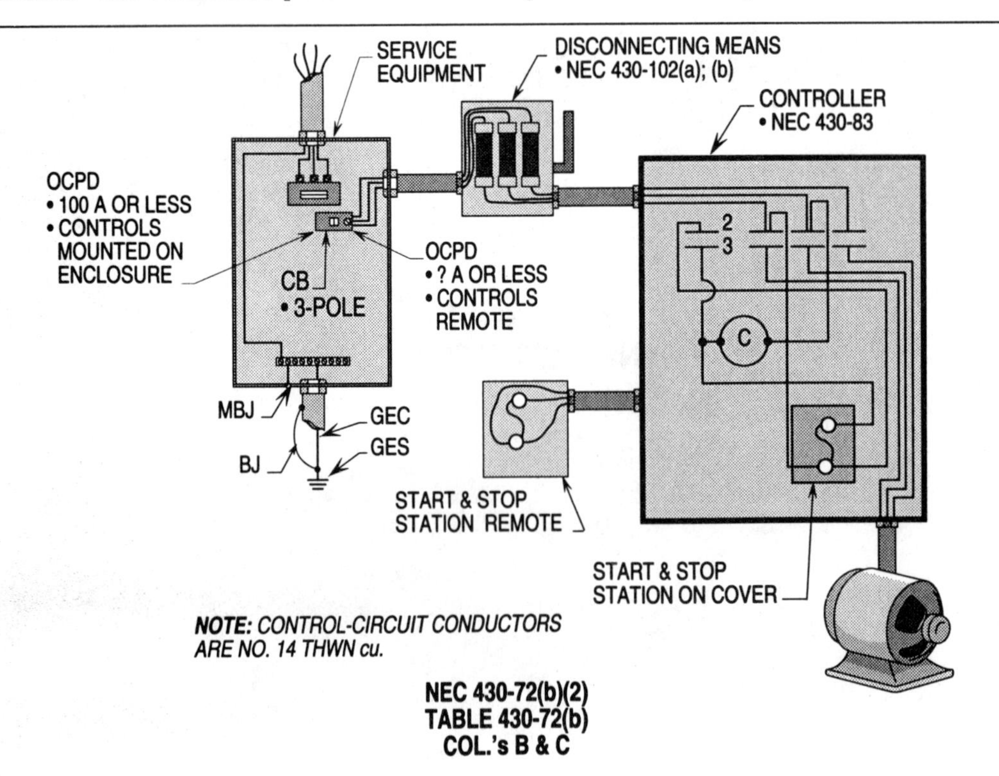

Ill. 13-25

14

A/C & Refrigerating Equipment

The conductors supplying power to HACR equipment are sized from the full-load amp (FLA) ratings of the compressor and condenser motor. These FLA ratings are increased by 125 percent per NEC 440-32.

The overcurrent protection devices (OCPD's) protecting the branch-circuits from short-circuit currents are sized from the provisions listed in NEC 440-22(a), which requires the FLA ratings to be increased from 175 percent up to 225 percent to allow the HACR equipment to start and run.

The elements used to supply the branch-circuits to HACR equipment can be selected by the branch-circuit selection currents listed on the nameplate of such equipment per NEC 440-4(c) and NEC 110-3(b).

Quick Reference

Rating And Interrupting Of Disconnecting Means 440-12(a)

The disconnecting means used to disconnect a hermetic refrigerant motor compressor must be capable of disconnecting both full-load amps (FLA) and locked rotor amps (LRA).

Example 14-1(a). What size nonautomatic, nonfused, and horsepower rated disconnecting means is required to disconnect the A/C unit in Illustration 14-1(a)?

Sizing nonautomatic or nonfused disconnect

Step 1: Calculating disconnect
440-12(a)(1)
(29 A + 2.5 A) x 115% = 36.2 A

Note: To determine the HP rating of the compressor motor, see **Additional Help** 64 on page 13-62.

Step 2: Selecting disconnect
440-12(a)(1); 240-6(a); chart
36.2 A requires 40 A CB
36.2 A requires 60 A disconnect

Solution: The minimum size nonautomatic CB is 40 amps and nonfused disconnect is 60 amps.

Sizing HP of disconnect

Step 1: Calculating disconnect
440-12(a)(2); Table 430-150;
Table 430-151(B)
31.5 FLA requires 25 HP
183 LRA requires 25 HP

Solution: The disconnecting means is required to be rated at least 25 HP.

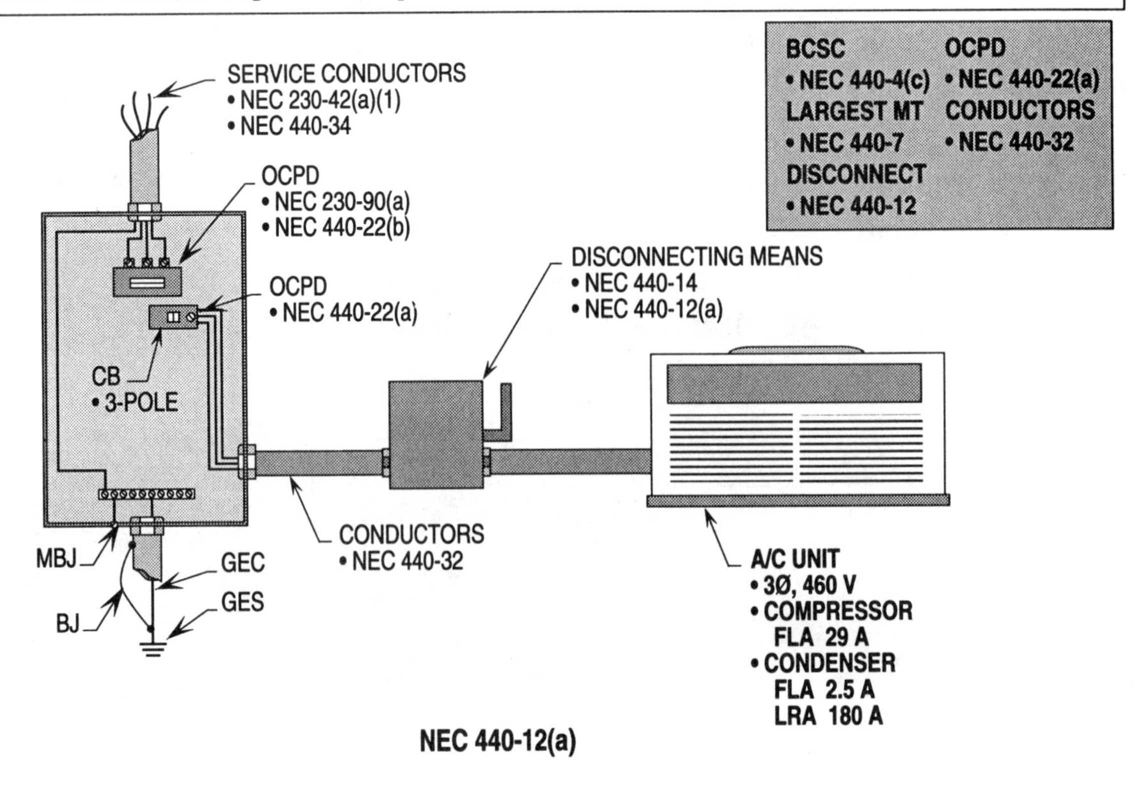

Ill. 14-1(a)

Rating And Interrupting Of Disconnecting Means 440-12(a)

Example 14-1(b). What size disconnecting means rated in horsepower is required to disconnect the A/C units in Illustration 14-1(b)?

Sizing disconnect for feeder-circuit

Step 1: Calculating disconnect
440-12(b); Table 430-150;
Table 430-151(B)

FLA = HP	LRA = HP
30 A = 15	200 A = 15
25 A = 10	160 A = 10
20 A = 7 1/2	140 A = 10
75 A = 32 1/2	500 A = 35

Step 2: Selecting disconnect
440-12(b)
FLA = HP
75 A = 32 1/2
LRA = HP
500 = 35

Solution: The disconnecting means must be rated at 35 HP and be capable of disconnecting 500 amps of locked rotor current.

Note 1: This procedure is to determine the size disconnect used even if the largest load was a motor and not a compressor. The disconnect shall be capable of disconnecting both the FLA and LRA of the A/C unit.

Note 2: The NEC, based upon LRC, requires at least a 35 HP rated disconnect.

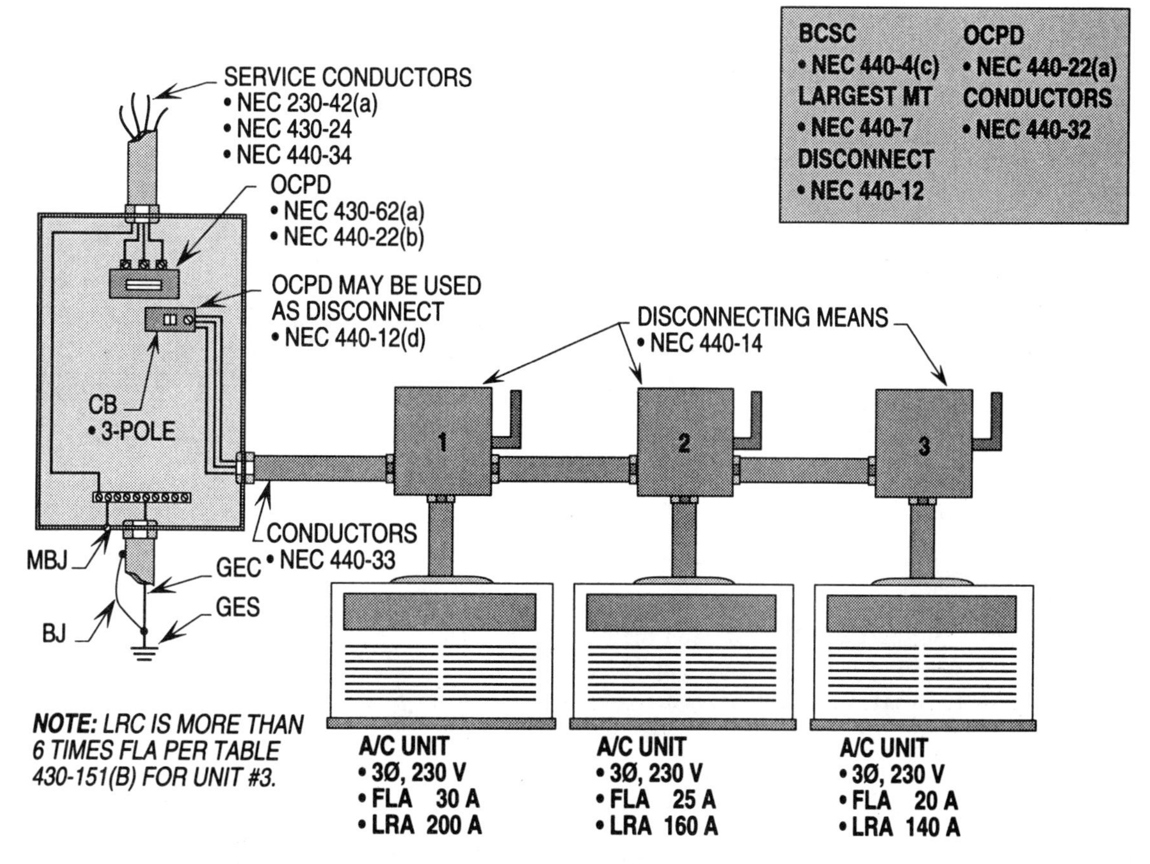

Ill. 14-1(b)

Rating Or Setting For Individual Motor-compressor 440-22(a)

The OCPD for hermetic seal compressors are selected at 175 percent or 225 percent of the compressor's FLA rating or the branch-circuit selection current, whichever is greater.

Example 14-2(a). What is the minimum and maximum size OCPD for the unit in Illustration 14-2(a)?

Sizing minimum OCPD

Step 1: Calculating OCPD
440-22(a)
19 A x 175% = 33.25 A

Step 2: Selecting OCPD
240-3(g); 240-6(a)
33.25 A requires 30 A OCPD

Solution: The minimum size OCPD is 30 amps.

Sizing maximum OCPD

Step 1: Calculating OCPD
440-22(a)
19 A x 225% = 42.75 A

Step 2: Selecting OCPD
240-3(g); 240-6(a)
42.75 A requires 40 A OCPD

Solution: The maximum size OCPD is 40 amps.

Example 14-2(b). What size OCPD is required for the A/C unit in Illustration 14-2(b)?

Sizing OCPD

Step 1: Selecting OCPD
440-4(c); 440-22(a)
BCSC requires 40 A OCPD

Solution: The size OCPD based upon the branch-circuit selection current is a 40 amp HACR CB or fuses rated at 40 amps each.

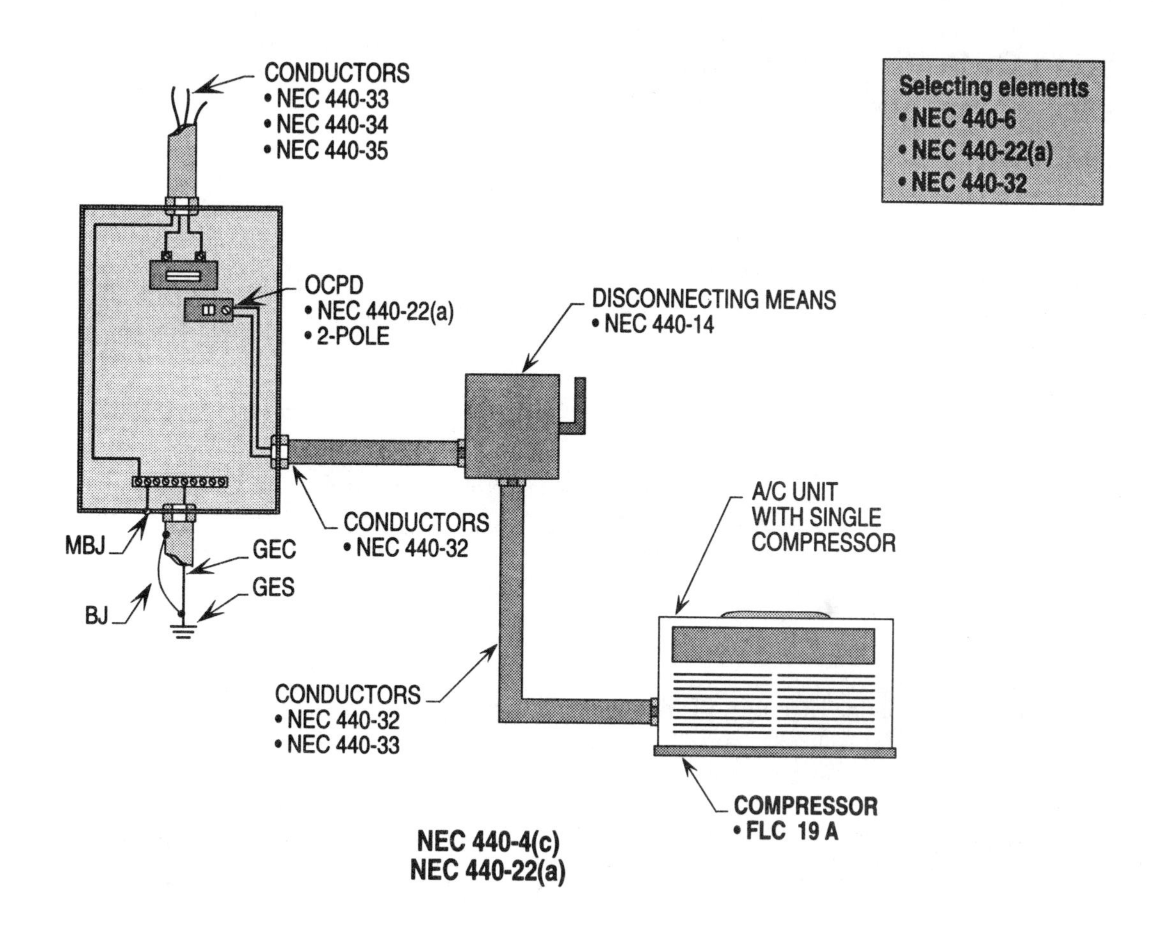

Ill. 14-2(a)

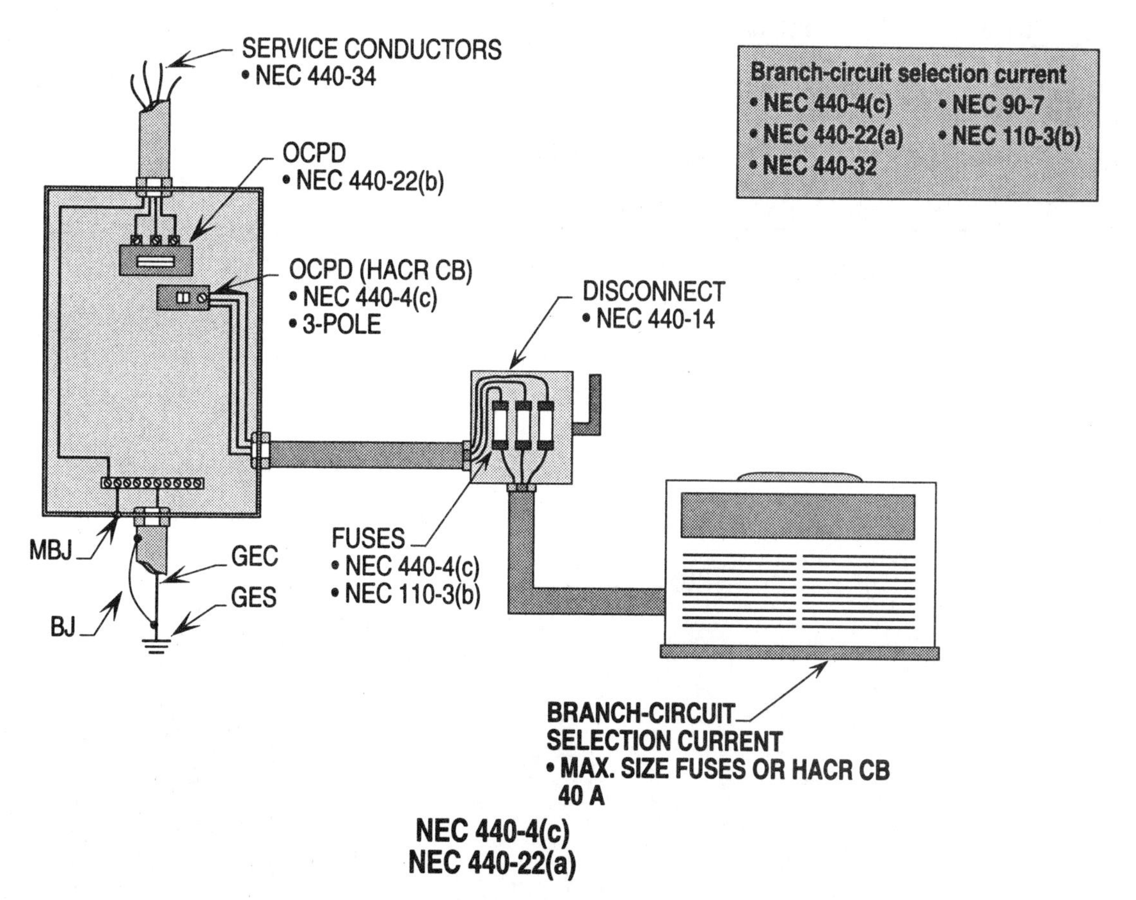

Ill. 14-2(b)

Rating Or Setting For Individual Motor-compressor 440-22(b)(1)

The OCPD for a feeder-circuit supplying two or more air conditioning or refrigerating units shall be sized to allow the largest unit to start and allow the other units to start at different intervals of time.

Example 14-3. What size OCPD is required for the feeder-circuit in Illustration 14-3 where the compressor is considered having the largest OCPD in the group?

Sizing OCPD for feeder-circuit

Step 1: Calculating largest OCPD rating for compressor
440-22(b)(1); 440-22(a)
28 A x 175% + 2.5 = 51.5

Step 2: Calculating OCPD for feeder-circuit
440-22(b)(1)
51.5 A + 25 A + 2.5 A + 23 A + 2.5 A = 104.5 A

Step 3: Selecting OCPD for feeder-circuit
240-3(g); 240-6(a)
104.5 A allows 100 A OCPD

Solution: **The size OCPD for the feeder-circuit using a circuit breaker is 100 amps.**

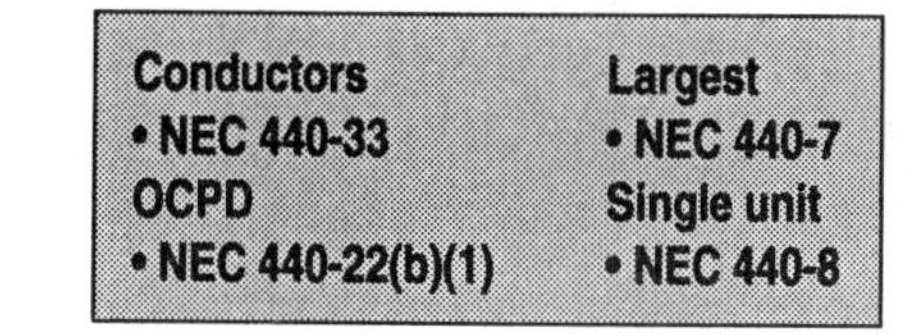

Conductors	Largest
• NEC 440-33	• NEC 440-7
OCPD	Single unit
• NEC 440-22(b)(1)	• NEC 440-8

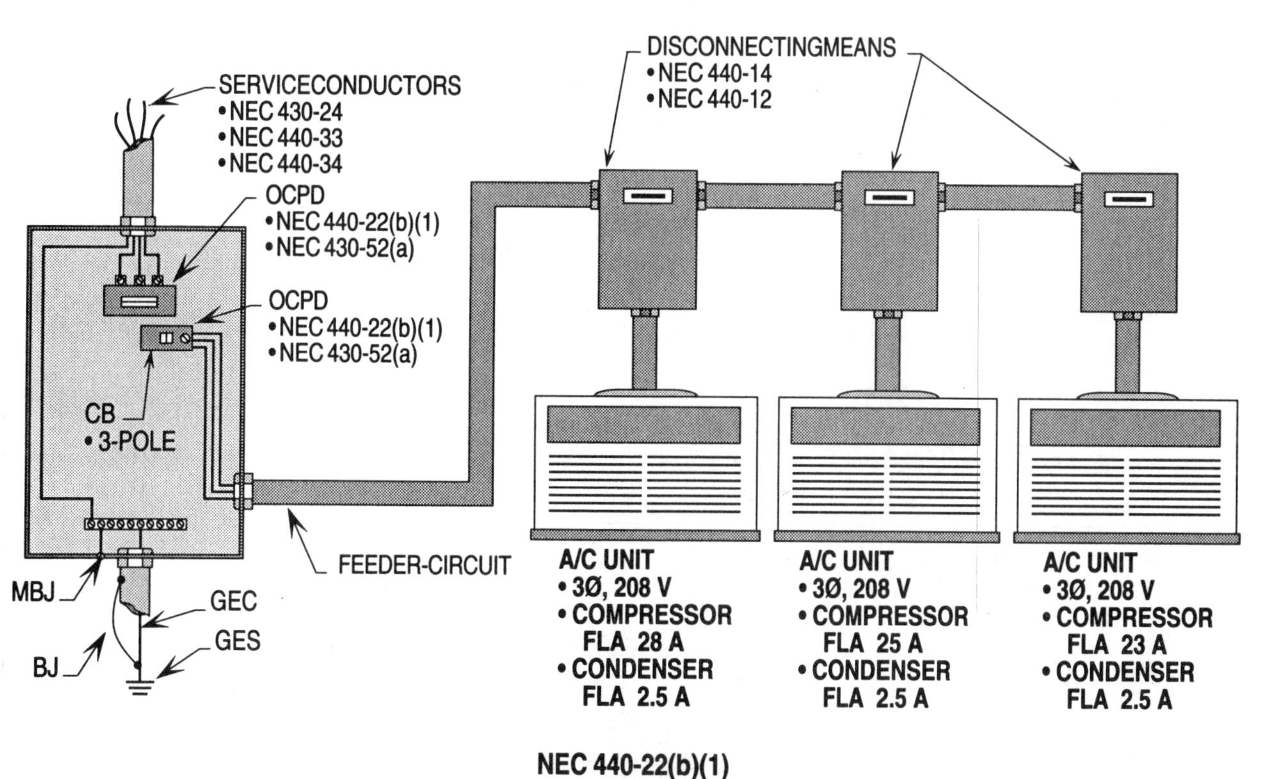

Ill. 14-3

Rating Or Setting For Equipment 440-22(b)(2)

The largest OCPD for a service load or feeder-circuit load containing motors and compressors is based upon the largest device for any motor or compressor in the group. The remaining loads are added to this single device rating.

Example 14-4(a). What size OCPD is required for the feeder-circuit in Illustration 14-4(a) where the motor has the largest OCPD?

Sizing OCPD for feeder-circuit

Step 1: Finding FLA for motor
440-22(b)(2); Table 430-150
15 HP = 46.2 A

Step 2: Calculating OCPD for motor
440-22(b)(2); Table 430-152
46.2 A x 250% = 115.5 A

Step 3: Selecting OCPD for motor
240-3(g); 430-52(c)(1); 240-6(a)
115.5 A allows 110 A OCPD

Step 4: Calculating OCPD for feeder-circuit
440-22(b)(2); 430-62(a)
110 A + 25 A + 23 A = 158 A

Step 5: Selecting OCPD for feeder-circuit
440-22(b)(2); 240-6(a)
158 A allows 150 A OCPD

Solution: The size OCPD for the feeder-circuit using a circuit breaker is 150 amps.

Note 1: The conductors supplying power to a compressor with other motors are computed at 125 percent times the largest FLA plus the FLA of the other motors.

Note 2: The same rule applies where only compressors are supplied by such conductors.

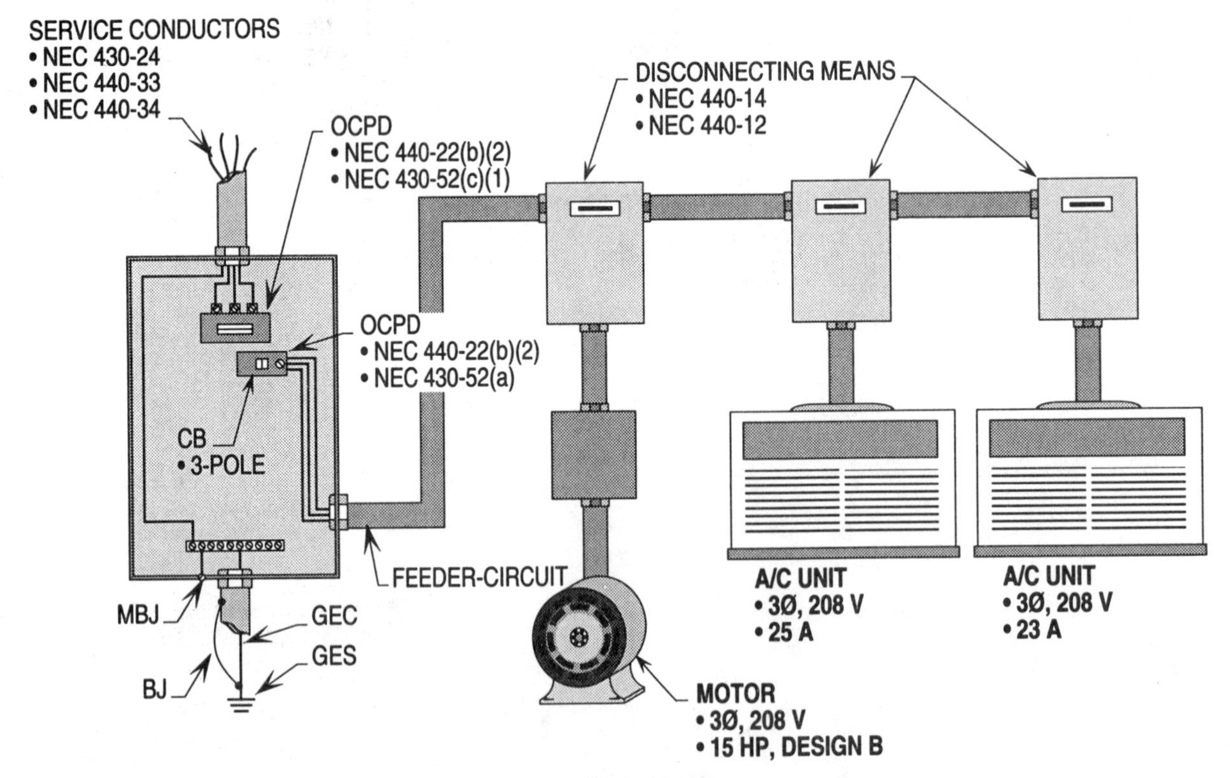

Ill. 14-4(a)

Single Motor For Branch-circuit
440-32

The conductors supplying power to an air-conditioning or refrigerating unit shall be sized to carry the load of the unit plus an overload for a short period of time.

Example 14-4(b). What size THHN copper conductors are required to supply the A/C unit in Illustration 14-4(b)?

Sizing conductors based on calculation

Step 1: Calculating FLA
440-32
25 A x 125% + 2.5 A = 33.75 A

Step 2: Selecting conductors
Table 310-16; 240-3(d)
33.75 A requires #10 cu.

Solution: The size conductors are No. 10 THHN copper.

Sizing conductors based on branch-circuit selection current

Step 1: Selecting conductors
440-4(c); 440-33
35 A requires #10 cu.

Solution: The branch-circuit selection current requires No. 10 THHN copper conductors.

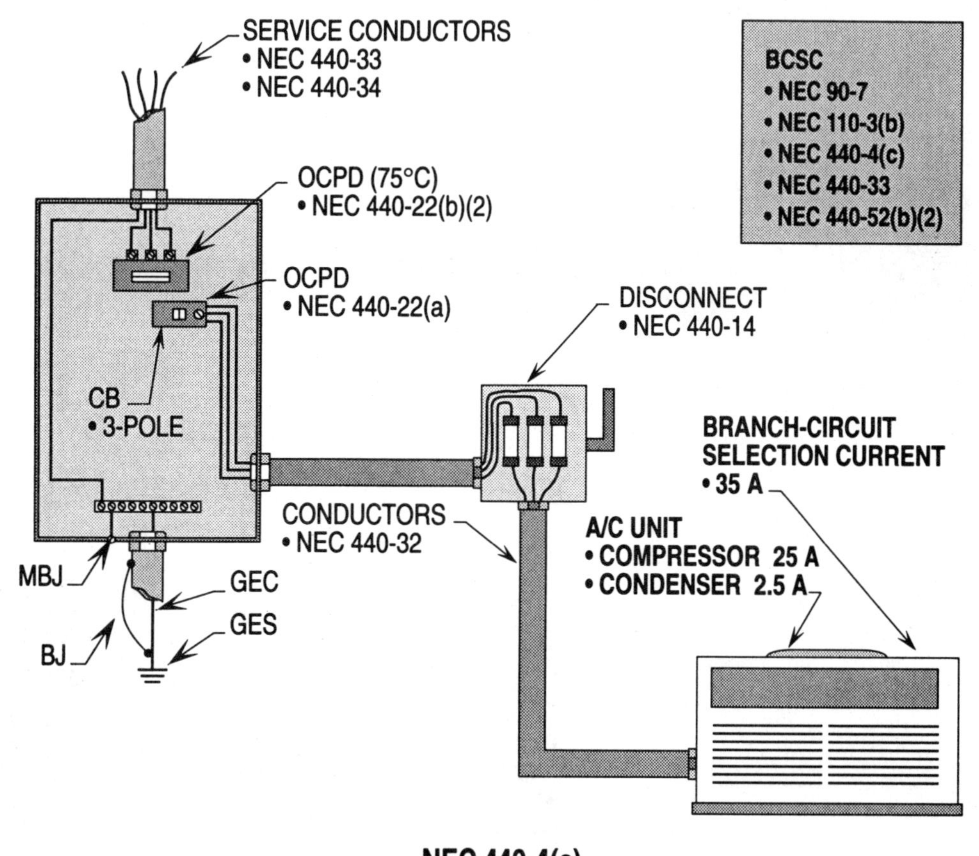

Ill. 14-4(b)

Two Or More A/C Units On A Feeder-circuit 440-33

Two or more compressors plus other motor loads can be connected to a feeder-circuit. The largest compressor is computed at 125 percent of its FLA and the remaining compressor loads are added to this total at 100 percent of their FLA ratings.

Example 14-5(a). What size THWN copper conductors are required for the feeder-circuit in Illustration 14-5(a)?

Sizing conductors for the feeder

Step 1: Finding FLA
A/C unit #1 = 40 A + 3 A
A/C unit #2 = 25 A + 2.5 A
A/C unit #3 = 23 A + 2.5 A

Step 2: Calculating FLA
440-33

40 A x 125% + 3 A	=	53 A
25 A x 100% + 2.5 A	=	27.5 A
23 A x 100% + 2.5 A	=	25.5 A
Total	=	106 A

Step 3: Selecting conductors
310-10(2); Table 310-16
106 A requires #2 cu.

Solution: The size THWN conductors are No. 2 copper.

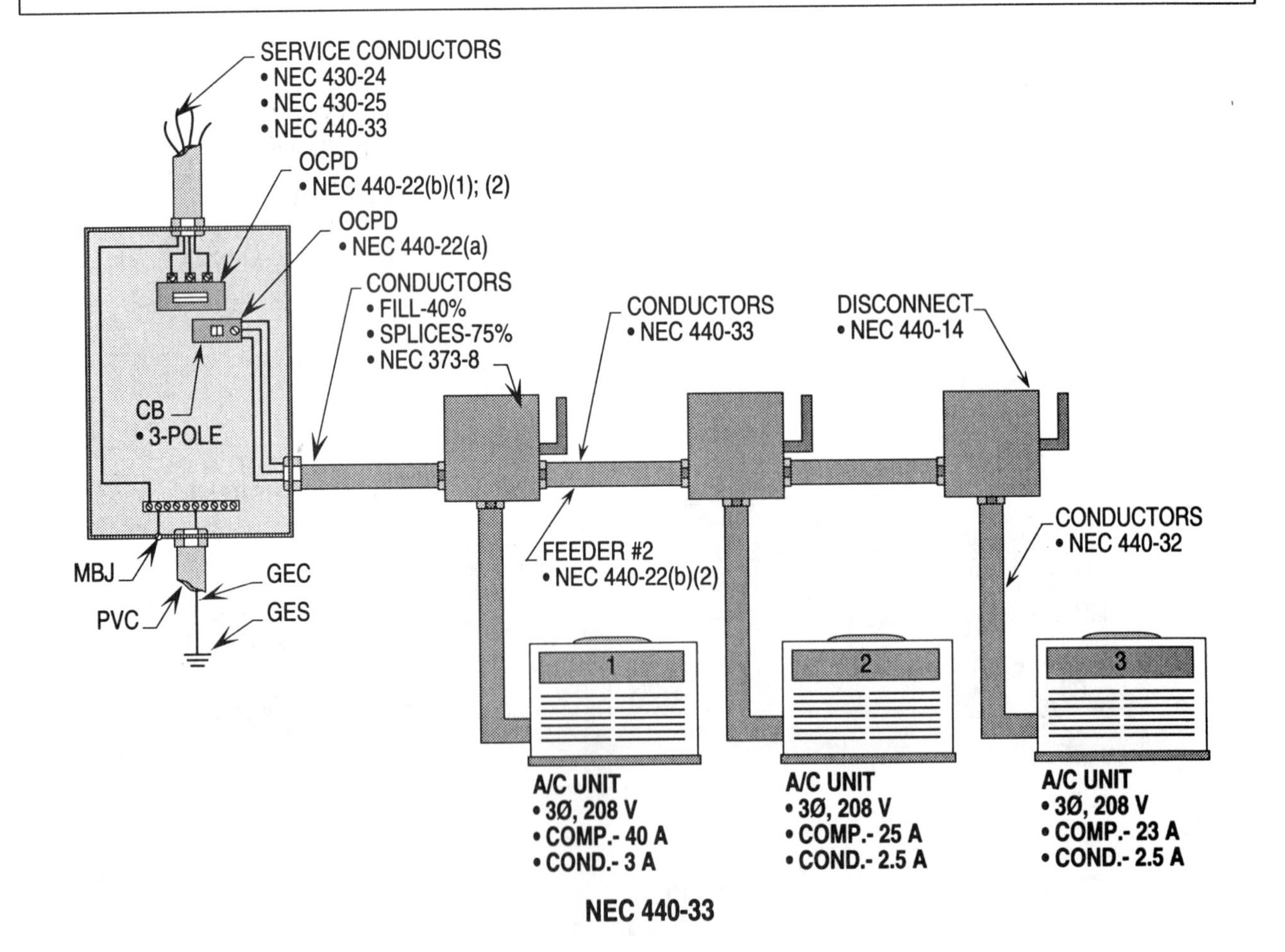

Ill. 14-5(a)

Two Or More A/C Units On A Feeder-circuit 440-33

For units with a branch-circuit selection current, the circuit conductors are selected and based upon the nameplate values.

Example 14-5(b). What size THHN copper conductors are required for the feeder-circuit in Illustration 14-5(b)?

Sizing conductors for feeder-circuit

Step 1: Finding FLA
440-33; 440-4(c); 110-3(b)
Branch-circuit selection current = 115 A

Step 2: Selecting conductors
310-10(2); Table 310-16
Branch-circuit selection current requires #2 cu.

Solution: **The size copper conductors are No. 2 THHN copper.**

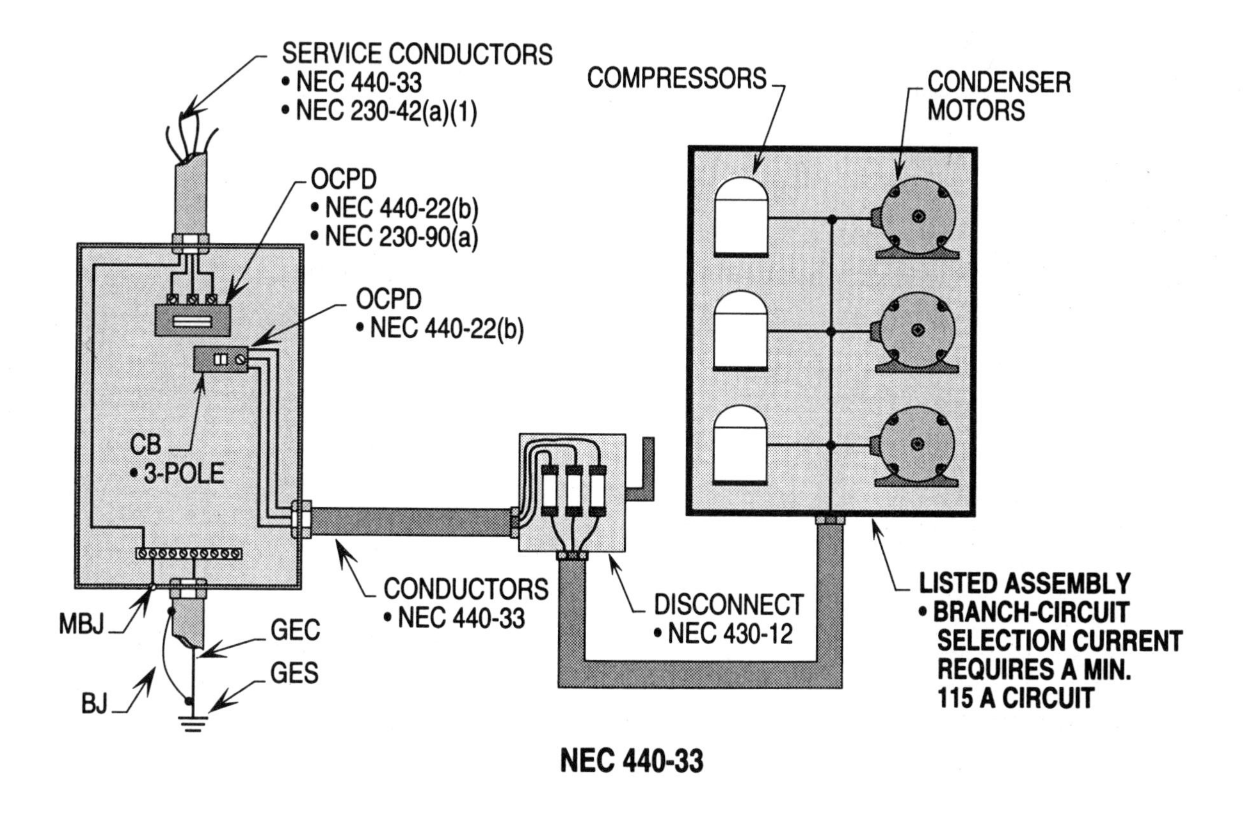

Ill. 14-5(b)

Two Or More A/C Units With Other Loads 440-34

Two or more compressors with motor loads plus other loads may be connected to a feeder-circuit or service conductors. The largest compressor or motor load is calculated at 125 percent plus 100 percent of the remaining compressors and motors. The other loads are computed at 125 percent for continuous and 100 percent for noncontinuous operation per NEC 215-2(a) and 230-42(a)(1).

Example 14-6. What size THWN copper conductors are required to supply power to the loads in Illustration 14-6?

Sizing conductors for feeder-circuit

Step 1: Finding FLA
440-34
A/C unit #1 = 33 A + 2.5 A
A/C unit #2 = 27 A + 2.5 A
A/C unit #3 = 21 A + 2.5 A
Other loads = 82 A

Step 2: Calculating FLA
440-34; 220-10; 215-2(a)

33 A x 125% + 2.5 A	=	43.75 A
27 A x 100% + 2.5 A	=	29.5 A
21 A x 100% + 2.5 A	=	23.5 A
65.6 A x 125%	=	82 A
Total load	=	178.75 A

Step 3: Selecting conductors
310-10(2); Table 310-16
178.75 A requires #3/0 cu.

Solution: **The size THWN conductors for the feeder-circuit are No. 3/0 copper.**

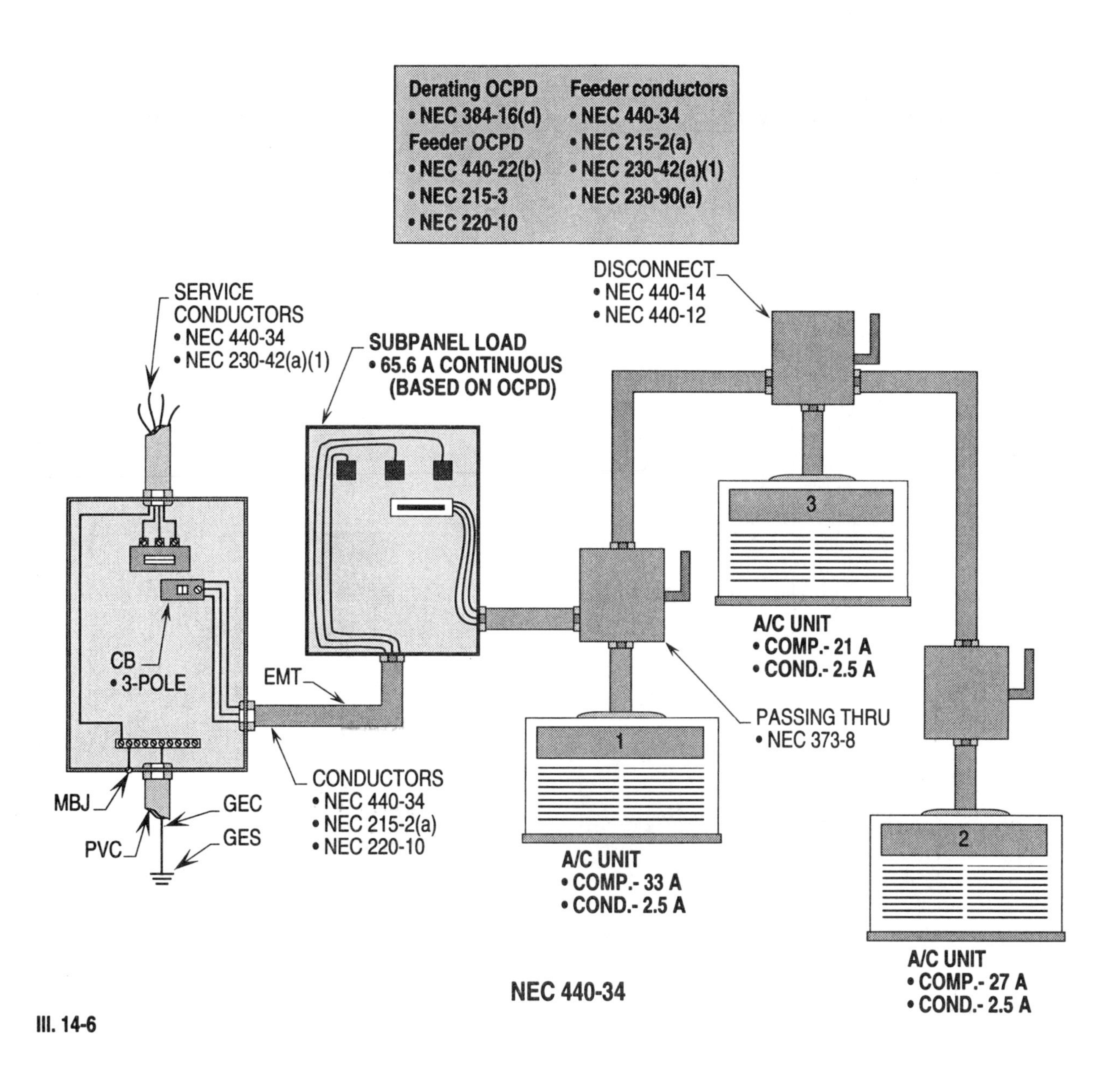

NEC 440-34

Ill. 14-6

Overload Protection
440-52(a)(1); (2)

The overload (OL) protection for compressors may be accomplished by using OCPD's in separate enclosures, separate overload relays or thermal protectors which are an integral part of the compressor.

Example 14-7. What size overload protective devices are required to prevent excessive overloading for the A/C units in Illustration 14-7?

Sizing OL's for A/C unit #1

Step 1: Finding FLA
440-52(a)(3)
FLA = 25 A

Step 2: Calculating FLA
440-52(a)(3)
25 A x 125% = 31.25 A

Step 3: Selecting TDF's
240-6(a)
31.25 A requires 30 A TDF's

Solution: The size time delay fuses are 30 amps.

Sizing OL's for A/C unit #2

Step 1: Finding FLA
440-52(a)(1)
FLA = 25 A

Step 2: Calculating FLA
440-52(a)(1)
25 A x 140% = 35 A OL's

Solution: The size of the overloads must not exceed 35 amps.

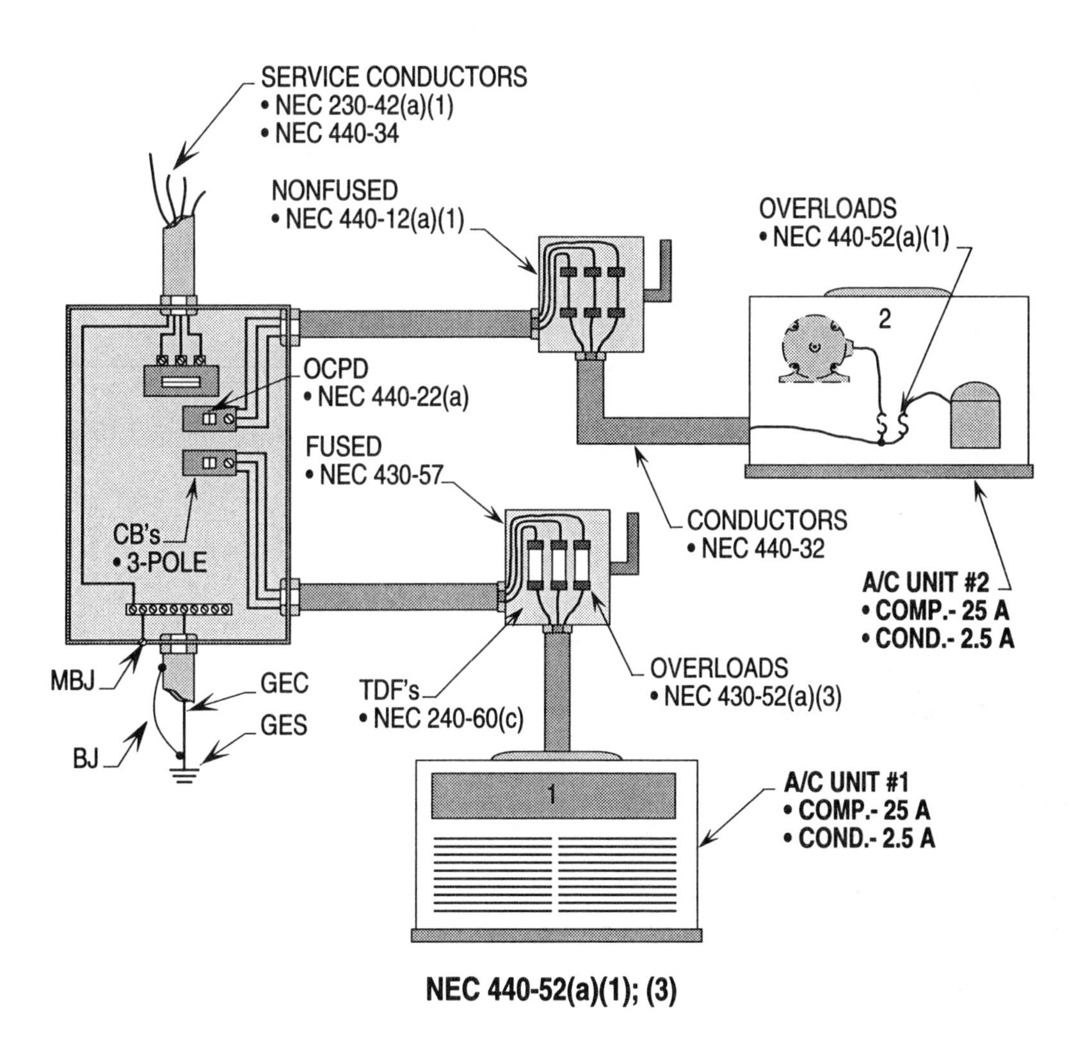

NEC 440-52(a)(1); (3)

Ill. 14-7

Branch-circuit Requirements
440-62(b); (c); 210-23(a)

The ampacity of a cord-and-plug connected A/C window unit must not exceed 80 percent of the branch-circuit where no other loads are served. If other loads are served by the branch-circuit, the cord-and-plugged A/C unit must not exceed 50 percent of the branch-circuit.

Example 14-8. What size OCPD is required for the A/C window units in Illustration 14-8?

Sizing OCPD for Unit #1

Step 1: Finding FLA
440-62(b); 210-23(b)(1)
FLA = 29 A + 2.5 A = 31.5 A

Step 2: Loading OCPD
440-62(b); 384-16(d)
40 A OCPD x 80% = 32 A

Step 3: Verifying loading
440-62(b); 384-16(d)
31.5 A does not exceed 32 A

Solution: A 40 amp OCPD is required to limit the load to 32 amps or less.

Sizing OCPD for Unit #2

Step 1: Finding FLA of OCPD
440-62(c); 210-23(a)
20 A OCPD = 20 A

Step 2: Calculating FLA allowed
440-62(c); 210-23(a)
20 A OCPD x 50% = 10 A

Step 3: Load of Unit #2
440-32; 440-33
7 A x 125% + 1.2 A = 9.95 A

Step 4: Verifying load to be added
440-62(c); 210-23(a)
9.95 A is less than 10 A

Solution: A 20 A OCPD is required to permit the 9.95 amps window unit to be connected to the No. 12 copper conductors.

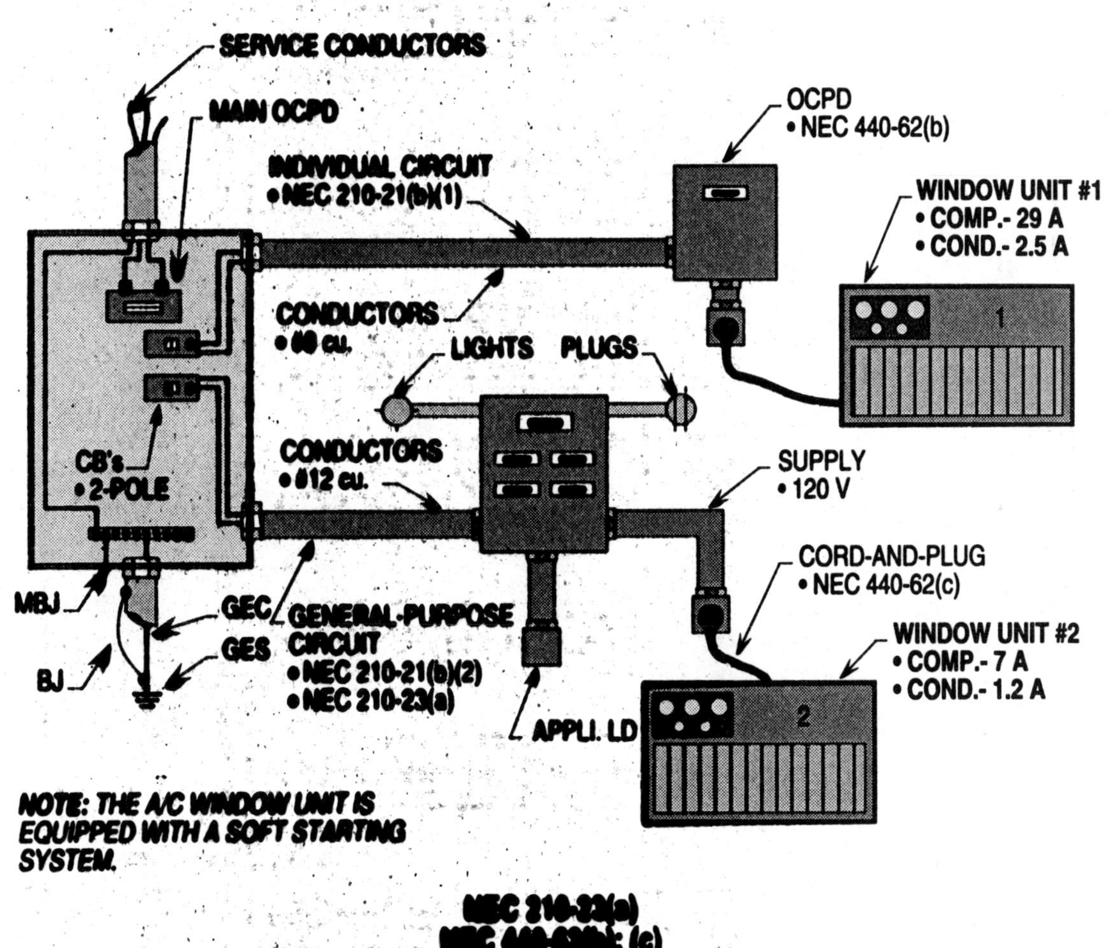

NEC 210-22(a)
NEC 440-62(b); (c)

NEC [illegible]	NEC 440-63
NEC [illegible]	NEC 440-64
NEC [illegible]	

Ill. 14-8

15 Generators and Transformers

Generators and transformers are required to be sized to supply power to the loads served and allow high inertia loads to start and run. In addition, they are to be protected by properly sized OCPD's and be equipped with conductors having enough capacity to supply the loads. OCPD's and conductors must be designed and installed in such a manner to safely protect the windings of such power sources from dangerous overloads.

The OCPD's and conductors are sometimes required to be adjusted in size in order to protect the transformer windings or the conductors from overload conditions.

Generators and transformers may be utility owned, which falls under the provision of the National Electrical Safety Code. If customer owned, they fall under the National Electrical Safety Code as well as the National Electrical Code.

Quick Reference

Generators
445-5

If conductors are fully sized and selected using certain percentages times the output of the generator, the OCPD may be eliminated at the generator. Care should be taken when applying this rule. In other words, a particular type generator must be selected and installed.

Example 15-1. What size THWN copper conductors are required to supply buildings #2 and #4 in Illustration 15-1 with an OCPD used for building #2?

Sizing conductors and OCPD for building #2

Step 1: Finding FLA of generator
445-5
FLA = kVA x 1,000 / V x √3
FLA = 200 x 1,000 / 480 V x 1.732
FLA = 240.7 A

Step 2: Calculating FLA for conductors
445-5; 220-10; 215-2(a); 215-3
160 A x 125% = 200 A

Step 3: Selecting conductors
310-10(2); Table 310-16
200 A requires #3/0 cu.

Solution: **The size THWN conductors are No. 3/0 copper protected by a 200 amp OCPD.**

Note: The conductors to building #2 do not have to be sized at 115 percent of the generator output because they are protected by the 200 amp OCPD.

Sizing conductors to building #4

Step 1: Finding FLA of generator
445-5
FLA = 240.7 A (See Step 1 above)

Step 2: Calculating FLA for conductors
445-5
240.7 A x 115% = 276.8 A

Step 3: Selecting conductors
310-10(2); Table 310-16
276.8 A requires #300 KCMIL cu.

Solution: **The size THWN conductors are required to be No. 300 KCMIL copper because there is no OCPD at the generator.**

Note: The conductors could be No. 4/0 THWN copper conductors if they were protected by a 225 A OCPD (184 A x 125% = 230 A).

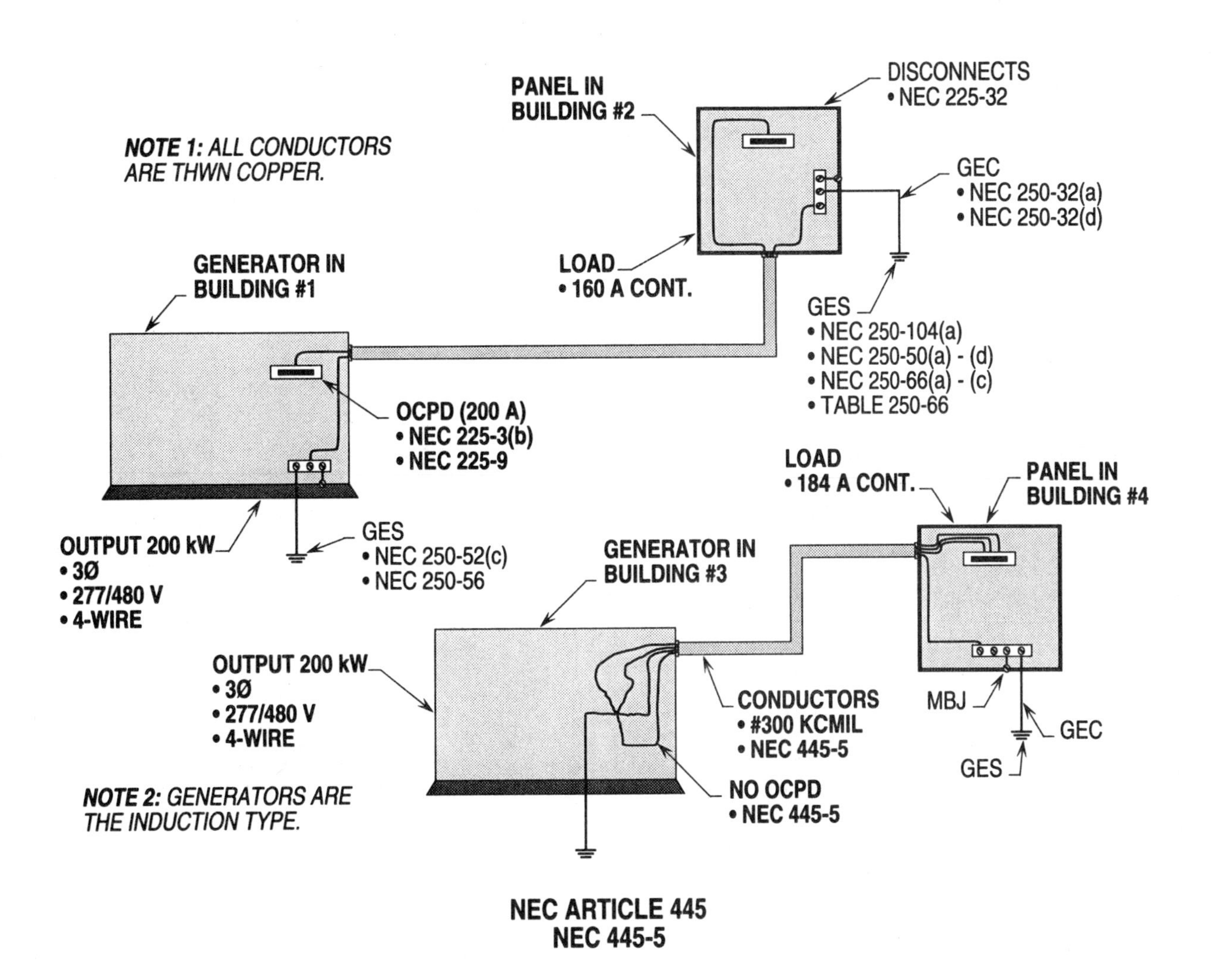

PANEL IN
BUILDING #2
DISCONNECTS
• NEC 225-32
NOTE 1: ALL CONDUCTORS
ARE THWN COPPER.
GEC
• NEC 250-32(a)
• NEC 250-32(d)
GENERATOR IN
BUILDING #1
LOAD
• 160 A CONT.
GES
• NEC 250-104(a)
• NEC 250-50(a) - (d)
• NEC 250-66(a) - (c)
• TABLE 250-66
OCPD (200 A)
• NEC 225-3(b)
• NEC 225-9
LOAD
• 184 A CONT.
PANEL IN
BUILDING #4
OUTPUT 200 kW
• 3Ø
• 277/480 V
• 4-WIRE
GES
• NEC 250-52(c)
• NEC 250-56
GENERATOR IN
BUILDING #3
OUTPUT 200 kW
• 3Ø
• 277/480 V
• 4-WIRE
CONDUCTORS
• #300 KCMIL
• NEC 445-5
MBJ
GEC
GES
NOTE 2: GENERATORS ARE
THE INDUCTION TYPE.
NO OCPD
• NEC 445-5
NEC ARTICLE 445
NEC 445-5

Ill. 15-1

Generators
Article 445

The size generator required to supply a number of loads may be sized by applying a rule of thumb method.

Example 15-2. What size generator is required to supply the loads in Illustration 15-2?

Sizing generator using the rule of thumb method

Step 1: Finding FLA of loads
430-6(a)(1); Table 430-150;
220-10

100 HP	= 124 A
75 HP	= 96 A
60 HP	= 77 A
Ltg. ld.	= 62 A
Plug ld.	= 25 A
Other lds.	= 122 A

Step 2: Calculating Running A

MT lds.	= 124 A
	= 96 A
	= 77 A
Ltg. ld.	= 62 A
Plug ld.	= 25 A
Other lds.	= 122 A
Running A	= 506 A

Step 3: Calculating Starting A
(Applying 125% multiplier)

Motor load	
124 A x 125%	= 155 A
96 A x 125%	= 120 A
77 A x 125%	= 96 A
Lighting load	
62 A x 100%	= 62 A
Plug load	
25 A x 100%	= 25 A
Other loads	
122 A x 100%	= 122 A
Starting A	= 580 A

Step 4: Selecting kW of generator
Running kW
kW = 506 A x 480 V x 1.732 / 1,000
kW = 421
Starting kW
kW = 580 A x 480 V x
1.732 / 1,000
kW = 482

Solution: The size generator must have a starting capacity of 482 kW and a running capacity of 421 kW respectively.

Note: Largest motors are connected to start in sequence at 125 percent and the all other loads at 100 percent.

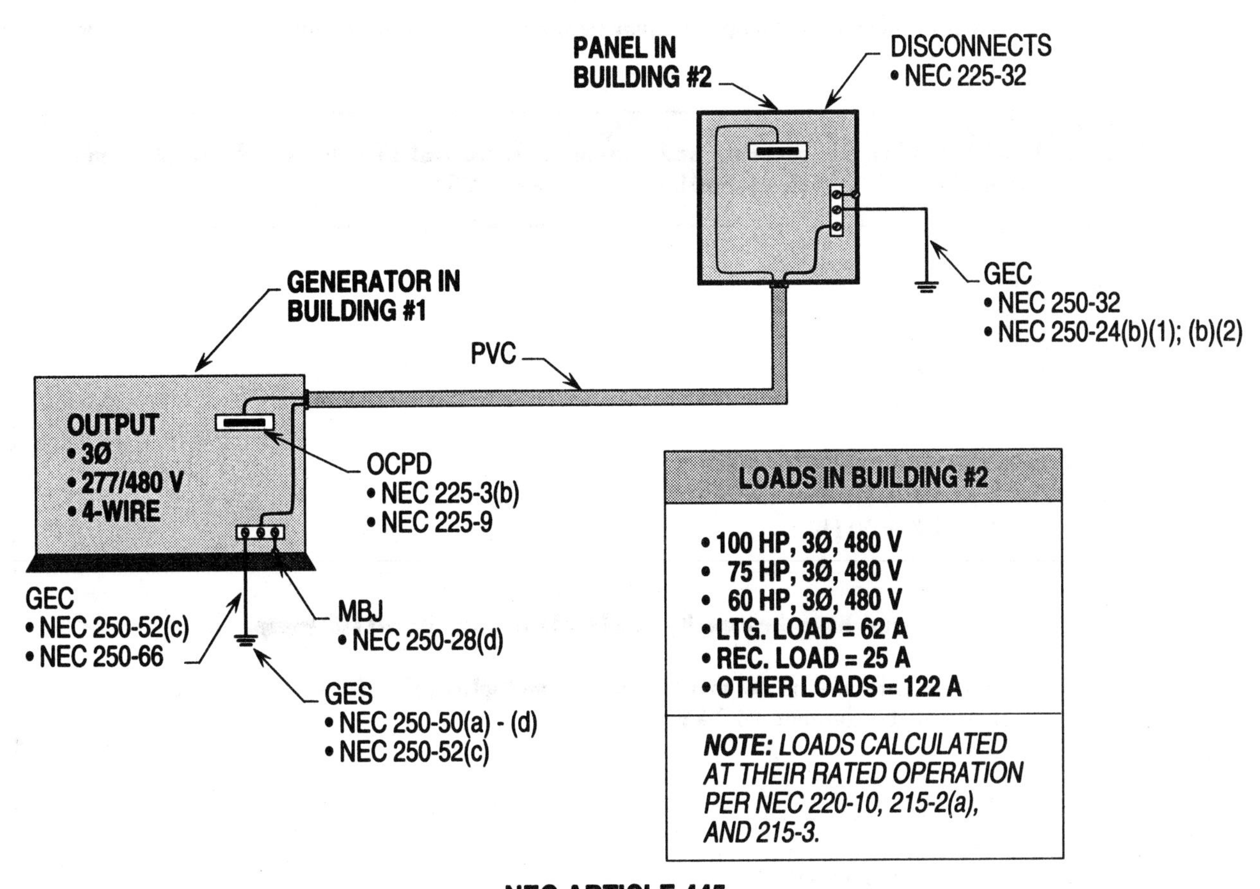

NEC ARTICLE 445

Ill. 15-2

Generators
Article 445

Generators that supply power to submersible pumps, etc. may be sized using the rule of thumb method which provides a fast and easy procedure.

Example 15-3(a) and (b). What size generator and transformer is required to start the submersible pump in Illustrations 15-3(a) and (b)? Use the rule of thumb method to derive kW.

Sizing generator

Step 1: Finding multiplier
3 1/2 kW x HP

Step 2: Calculating kW of generator
kW = 3 1/2 kW x HP
kW = 3 1/2 kW x 50 HP
kW = 175

Solution: The generator must have the capacity of 175 kW to start the 50 HP pump.

Note: The size of the generator in kW may be found by multiplying the HP of the pump motor by the constant 3 1/2 kW.

Sizing a Transformer

Step 1: Finding HP of pump motor
Ill. 15-3(b)
HP = 50

Step 2: Finding multiplier
Ill. 15-3(b)
multiplier = 1 kVA per HP

Step 3: Sizing Transformer
50 HP x 1 kVA = 50 kVA

Solution: The size transformer to supply the pump motor is 50 kVA.

Note: When using a transformer, multiply the HP of the pump motor by 1 kVA to obtain size of transformer.

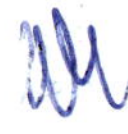

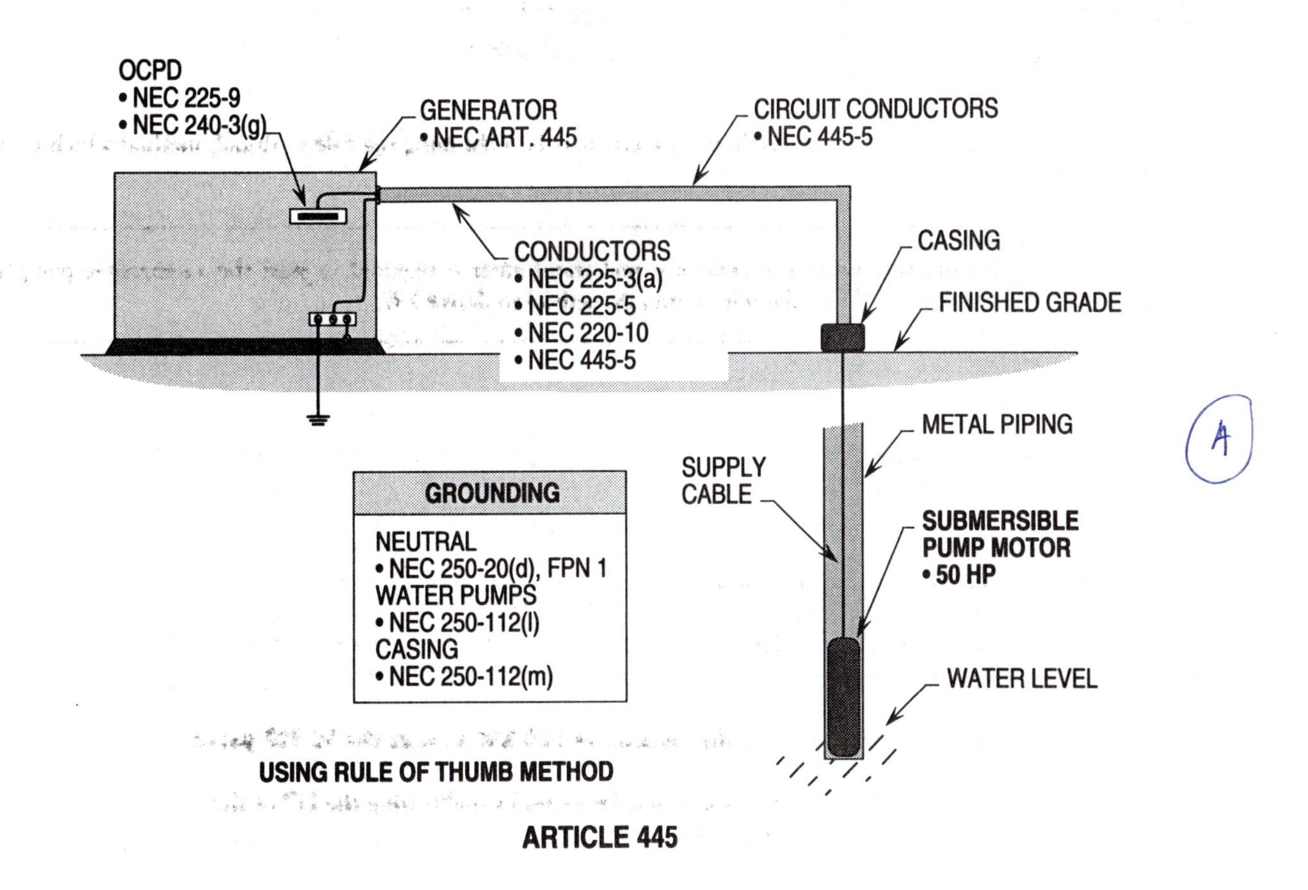

ARTICLE 445

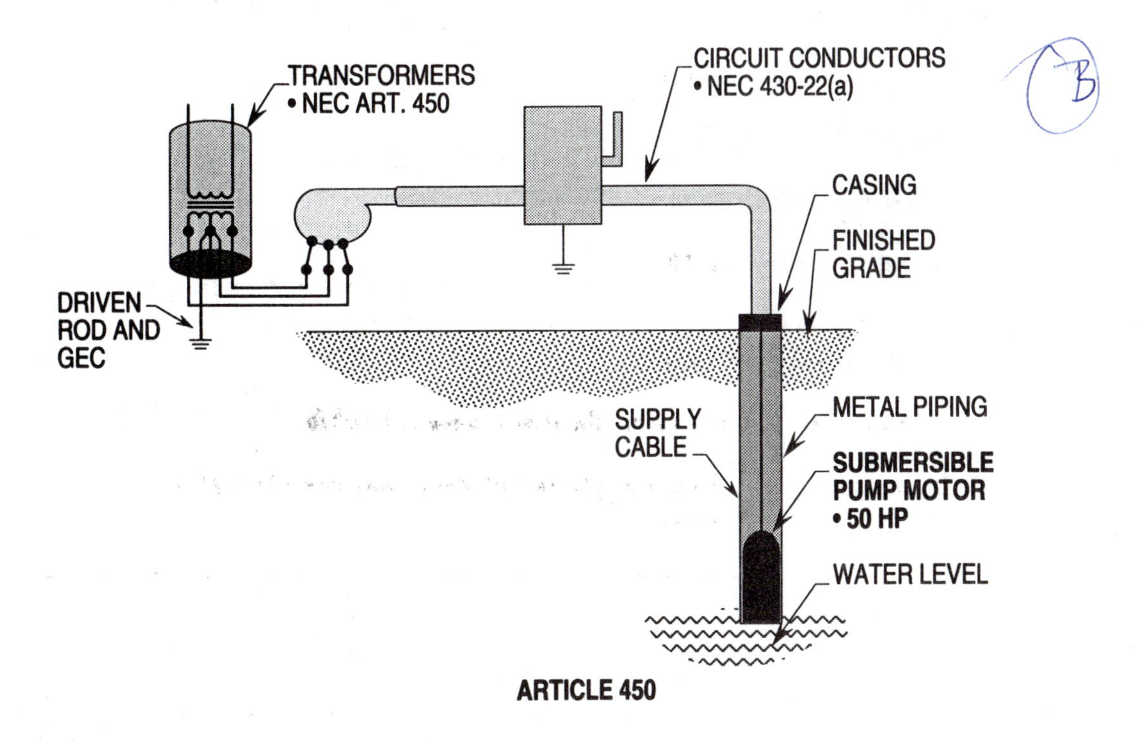

ARTICLE 450

Ill. 15-3(a); (b)

Overcurrent Protection Over 600 Volt 450-3(a); Table 450-3(a)

Overcurrent protection devices protecting the windings of transformers rated over 600 volts may be placed in the primary or primary and secondary sides of the circuits. OCPD's are sized and selected based upon voltage levels related to primary and secondary inputs and outputs of the windings.

Example 15-4. What size OCPD's are required to provide individual overcurrent protection in the primary side of the transformer in Illustration 15-4?

Sizing OCPD using CB's

Step 1: Finding FLA of primary
450-3(a)
FLA = kVA x 1,000 / V x √3
FLA = 1,000 x 1,000 / 12,470 V x 1.732
FLA = 46.3 A

Step 2: Calculating FLA for OCPD
450-3(a); Table 450-3(a)
46.3 A x 300% = 138.9 A

Step 3: Selecting OCPD
Table 450-3(a), Note 1; 240-6(a)
138.9 A requires 150 A OCPD

Solution: The size circuit breaker is 150 amps.

Sizing OCPD using TDF's

Step 1: Calculating FLA for TDF's
450-3(a); Table 450-3(a)
46.3 A x 250% = 115.8 A

Step 2: Selecting OCPD
Table 450-3(a), Note 1; 240-6(a)
115.8 A allows 125 A OCPD

Solution: The size time delay fuses are 125 amps.

ADDITIONAL HELP 67:

Calculating the primary voltage when the current and secondary voltage are known.

- Setting up formula

$$Ep = \frac{Es \times Is}{Ip}$$

- Calculating Pri. V

$$Ep = \frac{480\text{ V} \times 1{,}203\text{ A}}{46.3\text{ A}}$$

Ep = 12,471 V

Solution: Voltage rating is rounded down to 12,740 volts.

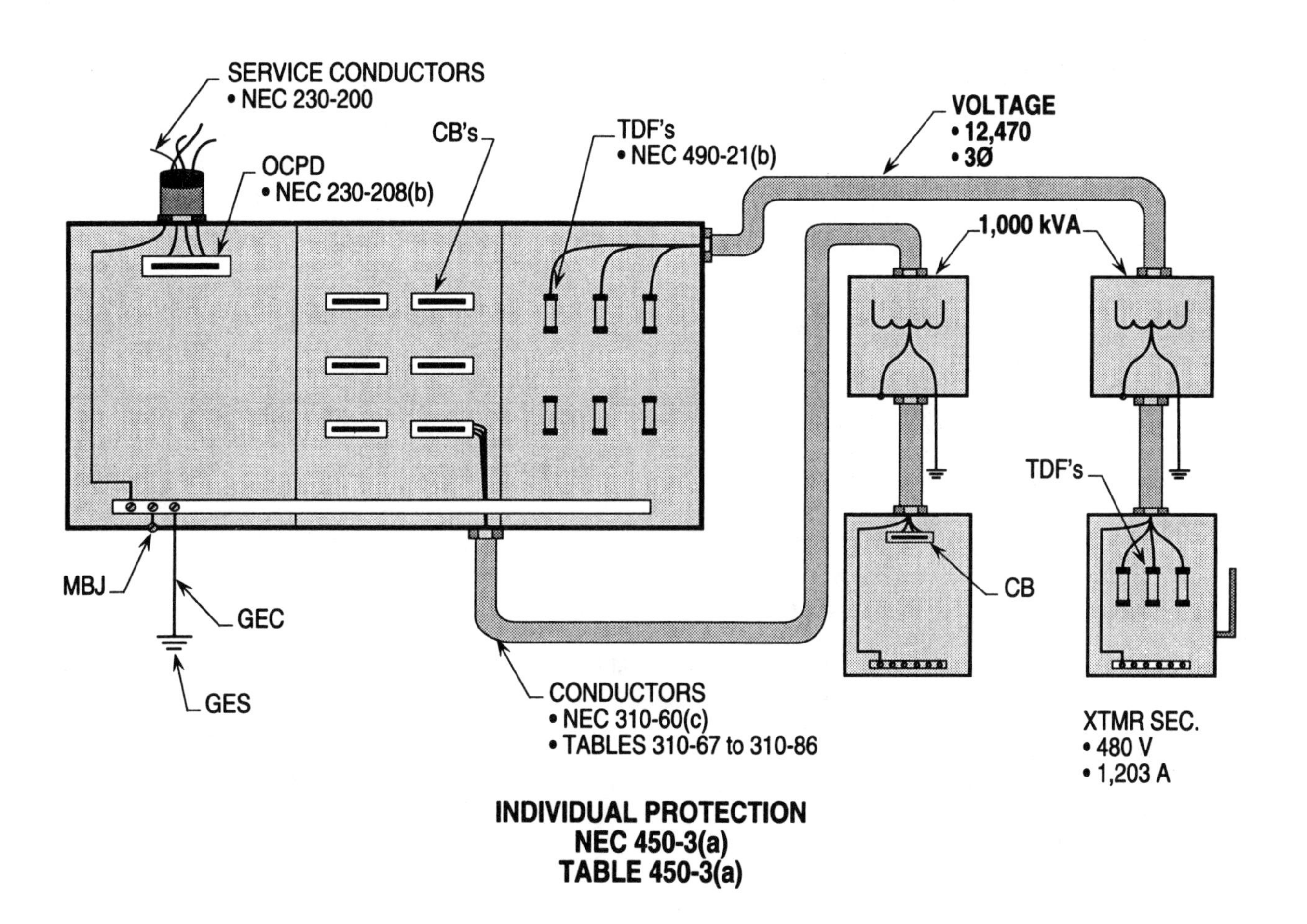
SERVICE CONDUCTORS
• NEC 230-200
OCPD
• NEC 230-208(b)
CB's
TDF's
• NEC 490-21(b)
VOLTAGE
• 12,470
• 3Ø
1,000 kVA
TDF's
CB
MBJ
GEC
GES
CONDUCTORS
• NEC 310-60(c)
• TABLES 310-67 to 310-86
XTMR SEC.
• 480 V
• 1,203 A
INDIVIDUAL PROTECTION
NEC 450-3(a)
TABLE 450-3(a)

Ill. 15-4

Primary And Secondary Protection (Over 600 V) 450-3(a); Table 450-3(a)

Overcurrent protection for a nonsupervised location may be placed in the primary and secondary side of high-voltage transformers, if the OCPD's are designed and installed according to the provisions listed in Table 450-3(a).

Example 15-5. What size OCPD is required to protect the primary and secondary side of the transformer in Illustration 15-5?

ADDITIONAL HELP 68:

Calculating the secondary voltage when the current and primary voltage are known.

- Setting up formula

$$Es = \frac{Ep \times Ip}{Is}$$

$$Es = \frac{4{,}160 \text{ V} \times 69.4 \text{ A}}{601.7 \text{ A}}$$

$$Es = 479.8 \text{ V}$$

Solution: Voltage rating is rounded up to 480 volts.

Sizing OCPD for primary side

Step 1: Finding FLA of transformer
FLA = kVA x 1,000 / V x √3
FLA = 500 x 1,000 / 4,160 V x 1.732
FLA = 69.4 A

Step 2: Calculating FLA for OCPD
450-3(a); Table 450-3(a)
FLA = 69.4 A x 600%
FLA = 416.4 A

Step 3: Selecting OCPD
Table 450-3(a), Note 1; 240-6(a)
416.4 A allows 450 A OCPD

Solution: The size OCPD for primary side is 450 amps.

Note 1: Higher percentages than 125 percent per Table 450-3(a) may be used to size the OCPD in the secondary side where the secondary voltage is greater than 600 volts.

Sizing OCPD for secondary side

Step 1: Finding FLA of transformer
FLA = 500 x 1,000 / 480 V x 1.732
FLA = 601.7 A

Step 2: Calculating FLA for OCPD
450-3(a); Table 450-3(a)
FLA = 601.7 A x 125%
FLA = 752 A

Step 3: Selecting OCPD
Table 450-3(a)
752 A allows 800 A OCPD

Solution: The size OCPD for the secondary side is 800 amps.

Note 2: If the secondary voltage is 4,160, the OCPD using a CB can be sized at 300 percent and a fuse can be sized at 250 percent of transformers.

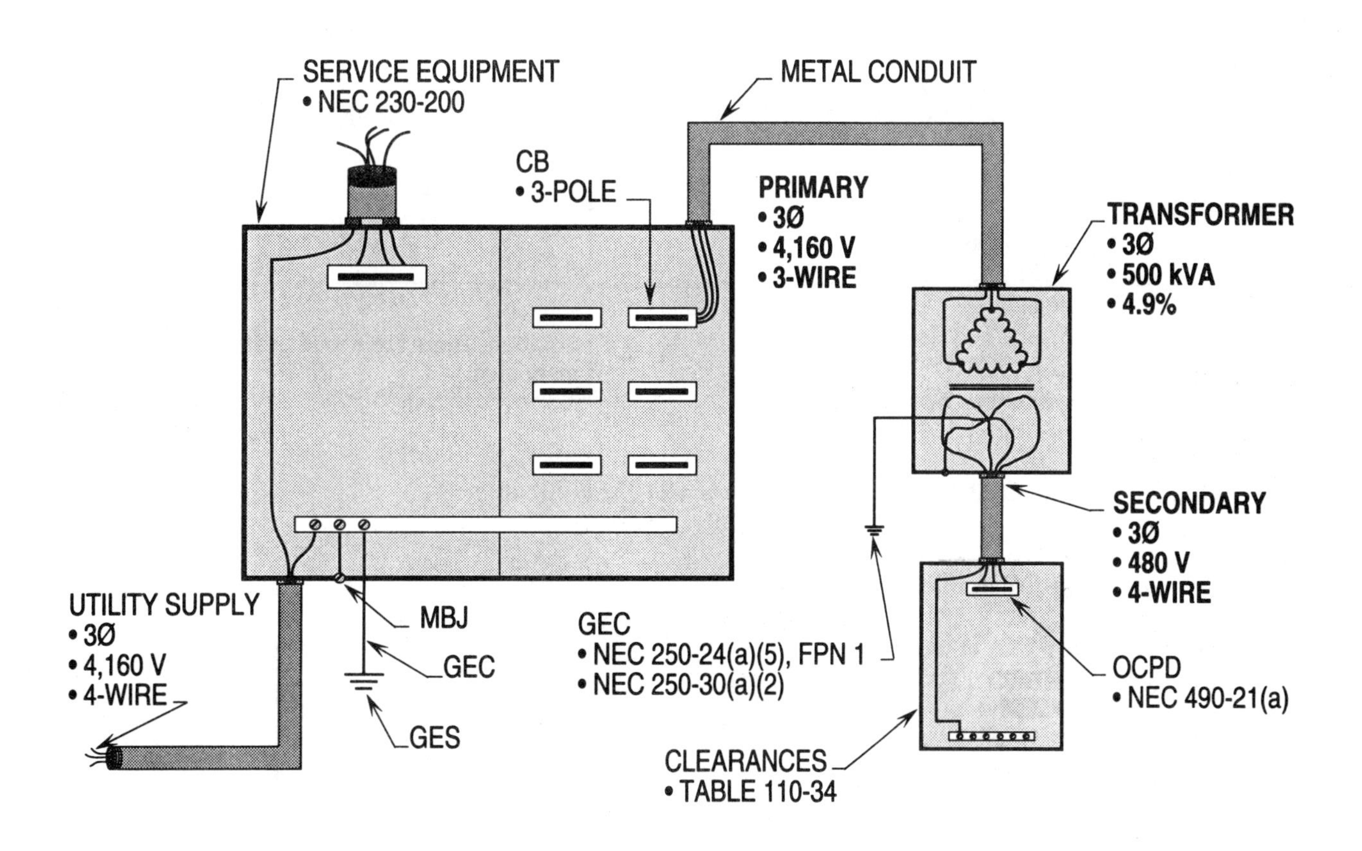

NONSUPERVISED LOCATION
NEC 450-3(a); TABLE 450-3(a)

Ill. 15-5

Primary And Secondary Protection (Over 600 V) 450-3(a); Table 450-3(a)

Overcurrent protection may be placed in the primary and secondary side of high-voltage transformers if the OCPD's are designed and installed according to the provisions listed in Table 450-3(a).

Example 15-6(a). What size OCPD is required to protect the primary and secondary side of the transformer in Illustration 15-6(a)?

Sizing OCPD for primary side

Step 1: Finding FLA of transformer
FLA = kVA x 1,000 / V x √3
FLA = 450 x 1,000 / 13,800 V x 1.732
FLA = 18.83 A

Step 2: Calculating FLA for OCPD
450-3(a); Table 450-3(a)
FLA = 18.83 A x 600%
FLA = 112.9 A

Step 3: Selecting OCPD
Table 450-3(a), Note 3; 240-6(a)
112.9 A allows 110 A OCPD

Solution: The size OCPD for the primary side is a 110 amp circuit breaker.

ADDITIONAL HELP 69:

Calculating the number of turns in the primary of the XTMR in Ill. 15-6(a) when the turns in the secondary are 600.

- Setting up formula

$$Tp = \frac{Ep \times Ts}{Es}$$

$$Tp = \frac{13{,}800 \times 600}{4{,}160}$$

Tp = 1,990 Turns

Solution: There are about 1,990 turns in the primary side.

Sizing OCPD for secondary side

Step 1: Finding FLA of transformer
FLA = 450 x 1,000 / 4,160 V x 1.732
FLA = 62.5 A

Step 2: Calculating FLA for OCPD
450-3(a); Table 450-3(a)
FLA = 62.5 A x 250%
FLA = 156.3 A

Step 3: Selecting OCPD
Table 450-3(a), Note 3; 240-6(a)
156.3 A allows 150 A OCPD

Solution: The size OCPD for secondary side is 150 amp fuses.

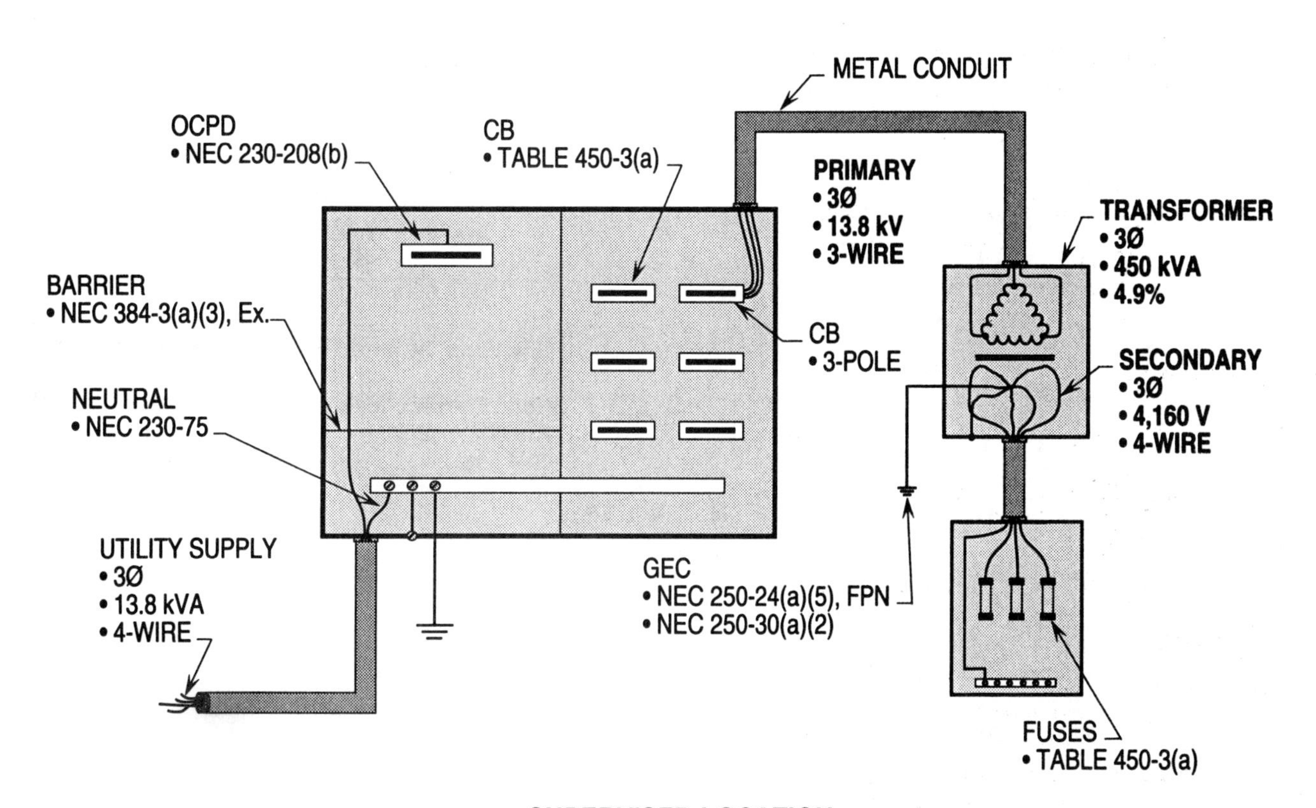

SUPERVISED LOCATION
NEC 450-3(a); TABLE 450-3(a)

Primary And Secondary Protection (Over 600 V) 450-3(a); Table 450-3(a)

Overcurrent protection for a supervised location may be placed in the primary and secondary side of high-voltage transformers, if the OCPD's are designed and installed according to the provisions listed in Table 450-3(a).

Example 15-6(b). What size OCPD is required to protect the primary and secondary side of the transformer in Illustration 15-6(b)?

Sizing OCPD for primary side

Step 1: Finding FLA of transformer
FLA = kVA x 1,000 / V x √3
FLA = 500 x 1,000 / 4160 V x 1.732
FLA = 69.4 A

Step 2: Calculating FLA for OCPD
450-3(a); Table 450-3(a)
FLA = 69.4 A x 600%
FLA = 416.4 A

Step 3: Selecting OCPD
Table 450-3(a), Note 3; 240-6(a)
416.4 A allows 400 A OCPD

Solution: The size OCPD for the primary side is 400 amps.

ADDITIONAL HELP 70:

Calculating the amps in the primary for the XTMR in Ill. 15-6(b) when the turns in the primary is 2,500 and secondary turns are 288.

• Setting up formula

$$Ip = \frac{Is \times Ts}{Tp}$$

$$Ip = \frac{602\ A \times 288}{2,500}$$

$$Ip = 69.4\ A$$

Solution: The amps in the primary are 69.4.

Note: The secondary amps are rounded up (601.7 A to 602 A) and the solution 69.35 is rounded up to 69.4 amps.

Sizing OCPD for secondary side

Step 1: Finding FLA of transformer
FLA = 500 x 1,000 / 480 V x 1.732
FLA = 601.7 A

Step 2: Calculating FLA for OCPD
450-3(a); Table 450-3(a)
601.7 A x 250% = 1,504.3 A

Step 3: Selecting OCPD
Table 450-3(a), Note 3; 240-6(a)
1,504.3 A allows 1,500 A OCPD

Solution: The size OCPD for the secondary side is 1,500 amps.

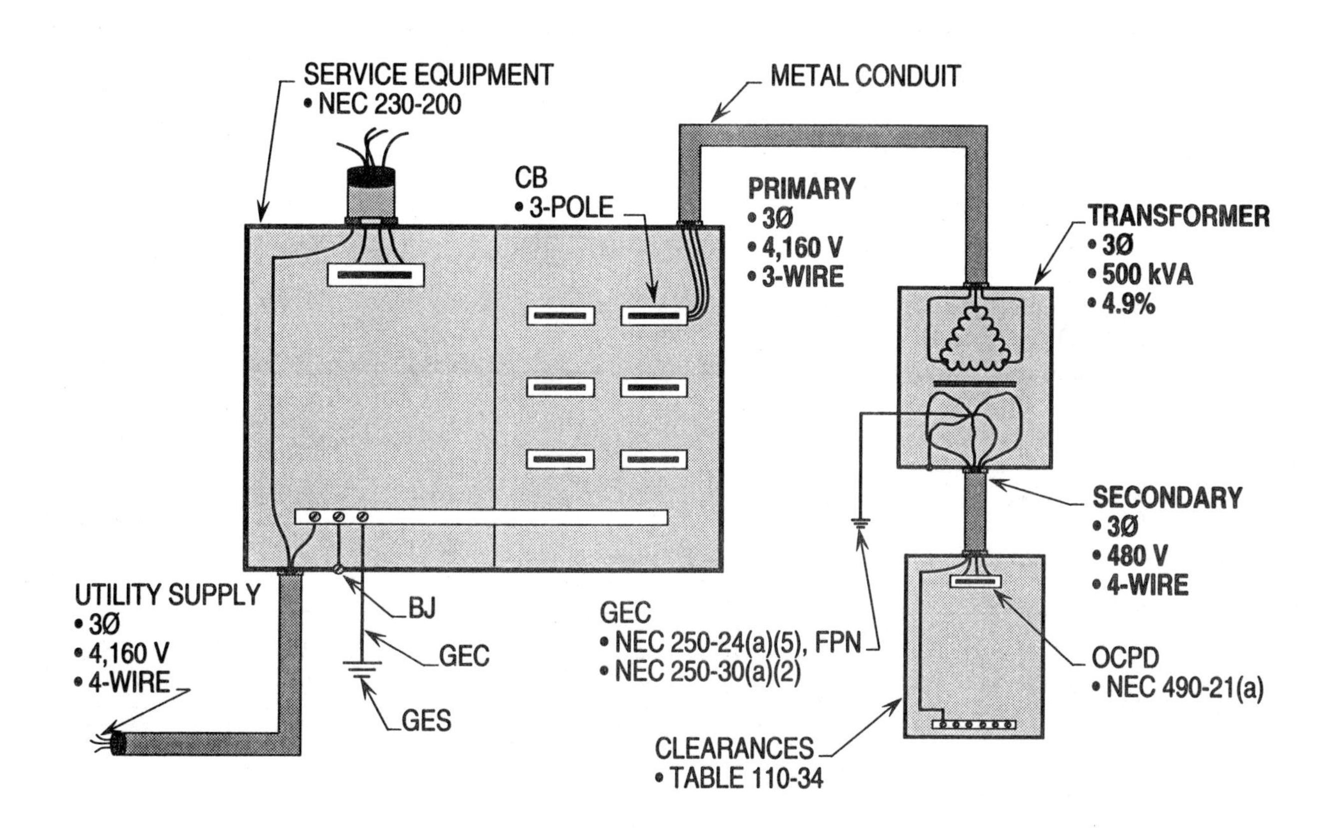

SUPERVISED LOCATION
NEC 450-3(a); TABLE 450-3(a)

Ill. 15-6(b)

Primary Protection (600 V Or Less) 450-3(b); Table 450-3(b)

To determine the proper size OCPD for the primary side of transformers rated at 600 volts or less, the primary FLC is multiplied by a percentage based upon the different levels of amps.

Example 15-7(a). What size OCPD is required in the primary side of the transformer in Illustration 15-7(a)? (Based upon first level of amps.)

Sizing primary OCPD (First Level)

Step 1: Finding FLA of primary
FLA = kVA x 1,000 / V x $\sqrt{3}$
FLA = 50 x 1,000 / 480 V x 1.732
FLA = 60.2 A

Step 2: Calculating OCPD
450-3(b); Table 450-3(b)
60.2 A x 125% = 75.3 A

Step 3: Selecting OCPD
Table 450-3(b), Note 1; 240-6(a)
75.3 A requires 80 A OCPD

Solution: The size OCPD in the primary side is 80 amps.

Example 15-7(b). What size OCPD is required in the primary side of the transformer in Illustration 15-7(b)? (Based upon second level of amps.)

Sizing primary OCPD (Second Level)

Step 1: Finding FLA of primary
FLA = kVA x 1,000 / V
FLA = 3 x 1,000 / 480 V
FLA = 6.25 A

Step 2: Calculating OCPD
450-3(b); Table 450-3(b)
6.25 A x 167% = 10.4 A

Step 3: Selecting OCPD
Table 450-3(b); 240-6(a)
10.4 A requires 10 A OCPD

Solution: The size OCPD in the primary side is 10 amps.

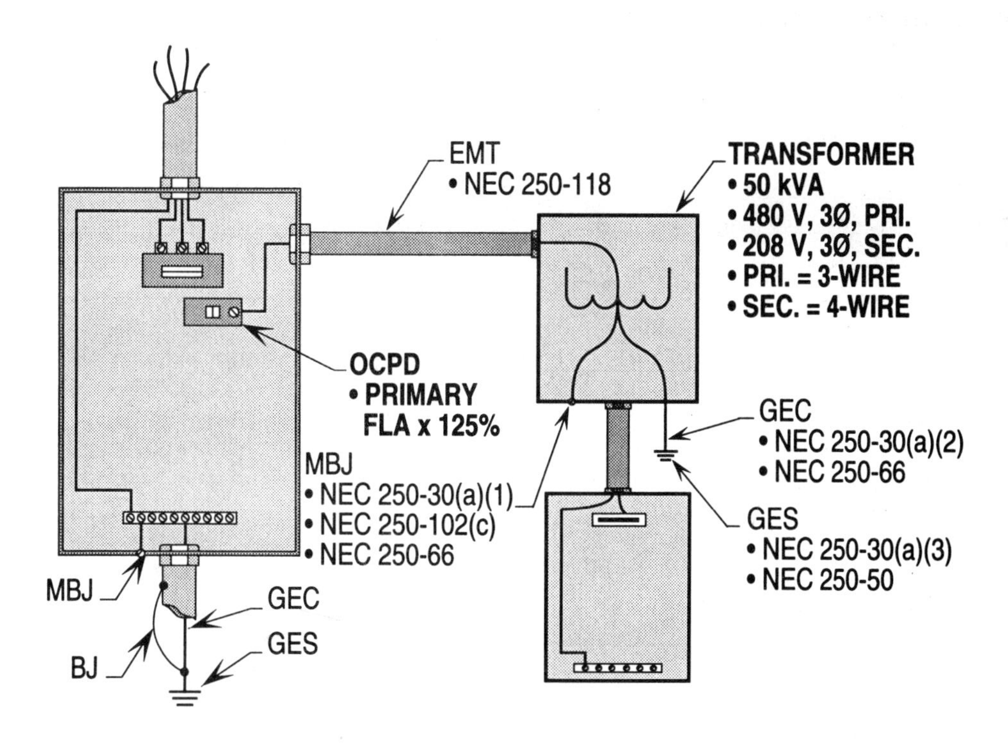

PRIMARY 9 AMPS OR MORE
NEC 450-3(b); TABLE 450-3(b)

Ill. 15-7(a)

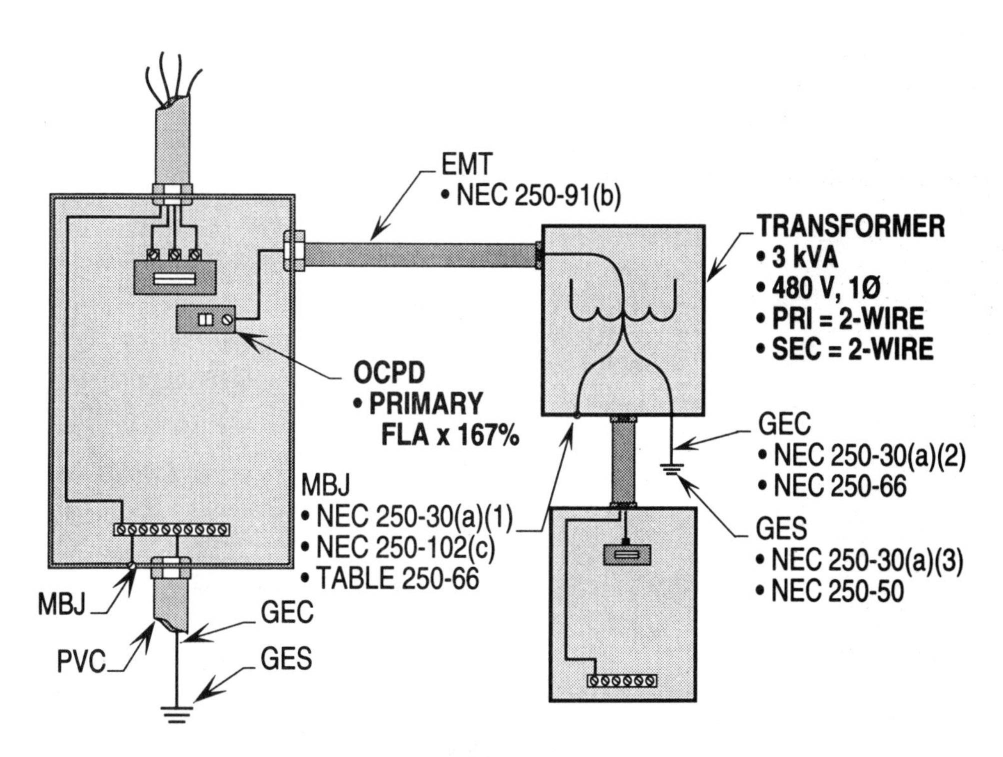

PRIMARY 2 AMPS OR MORE BUT LESS THAN 9 AMPS
NEC 450-3(b); TABLE 450-3(b)

Ill. 15-7(b)

Primary Protection (600 V Or Less) 450-3(b); Table 450-3(b)

There are three levels of primary full-load current in amps that are used to determine the percentage selected to size the primary OCPD.

Example 15-7(c). What size OCPD is required on the primary side of the transformer in Illustration 15-7(c)? (Based upon the third level of amps..)

Sizing minimum primary OCPD

Step 1: Finding FLA of primary
FLA = kVA x 1,000 / V
FLA = .8 x 1,000 / 480 V
FLA = 1.7 A

Step 2: Calculating OCPD
450-3(b); Table 450-3(b)
1.7 A x 300% = 5.1 A

Step 3: Selecting OCPD
Table 450-3(b); 240-6(a)
5.1 A allows 3 A OCPD

Solution: **The minimum size OCPD in the primary side is 3 amps.**

Sizing maximum primary OCPD

Step 1: Calculating OCPD
430-72(c)(4); Step 1
1.7 A x 500% = 8.5 A

Step 2: Selecting OCPD
430-72(c)(4); 240-6(a)
8.5 A allows 6 A OCPD

Solution: **The maximum size OCPD in the primary side is 6 amps.**

Note: The secondary side is a motor controlled circuit.

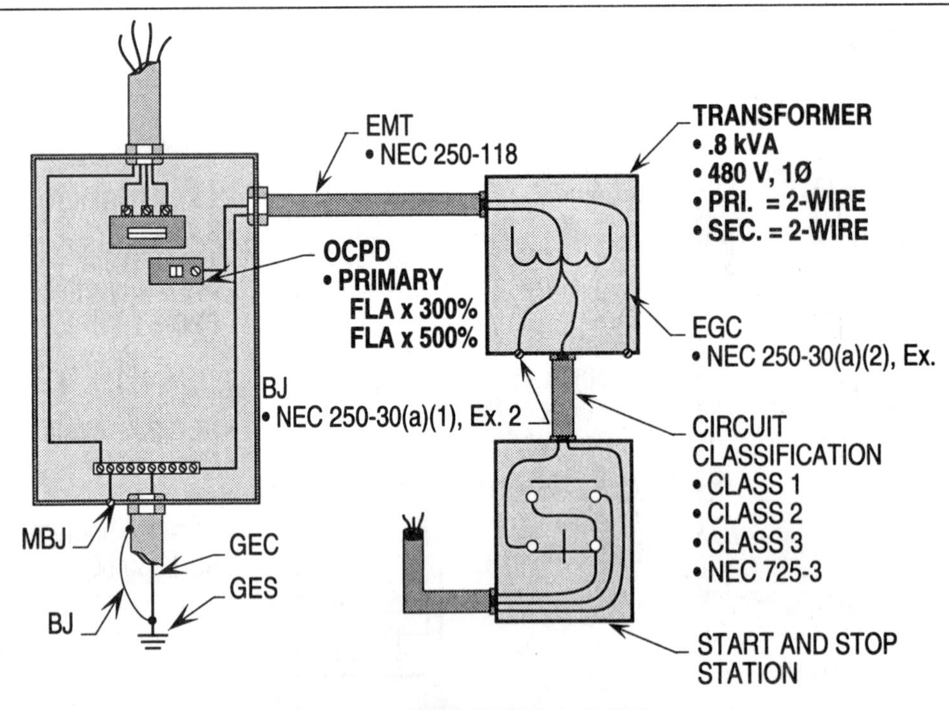

Ill. 15-7(c)

Primary And Secondary Protection (600 V Or Less) 450-3(b); Table 450-3(b)

Primary OCPD's of transformers which trip open due to high inrush currents of equipment connected to the secondary side may be increased up to 250 percent of FLA ratings.

Example 15-8. What is the maximum size OCPD permitted in the primary and secondary side of the transformer in Illustration 15-8?

Sizing OCPD in primary

Step 1: Finding FLA of primary
FLA = kVA x 1,000 / V x √3
FLA = 50 x 1,000 / 480 V x 1.732
FLA = 60.2 A

Step 2: Calculating OCPD
450-3(b); Table 450-3(b)
60.2 A x 250% = 150.5 A

Step 3: Selecting OCPD
Table 450-3(b); 240-6
150.5 A allows 150 A OCPD

Solution: The size OCPD in the primary side of the transformer is 150 amps.

Note: The size of the OCPD in primary must not exceed the 250 percent times the FLA of the primary side.

Sizing OCPD in secondary

Step 1: Finding FLA of primary
FLA = kVA x 1,000 / V x √3
FLA = 50 x 1,000 / 208 V x 1.732
FLA = 138.9 A

Step 2: Calculating OCPD
450-3(b); Table 450-3(b)
138.9 A x 125% = 173.6 A

Step 3: Selecting OCPD
Table 450-3(b), Note 1; 240-6(a)
173.6 A allows 175 A OCPD

Solution: The size OCPD in the secondary side of the transformer is 175 amps.

Note: Table 450-3(b), Note 1 allows the next higher size OCPD to be used.

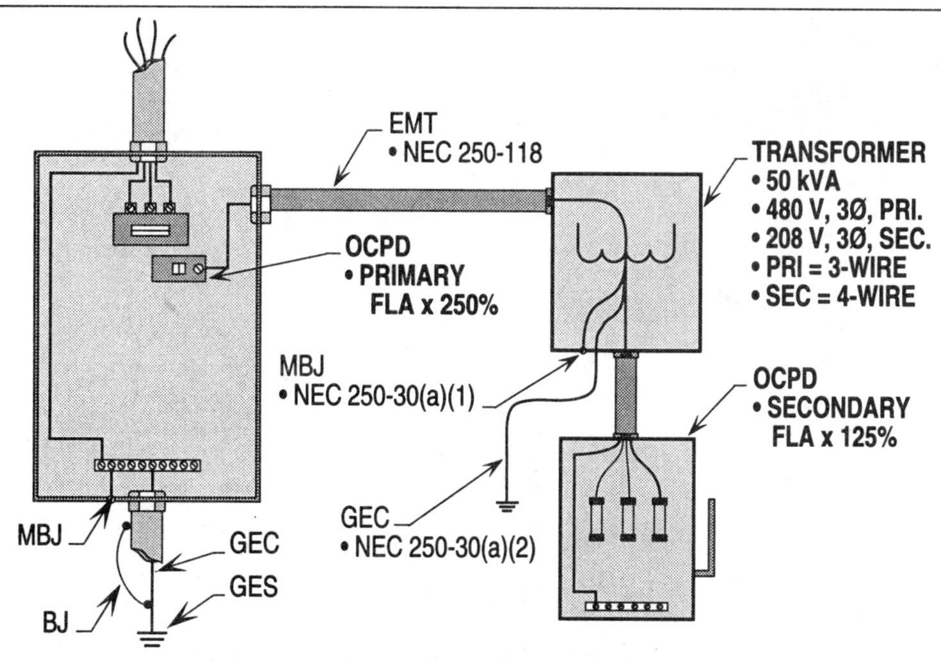

Ill. 15-8

Autotransformers Rated 600 V Or Less 450-4(a)

Autotransformers rated at 600 volts or less are required to be protected by OCPD installed on their input side.

Example 15-9. What size OCPD is required to be installed on the input side of the autotransformer in Illustration 15-9?

Sizing OCPD for 9 amps or more

Step 1: Calculating OCPD
450-4(a); 210-9
14 A x 125% = 17.5 A

Step 2: Selecting OCPD
450-4(a); 240-6(a)
17.5 A allows 20 A OCPD

Solution: **The size OCPD allowed in the input side of autotransformer is 20 amps.**

Sizing OCPD for less than 9 amps

Step 1: Calculating OCPD
450-4(a), Ex.; 210-9
7 A x 167% = 11.69 A

Step 2: Selecting OCPD
450-4(a), Ex.; 240-6(a)
11.69 A allows 10 A OCPD

Solution: **The size OCPD allowed in the input side of the autotransformer is 10 amps.**

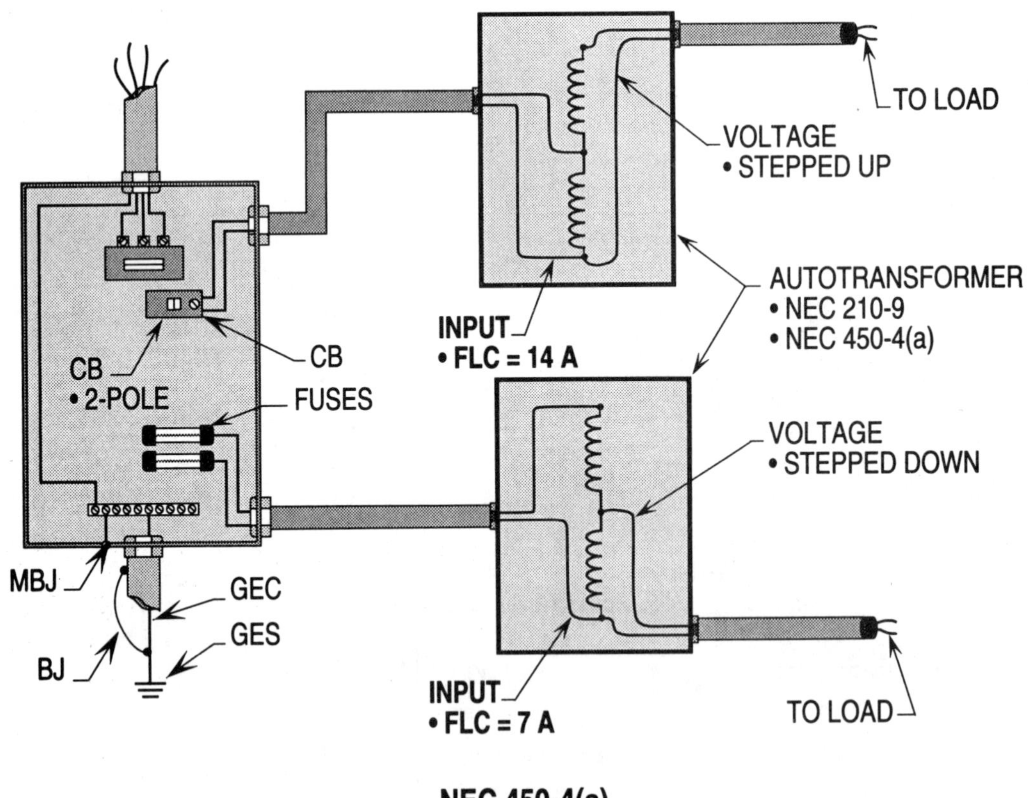

Ill. 15-9

Overcurrent Protection For Grounding Autotransformer 450-5(a)(2)

OCPD's must be sized in such a manner so they will cause the main switch or common trip OCPD to open if the load on the autotransformer reaches or exceeds certain values.

Example 15-10. What continuous current rating is required to trip open the service OCPD in Illustration 15-10?

Finding amount of current to trip open service OCPD

Step 1: Calculating current
450-5(a)(2)
100 A x 125% = 125 A

Solution: A current of 125 amps or greater will cause the service OCPD to trip open.

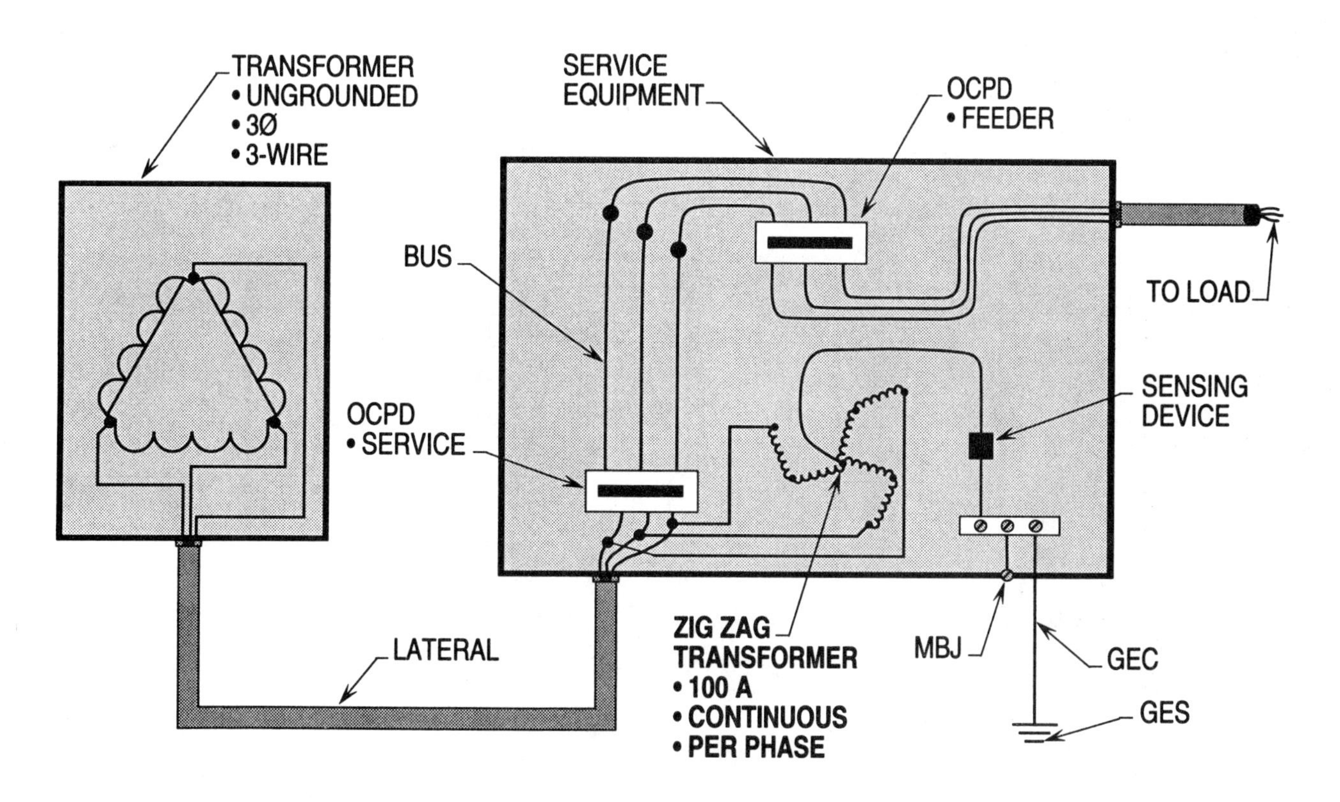

NEC 450-5(a)(2)

Ill. 15-10

Overcurrent Protection For Ground Reference For Fault Protection Devices 450-5(b)(2)

OCPD's must simultaneously open all ungrounded conductors when it operates. The percentages used to determine OCPD's ratings are based upon the continuous phase or neutral current values.

Example 15-11. What continuous current rating is required to trip open the service OCPD in Illustration 15-11?

Finding phase current to trip open service OCPD

Step 1: Calculating current per phase
450-5(b)(2)
100 A x 125% = 125 A

Solution: A current of 125 amps or greater will cause the service OCPD to trip open.

Finding neutral current to trip open service OCPD

Step 1: Calculating current per neutral
450-5(b)(2)
100 A x 42% = 42 A

Solution: A current of 42 amps or greater will cause the service OCPD to trip open.

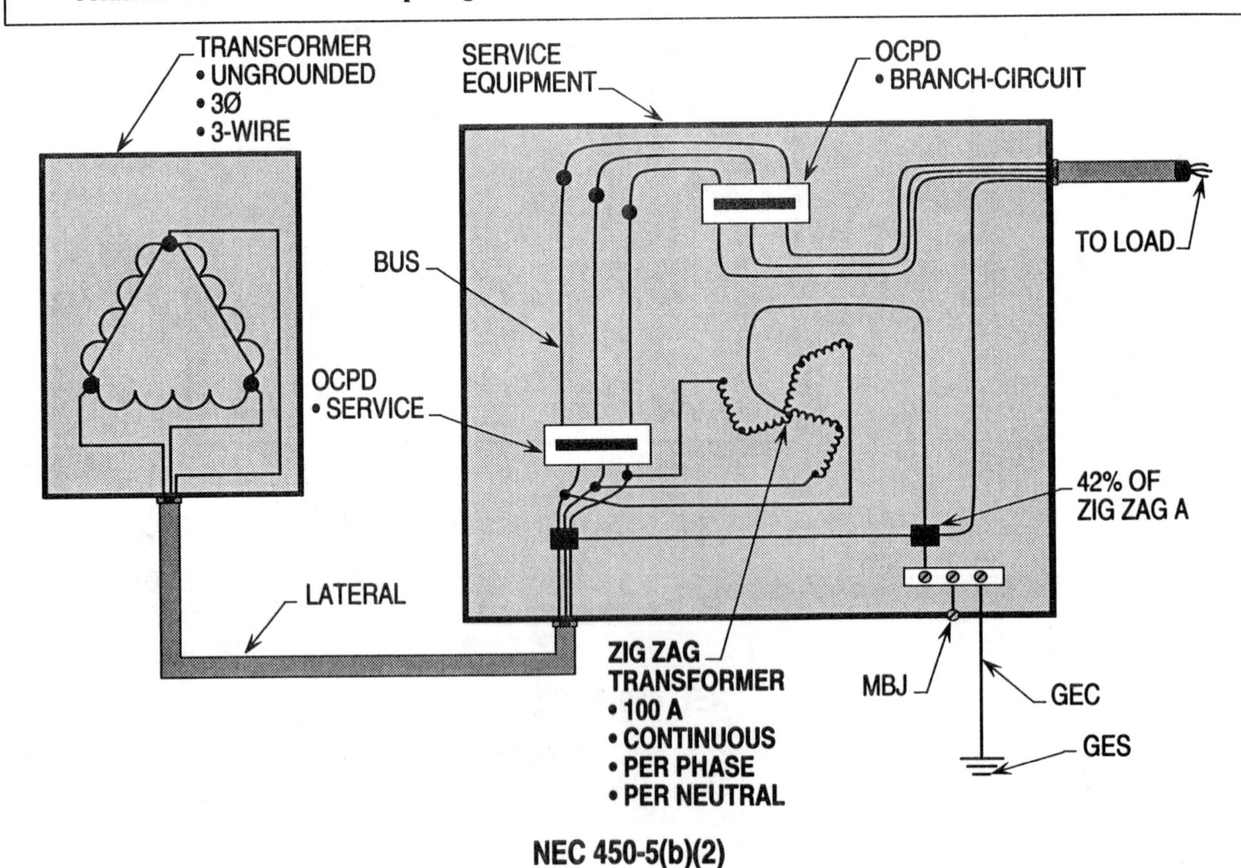

Ill. 15-11

Secondary Ties
450-6(a)(1)

Secondary ties are used to connect two power sources between secondary phases of two transformers to provide continuous power where one transformer is lost. Bus ties are based upon 67 percent or 100 percent of the secondary output.

Example 15-12(a). What size bus tie is required to tie together the secondaries of the transformers in Illustration 15-12(a)?

Sizing bus tie

Step 1: Finding FLA of transformer
FLA = kVA x 1,000 / V x √3
FLA = 2,000 x 1,000 / 480 V x 1.732
FLA = 2,406.7 A

Step 2: Calculating tie
450-6(a)(1)
2,406.7 A x 67% = 1,612.5 A

Solution: The secondary bus tie must be rated at least 2,000 amps.

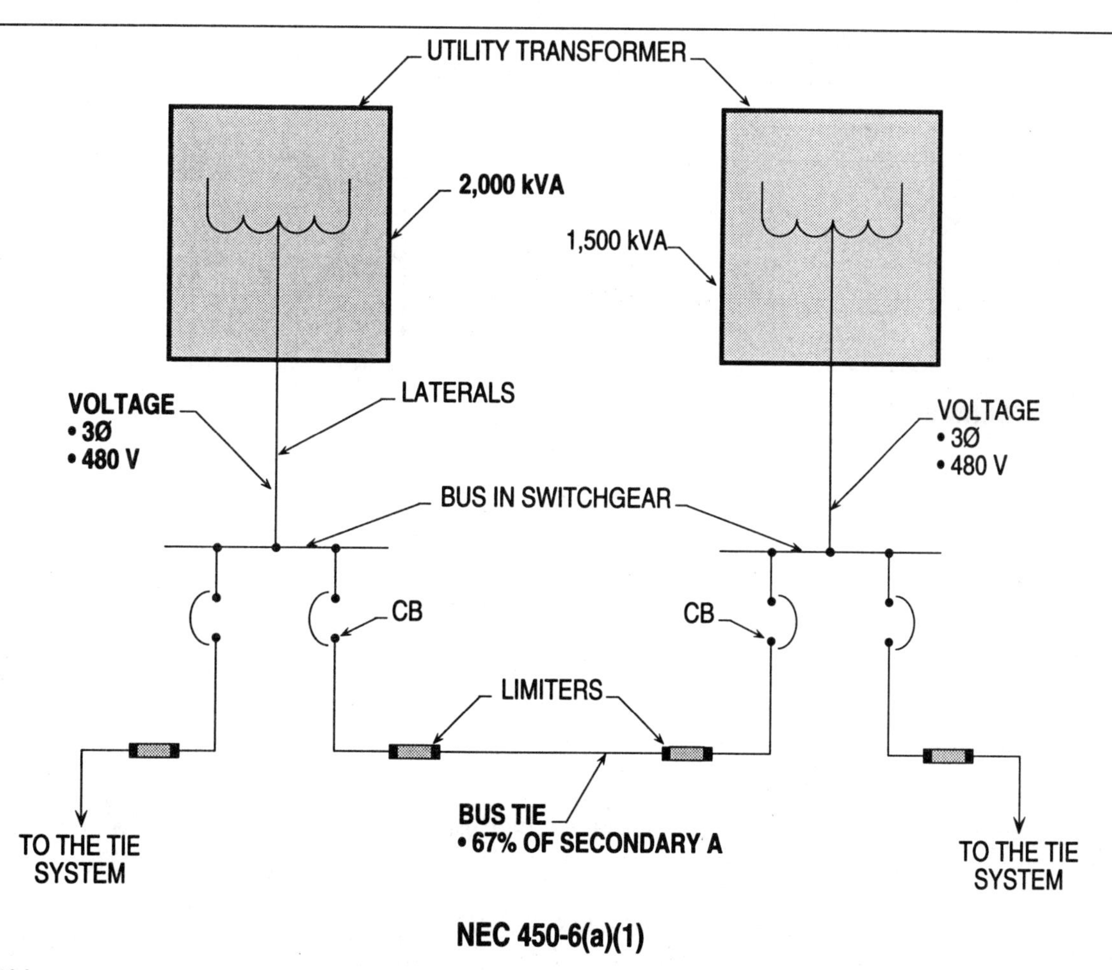

Ill. 15-12(a)

Secondary Ties
450-6(a)(2)

Secondary ties are used to connect two power sources between secondary phases of two transformers to provide continuous power where one transformer is lost. Bus ties are based upon 67 percent or 100 percent of the secondary output.

Example 15-12(b). What size bus tie is required to tie together the secondaries of the transformers in Illustration 15-12(b)?

Sizing bus tie

Step 1: Finding FLA of transformer
FLA = kVA x 1,000 / V x √3
FLA = 2,000 x 1,000 / 480 V x 1.732
FLA = 2,406.7 A

Step 2: Calculating tie
450-6(a)(2)
2,406.7 A x 100% = 2,406.7 A

Solution: The secondary bus tie must be rated at least 2,500 amps.

Note: The 2,500 amps is the next available size.

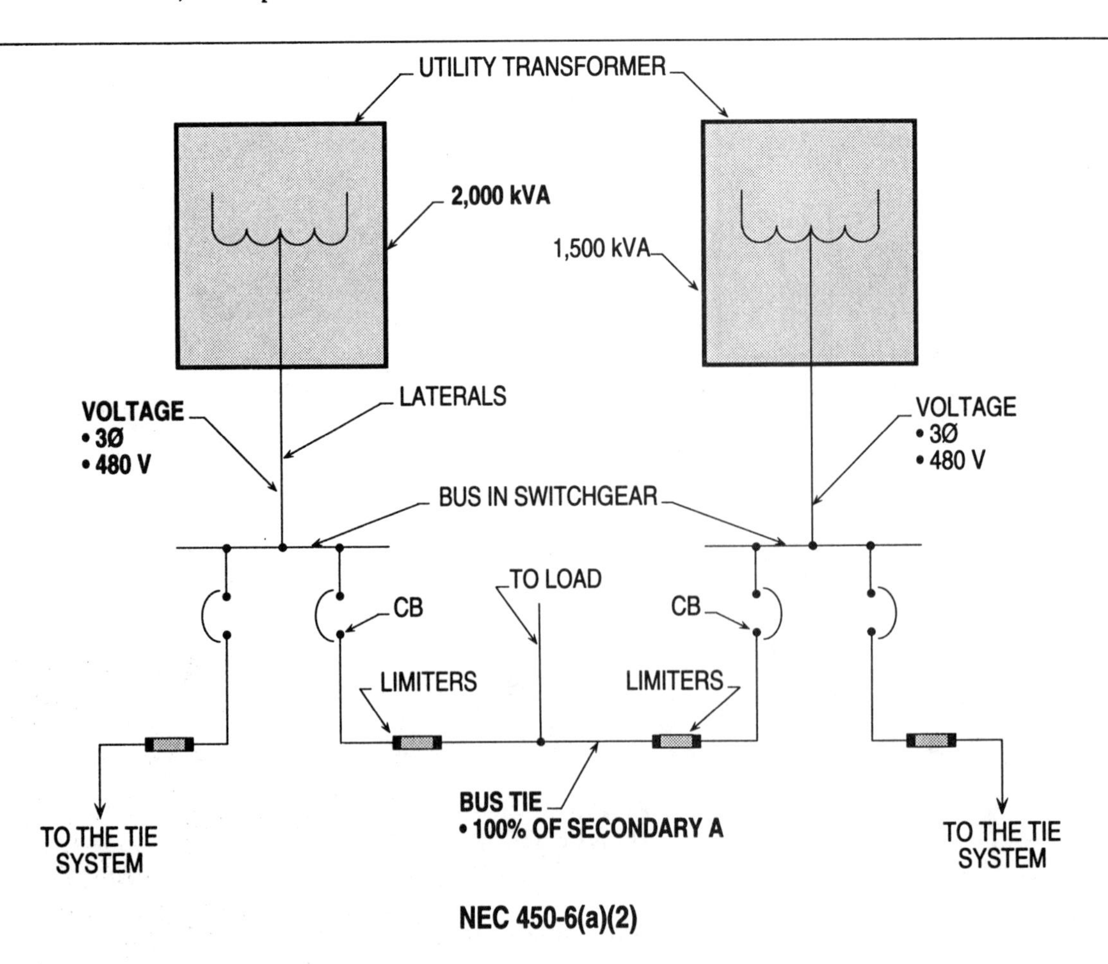

Ill. 15-12(b)

Secondary Ties
450-6(a)(4)(b)

Secondary ties are used to connect two power sources between secondary phases of two transformers to provide continuous power, where one transformer is lost. The bus tie is based upon 133 percent and the OCPD upon 250 percent of the secondary output.

Example 15-12(c). What size bus tie is required to tie together the secondaries of the transformers in Illustration 15-12(c)?

Sizing bus tie

Step 1: Finding FLA of transformers
FLA = kVA x 1,000 / V x √3
FLA = 2,000 x 1,000 / 480 V x 1.732
FLA = 2,406.7 A

Step 2: Calculating tie
450-6(a)(4)(b)
2,406.7 A x 133% = 3,200.9 A

Solution: **The secondary bus ties are required to be rated at 3,500 amps.**

Note: The 3,500 amps is the next available size.

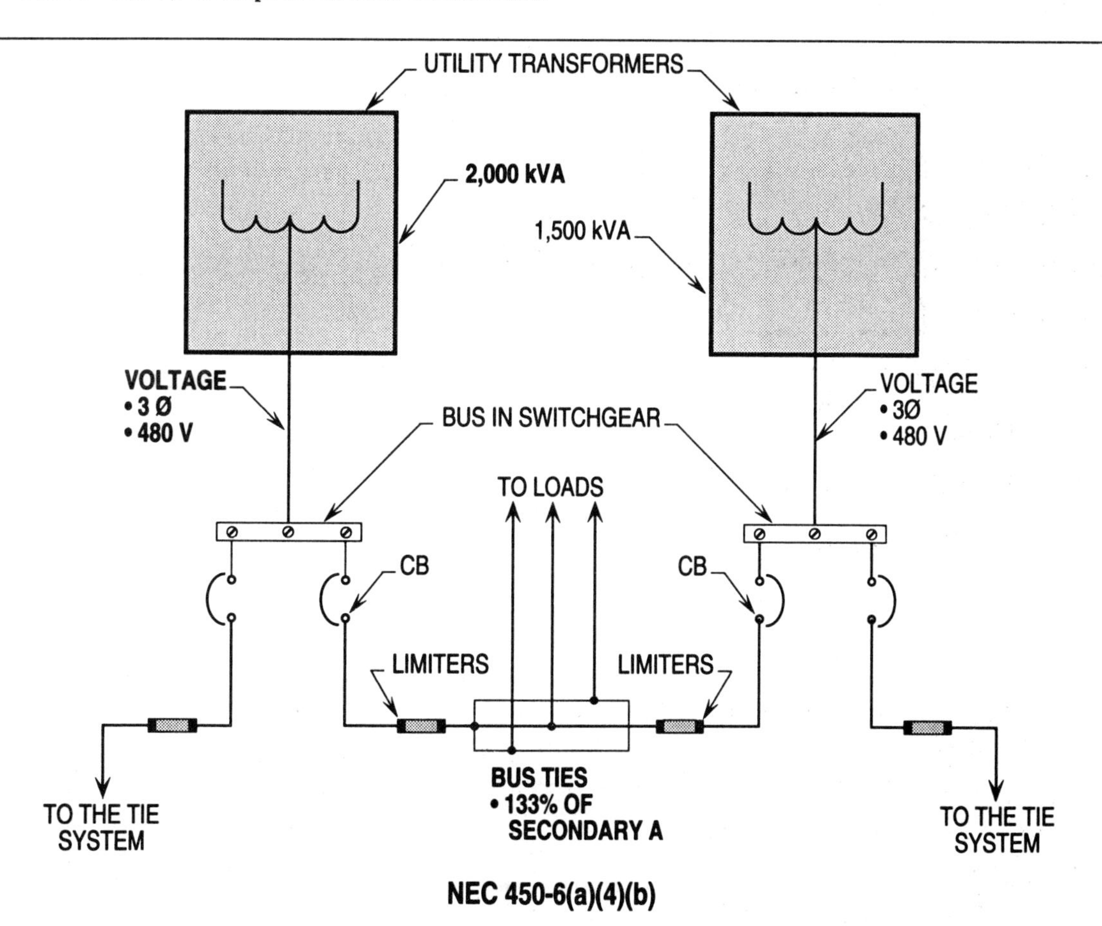

Ill. 15-12(c)

Secondary Ties
450-6(b)

Secondary ties are used to connect two power sources between secondary phases of two transformers to provide continuous power, where one transformer is lost. The bus tie is based upon 133 percent and the OCPD upon 250 percent of the secondary output.

Example 15-12(d). What size circuit breaker is required to protect the secondary tie in Illustration 15-12(d)?

Sizing bus protection

Step 1: Finding FLA of transformers
FLA = kVA x 1,000 / V x √3
FLA = 2,000 x 1,000 / 480 V x 1.732
FLA = 2,406.7 A

Step 2: Calculating protection
450-6(b)
2,406.7 A x 250% = 6,016.8 A

Step 3: Selecting OCPD
450-6(b)
6,016.8A allows 7,000 A

Solution: **The OCPD for the secondary tie is required to be 7,000 amps.**

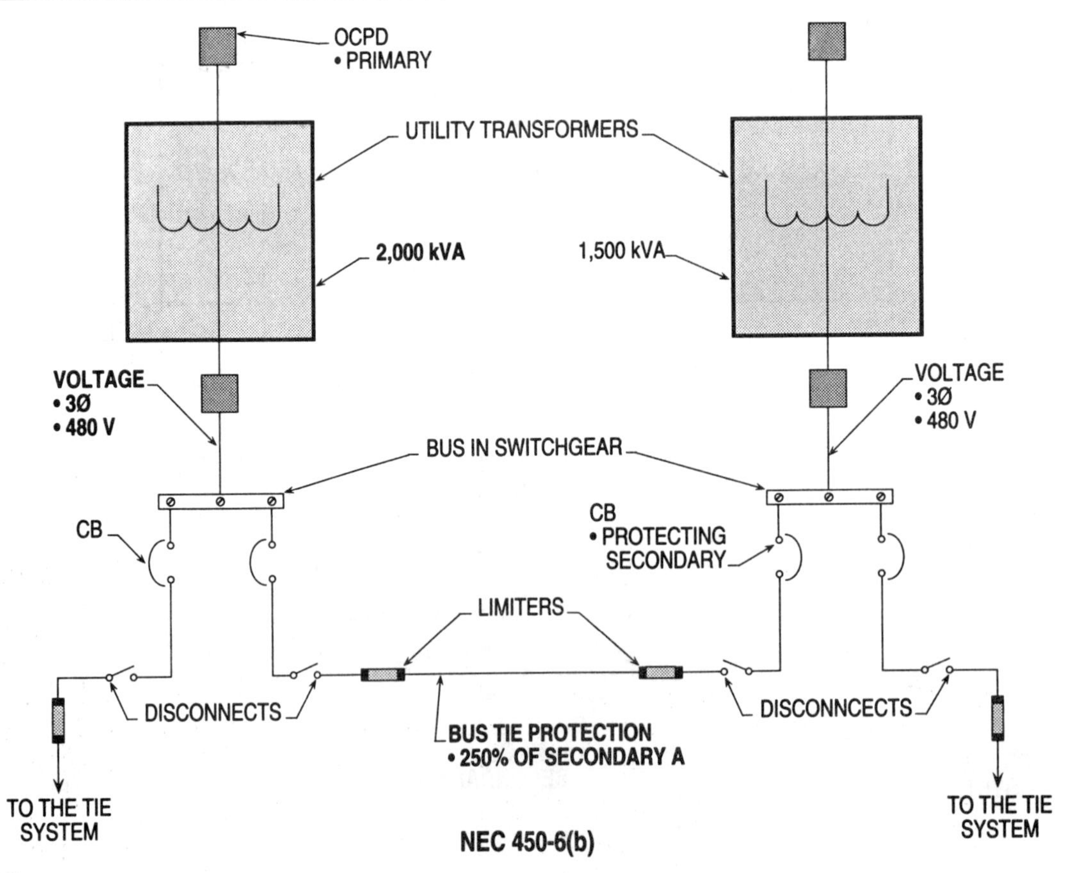

Ill. 15-12(d)

16

Phase Converters and Capacitors

Phase converters are used to convert single-phase power to three-phase power.

Branch-circuit conductors are sized per NEC 455-6 at 125 percent times the phase converters nameplate single-phase input full-load amperage. The OCPD is sized per NEC 455-7 at 125 percent times the phase converters nameplate single-phase input full-load amperage. The OCPD shall not exceed the 125 percent, but shall be equal to or lower than 125 percent.

Branch-circuit elements such as OCPD's and conductors supplying specific loads are calculated at 250 percent of the equipment's FLA ratings.

Feeder-circuit conductors that convert single-phase power to three-phase power to supply power to two or more phase converters are sized at 250 percent times the three-phase amperage of all motors and other loads served. The OCPD is sized per NEC 455-7(b) at 250 percent times the full-load three-phase amperage of all motors and other loads. The OCPD is selected per NEC 240-6(a). If the percentage does not correspond to a standard size, the next size OCPD above this percentage can be selected per NEC 455-7(b) and 455-6(a)(2).

Quick Reference

Phase Converters
455-6; 455-6(a)(1); 455-7; 455-7(a)

Phase converters with single-phase input and FLA's listed on their nameplates are required to have supply elements sized and selected based upon such ratings.

Example 16-1. What is the size THWN cu. conductors and OCPD required to supply power to the phase converter in Illustration 16-1?

Sizing conductors

Step 1: Finding amperage
455-4
Nameplate = 40 A

Step 2: Calculating amperage
240-3(g); 455-6(a)
40 x 125 % = 50 A

Step 3: Selecting conductors
Table 310-16
50 A = #8 THWN cu.

Solution: The size THWN cu. copper conductors are required to be No. 8.

Sizing OCPD

Step 1: Finding amperage
455-4
Nameplate = 40 A

Step 2: Calculating amperage
455-7(a)
40 x 125 % = 50 A

Step 3: Selecting OCPD
455-7; 240-6(a)
50 A = 50 A OCPD

Solution: A 50 amp OCPD is required.

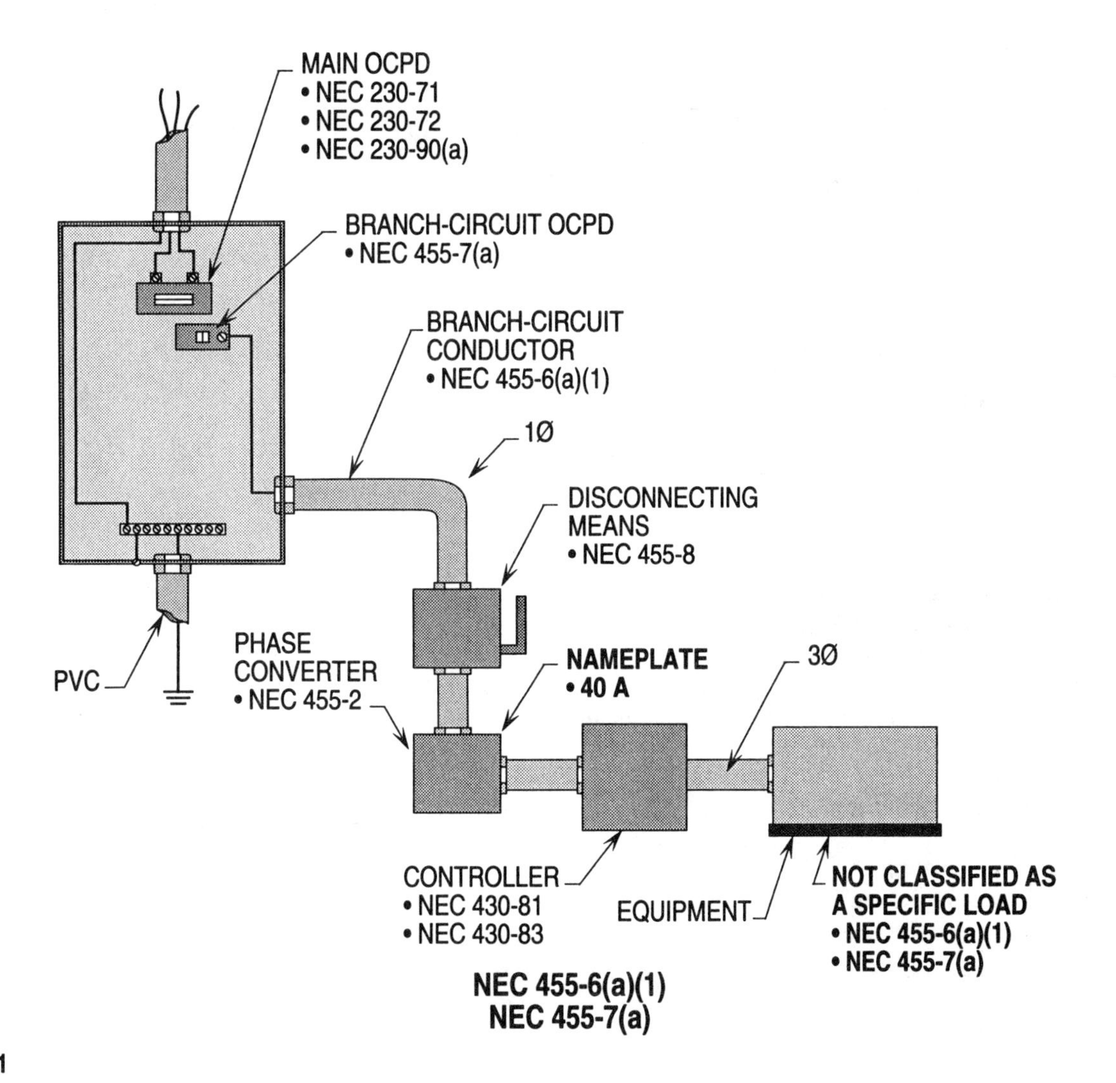
MAIN OCPD
• NEC 230-71
• NEC 230-72
• NEC 230-90(a)
BRANCH-CIRCUIT OCPD
• NEC 455-7(a)
BRANCH-CIRCUIT CONDUCTOR
• NEC 455-6(a)(1)
1Ø
DISCONNECTING MEANS
• NEC 455-8
PHASE CONVERTER
• NEC 455-2
NAMEPLATE
• 40 A
3Ø
PVC
CONTROLLER
• NEC 430-81
• NEC 430-83
EQUIPMENT
NOT CLASSIFIED AS A SPECIFIC LOAD
• NEC 455-6(a)(1)
• NEC 455-7(a)
NEC 455-6(a)(1)
NEC 455-7(a)

Ill. 16-1

Phase Converters
455-6(a)(2); 455-7(b)

Phase converters are used to convert single-phase to three-phase electrical power. The two types utilized in modern electrical systems are rotary and static respectively. The elements of circuits supplying phase converters and specific loads are sized at 250 percent of the FLA ratings, where the input and output voltage levels are the same.

Example 16-2. What size OCPD and THWN copper conductors are required to supply the input side of the phase converter in Illustration 16-2?

Sizing conductors for input side

Step 1: Finding FLA of motor
430-6(a)(1); Table 430-150
25 HP = 68 A

Step 2: Calculating conductors
455-6(a)(2)
68 A x 250% = 170 A

Step 3: Selecting conductors
310-10(2); Table 310-16
170 A requires #2/0 cu.

Solution: The size THWN copper conductors are No. 2/0.

Sizing OCPD for input side

Step 1: Calculating OCPD
455-7(b); Step 1 above
68 A x 250% = 170 A

Step 2: Selecting OCPD
455-7; 240-3(g); 240-6(a)
170 A requires 175 A OCPD

Solution: The size OCPD is 175 amps.

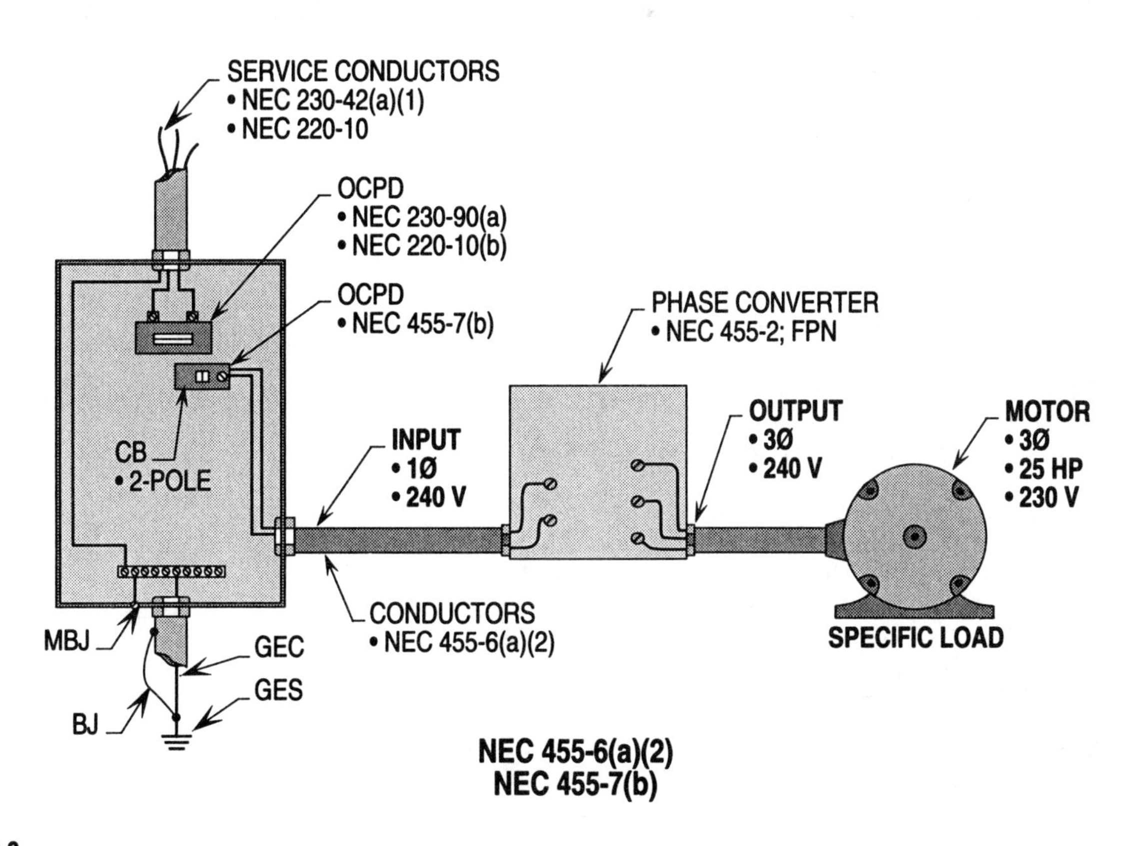
SERVICE CONDUCTORS
• NEC 230-42(a)(1)
• NEC 220-10
OCPD
• NEC 230-90(a)
• NEC 220-10(b)
OCPD
• NEC 455-7(b)
PHASE CONVERTER
• NEC 455-2; FPN
CB
• 2-POLE
INPUT
• 1Ø
• 240 V
OUTPUT
• 3Ø
• 240 V
MOTOR
• 3Ø
• 25 HP
• 230 V
CONDUCTORS
• NEC 455-6(a)(2)
SPECIFIC LOAD
MBJ
GEC
GES
BJ
NEC 455-6(a)(2)
NEC 455-7(b)

Ill. 16-2

Phase Converters
455-6(a)(2); 455-7(b)

Phase converters are used to convert single-phase power to three-phase electrical power. The two types utilized in modern electrical systems are rotary and static respectively. The elements of circuits supplying phase converters and specific loads are sized at 250 percent of the FLA's of such loads, plus the ratio of the input and output voltage levels.

Example 16-3. What size OCPD and THWN copper conductors are required to supply the input side of the phase converter in Illustration 16-3?

Sizing conductors for input side

Step 1: Finding FLA of motor
430-6(a)(1); Table 430-150
25 HP = 34 A

Step 2: Calculating ratio
455-6(a)(2)
Ratio = 480 V / 240 V = 2

Step 3: Calculating conductors
455-6(a)(2)
34 A x 250% x 2 = 170 A

Step 4: Selecting conductors
310-10(2); Table 310-16
170 A requires #2/0 cu.

Solution: The size THWN copper conductors are No. 2/0.

Sizing OCPD for input side

Step 1: Calculating OCPD
455-7(b); Step 1 above
34 A x 250% x 2 = 170 A

Step 2: Selecting OCPD
455-7(b); 240-3(g); 240-6(a)
170 A requires 175 A OCPD

Solution: The size OCPD is 175 amps.

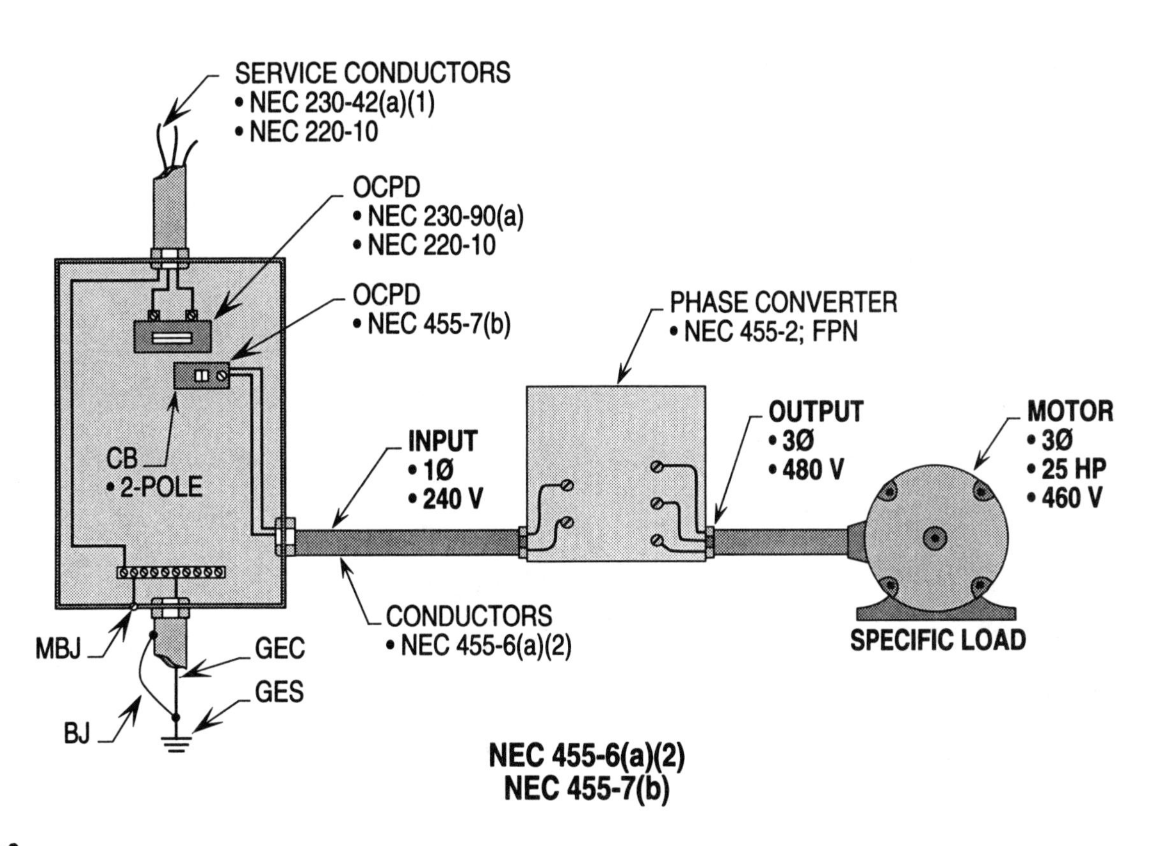
SERVICE CONDUCTORS
• NEC 230-42(a)(1)
• NEC 220-10
OCPD
• NEC 230-90(a)
• NEC 220-10
OCPD
• NEC 455-7(b)
PHASE CONVERTER
• NEC 455-2; FPN
CB
• 2-POLE
INPUT
• 1Ø
• 240 V
OUTPUT
• 3Ø
• 480 V
MOTOR
• 3Ø
• 25 HP
• 460 V
MBJ
GEC
GES
BJ
CONDUCTORS
• NEC 455-6(a)(2)
SPECIFIC LOAD
NEC 455-6(a)(2)
NEC 455-7(b)

Ill. 16-3

PHASE CONVERTERS
455-6; 455-6(a)(2); 455-7; 455-7(b)

The elements of the circuits supplying phase converters and specific loads are sized at 250 percent of the FLA of the equipment served, where the input and output voltage are the same.

Example 16-4. What is the size THWN cu. conductors and OCPD required to supply power to the group of motors in Illustration 16-4?

Sizing conductors

Step 1: Finding amperage
430-6(a)(1); 455-6(a)(2); Table 430-150
10 HP = 28 A
25 HP = 68 A
30 HP = 80 A

Step 2: Calculating amperage
455-6(a)(2)
28 A + 68 A + 80 A = 176 A
176 A x 250% = 440 A

Step 3: Selecting conductors
310-10(2); Table 310-16
440 A = #700 KCMIL THWN cu.

Solution: **The size THWN cu. conductors are required to be No. 700 KCMIL.**

Note: These conductors may be connected in parallel per NEC 310-4.

Sizing OCPD

Step 1: Selecting amperage
240-3(g); 455-7(b); Step 2 above
440 A = 450 A OCPD

Solution: **A 450 amp OCPD is required.**

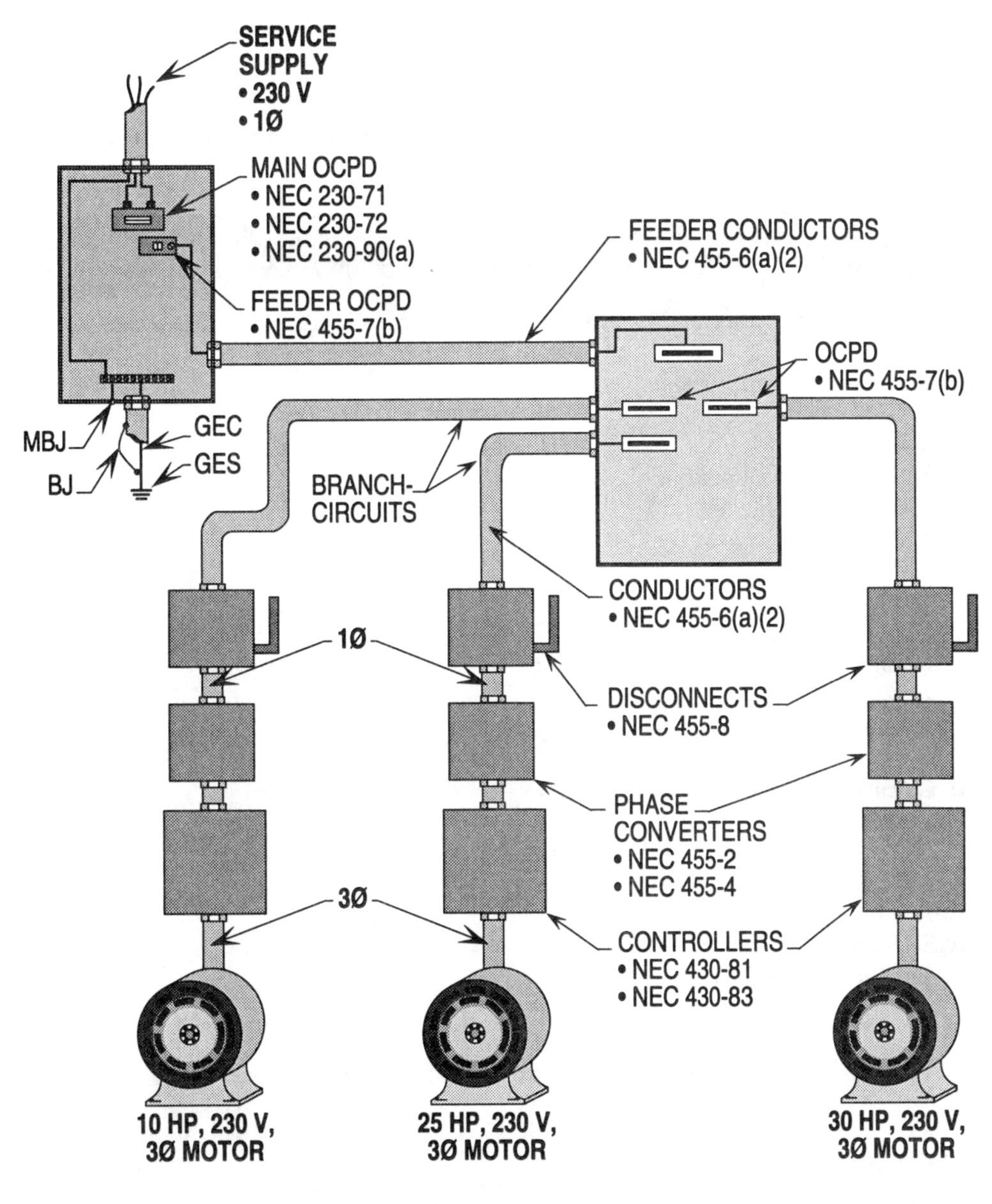
SERVICE SUPPLY
• 230 V
• 1Ø
MAIN OCPD
• NEC 230-71
• NEC 230-72
• NEC 230-90(a)
FEEDER OCPD
• NEC 455-7(b)
FEEDER CONDUCTORS
• NEC 455-6(a)(2)
OCPD
• NEC 455-7(b)
MBJ
BJ
GEC
GES
BRANCH-CIRCUITS
CONDUCTORS
• NEC 455-6(a)(2)
1Ø
DISCONNECTS
• NEC 455-8
PHASE CONVERTERS
• NEC 455-2
• NEC 455-4
3Ø
CONTROLLERS
• NEC 430-81
• NEC 430-83
10 HP, 230 V, 3Ø MOTOR
25 HP, 230 V, 3Ø MOTOR
30 HP, 230 V, 3Ø MOTOR
NEC 455-6(a)(2)
NEC 455-7(b)

Ill. 16-4

Phase Converters 455-6(a)(2); 455-7(b)

The elements of circuits supplying phase converters and specific loads plus other loads are sized at 250 percent of the FLA of such loads, where the input and output voltage levels are the same.

Example 16-5. What size OCPD and THHN copper conductors are required for the feeder-circuit in Illustration 16-5?

Sizing conductors for input side

Step 1: Finding FLA for input side
430-6(a)(1); Table 430-150
5 HP = 15.2 A
7 1/2 HP = 22 A
10 HP = 28 A

Step 2: Finding other loads
215-2(a); 220-10
35 A x 100% = 35 A

Step 3: Calculating conductors
455-6(a)(2)
15.2 A + 22 A + 28 A + 35 A = 100.2 A
100.2 A x 250% = 250.5 A

Step 4: Selecting conductors
310-10(2); Table 310-16
250.5 A requires #250 KCMIL cu.

Solution: The size THWN copper conductors are No. 250 KCMIL copper.

Sizing OCPD for input side

Step 1: Calculating OCPD
455-7(b); Step 3 above
100.2 A x 250% = 250.5 A

Step 2: Selecting OCPD
455-7; 240-6(a)
250.5 A allows 250 A OCPD

Solution: The size OCPD is 250 amps.

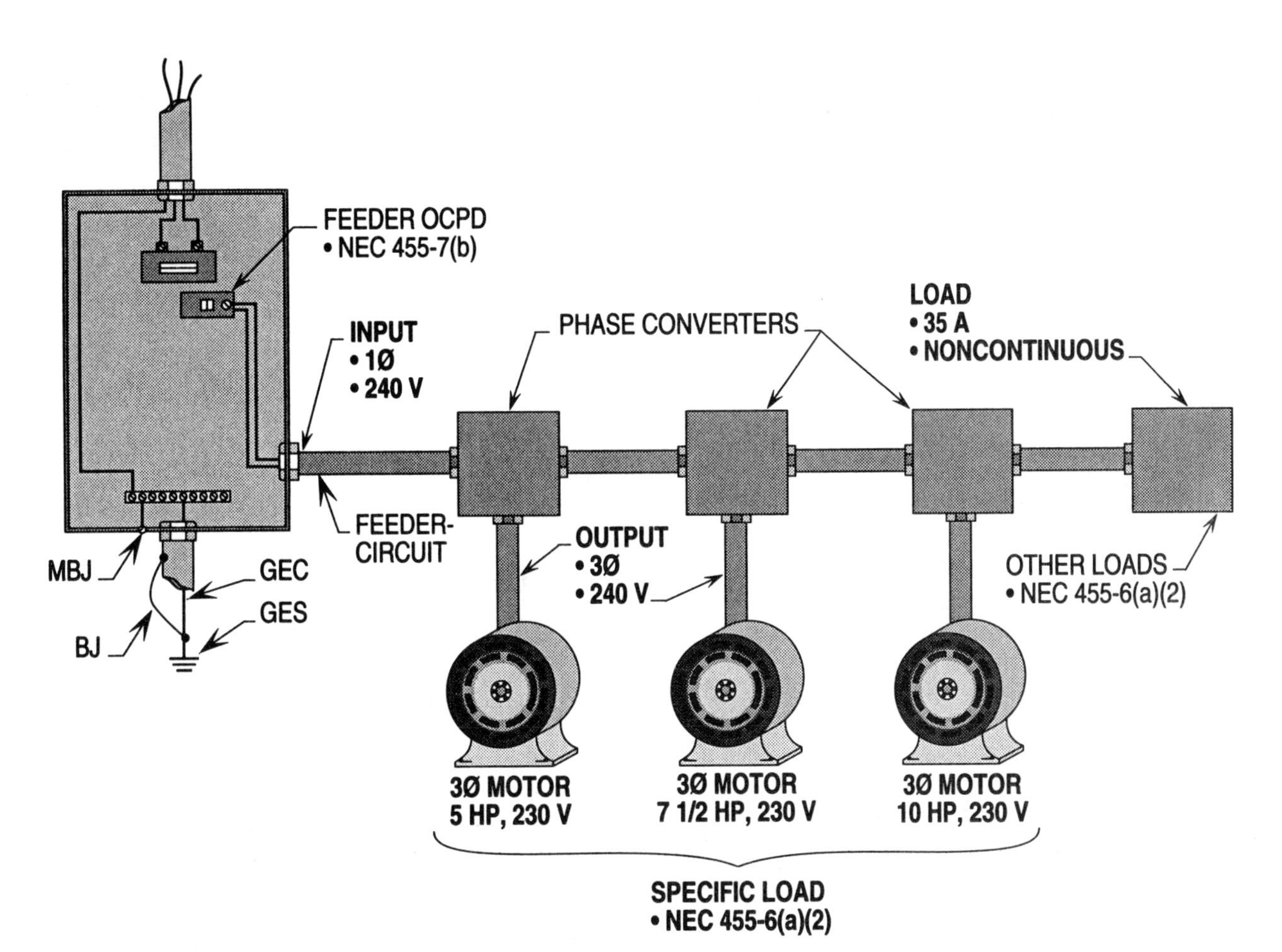

NEC 455-6(a)(2)
NEC 455-7(b)

Ill. 16-5

Phase Converters
455-6(a)(2); 455-7(b)

The elements of circuits supplying phase converters and specific loads plus other loads are sized at 250 percent of the FLA of such loads plus the ratio of the input and output voltage levels.

Example 16-6. What size OCPD and THWN copper conductors are required for the feeder-circuit in Illustration 16-6?

Sizing conductors for input side

Step 1: Finding FLA of motors
430-6(a)(1); Table 430-150
5 HP = 7.6 A
7 1/2 HP = 11 A
10 HP = 14 A

Step 2: Finding other loads
215-2(a); 220-10
34 A x 100% = 34 A

Step 3: Finding ratio
Ratio = Sec. V / Pri. V
Ratio = 480 V / 240 V
Ratio = 2

Step 4: Calculating conductors
455-6(a)(2)
7.6 A + 11 A + 14 A + 34 A = 66.6 A
66.6 A x 2 x 250% = 333 A

Step 5: Selecting conductors
310-10(2); Table 310-16
333 A requires #400 KCMIL cu.

Solution: The size THWN copper conductors are No. 400 KCMIL.

Sizing OCPD for input side

Step 1: Calculating OCPD
455-7(b); Step 4 above
66.6 A x 2 x 250% = 333 A

Step 2: Selecting OCPD
240-3(g); 455-7(b); 240-6(a)
333 A requires 350 A OCPD

Solution: The size OCPD is 350 amps.

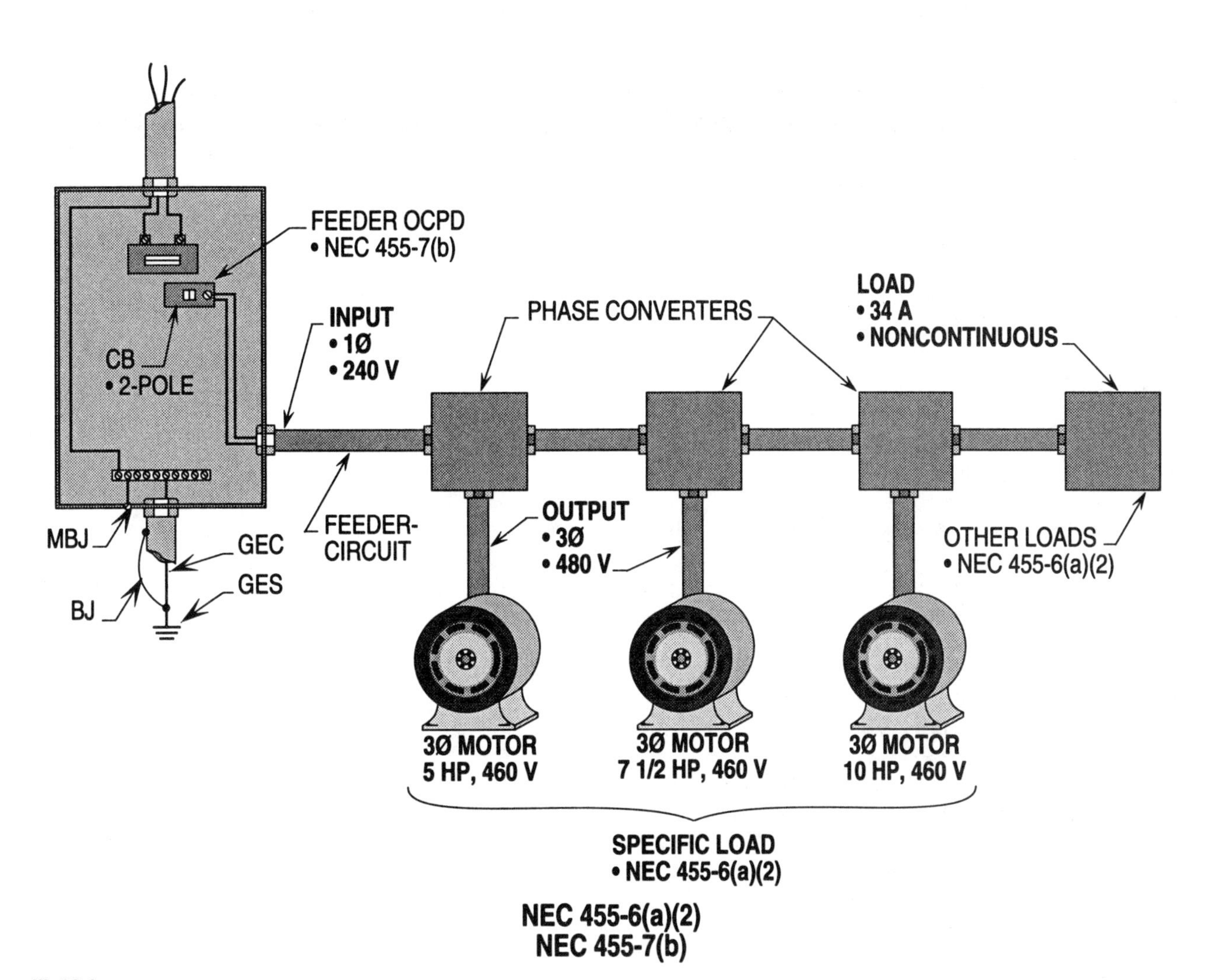
FEEDER OCPD
• NEC 455-7(b)
LOAD
• 34 A
• NONCONTINUOUS
INPUT
• 1Ø
• 240 V
PHASE CONVERTERS
CB
• 2-POLE
MBJ
GEC
BJ
GES
FEEDER-CIRCUIT
OUTPUT
• 3Ø
• 480 V
OTHER LOADS
• NEC 455-6(a)(2)
3Ø MOTOR
5 HP, 460 V
3Ø MOTOR
7 1/2 HP, 460 V
3Ø MOTOR
10 HP, 460 V
SPECIFIC LOAD
• NEC 455-6(a)(2)
NEC 455-6(a)(2)
NEC 455-7(b)

Ill. 16-6

Phase Converters
455-8(c)

The disconnecting means for phase converters shall be sized with enough capacity to deenergize the entire load without rupturing.

Example 16-7. What size circuit breaker is required to disconnect the phase converters in Illustration 16-7?

Sizing disconnect for not a specific load

Step 1: Calculating load
455-8(c)
34 A x 115% = 39.1 A

Step 2: Selecting disconnect
455-8(c); 240-6(a)
39.1 A requires 40 A

Solution: The size disconnecting means using a CB is 40 amps.

Sizing disconnect for a specific load

Step 1: Finding FLA of motor
430-6(a)(1); Table 430-150
20 HP = 54 A

Step 2: Calculating load
455-8(c)(1)
54 A x 250% = 135 A

Step 3: Selecting disconnect
455-8(c)(1); 240-6(a)
135 A requires 150 A

Solution: The size disconnecting means using a CB is 150 amps.

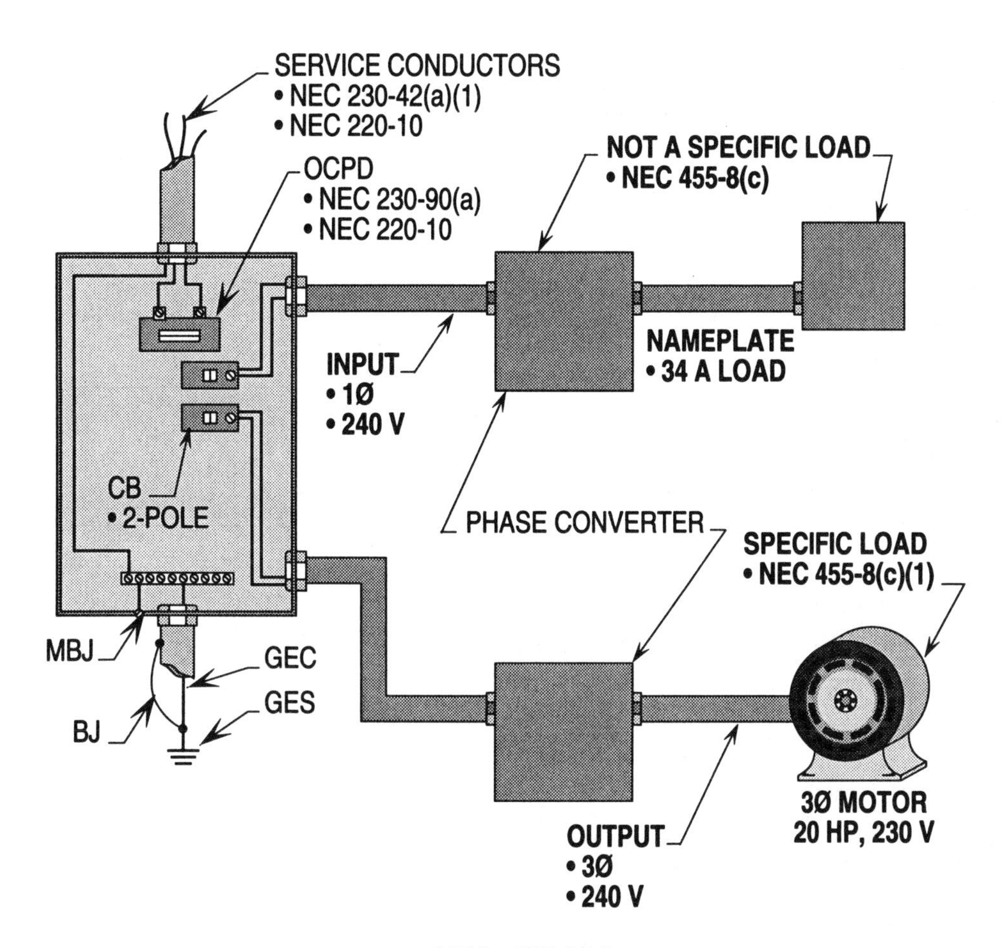
SERVICE CONDUCTORS
• NEC 230-42(a)(1)
• NEC 220-10
OCPD
• NEC 230-90(a)
• NEC 220-10
NOT A SPECIFIC LOAD
• NEC 455-8(c)
INPUT
• 1Ø
• 240 V
NAMEPLATE
• 34 A LOAD
CB
• 2-POLE
PHASE CONVERTER
SPECIFIC LOAD
• NEC 455-8(c)(1)
MBJ
GEC
GES
BJ
3Ø MOTOR
20 HP, 230 V
OUTPUT
• 3Ø
• 240 V
NEC 455-8(c)
NEC 455-8(c)(1)

Ill. 16-7

Phase Converters
455-8(c)(1); 455-8(c)(2)

If the disconnecting means contains CB's or fuses, the disconnecting enclosure will be larger in size than an enclosure housing a nonautomatic device.

Example 16-8. What size circuit breaker and HP rated switch is required to disconnect the motors in Illustration 16-8?

Sizing disconnect using CB

Step 1: Finding FLA of motor
430-6(a)(1); Table 430-150
20 HP = 27 A

Step 2: Calculating load
455-8(c)(1)
27 A x 2 x 250% = 135 A

Step 3: Selecting OCPD
240-3(g); 455-8(c)(1); 240-6(a)
135 A requires 150 A OCPD

Solution: The size disconnecting means using a CB is 150 amps.

Sizing disconnect using HP rated switch

Step 1: Finding FLA of motor
430-6(a)(1); Table 430-150
30 HP = 40 A

Step 2: Calculating other load
215-2(a); 220-10
26 A x 100% = 26 A

Step 3: Calculating total load
455-8(c)(2)
40 A + 26 A = 66 A
66 A x 200% = 132 A

Step 4: Selecting disconnect
240-3(g); 455-8(c)(2); 240-6(a)
132 A requires 150 A

Solution: The size HP rated switch is 150 amps.

Sizing capacitor in microfarads rule of thumb method.

Example 16-9. What size capacitor in microfarads (mf) is required for the motor in Illustration 16-9?

Sizing capacitor in mf

Step 1: Applying formula
mf = 159,300 / hertz x amps / volts
mf = 159,300 / 60 Hz x 21 A / 480 V x 1.732
mf = 2,655 x 21 A ÷ 831 V
mf = 67

Solution: The size capacitor in microfarads (mf) is 67 mf.

Note: The constant 159,300 is used in this formula when sizing microfarads.

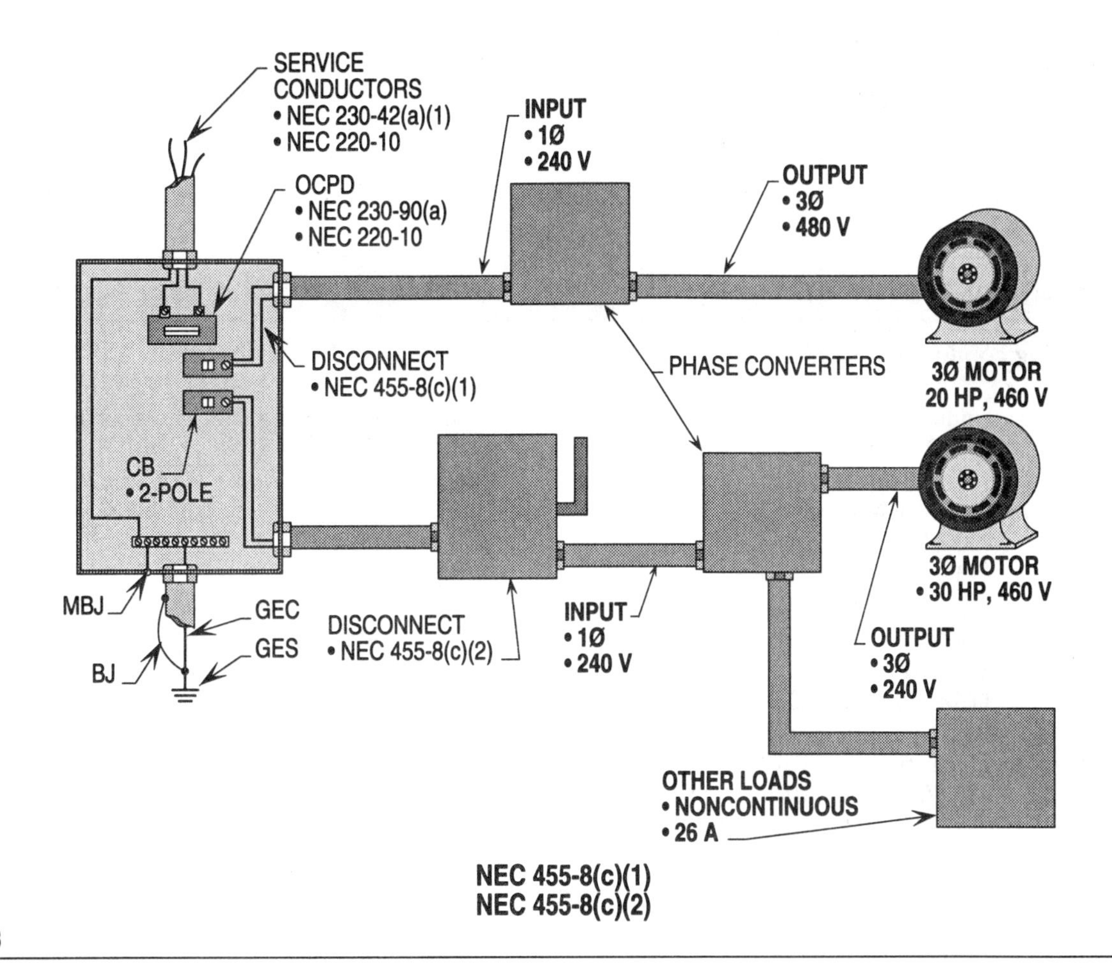

Ill. 16-8

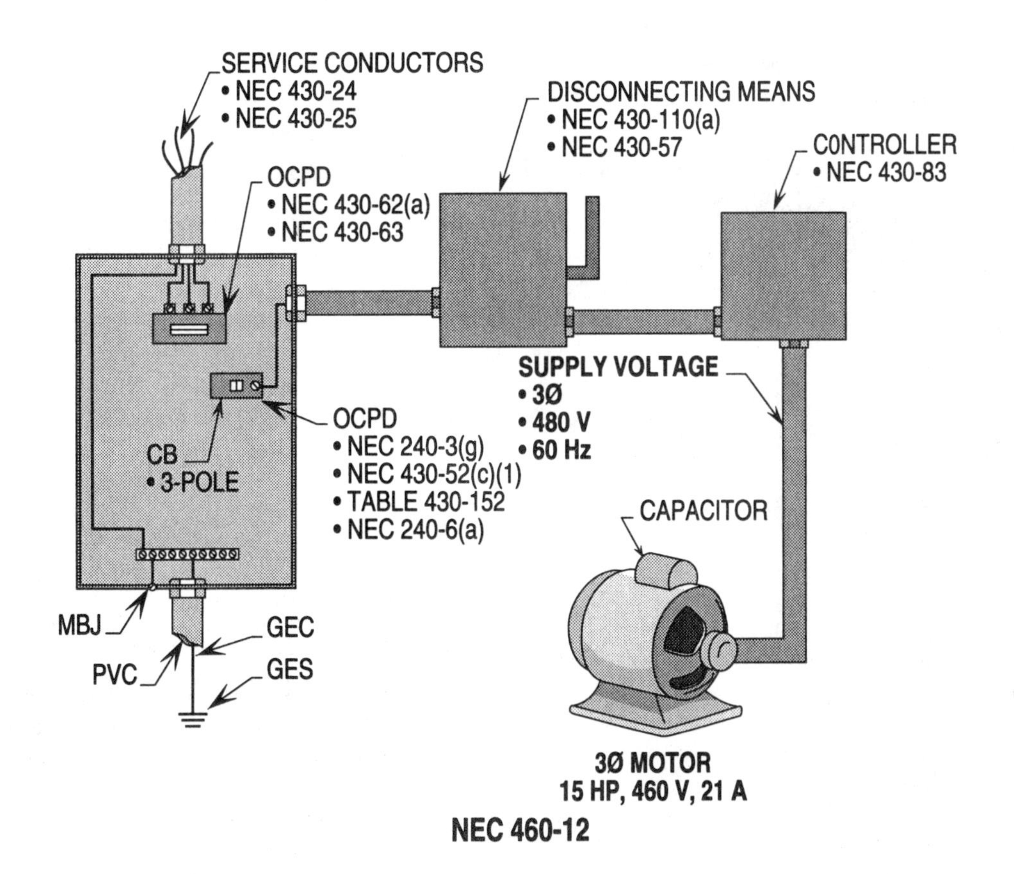

Ill. 16-9

Capacitor
460-8(a)

The size conductors supplying power to capacitors are sized and selected, based either upon the FLA of the capacitor times 135 percent or 1/3 of the conductors ampacity servicing the motor.

Example 16-10. What size THWN copper conductors are required for the capacitor in Illustration 16-10?

Sizing conductors based on 1/3

Step 1: Finding FLA of motor
50 HP = 143 A

Step 2: Calculating conductors
430-22(a)
143 A x 125% = 178.8 A

Step 3: Selecting conductors
310-10(2); Table 310-16
178.8 A requires #3/0 THWN cu.

Step 4: Calculating conductors at 1/3 of #3/0 cu.
460-8(a); Table 310-16
#3/0 THHN cu. = 200 A
1/3 of 200 A = 66.7 A

Step 5: Selecting conductors
Table 310-16
66.7 A requires #4 cu.

Solution: **The size THWN copper conductors based on 1/3 of branch-circuit is No. 4.**

Sizing conductors based on 135%

Step 1: Calculating FLA of capacitor
460-8(a)
FLA = kVA x 1,000 / V x √3
FLA = 25 x 1,000 / 208 V x 1.732
FLA = 69.4

Step 2: Calculating conductors
460-8(a)
69.4 A x 135% = 93.7

Step 3: Selecting conductors
310-10(2); Table 310-16
93.7 A requires #3 cu.

Solution: **The size THWN copper conductors based on 135 percent of FLA of capacitor are No. 3.**

Note: NEC 460-8(a) requires the largest conductors calculated which are the No. 3 THHN copper.

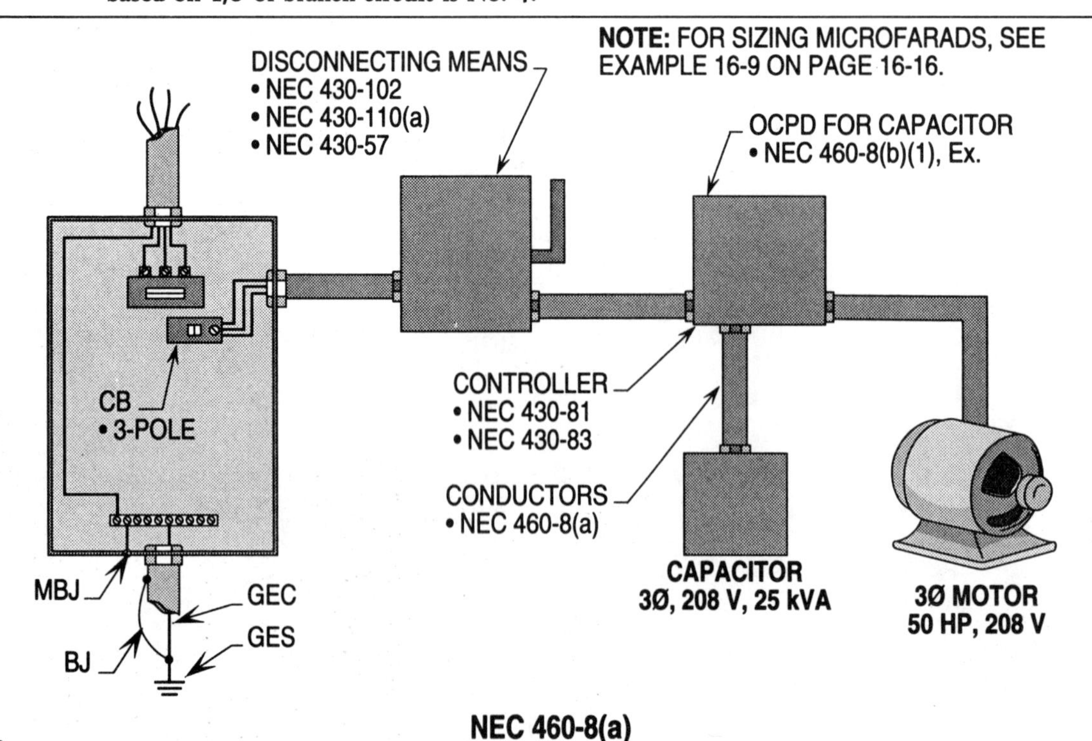

NEC 460-8(a)

Ill. 16-10

Capacitors
460-8(b); (c)

The disconnecting means for a capacitor that is located on the line side of the motor controller is sized at 135 percent of the capacitor's FLA and the OCPD's are selected at 150 percent to 250 percent of the capacitor's FLA.

Example 16-11. What size disconnect and OCPD is required for the capacitor in Illustration 16-11?

Sizing disconnect for capacitor

Step 1: Calculating FLA of capacitor
460-8(c)(4)
FLA = kVA x 1,000 / V x √3
FLA = 25 x 1,000 / 208 V x 1.732 = 69.4 A
FLA = 69.4 x 135% = 93.69
FLA = 93.69 A

Step 2: Calculating disconnect
460-8(c)(4); 240-6(a)
93.69 A requires 100 A

Solution: **The size disconnect switch is 100 amps.**

Sizing OCPD for capacitor

Step 1: Calculating OCPD
460-8(b)(2)
69.4 A x 150% = 104.1 A

Step 2: Selecting OCPD
460-8(b)(2); 240-6(a)
104.1 A allows 110 A OCPD

Solution: **The next size OCPD above 104.1 amps is used, but the OCPD can be sized up to 250 percent of the capacitors FLA rating.**

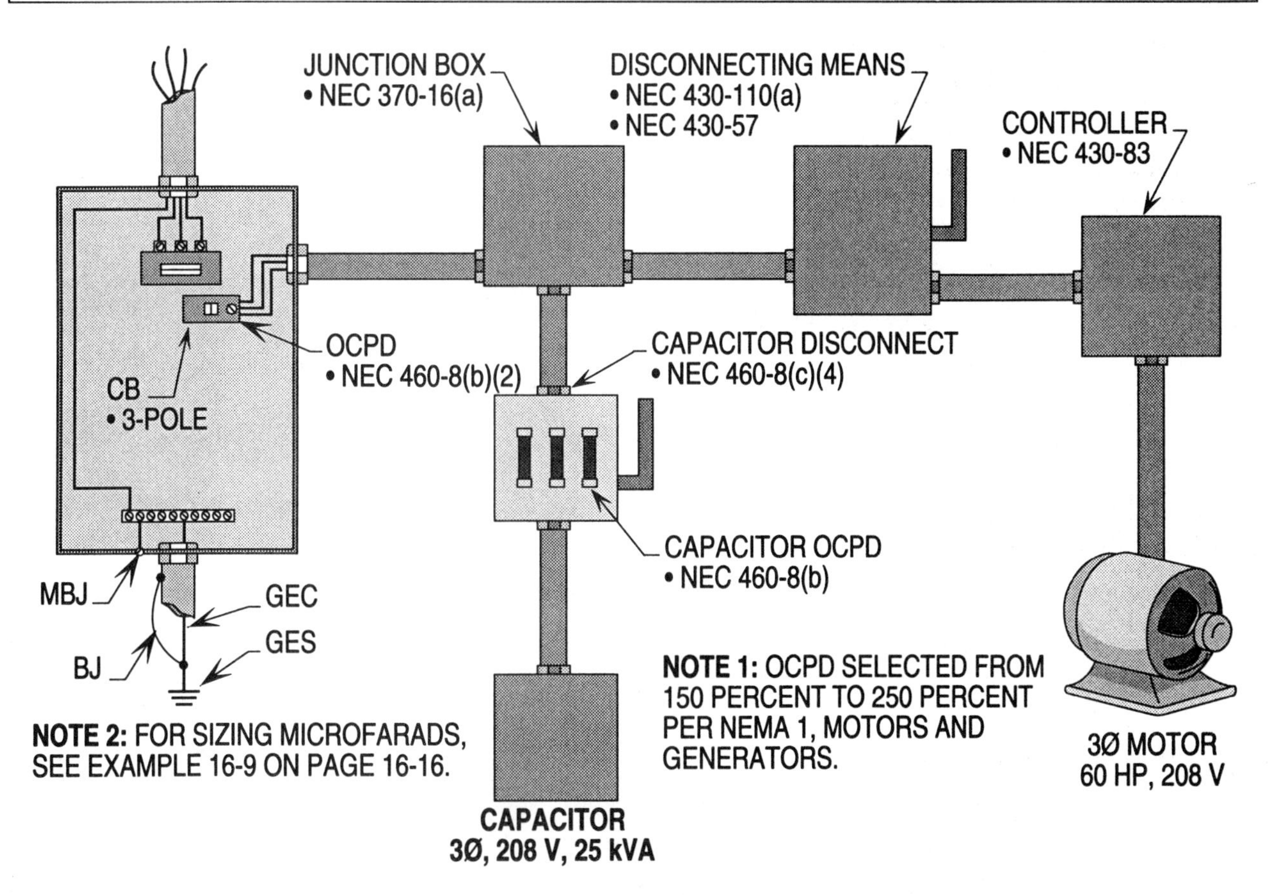

Ill. 16-11

Health Care Facilities

Health care facilities are loaded with special equipment that requires comprehensive procedures to calculate and determine load values. This chapter deals with the methods utilized to compute such values to be used for sizing the elements to X-ray equipment and settings for ground-fault protection of equipment and other sophisticated loads.

Diagnostic and therapeutic X-ray equipment is usually the type utilized in health care facilities to examine patients for various medical problems.

Most manufacturers specify the OCPD's and conductors to supply power to such sensitive electronic operated equipment. The NEC requires designers to calculate such load values where manufacturers fail to provide specific information for the selection of such wiring elements.

Quick Reference

Rating of Supply Conductors and OCPD's 517-73(a)(2), FPN; 517-73(b), FPN

Sizing and selecting conductors and OCPD's for X-ray equipment is usually based upon manufacturer's nameplate data calling for specific sizes.

Example 17-1. What size OCPD and THHN copper conductors are required for the X-ray equipment in Illustration 17-1?

Sizing conductors and OCPD

Step 1: Finding A for power supply
517-73(a), FPN; 517-73(b), FPN
Circuit rating = 100 A
OCPD = 100 A

Step 2: Selecting conductors and OCPD
310-10(2); Table 310-16; 240-6(a)
100 A requires #3 cu.
100 A requires 100 A OCPD

Solution: The size THHN copper conductors are No. 3 and the size OCPD is 100 amps.

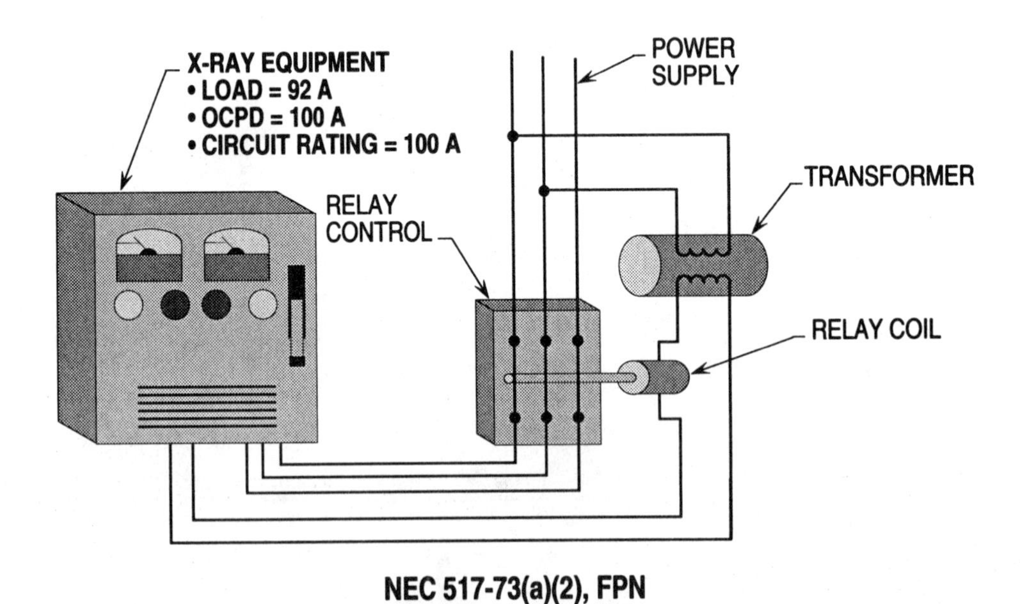

Ill. 17-1

Disconnecting Means (Capacity) 517-72(a)

Disconnects used for X-ray equipment are sized and selected based upon the momentary or long time rating which is based upon operation of the particular unit.

Example 17-2. What size disconnect is required for the X-ray equipment in Illustration 17-2?

Sizing disconnect for momentary rating

Step 1: Calculating amps
517-72(a)
A = MA / 1,000 x Sec. V / Pri. V
A = 175 x 125,000 V ÷ 208 V x 1,000
A = 21,875,000 / 208,000
A = 105 A

Step 2: Applying demand factor
517-72(a)
105 A x 50% = 52.5 A

Step 3: Selecting disconnect
517-72(a); 240-6(a)
52.5 A requires 60 A

Solution: **The disconnecting means is 60 amps.**

Sizing disconnect for long time rating

Step 1: Calculating amps
517-72(a)
A = MA / 1,000 x Sec. V / Pri. V
A = 20 x 250,000 V ÷ 208 V x 1,000
A = 5,000,000 / 208,000
A = 24 A

Step 2: Applying demand factor
517-72(a)
24 A x 100% = 24 A

Step 3: Selecting disconnect
517-72(a)
24 A requires 25 A

Solution: **The size disconnect is a 25 amp CB or 30 amp switch.**

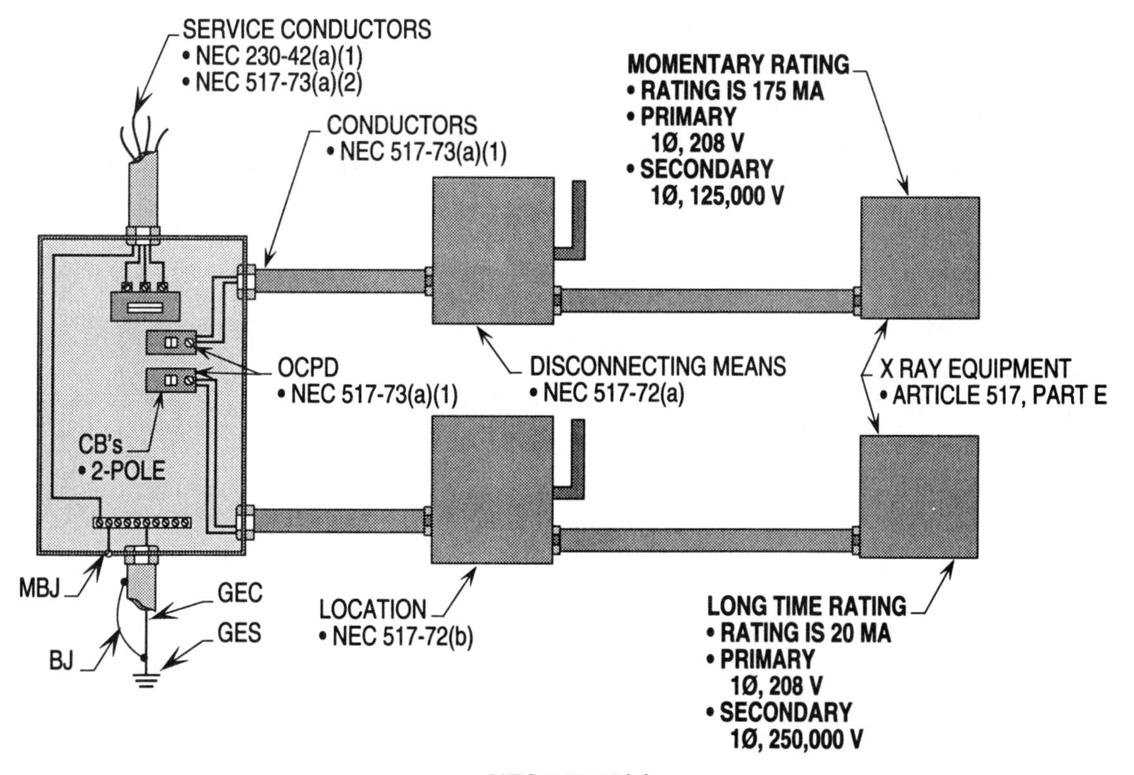

NEC 517-72(a)

Ill. 17-2

Rating of Supply Conductors and OCPD's 517-73(a)(1)

Sizing and selecting conductors and OCPD's for diagnostic X-ray equipment based upon nameplate ratings and not manufacturer's data which calls for specific sizes.

Example 17-3. What size OCPD and THWN copper conductors are required to supply the X-ray equipment in Illustration 17-3?

Sizing OCPD for momentary rating

Step 1: Calculating amps
517-73(a)(1)
A = MA / 1,000 x Sec. V / Pri. V
A = 175 x 125,000 V ÷ 208 V x 1,000
A = 105 A

Step 2: Applying demand factor
517-73(a)(1)
105 A x 50% = 52.5 A

Step 3: Selecting OCPD
517-73(a)(1); 240-6(a)
52.5 A requires 60 A OCPD

Solution: The size OCPD using a CB or fuses is 60 amps.

Sizing OCPD to long time rating

Step 1: Calculating amps
517-73(a)(1)
A = MA / 1,000 x Sec. V / Pri. V
A = 20 x 250,000 V ÷ 208 V x 1,000
A = 24 A

Step 2: Applying demand factor
517-73(a)(1)
24 A x 100% = 24 A

Step 3: Selecting OCPD
517-73(a)(1); 240-6(a)
24 A requires 25 A OCPD

Solution: The size OCPD using a CB or fuses is 25 amps.

Sizing conductors for momentary rating

Step 1: Selecting conductors
517-73(a)(1); Table 310-16;
Step 2 above
52.5 A requires #6 cu.

Solution: The size THWN copper conductors are No. 6.

Sizing conductors for long time rating

Step 1: Selecting conductors
517-73(a)(1); Table 310-16;
Step 2 above
24 A requires #10 cu.

Solution: The size THWN copper conductors are No. 10.

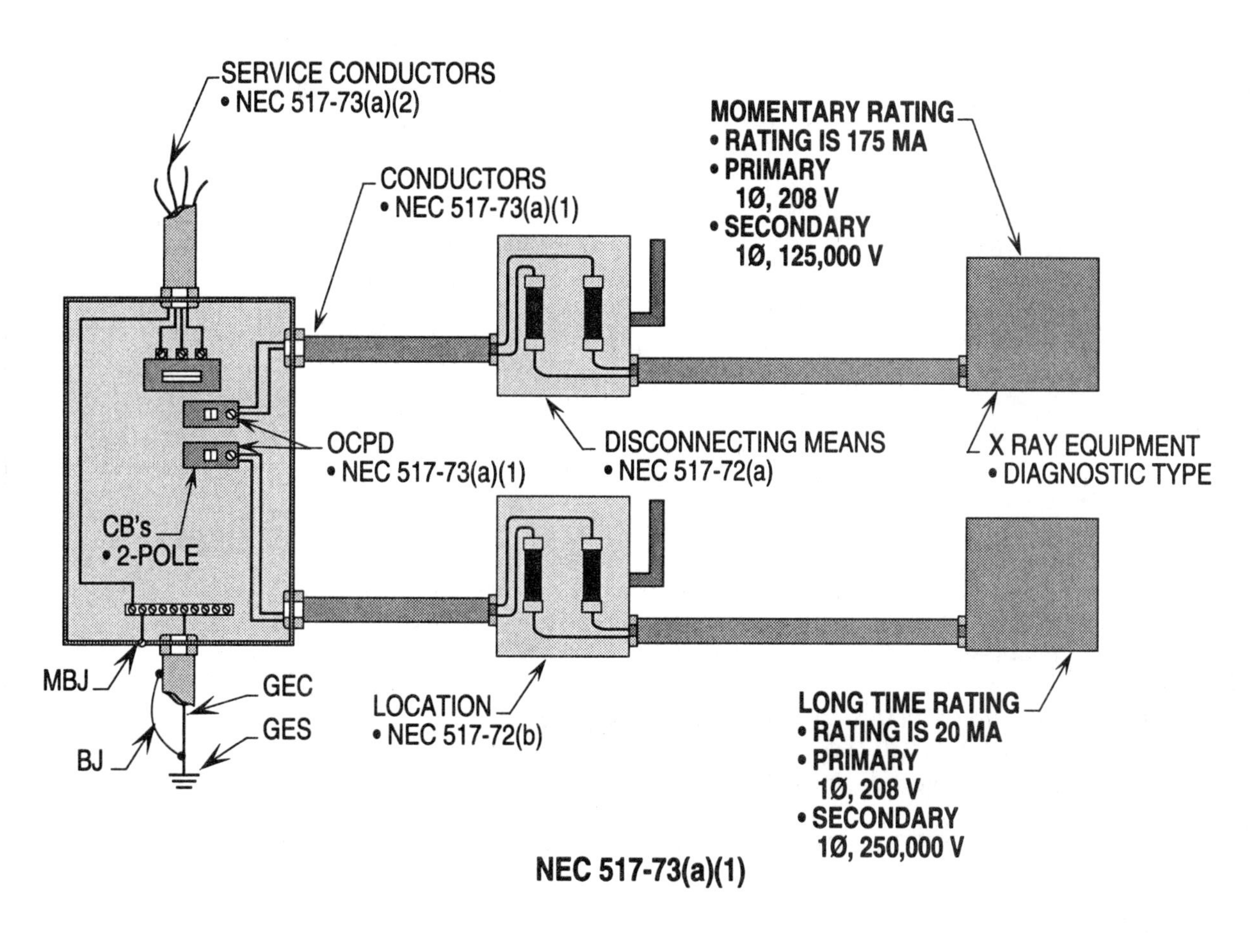
SERVICE CONDUCTORS
• NEC 517-73(a)(2)
CONDUCTORS
• NEC 517-73(a)(1)
MOMENTARY RATING
• RATING IS 175 MA
• PRIMARY
1Ø, 208 V
• SECONDARY
1Ø, 125,000 V
OCPD
• NEC 517-73(a)(1)
DISCONNECTING MEANS
• NEC 517-72(a)
X RAY EQUIPMENT
• DIAGNOSTIC TYPE
CB's
• 2-POLE
MBJ
GEC
GES
BJ
LOCATION
• NEC 517-72(b)
LONG TIME RATING
• RATING IS 20 MA
• PRIMARY
1Ø, 208 V
• SECONDARY
1Ø, 250,000 V
NEC 517-73(a)(1)

Ill. 17-3

Rating of Supply conductors and OCPD's 517-73(a)(2)

Sizing elements for diagnostic X-ray equipment based upon nameplate information and not manufacturer data which calls for specific sizes.

Example 17-4. What size OCPD and THHN copper conductors are required for the feeder-circuit in Illustration 17-4?

Sizing OCPD for feeder-circuit

Step 1: Calculating amps
517-73(a)(2)
Unit #1
A = MA / 1,000 x Sec. V / Pri. V
A = 25 x 150,000 V ÷ 208 V x 1,000
A = 18.03 A
Unit #2
A = 30 x 175,000 V ÷ 208 V x 1,000
A = 25.24 A
Unit #3
A = 35 x 200,000 V ÷ 208 V x 1,000
A = 33.65 A

Step 2: Applying demand factors
517-73(a)(2)

33.65 A x 50%	= 16.83 A
25.24 A x 25%	= 6.31 A
18.03 A x 10%	= 1.80 A
Total load	= 24.94 A

Step 3: Selecting OCPD
517-73(a)(2); 240-6(a)
24.94 A requires 30 A OCPD

Solution: The size OCPD for the feeder-circuit is 30 amps.

Sizing conductor for feeder-circuit

Step 1: Selecting conductors
517-73(a)(2); Table 310-16;
Step 2 above
24.94 A requires #10 cu.

Solution: The size THHN copper conductors for the feeder-circuit is No. 10.

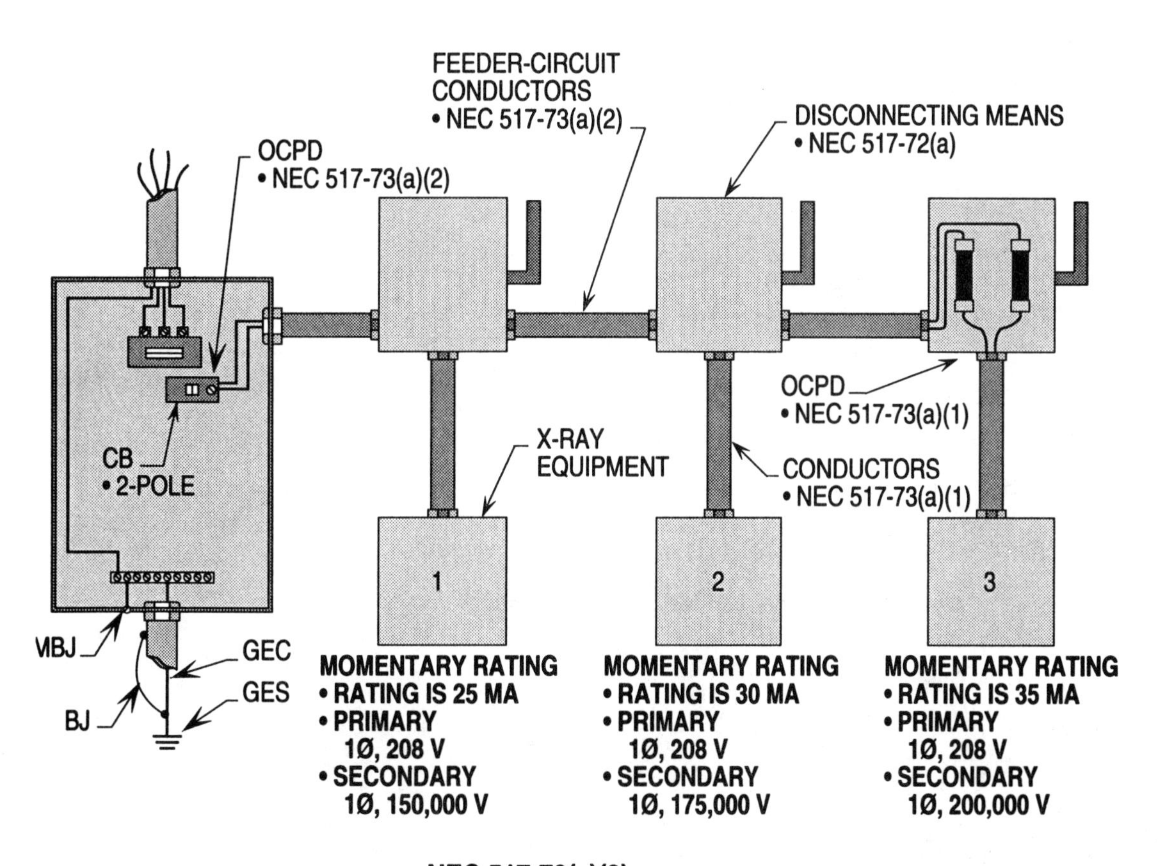

NEC 517-73(a)(2)

Ill. 17-4

Rating of Supply Conductors and OCPD's 517-73(b)

Sizing the OCPD's and conductors for therapeutic X-ray equipment based upon nameplate ratings and not manufacturer data which calls for specific sizes.

Example 17-5. What size OCPD and THWN copper conductors are required for the branch-circuit in Illustration 17-5?

Sizing OCPD for branch-circuit

Step 1: Calculating amps
517-73(b)
40 A x 100% = 40 A

Step 2: Selecting OCPD
517-73(b); 240-6(a)
40 A requires 40 A OCPD

Solution: The size OCPD for the branch-circuit is 40 amps.

Sizing conductors for branch-circuit

Step 1: Finding amps
517-73(b)
40 A x 100% = 40 A

Step 2: Selecting conductors
517-73(b); Table 310-16
40 A requires #8 cu.

Solution: The size THWN copper conductors for the branch-circuit is No. 8.

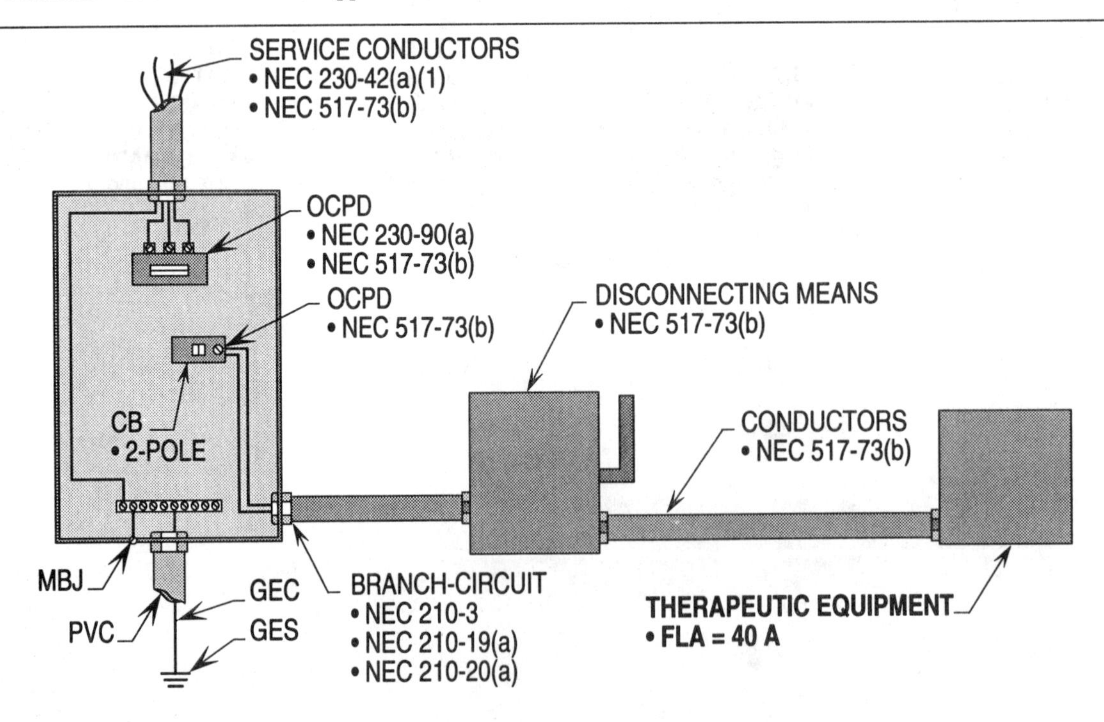

NEC 517-73(b)

Ill. 17-5

18

Theaters and Motion Picture Studios

Special rules regulate the design and layout of electrical systems installed in theaters and motion picture studios. Overcurrent protection devices and conductors protecting and supplying power to equipment such as dimmers, footlights, and receptacles are required to be sized with capacity capable of carrying the heating effects of such loads.

Feeder-circuit elements serving stage switchboards and portable patch panels for road shows must be sized and selected to handle all or parts of the load required during performances.

Overcurrent protection devices and conductors used on the ungrounded or grounded legs of feeders are sometimes oversized to accommodate the varying loads without overheating and tripping open the circuits.

The elements for circuits run in theaters and motion picture studios are calculated and designed based upon continuous operation, noncontinuous operation, or demand factors which are selected according to the conditions of use for such equipment.

Quick Reference

Dimmers (Disconnect and Overcurrent Protection) 520-25(a)

The elements of circuits supplying power to a dimmer located in a bank of dimmers are calculated at 125 percent of the ampacity of such dimmers.

Example 18-1. What size OCPD and disconnect is required for one of the dimmers in Illustration 18-1?

Sizing OCPD and disconnect for dimmers

Step 1: Finding A of dimmers
520-25(a)
Dimmers = 23 A each

Step 2: Calculating A for OCPD and disconnect
520-25(a)
23 A x 125% = 28.75 A

Step 3: Selecting OCPD and disconnect
520-25(a); 240-6(a)
28.75 A requires 30 A OCPD
28.75 A requires 30 A disconnect

Solution: The size OCPD and disconnect is required to be 30 amps.

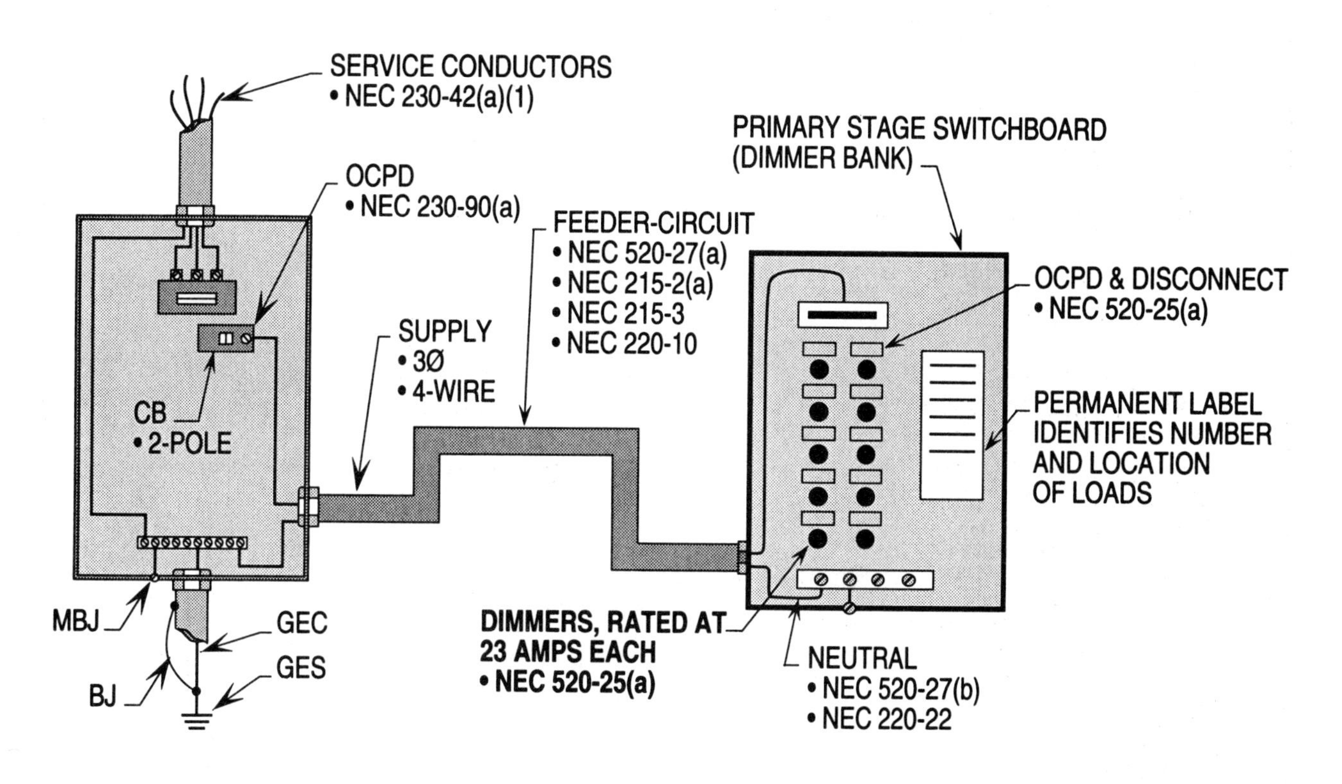

NEC 520-25(a)

Ill. 18-1

Stage Switchboard Feeders (Dimmer Bank) 520-27(c)

The elements of circuits supplying power to dimmer banks in stage switchboards are sized at 125 percent of the load that is intended to be controlled.

Example 18-2. What size OCPD and THWN copper conductors are required for the feeder-circuits in Illustration 18-2?

Sizing conductors to dimmer bank

Step 1: Finding A for dimmer bank
520-25(a); 520-27(c)
Load = 400 A

Step 2: Calculating A for dimmer bank
520-27(c), FPN; 220-10; 215-2(a)
400 A x 125% = 500 A

Step 3: Paralleling 3 times per phase
310-4; 300-5(i); 300-20
500 A / 3 per phase = 166.7 A

Step 4: Selecting conductors for dimmer bank
310-10(2); Table 310-16
166.7 A requires #2/0 cu.

Solution: The size THWN copper conductors are No. 2/0.

Sizing OCPD for dimmer bank

Step 1: Selecting OCPD for dimmer bank
240-3(b); 240-6(a); Step 4 above
#2/0 THWN cu. = 175 A
175 A requires 175 OCPD

Solution: The size OCPD is 175 amps.

Sizing conductors and OCPD's to patch panel

Step 1: Finding A for patch panel
520-27(b)
Load = 40 A per circuit

Step 2: Calculating A for conductors and OCPD's
520-27(c); 220-10; 215-2(a)
40 A x 125% = 50 A

Step 3: Selecting conductors and OCPD's
240-6(a); Table 310-16
50 A requires #8 cu.
50 A requires 50 A OCPD

Solution: The size conductors are No. 8 THWN copper conductors and the OCPD's are 50 amp each.

Note: The neutral is considered current-carrying if the dimmers are solid state.

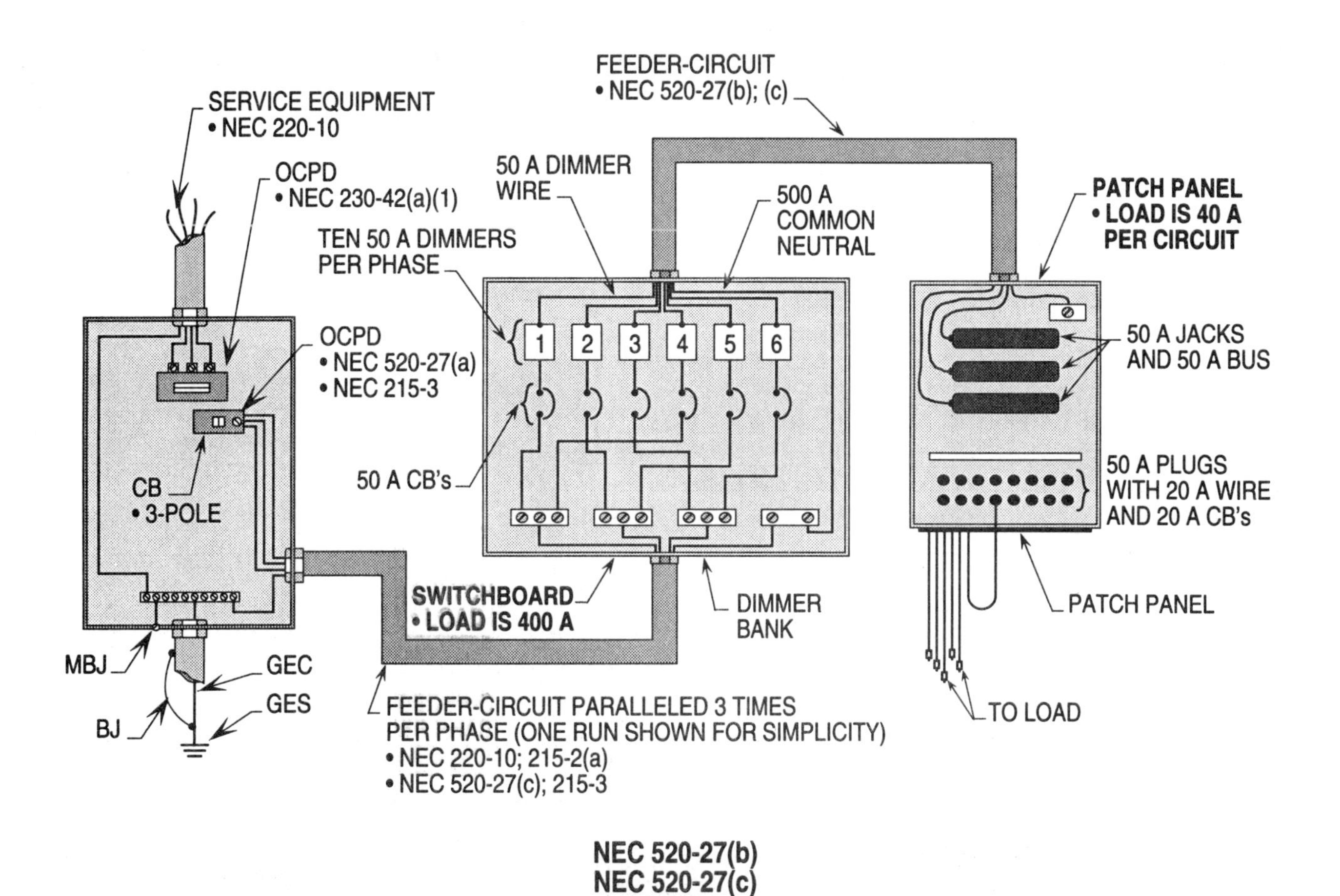

NEC 520-27(b)
NEC 520-27(c)
NEC 220-10
NEC 215-2(a)

Ill. 18-2

Location Boards (OCPD) 530-18(b); (d)

The OCPD protecting the circuit conductor supplying power to location boards may be sized up to 400 percent of the feeder-circuit conductor's ampacity.

Example 18-3. What is the setting of the OCPD for the feeder-circuit between the substation and location board in Illustration 18-3?

Sizing the setting of the OCPD for the feeder-circuit

Step 1: Selecting the conductors for feeder-circuit
310-10(2); Table 310-16
267 A requires #300 KCMIL cu.

Step 2: Calculating the setting for OCPD
530-18(b); (d)
285 A (#300 KCMIL cu.) x 400% = 1,140 A

Solution: The setting of the OCPD for the feeder-circuit is 1,140 amps.

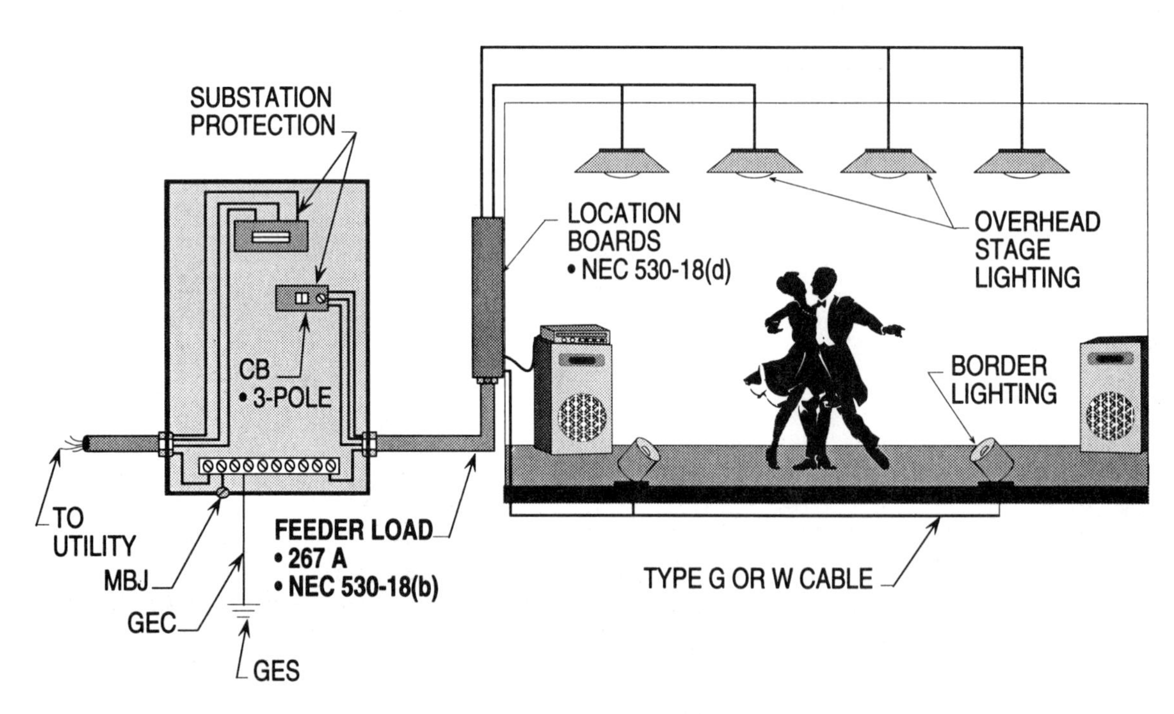

NEC 530-18(b); (d)

Ill. 18-3

Television Studio Sets
530-19(a)

Circuits and elements used to supply stage set lighting loads are sized and selected, based upon demand factors times the connected loads.

Example 18-4. What is the demand load in amps for the feeder-circuit in Illustration 18-4?

Sizing the demand load in amps

Step 1: Finding VA of feeder-circuit
Load = 250,000 VA

Step 2: Applying demand factors
530-19(a); Table 530-19(a)

First 50,000 VA x 100%	= 50,000 VA
From 50,000 VA x 75%	= 37,500 VA
From 100,000 VA x 60%	= 60,000 VA
Remaining 50,000 VA x 50%	= 25,000 VA
Total A	=172,500 VA

Step 3: Calculating A of feeder-circuit
A = 172,500 VA / 208 V x 1.732 (360 V)
A = 479.2 A

Solution: The demand load in amps for the feeder-circuit is 479.2 amps.

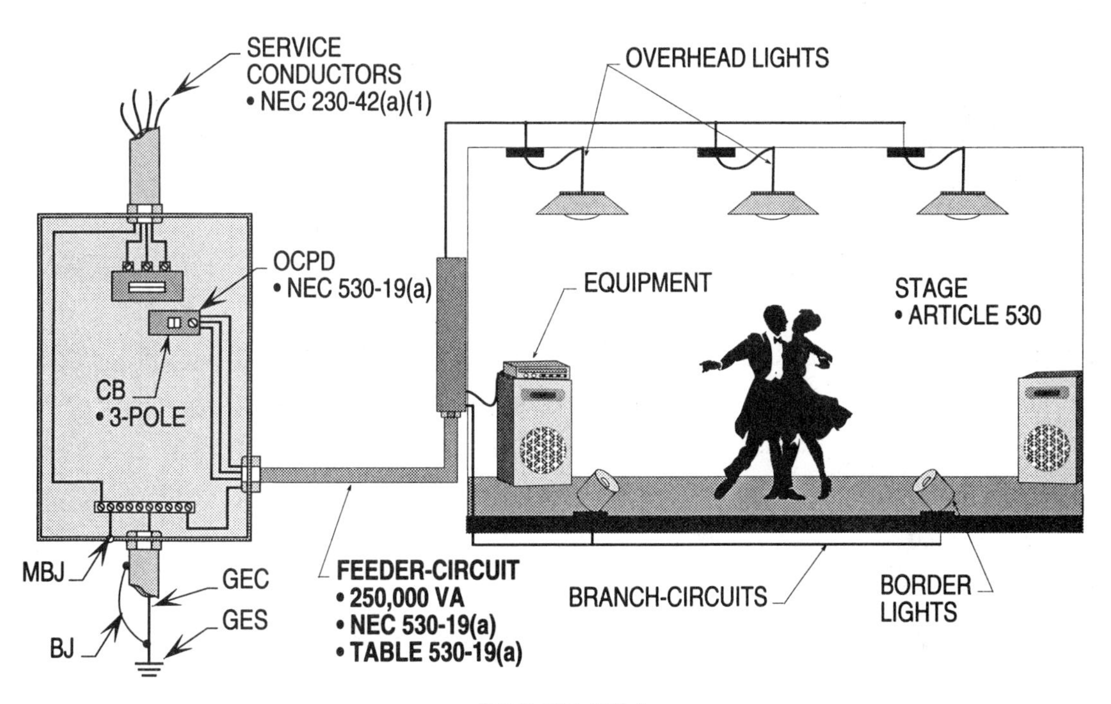

NEC 530-19(a)
TABLE 530-19(a)

Ill. 18-4

19

Mobile Homes, RV's and Marinas

The elements used in electrical systems to connect mobile homes and RV's to a supply of electricity which includes feeder-circuit conductors and service equipment components are calculated using the standard or optional method.

Wiring and equipment elements utilized to serve fixed or floating piers, wharfs, docks, and other areas in marinas, boatyards, boat basins, and similar establishments must be sized to carry the loads. Such loads are computed by applying either the standard or optional calculation or both. Designers and installers must consider areas of extreme temperatures before demand factors are applied.

There are certain electrical components that are required by the NEC to be supplied by GFCI circuits, for the protection of personnel using such systems.

Note: When performing computations for park trailers use the same calculation procedure as for mobile homes found in NEC 552-47. See NEC 550-13 and page 19-6 of this chapter.

Quick Reference

Lighting Circuits Required
550-7(a)

When determining the number of lighting circuits required for a mobile home, the square footage is multiplied by 3 VA. When using 15 amp circuit breakers to determine the volt-amps, an 1,800 volt-amp value may be used (15 A x 120 V = 1,800 VA). The square footage volt-amps are divided by the circuit breaker voltage to obtain the number of lighting circuits required to be installed.

Example 19-1. Determine the number of lighting circuits required for the mobile home in Illustration 19-1?

# of lighting circuits

Step 1: Finding sq. ft. VA
550-7(a)
900 sq. ft. x 3 VA = 2,700 VA

Step 2: Finding circuit breaker VA
550-7(a)
15 A x 120 V = 1,800 VA

Step 3: Calculating # of lighting circuits
2,700 VA ÷ 1,800 VA = 1.5

Step 4: Selecting # of lighting circuits
550-7(a)
1.5 = 2 - 15 amp, 2-wire circuits

Solution: **2 - 15 amp, 2-wire circuits are required.**

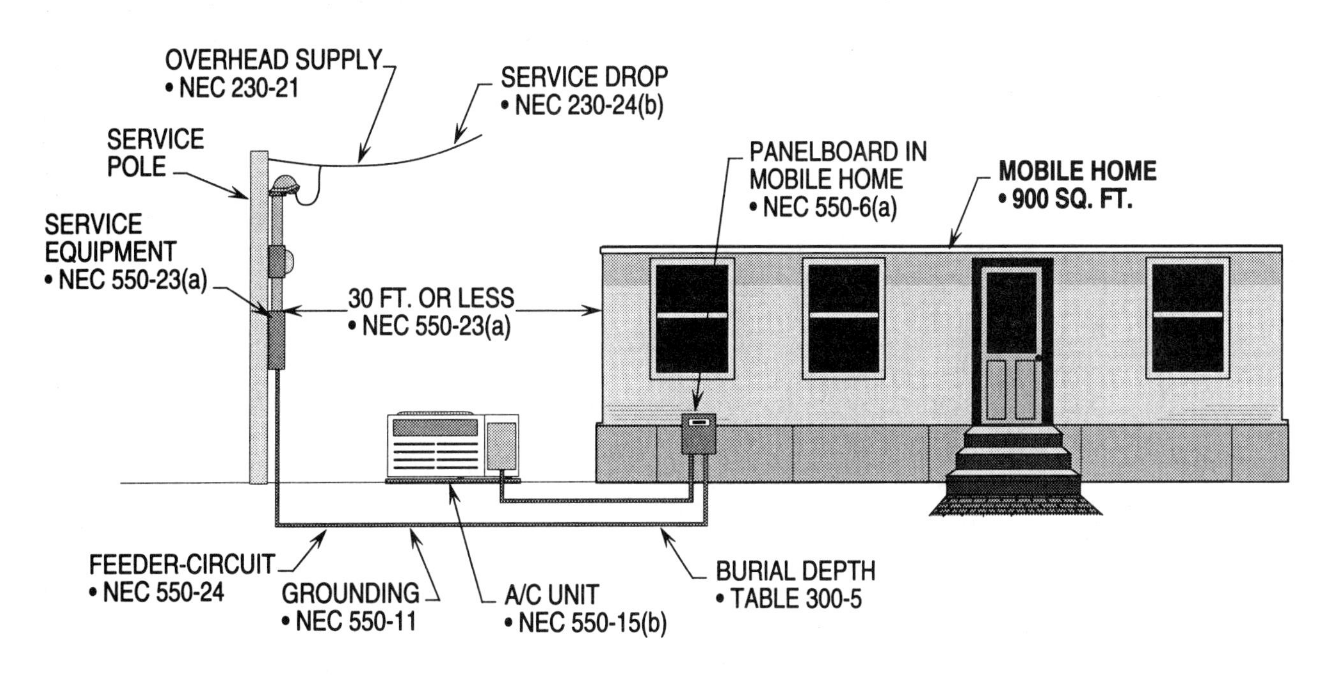

NEC 550-7(a)

Ill. 19-1

General Appliance Loads 550-7(c)(1) thru (c)(4)

When sizing the ampere rating for fixed appliances connected on lighting circuits, the branch-circuit rating shall not exceed 50 percent.

When sizing the fixed appliances connected to an individual branch-circuit, the branch-circuit rating may be loaded to 80 percent.

When sizing for range circuits, the branch-circuit is based on the range nameplate rating and may be derated by applying the demand factors listed in NEC 550-13(b)(5).

Example 19-2. Determine the total VA for the fixed appliance loads in Illustration 19-2?

Finding range VA

Step 1: Finding range VA
8.5 kVA = 8,500 VA

Note: Review NEC 550-13(b)(2), (3), and (4) very carefully before performing this calculation.

Step 2: Calculating range VA
550-7(c)(4); 550-13(b)(5)
8,500 VA x 80% = 6,800 VA

Solution: The range is 6,800 volt-amps.

Finding fixed appliance VA

Step 1: Calculating load
550-13(b)(2); (3); (4)

water heater	=	6,000 VA
disposal	=	450 VA
dishwasher	=	700 VA
heating	=	5,500 VA
largest motor (450 VA x 25%)	=	113 VA
Total load	=	12,763 VA

Solution: The fixed appliance load is 12,763 volt-amps.

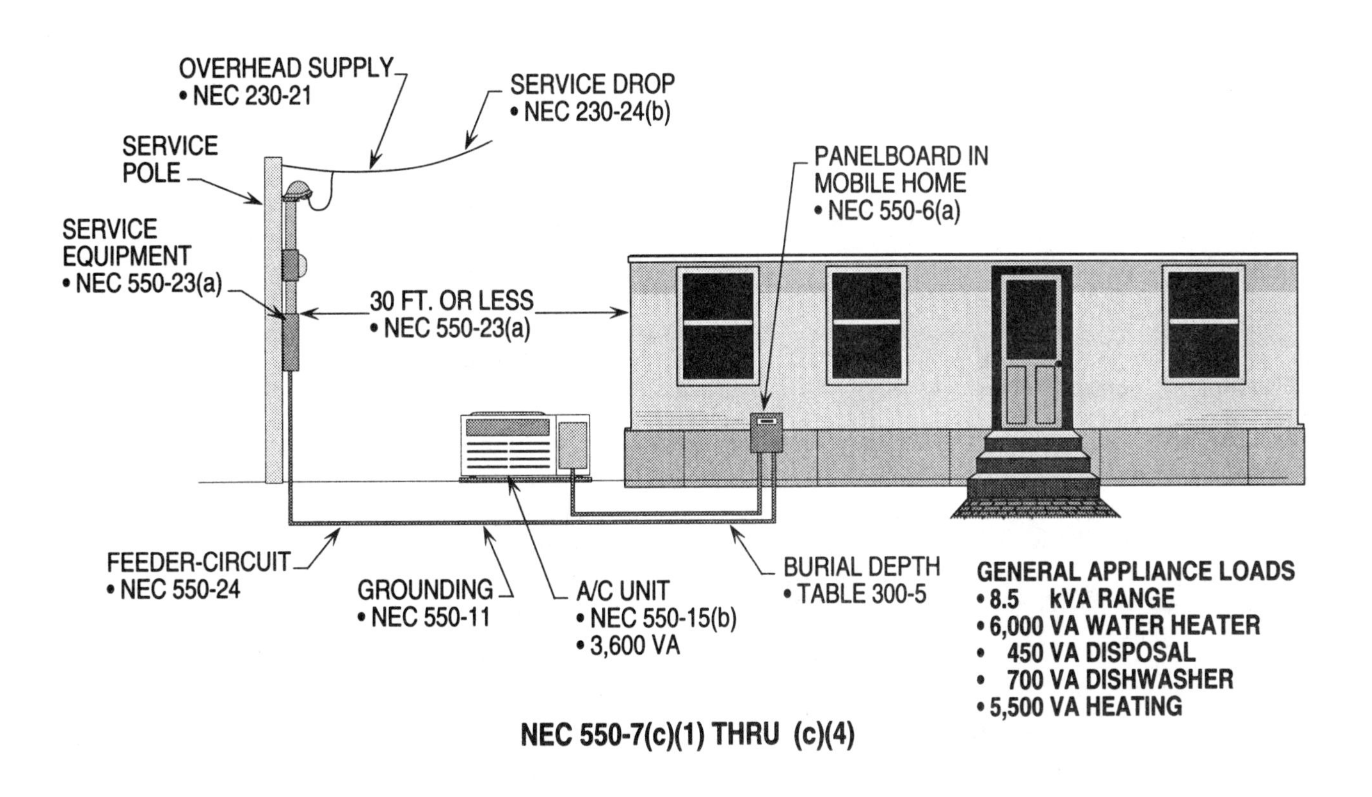

NEC 550-7(c)(1) THRU (c)(4)

Ill. 19-2

Feeder to Mobile Home 550-13

The standard calculation may be used to determine amps to size circuit elements. Use this same calculation procedure for park trailers.

Example 19-3. Determine the size service-entrance conductors required for Phases A and B and the neutral for the mobile home in Illustration 19-3?

General lighting and receptacle load

550-13(a)
800 sq. ft. x 3 VA = 2,400 VA

Small appliance loads

550-13(a)
Small appliance load
1,500 VA x 2 = 3,000 VA
Laundry load
1,500 VA x 1 = 1,500 VA
Total load = 4,500 VA

Applying demand factor

Finding load
550-13(a)
General lighting and receptacle load = 2,400 VA
Small appliance loads = 4,500 VA
Total load = 6,900 VA
First 3,000 VA at 100% = 3,000 VA
Next 3,900 VA at 35% = 1,365 VA
Total load = **4,365 VA•√**

Special appliance loads (other loads)

550-13(b)(2)(3)(4)
Water heater = 6,000 VA
Dishwasher = 800 VA √
Disposal = 540 VA √
Heating = 5,500 VA
Largest motor = 135 VA √
(540 VA x 25%)
Total load = **12,975 VA •**

Range load

550-13(b)(5)
8.5 kVA range = 8,500 VA
8,500 VA x 80% = **6,800 VA •√**

Total VA loads (phases)

General lighting load = 4,365 VA •
Small appliance load = 12,975 VA •
Range load = 6,800 VA •
Total load = 24,140 VA

Finding amps

I = VA / V
I = 24,140 VA ÷ 240 V
I = 101 A

Total VA loads (neutral)

220-22
General lighting load = 4,365 VA √
Dishwasher load = 800 VA √
Disposal load = 540 VA √
Largest motor = 135 VA √
Range = 4,760 VA √
(6,800 VA x 70%)
Total load = 10,600 VA

Finding amps

I = VA / V
I = 10,600 VA / 240 V
I = 44 A

Sizing conductors

Table 310-16
Phases - 101 A = #2 THWN copper
Neutral - 44 A = #8 THWN copper

Table 310-15(b)(6)
Phases - 110 A = #3 THWN copper
Neutral - 44 A = #8 THWN copper

Note: Use this same procedure for park trailers.

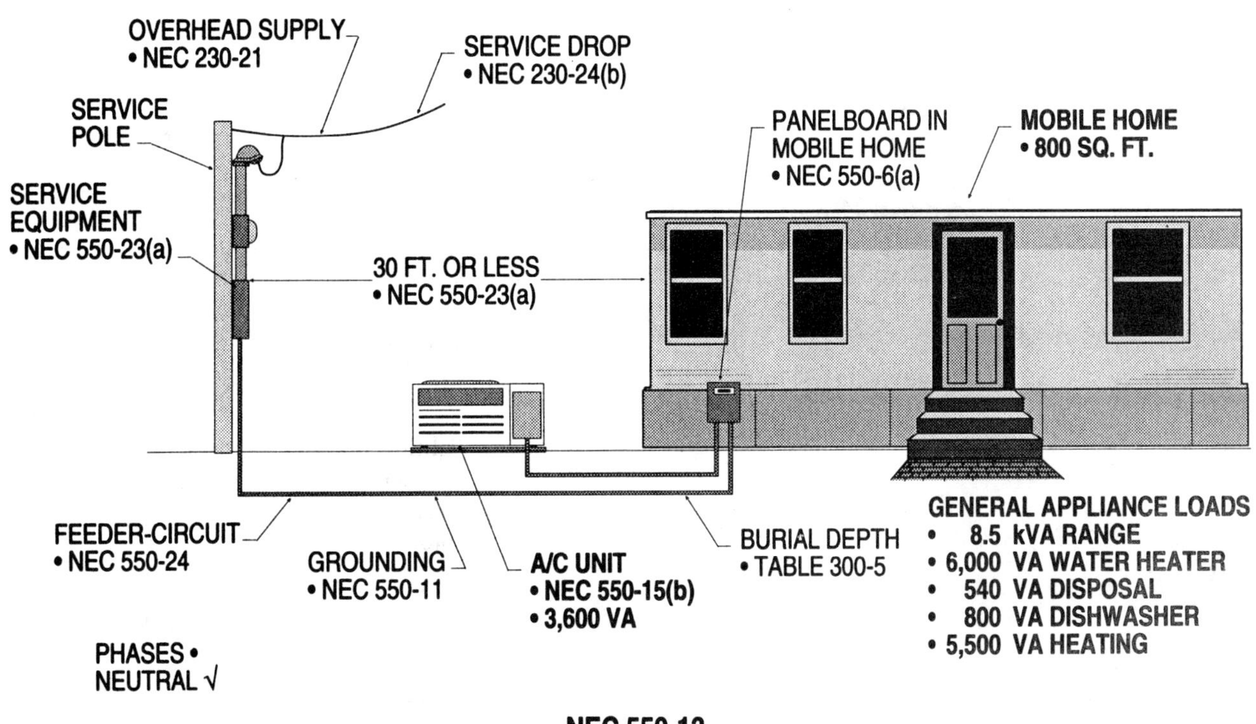

NEC 550-13

Ill. 19-3

Mobile Home Park Service and Feeders
550-22; Table 550-22

The optional calculation may be used to determine the amps to size the circuit elements.

Example 19-4. Determine the size service-entrance conductors required for phases A and B and the neutral for the mobile home in Illustration 19-4?

Finding total VA load

Step 1: Finding VA load
550-22; Table 550-22
Mobile home = 17,000 VA

Step 2: Calculating VA load
Table 550-22
17,000 VA x 28 x 24% = 114,240 VA

Solution: **The total VA required is 114,240 VA.**

Sizing conductors

Step 1: Finding VA
550-22
Total VA = 114,240 VA

Step 2: Calculating A
220-2(a)
A = 114,240 VA / 240 V
A = 476 A

Step 3: Selecting conductors
Table 310-16; 310-4
Phases A and B (Parallel)
A = 476 ÷ 3 (no. runs per phase)
A = 158.7 A
158.7 A requires #2/0 THWN cu.

Solution: **The size THWN copper conductors required are No. 2/0.**

Sizing neutral

Step 1: Finding VA
550-22
Total VA = 114,240 VA

Step 2: Calculating A
A = 114,240 VA / 240 V
A = 476 A

Step 3: Applying demand factors
220-22
First 200 A at 100% = 200 A
Next 276 A at 70% = 193 A
Total load = 393 A

Step 4: Selecting conductors
Table 310-16; 310-4
Neutral (Parallel)
A = 393 A ÷ 3 (no. runs per phase)
A = 131 A
131 A requires #1/0 THWN

Solution: **The size THWN copper conductors required are No. 1/0.**

Note: A No. 1/0 THWN copper conductor is required when paralleling the neutral per 310-4.

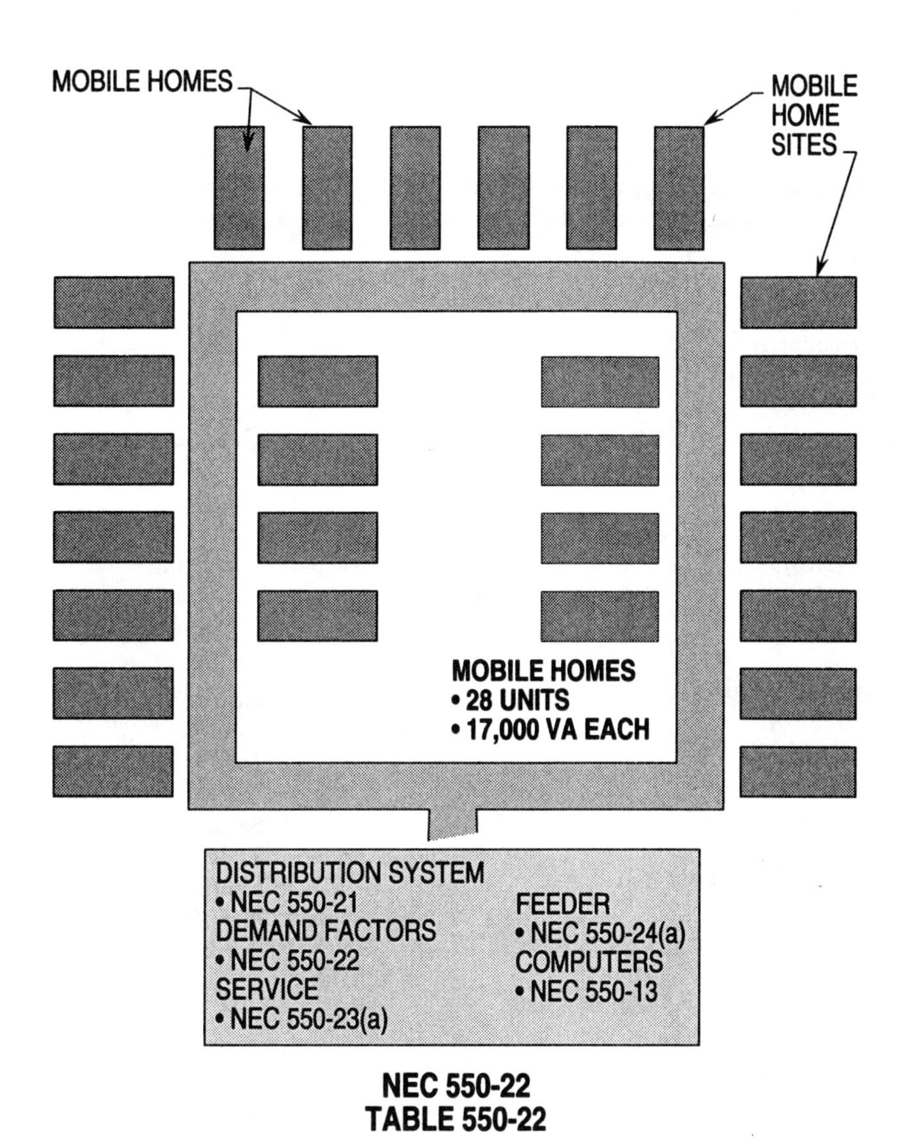
MOBILE HOMES
MOBILE HOME SITES
MOBILE HOMES
• 28 UNITS
• 17,000 VA EACH
DISTRIBUTION SYSTEM
• NEC 550-21
DEMAND FACTORS
• NEC 550-22
SERVICE
• NEC 550-23(a)
FEEDER
• NEC 550-24(a)
COMPUTERS
• NEC 550-13
NEC 550-22
TABLE 550-22

Ill. 19-4

RV (OCPD) Per 93 NEC 551-42(a) thru (d)

The standard calculation may be used to size the circuit elements serving RV's.

Example 19-5. What amp value is required for the OCPD's and conductors for the RV in Illustration 19-5?

Sizing amps for OCPD's and conductors

Step 1: Calculating A for RV
551-42(a) thru (d)
551-42(d)(1)
Lighting load = 200 sq. ft. x 3 = 600 VA
551-42(d)(2)

Small appliance load = 1,500 VA x 2 = 3,000 VA
551-42(d)(3)
Total VA = 3,600 VA

Step 2: Applying demand factors
551-42(d)(4)
First 3,000 VA x 100% = 3,000 VA
Next 600 VA x 35% = 210 VA
Total = 3,210 VA

Step 3: Calculating A per leg A & B
551-42(d)(4)
A = 3,210 VA / 240 V
A = 13.38 A•

Step 4: Calculating A/C and Htg. load
551-42(d)(5)
A/C unit = 15 A•

Step 5: Calculating largest motor load
551-42(d)(6)
A/C unit = 15 A x 25% = 3.75 A•

Step 6: Calculating appliance load
551-42(d)(7)
Disposal = 6 A•
Water heater = 15 A•
Dishwasher = 10 A•
Total = 31 A

Step 7: Calculating range load
551-42(d)(8); Table 551-42(8)
A = 6 kW x 1,000% x 80% / 240 V
A = 20 A•

Step 8: Adding loads for legs A & B
551-42(a) thru (d)
Amperes per leg

	A	B
	• 13.38 A	• 13.38 A
	• 15 A	• 15 A
	• 3.75 A	• 3.75 A
	• 6 A	• 0 A
	• 15 A	• 15 A
	• 0 A	• 10 A
	• 20 A	• 20 A
Total	73.13 A	77.13 A

Solution: The OCPD's and conductors are required to be sized based upon 77.13 amps per leg B.

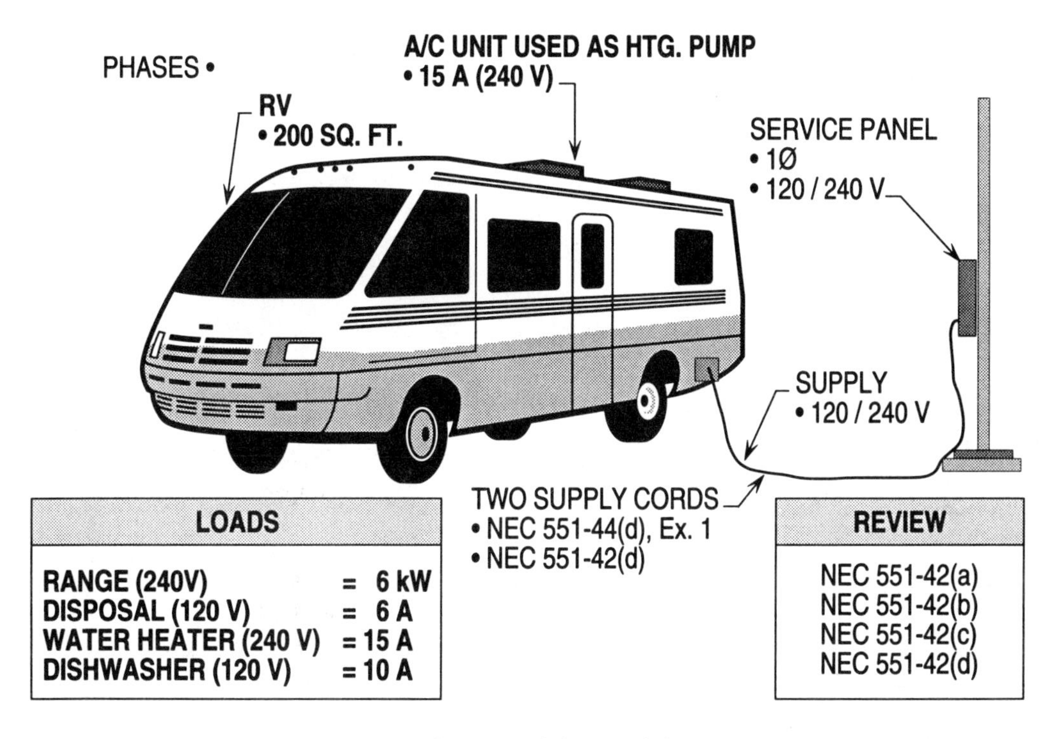

NEC 551-42(a) thru (d)
PER 93 NEC

Ill. 19-5

Power Supply Assembly
551-42(c); 551-44(c)

Power supply assemblies such as rubber cords may be used to serve RV's.

Example 19-6. What size power supply assembly is required for the RV in Illustration 19-6?

Sizing power supply assembly

Step 1: Finding A for RV
551-42(c); 551-44(c)
Load = 28 A

Step 2: Calculating PSA for RV
551-42(c); 551-44(c)
Circuits = 2 #15 and 1 #20 A
Load = 28 A

Step 3: Selecting PSA for RV
551-42(c); 551-44(c)
28 A requires 30 A

Solution: The power supply assembly (PSA) is required to be rated at least 30 amps.

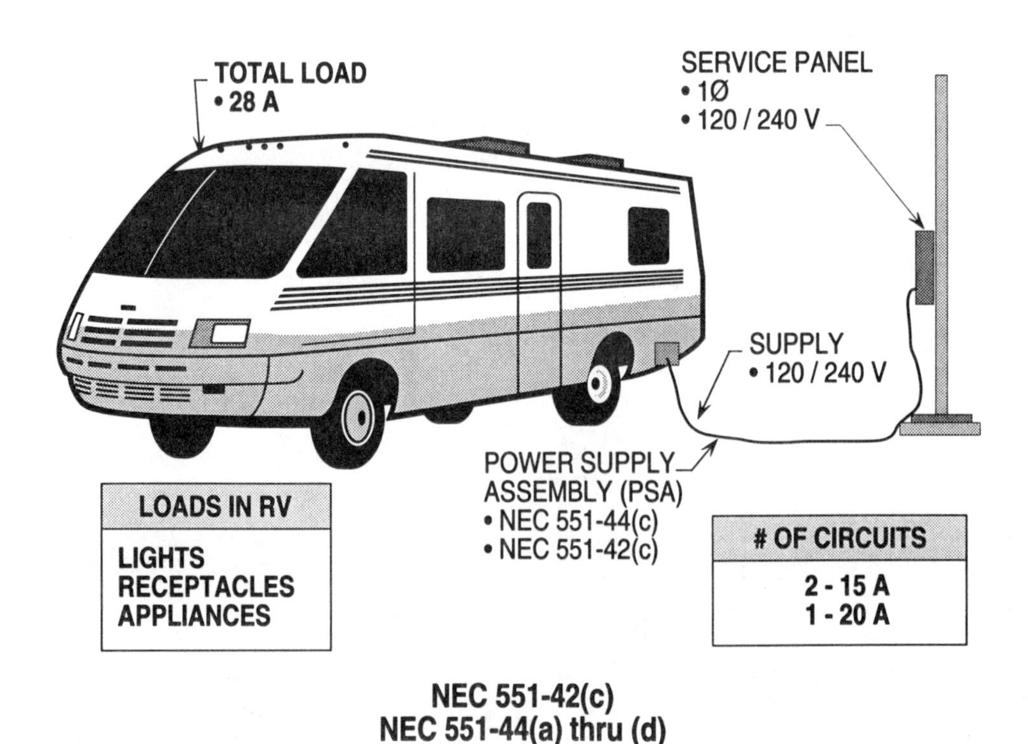

Ill. 19-6

Recreational Vehicle Parks
551-73(a)

The elements in feeder-circuits and service components may be sized by the amps derived from applying demand factors to the total connected loads.

Example 19-7. What is the demand load in amps for the number of RV sites in Illustration 19-7?

Sizing demand load in amps

Step 1: Finding load in VA
551-73(a)
9,600 VA x 10 sites = 96,000 VA
3,600 VA x 15 sites = 54,000 VA
2,400 VA x 8 sites = 19,200 VA
Total VA for sites = 169,200 VA

Step 2: Finding demand factors
551-73(a); Table 551-73
33 sites allows 42%

Step 3: Applying demand factors
551-73(a); Table 551-73
169,200 VA x 42% = 71,064 VA

Step 4: Calculating A of sites
551-72
A = 71,064 / 240 V
A = 296.1 A

Solution: **The demand load in amps for the sites is 296.1 amps.**

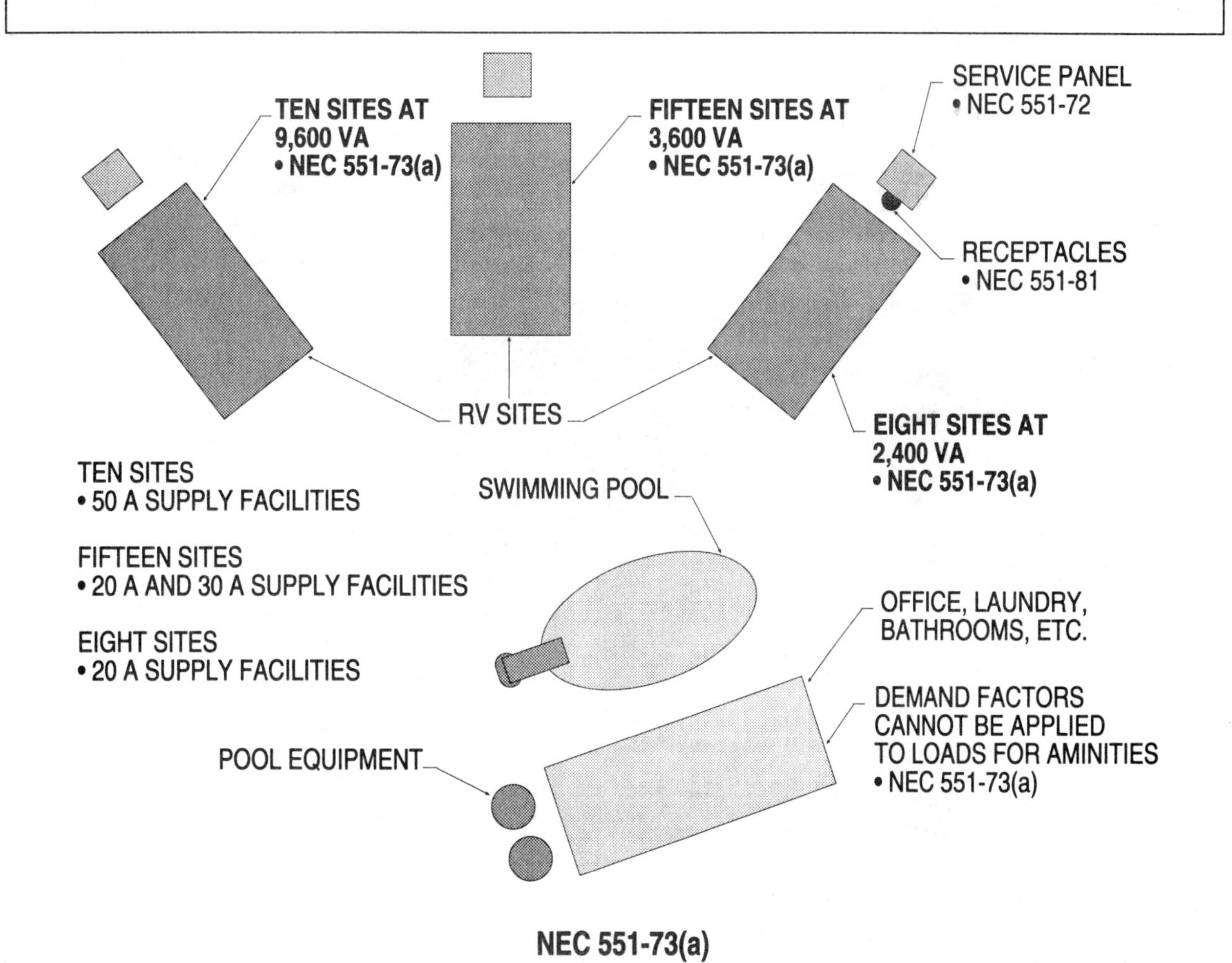

Ill. 19-7

Marinas (Feeders and Services) 555-6

The ampacity used to select elements of circuits supplying shore power receptacles may be determined from demand factors based upon the number of receptacles.

Example 19-8. What is the demand load for the 100 shore power receptacles in Illustration 19-8?

Sizing demand load for shore power receptacles

Step 1: Finding A of receptacles
555-6
20 A plugs x 50 = 1,000 A
30 A plugs x 50 = 1,500 A
Total A = 2,500 A

Step 2: Finding demand factor
555-6
100 allows 30%

Step 3: Calculating A for receptacles
555-6
2,500 A x 30% = 750 A

Solution: The demand load for the shore power receptacles is 750 amps.

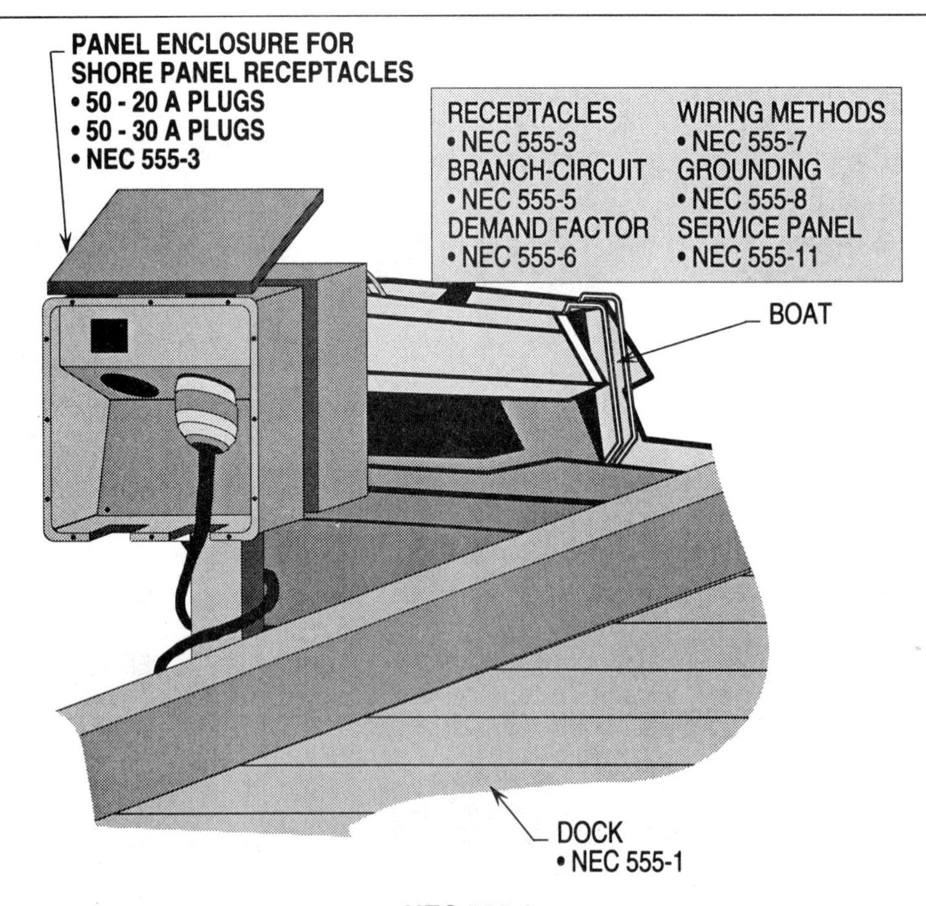

NEC 555-6

Ill. 19-8

20 Sign and Outline Lighting

The elements of circuits used to supply power to electric signs and outline lighting systems are sized at continuous or noncontinuous operation. Because the electrical system supplying such equipment must be serviced, disconnects and lockout procedures must be provided to ensure proper safety for service personnel.

There are certain types of signs that are allowed to have demand factors applied, based upon their condition of operation.

Special rules must be applied for neon installations to prevent shock and fire hazards to the general public.

Branch-circuits supplying ballasts, transformers, and incandescent lighting systems are limited to 15, 20, or 30 amp circuits, based upon how they are installed.

Quick Reference

Computed Load 600-5(b)

Special rules and wiring techniques must be used when designing and installing neon systems. Loads are calculated at continuous or noncontinuous operation.

Example 20-1(a). What size OCPD and THWN copper conductors are required for the circuit in Illustration 20-1(a)?

Sizing OCPD for BC

Step 1: Finding A for BC
600-5(b); 210-20(a)
1.5 A x 10 = 15 A

Step 2: Calculating A for BC
600-5(b); 210-20(a)
15 A x 125% = 18.75 A

Step 3: Selecting OCPD for BC
Table 310-16; 240-3(d)
240-3(b); 240-6(a)
18.75 A requires 20 A OCPD

Solution: The size OCPD for the BC is 20 amps.

Sizing conductors for BC

Step 1: Selecting conductors for BC
310-10(2); 210-19(a); 240-3(d)
Step 2 above
18.75 A requires #12 cu.

Solution: The size THWN copper conductors are No. 12.

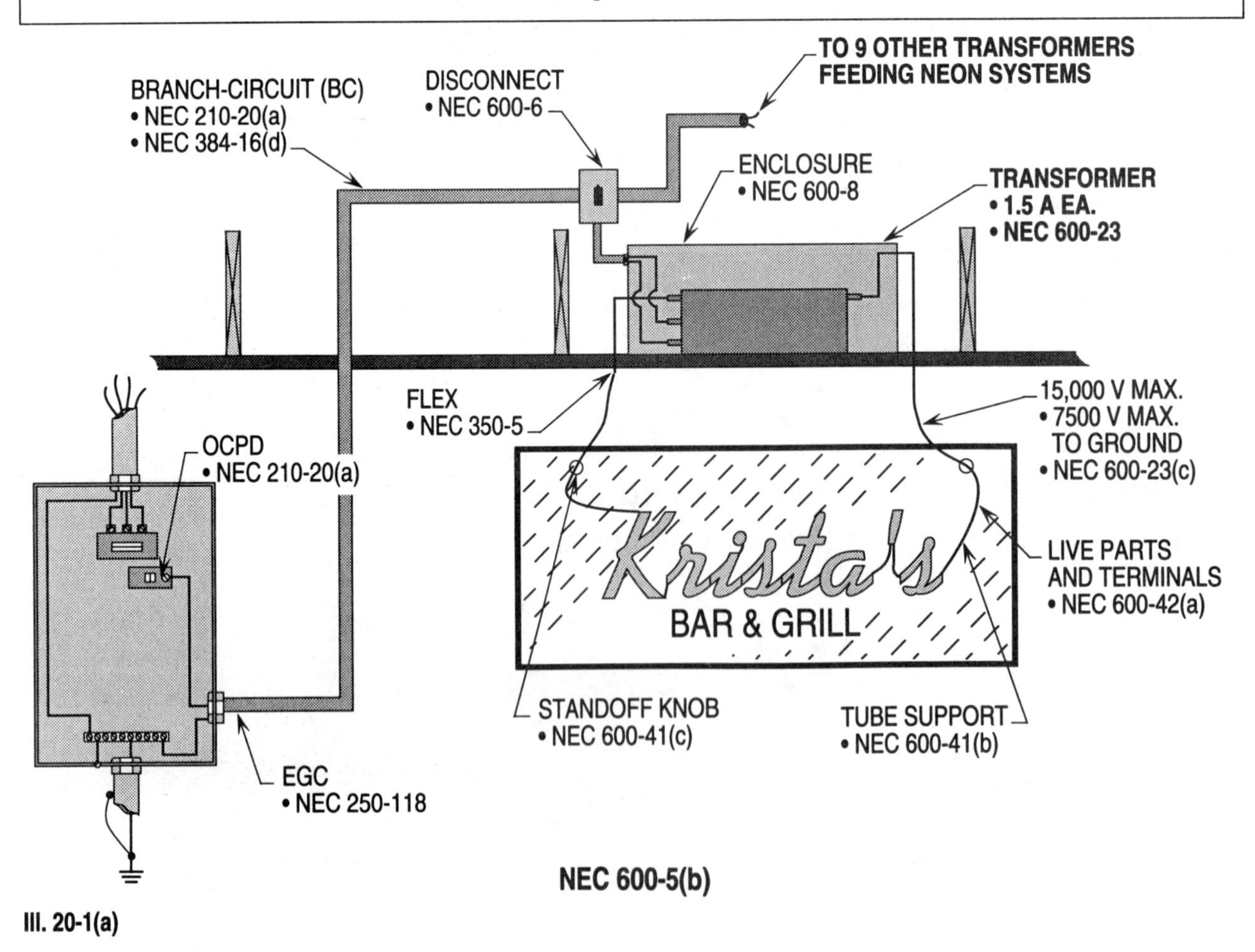

Ill. 20-1(a)

Computed Load
600-5(b)(3)

When sizing the service or feeder calculation for a sign computed at a minimum of 1,200 volt-amps, the load is multiplied by 125 percent for continuous duty when the sign burns for three hours or more. When sizing the load from the overcurrent protection device, the load is multiplied by 80 percent to obtain the continuous load to be used.

Example 20-1(b). Find the total size load and OCPD for the sign in Illustration 20-1(b)? In addition, find the VA for the service and feeder load?

Finding load and OCPD

Step 1: Finding A
600-5(b)(3)
OCPD = 20 A

Step 2: Calculating A
600-5(b)(3); 384-16(d); 210-20(a)
20 A x 80% = 16 A
16 A x 125% = 20 A

Step 3: Selecting OCPD
600-5(b)(3); 240-3(d); 240-6(a)
20 A requires 20 A OCPD

Solution: The size OCPD required is 20 amp for a sign that burns three hours or more.

Finding VA for service or feeder load

Step 1: Finding VA
600-5(b)(3)
1,200 VA is the minimum

Step 2: Finding continuous load
215-2(a); 230-42(a)(1)
1,200 VA x 125% = 1,500 VA

Solution: The load for a service or feeder is 1,500 VA.

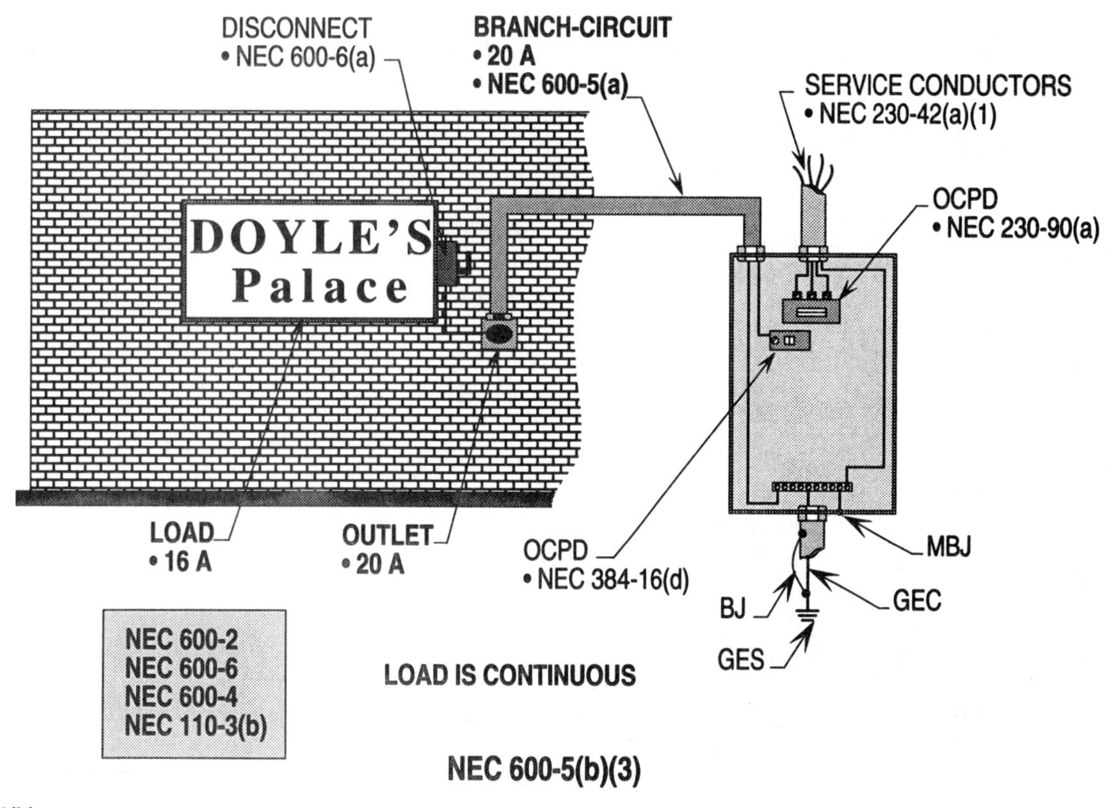

Ill. 20-1(b)

Messenger Signs
600-5(b)(3)

The components of circuits supplying power to messenger signs are determined by the number of lamps and their condition of use during operation.

Example 20-2. What size OCPD and THWN copper conductors are required for the circuits in Illustration 20-2?

Sizing OCPD and conductors for BC

Step 1: Finding VA for BC
600-5(b)(3)
33 W x 56 = 1,848 VA

Step 2: Finding A for BC
600-5(b)(3)
A = 1,848 VA / 120 V
A = 15.4

Step 3: Calculating A for BC
600-5(b)(3); 210-19(a); 210-20(a)
15.4 A x 125% = 19.25 A

Step 4: Selecting OCPD and conductors for BC
310-10(2); Table 310-16;
240-3(b); 240-6(a)
19.25 A requires #12 cu.
19.25 A requires 20 A OCPD

Solution: The size OCPD is 20 amps and the size THWN copper conductors are No. 12.

Sizing OCPD and conductors for FC

Step 1: Finding VA and A for FC
600-5(b)(3); 600-4(a)
33 W x 1,500 = 49,500 VA
A = 49,500 VA / 208 V x 1.732
A = 137.5

Step 2: Calculating A for FC
600-4(a)
137.5 A x 125% x 85% = 146 A

Step 3: Selecting OCPD and conductors
310-10(2); Table 310-16;
240-3(b); 240-6(a)
146 A requires #1/0 cu.
146 A requires 150 A OCPD

Solution: The size OCPD is 150 amps and the size THWN copper conductors are No. 1/0.

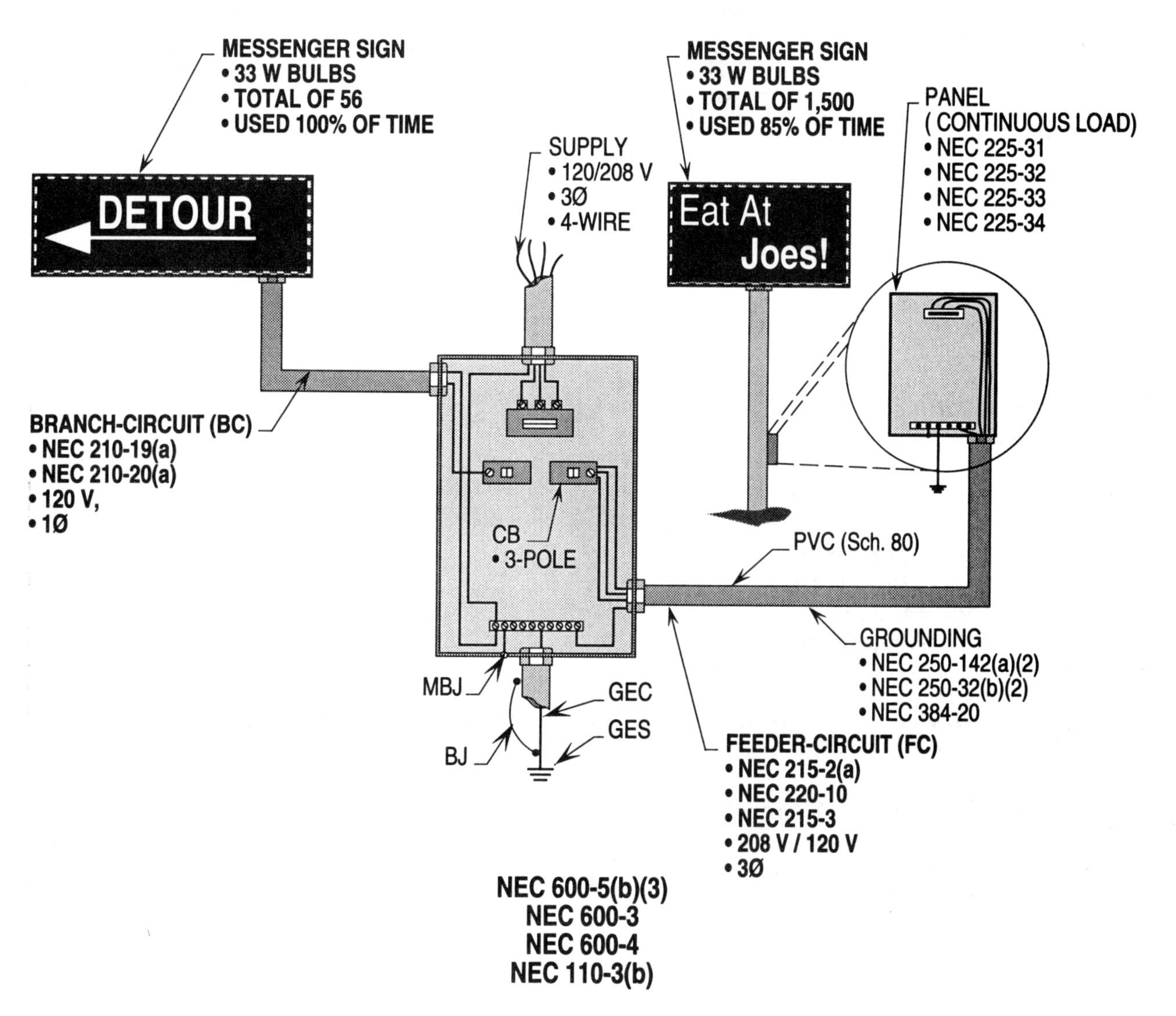
MESSENGER SIGN
• 33 W BULBS
• TOTAL OF 56
• USED 100% OF TIME
DETOUR
MESSENGER SIGN
• 33 W BULBS
• TOTAL OF 1,500
• USED 85% OF TIME
Eat At
Joes!
PANEL
(CONTINUOUS LOAD)
• NEC 225-31
• NEC 225-32
• NEC 225-33
• NEC 225-34
SUPPLY
• 120/208 V
• 3Ø
• 4-WIRE
BRANCH-CIRCUIT (BC)
• NEC 210-19(a)
• NEC 210-20(a)
• 120 V,
• 1Ø
CB
• 3-POLE
PVC (Sch. 80)
GROUNDING
• NEC 250-142(a)(2)
• NEC 250-32(b)(2)
• NEC 384-20
MBJ
GEC
GES
BJ
FEEDER-CIRCUIT (FC)
• NEC 215-2(a)
• NEC 220-10
• NEC 215-3
• 208 V / 120 V
• 3Ø
NEC 600-5(b)(3)
NEC 600-3
NEC 600-4
NEC 110-3(b)

Ill. 20-2

Control Switch Rating 600-6(b)

When sizing the control switch rating for a transformer, the rating of an AC general-use snap switch must be equivalent to the output of the transformer. The rating of an AC-DC general-use snap switch must be twice the output of the transformer. General-use snap switches are not rated for inductive loads. The rating for flashers to control transformers must be at least twice the output of the transformer.

Example 20-3(a). What is the control switch rating for the AC general-use snap switch in Illustration 20-3(a)?

Sizing switch

Step 1: Finding load and switch A
600-6(b)
Load = 9 A + 9 A = 18 A
Switch = 20 A

Step 2: Calculating A
600-6(b); 380-14(a)(1)
20 A x 100% = 20 A

Step 3: Selecting switch
600-6(b); 380-14(a)(1)
18 A requires 20 A

Solution: The size switch required is 20 amps to disconnect the 18 amp load of the sign.

Example 20-3(b). What is the control switch rating for an AC-DC general-use snap switch in Illustration 20-3(b)?

Sizing switch

Step 1: Finding switch A
600-6(b)
Switch = 20 A

Step 2: Calculating A
600-6(b); 380-14(b)(2)
20 A x 50% = 10 A

Step 3: Selecting switch
600-6(b); 380-14(b)(2)
10 A requires 20 A

Solution: The size switch required is 20 amps but cannot serve a transformer load rated over 10 amps.

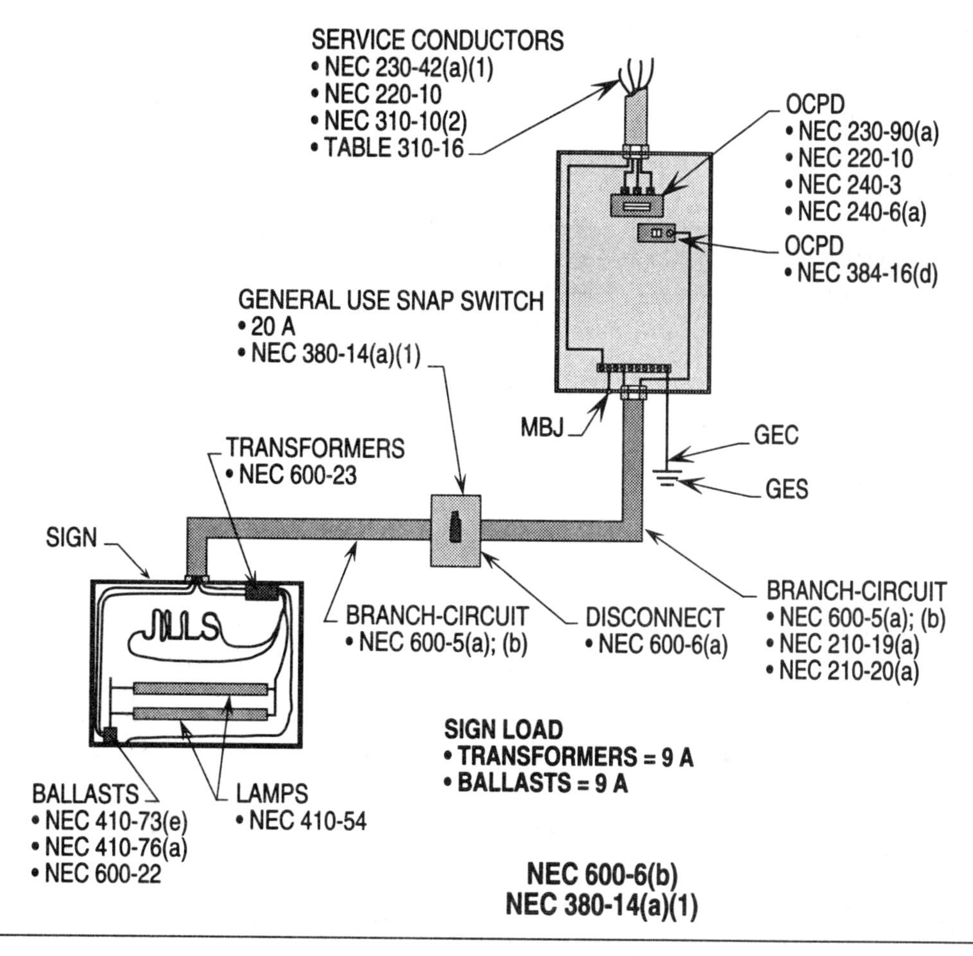

Ill. 20-3(a)

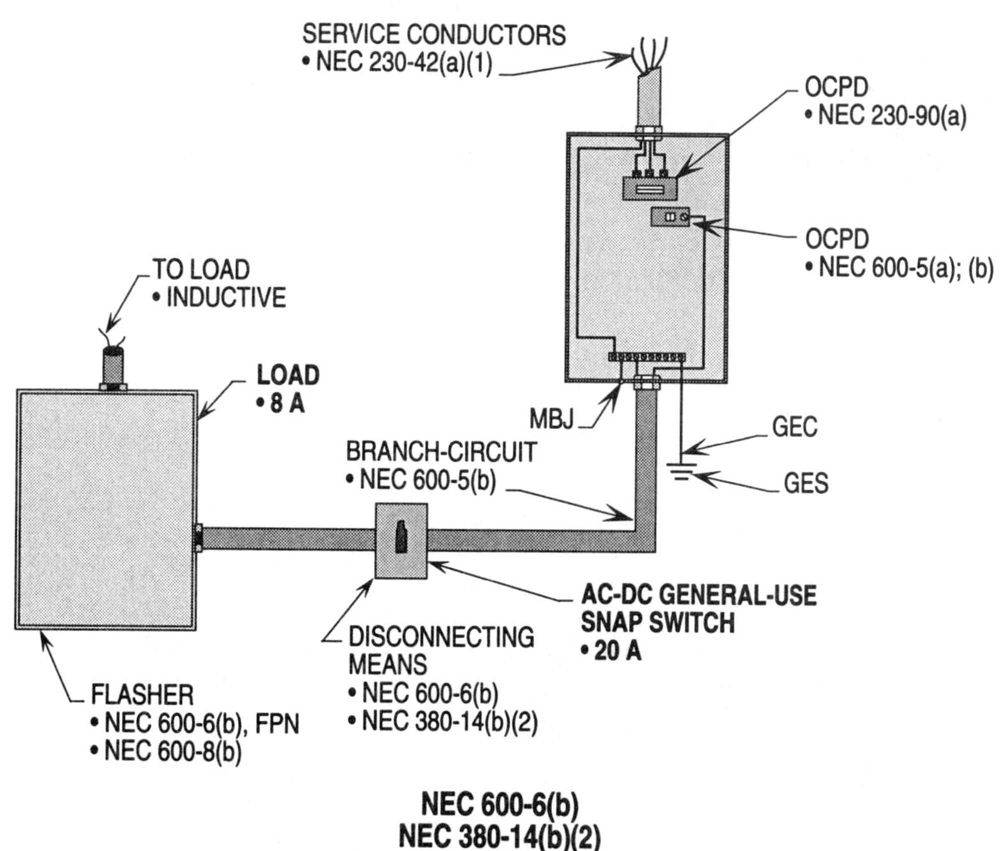

Ill. 20-3(b)

Control Switch Rating (Flasher)
600-6(b)

Flashers controlling an inductive load, such as transformers, must be sized at least twice the load served.

Example 20-3(c). What is the flasher rating to control the transformer in Illustration 20-3(c)?

Sizing flasher

Step 1: Finding flasher A
600-6(b)
Flasher = 30 A

Step 2: Calculating A
600-6(b); 380-14(b)(2)
30 A x 50% = 15 A
(15 A Ld. x 2 = 30 A)

Step 3: Selecting flasher
600-6(b)
15 A requires 30 A

Solution: The size flasher required is 30 amps but cannot serve a transformer load rated over 15 amps.

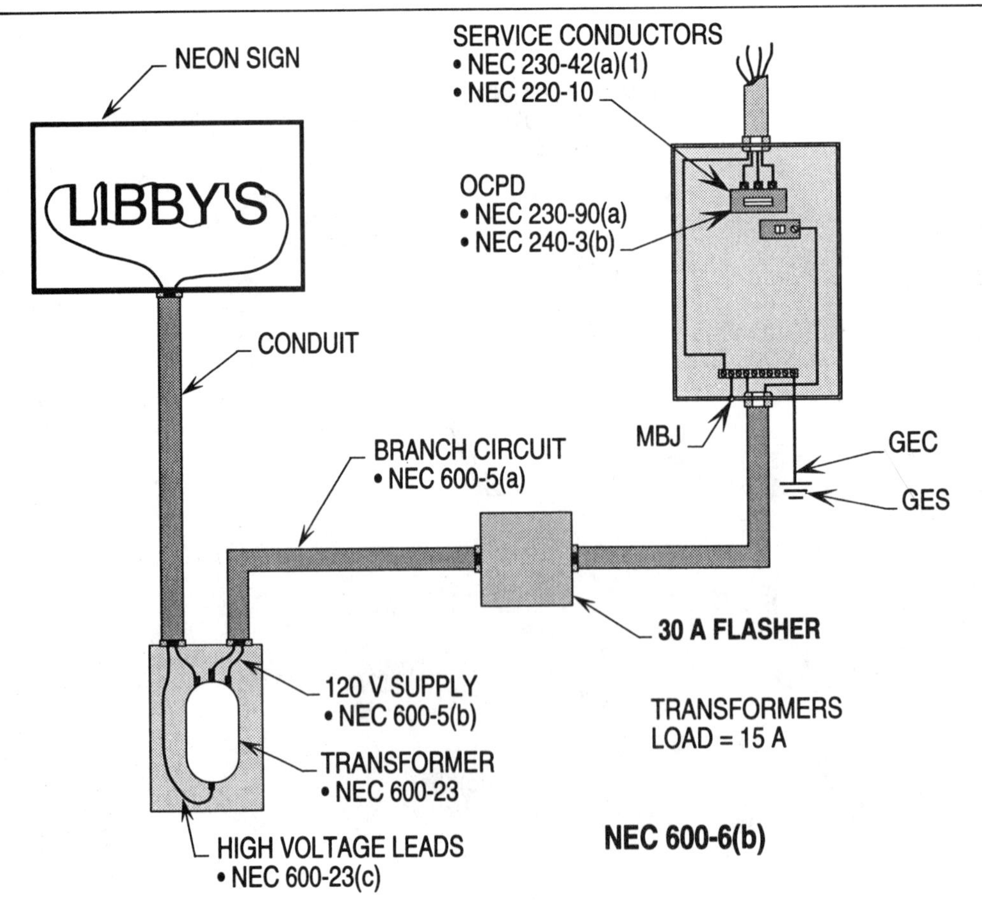

Ill. 20-3(c)

Number of Conductors in Raceway 600-31(c)

When sizing the same size conductors to be installed in a raceway, use Tables C1 through C12(A) of Appendix C to select conduit size. When sizing combination conductors to be installed in a raceway, use Tables 4 and 5 of Chapter 9 to select conduit based upon the total cross-sectional area of individual conductors.

Example 20-4(a). What is the size PVC (schedule 80) conduit required for the conductors in Illustration 20-4(a)?

Sizing PVC (Sch. 80) conduit

Step 1: Finding size PVC conduit
Table C9, Appendix C
4 #6 requires 1"

Solution: A 1 in. conduit is required.

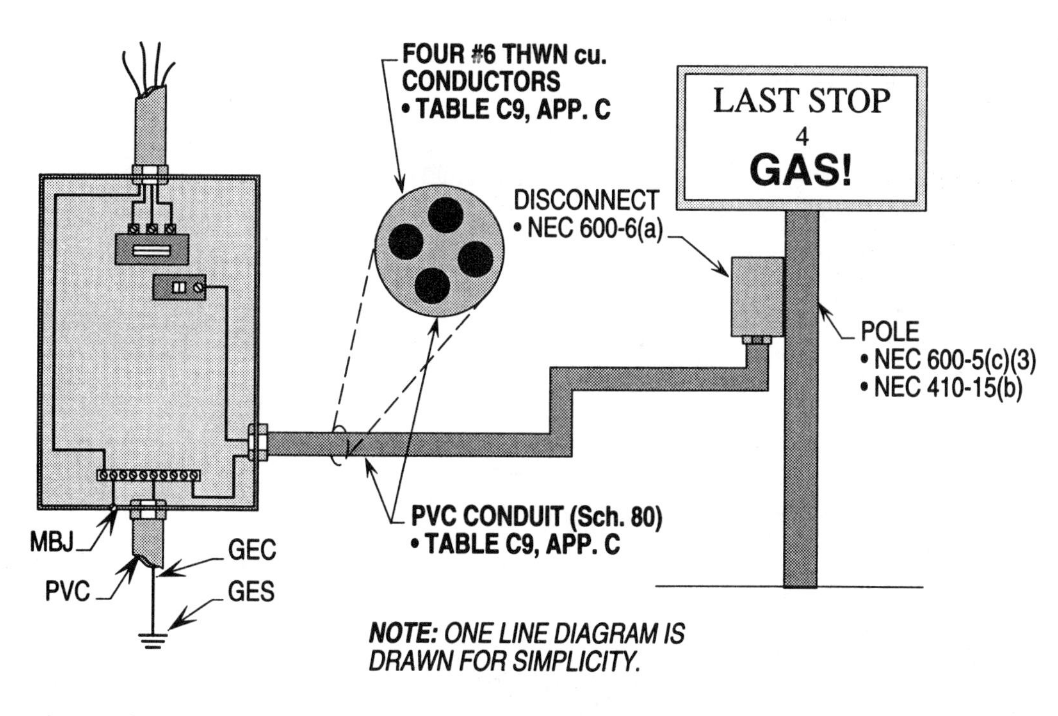

Ill. 20-4(a)

Number of Conductors in Raceway 600-31(c)

The number of conductors permitted in a raceway is determined by their size and insulation. The size of the raceway is selected and based upon the same or different types of conductors.

Example 20-4(b). What is the size conduit required for the conductors in Illustration 20-4(b)?

Sizing conduit

Step 1: Finding sq. in. area
Table 5, Ch. 9
#14 = .0097 sq. in. area
#12 = .0133 sq. in. area
#10 = .0211 sq. in. area

Step 2: Calculating sq. in. area
.0097 sq. in. x 12 = 0.1164 sq. in. area
.0133 sq. in. x 4 = 0.0532 sq. in. area
.0211 sq. in. x 4 = 0.0844 sq. in. area
Total = 0.2540 sq. in. area

Step 3: Selecting size conduit
Table 4, Ch. 9
0.254 sq. in. area requires
0.355 sq. in.

Solution: A 1 in. conduit is required.

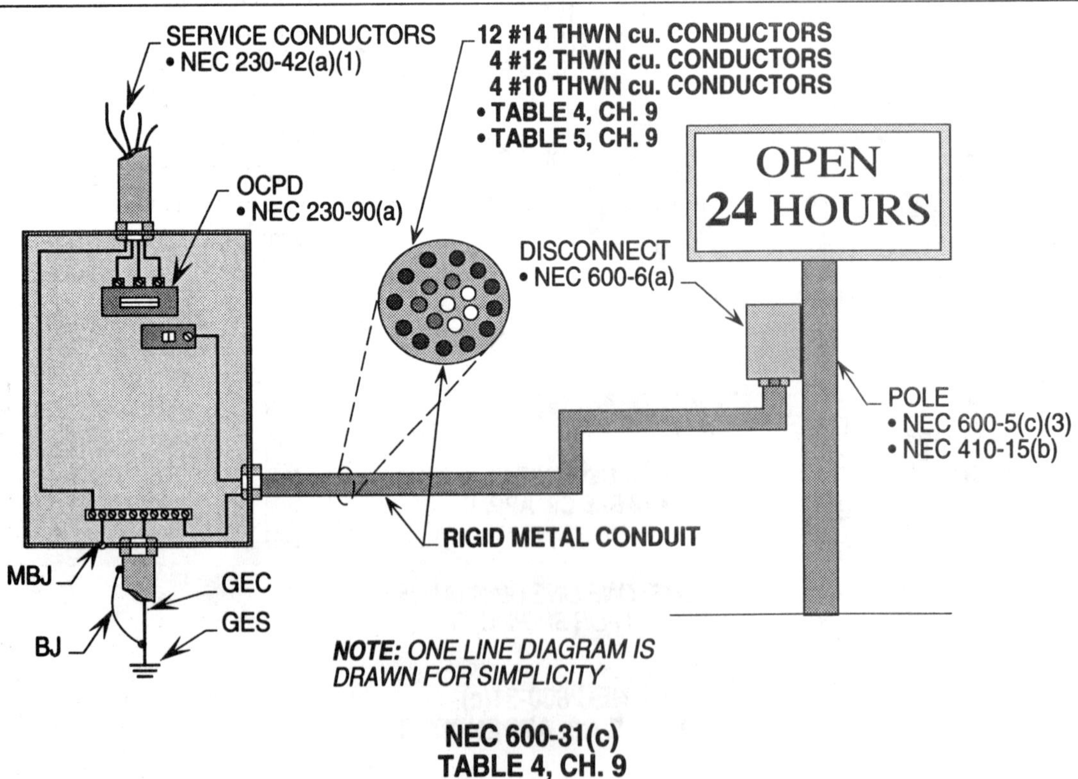

Ill. 20-4(b)

21

Cranes, Hoist, and Elevators

Cranes, hoists, and elevators are used primarily to move or lift people and equipment. The wiring methods and elements must be designed, sized and installed in such a manner to be reliable and dependable. The NEC has specific requirements and procedures to apply when calculating such loads to ensure safety.

Controls and disconnects are required to be installed and located in such a manner so that personnel may have easy access for operation.

The wiring methods and elements are to not only comply with NEC requirements, but must meet all the provisions of safety for the worker in his or her workplace.

Designers, installers, and inspectors must review the rules and regulations in the NEC and OSHA to ensure that both standards are complied with.

Note: The AHJ may require the Examples to the Appendix D to be applied when calculating ampacities for motors having a duty-cycle service.

Quick Reference

Sizing Conductors to One Motor 610-14(e)(1)

Conductors serving power to a motor used to operate a crane shall be sized and selected with enough capacity to lift and move the loads safely.

Example 21-1. What size THWN copper conductors are required to supply the motor in Illustration 21-1?

Sizing conductors to one motor

Step 1: Finding FLA
610-14(e)(1)
65 A x 100% = 65 A

Step 2: Selecting conductors
610-14(a); Table 610-14(a)
65 A requires #6 cu.

Solution: **The size THWN copper conductors are No. 6.**

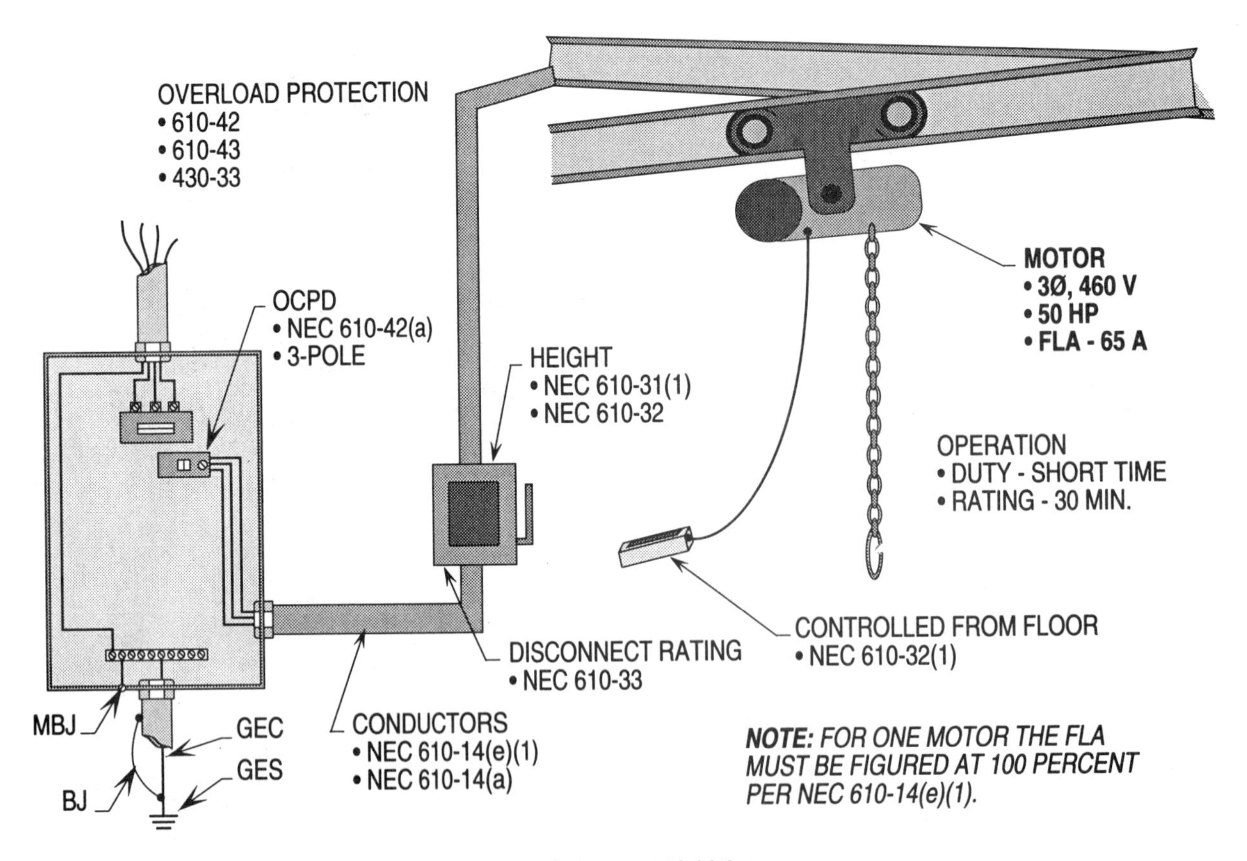

Ill. 21-1

Sizing Conductors to More Than One Motor 610-14(e)(2)

Conductors supplying power to a group of motors for a hoist are calculated and sized, based upon percentages being applied according to motor sizes.

Example 21-2. What size THWN copper conductors are required to supply the crane in Illustration 21-2?

Sizing conductors to more than one motor

Step 1: Finding FLA's of motors
610-14(e)(2)
60 min. rating = 52 A x 100% = 52 A
30 min. rating = 40 A x 50% = 20 A
Total load = 72 A

Step 2: Selecting conductors
610-14(e)(2); Table 610-14(a)
72 A requires #6 cu.

Solution: **The size THWN copper conductors are No. 6.**

Note: The size is based upon a 60 min. rating.

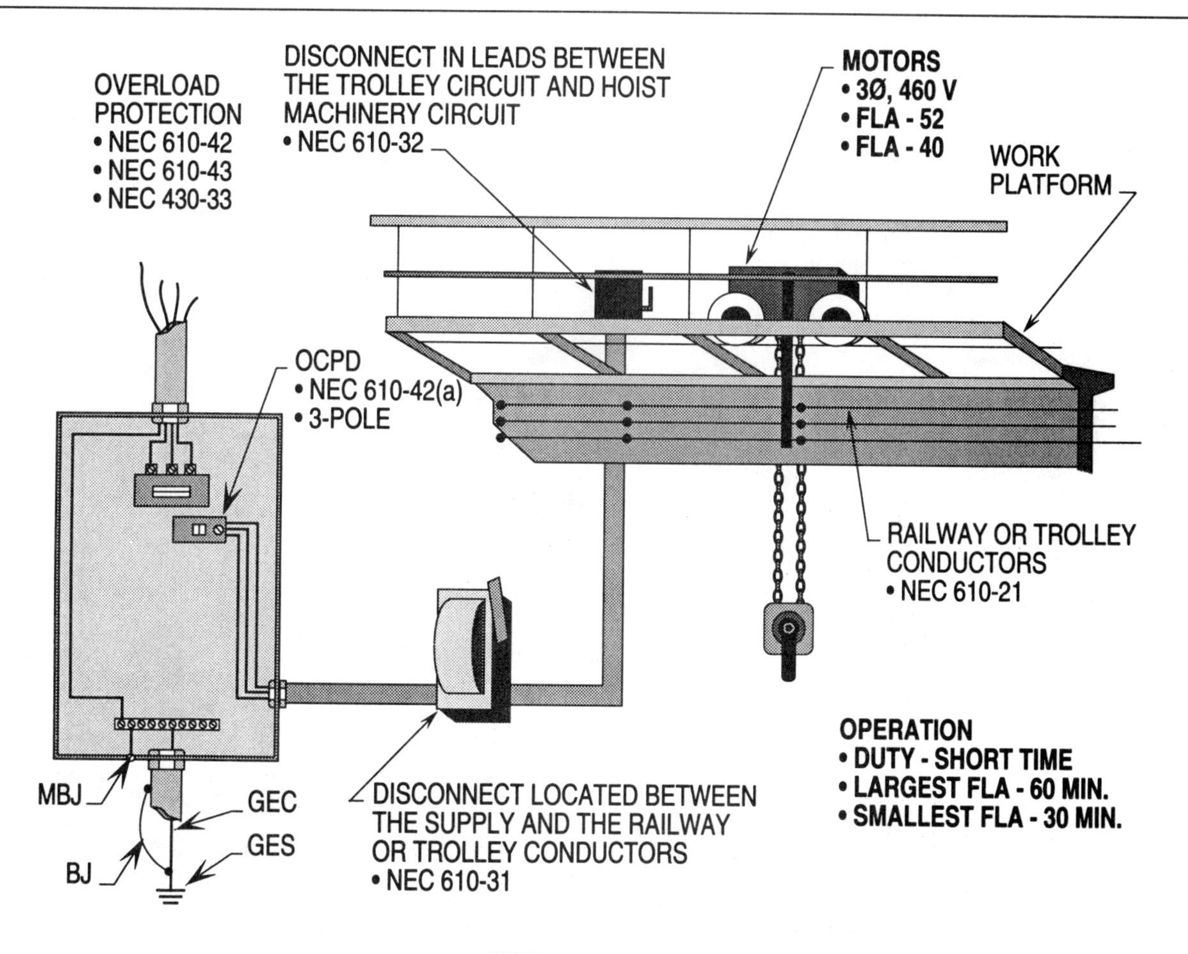

Ill. 21-2

Sizing Conductors for Feeders 610-14(e)(3)

Feeder-circuit conductors can be utilized to supply power to more than one crane. Conductors may be sized using demand factors based upon the number of cranes.

Example 21-3. What size THHN copper conductors are required to supply power for the feeder-circuit in Illustration 21-3?

Sizing conductors for feeder

Step 1: Finding FLA of motors
610-14(e)(3)
Crane 1 (FLC) = 124 A
Crane 2 (FLC) = 96 A
Crane 3 (FLC) = 77 A
Crane 4 (FLC) = 65 A
Total load = 362 A

Step 2: Finding demand factor
610-14(e)(3); Table 610-14(e)
4 cranes allows 87%

Step 3: Applying demand factors
610-14(e)(3); Table 610-14(e)
362 A x 87% = 314.9 A

Step 4: Selecting conductors
610-14(a); Table 610-14(a)
314.9 A requires #250 KCMIL cu.

Solution: The size THHN copper conductors are No. 250 KCMIL.

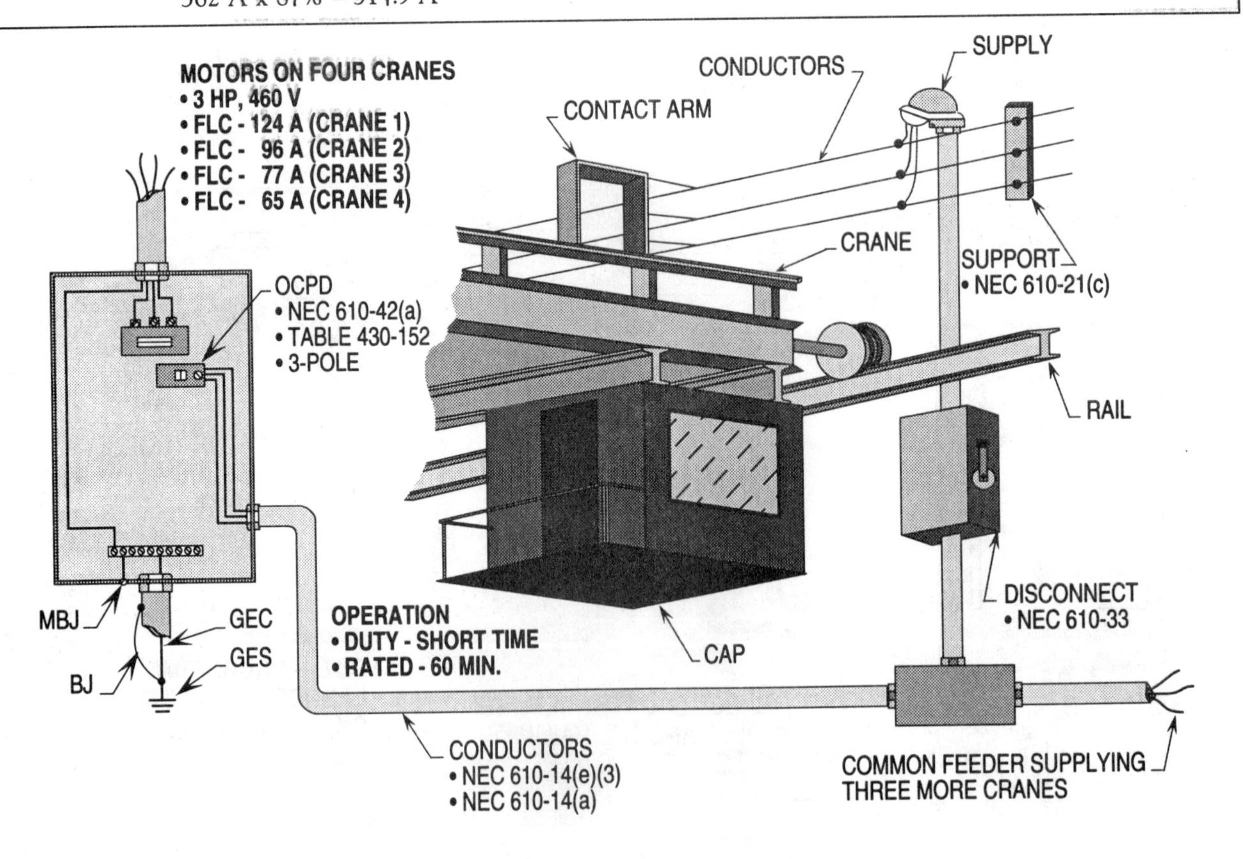

Ill. 21-3

Sizing Disconnects for Motors 610-31

The disconnecting means for a crane must be capable of deenergizing circuit conductors in a safe and reliable manner.

Example 21-4. What size disconnect is required to deenergize the motor for the crane in Illustration 21-4?

Sizing disconnect for motor

Step 1: Finding FLA of motor
610-31
Short time FLA = 65 A

Step 2: Calculating disconnect
610-14(e)(1)
65 A x 100% = 65 A

Step 3: Selecting a CB or switch for disconnect
240-6(a)
65 A requires 70 A CB
65 A requires 100 A switch

Solution: **The size disconnect using a CB is 70 amps or 100 amps for a switch.**

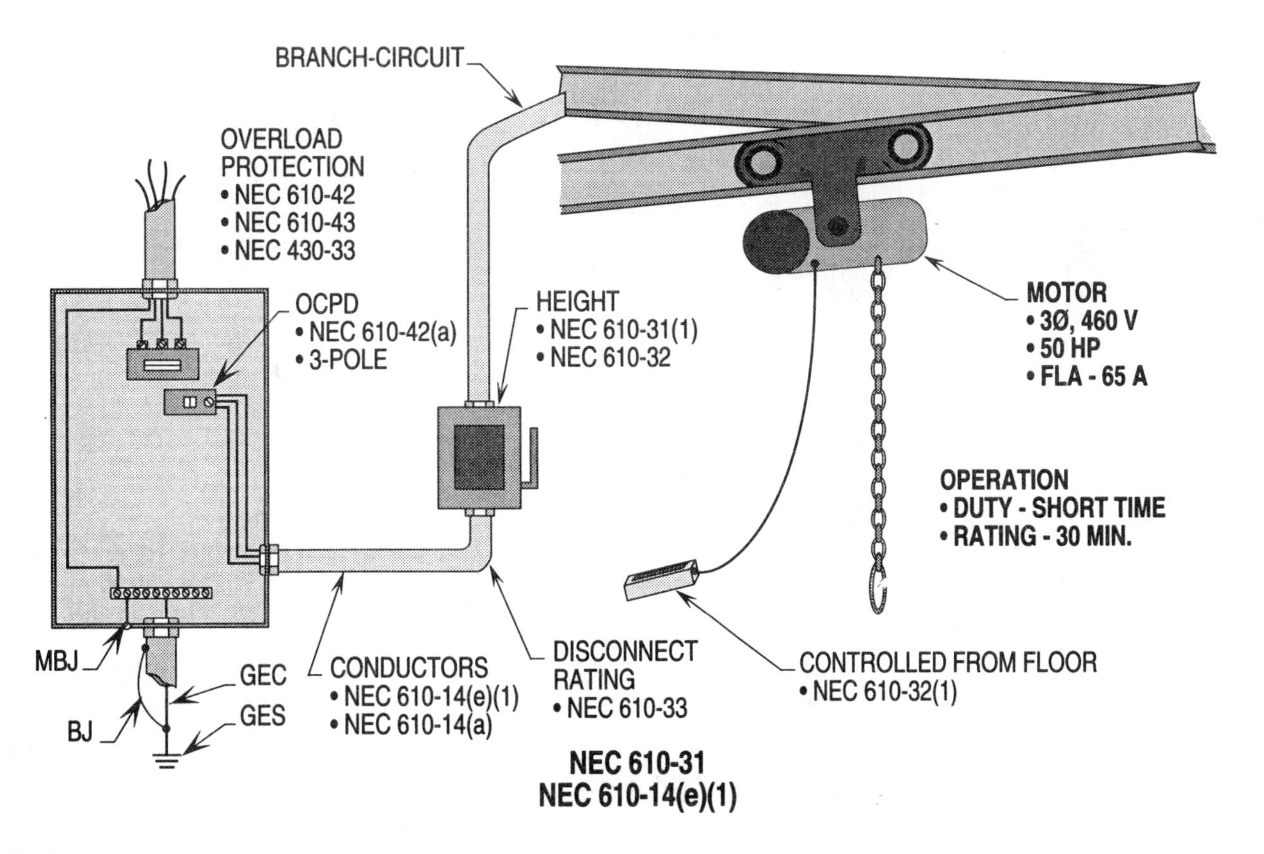

Ill. 21-4

Sizing Disconnects for Motors 610-33

When more than one motor is used for the operation of a crane, the disconnecting means is sized and based upon the operation of such motors.

Example 21-5. What size disconnect is required to deenergize the motors in Illustration 21-5?

Sizing disconnect for motors

Step 1: Finding FLA of motors
610-33; 610-32
STR = 52 A
STR = 40 A
Total = 92 A

Step 2: Calculating disconnect
610-33; 610-32
92 A x 50% = 46 A

Step 3: Selecting a CB or switch
610-33; 240-6(a)
46 A requires 50 A CB
46 A requires 60 A switch

Solution: The size disconnect using a CB is 50 amps or 60 amps for a switch.

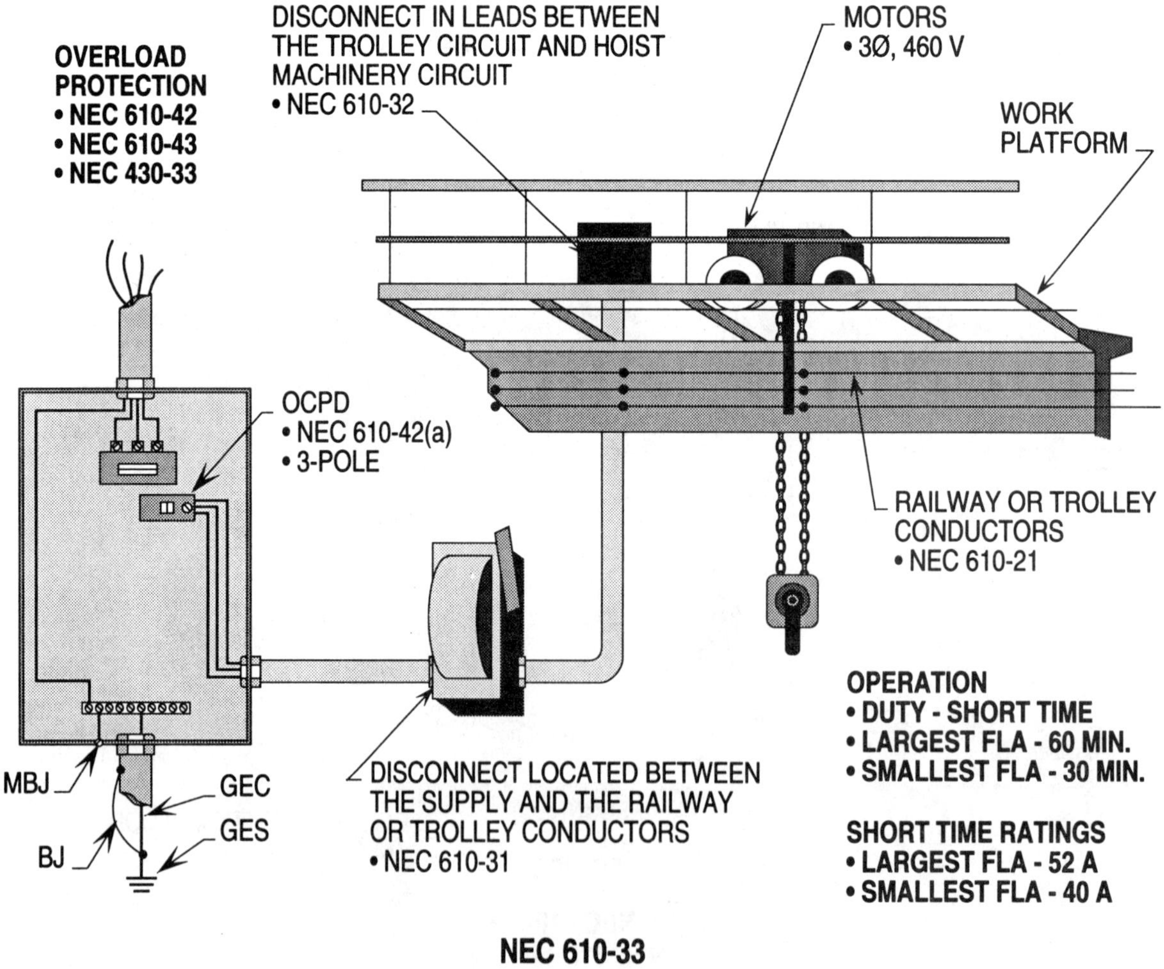

Ill. 21-5

Sizing Disconnects for Motors 610-33

The size of the disconnecting means required to disconnect a number of motors on a crane may be determined by demand factors which are selected according to the operation of the crane.

Example 21-6. What size disconnect is required to deenergize the motors in Illustration 21-6, based upon any single motion?

Sizing disconnect for motors

Step 1: Finding FLA for motors
610-33; 610-32
STR = 124 A
STR = 96 A
STR = 77 A
Total load = 297 A

Step 2: Calculating disconnect
610-33
297 A x 75% = 222.8 A

Step 3: Selecting CB or switch
610-33; 240-6(a)
222.8 A requires 225 A CB
222.8 A requires 400 A switch

Solution: The size disconnect using a CB is 225 amps or 400 amp switch.

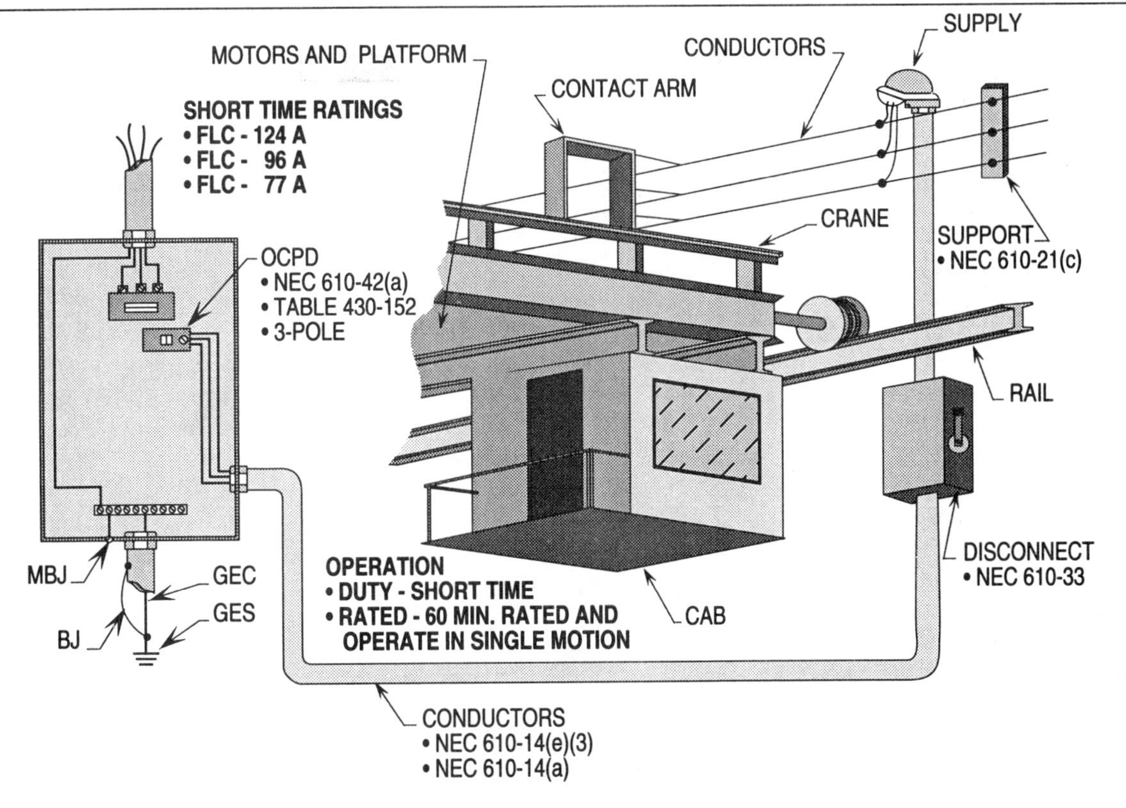

Ill. 21-6

Sizing OCPD to Start and Run Motor 610-42(a)

The size OCPD must be capable of holding the inrush current of a motor and allow the load to be lifted and moved by the crane.

Example 21-7. What size OCPD is required the motor in Illustration 12-7 to start and run?

Sizing OCPD to start and run motor

Step 1: Finding FLA for motor
430-6(a)(1); Table 430-150
FLA = 65 A

Step 2: Calculating OCPD (CB)
610-42(a); Table 430-152
65 A x 250% = 162.5 A

Step 3: Selecting OCPD
430-52(c)(1), Ex. 1; 240-6(a)
162.5 A requires 175 A OCPD

Solution: **The size OCPD using a CB is 175 amps.**

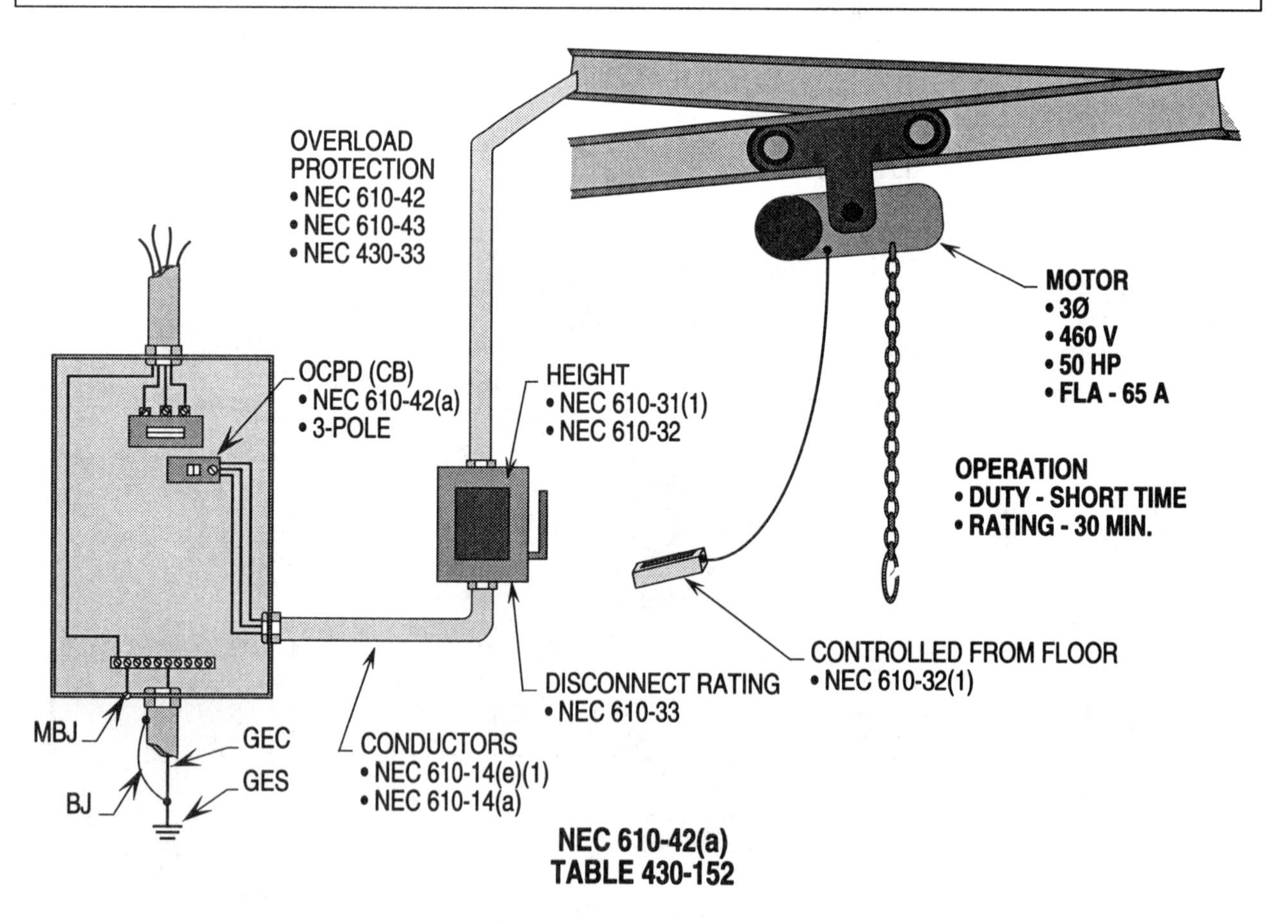

Ill. 21-7

Sizing CB to Allow Motors to Start and Run 610-42(a)

The OCPD serving a crane must be sized with enough capacity to allow the motors to start and operate properly.

Example 21-8. What size CB is required to allow the motors in Illustration 21-8 to start and operate the crane in a single operation?

Step 1: Finding FLA of motors
610-42(a); Table 430-152

Motor 1	= 124 A
Motor 2	= 96 A
Motor 3	= 77 A
Total load	= 297 A

Step 2: Calculating OCPD
430-52(a); Table 430-152
297 A x 250% = 742.5 A

Step 3: Selecting OCPD
430-52(c)(1), Ex. 1; 240-6(a)
742.5 A allows 800 A OCPD

Solution: The size OCPD using a CB is 800 amps.

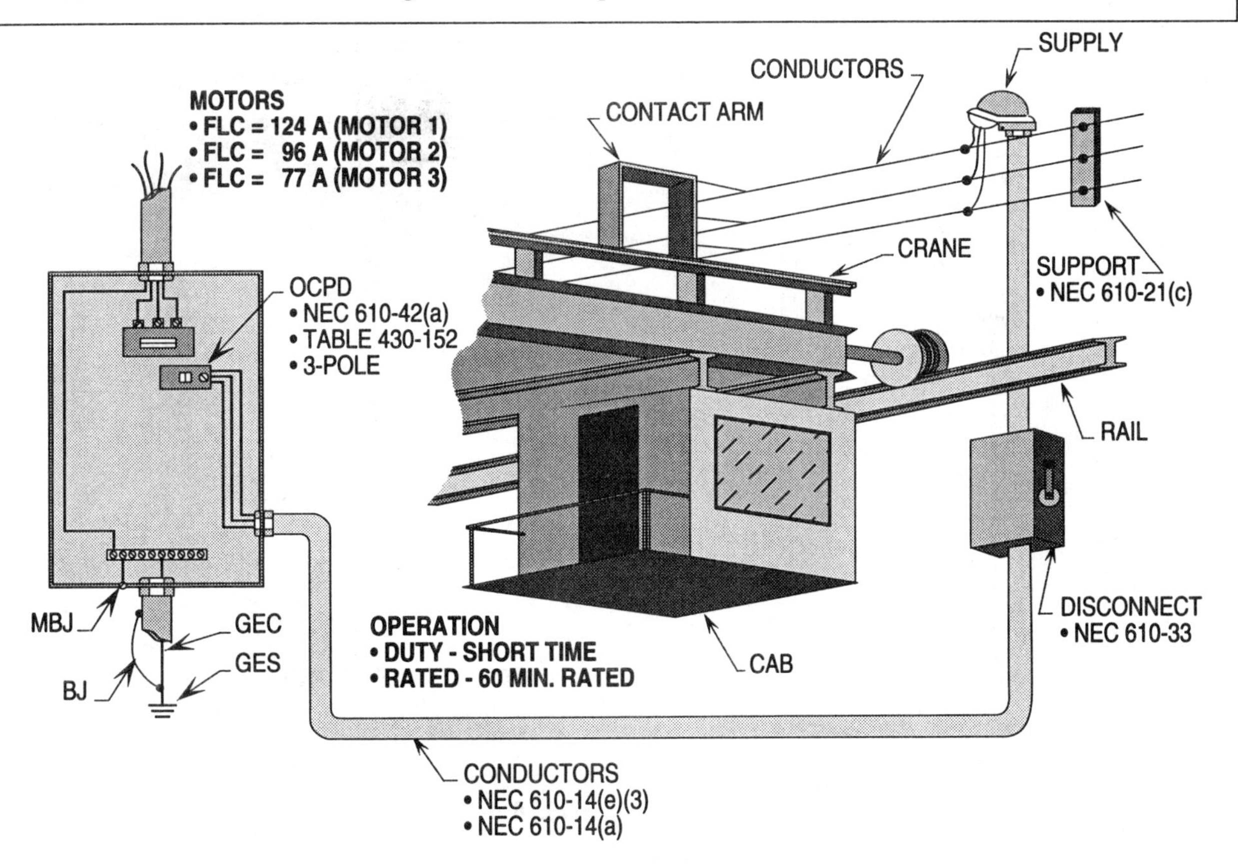

Ill. 21-8

Sizing OCPD for Control Circuits and Overloads 610-53

OCPD's used for the protection of control circuits and overloads are required to be sized and selected at ratings that provide safe and reliable operation.

Example 21-9. What size OCPD is allowed for the control circuit conductors and overloads in Illustration 21-9?

Sizing OCPD for control circuits

Step 1: Finding FLA of conductors
610-53; Table 310-16
#14 THWN cu. = 20 A

Step 2: Calculating OCPD
610-53
20 A x 300% = 60 A

Step 3: Selecting OCPD
60 A allows 60 A OCPD

Solution: The size OCPD for the control conductors is 60 amps.

Sizing OCPD for overloads

Step 1: Calculating conductors and OCPD
610-14(e)(1); Table 310-16; 240-6(a)
27 A x 100% = 27 A

Step 2: Selecting conductor and OCPD
610-14(e); Table 310-16; 240-6(a)
27 A requires #10 cu.
27 A requires 30 A CB

Step 3: Selecting OCPD for OL's
610-43(1)
OCPD = 30 A CB

Solution: The overloads are considered protected by the 30 amp CB in the panelboard.

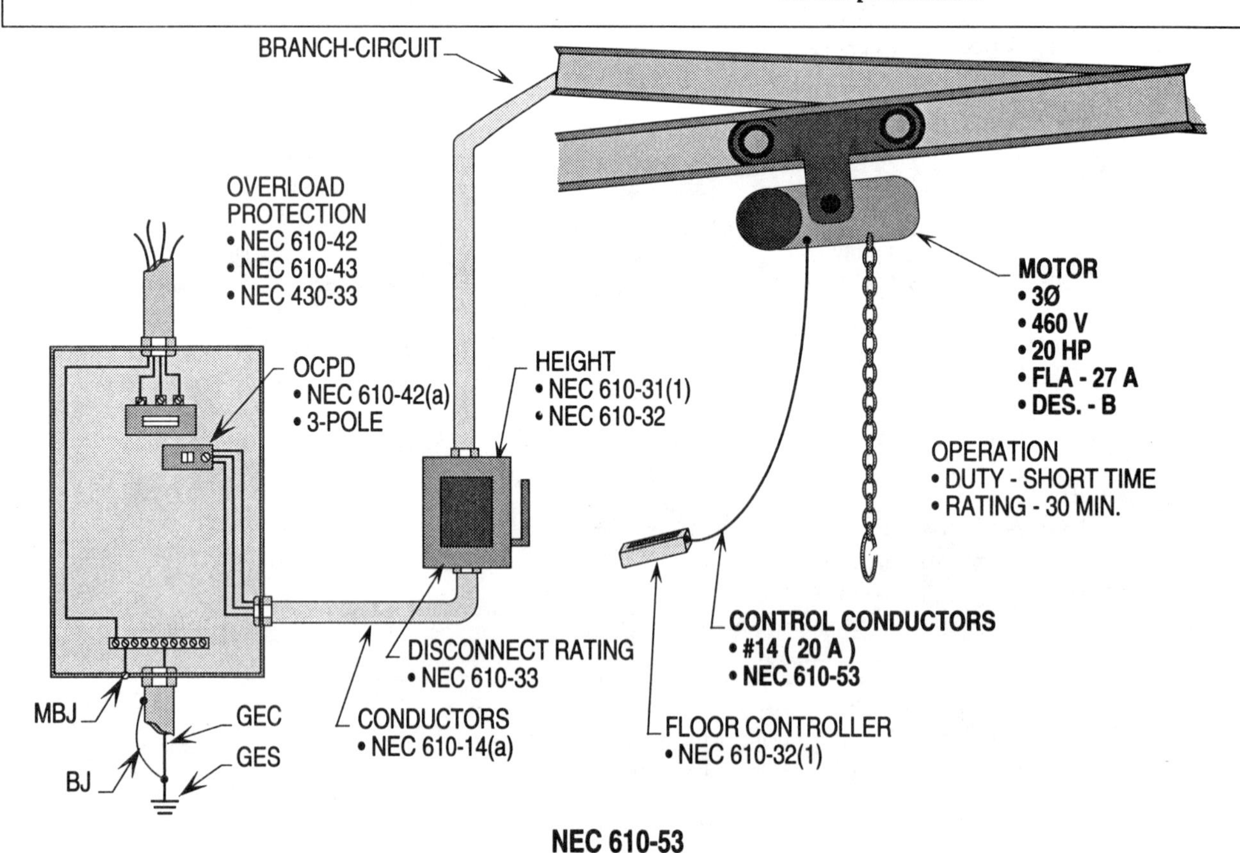

NEC 610-53

Ill. 21-9

Sizing OCPD for the Tap Control Conductors 610-53(a)

OCPD's must be sized and based upon the primary FLA's of the transformer or the ampacity of the control conductors, whichever provides the better protection.

Example 21-10. What size OCPD is required for the tap control conductors in Illustration 21-10?

Sizing tap control conductors

Step 1: Finding A of conductors
610-53(a); Table 310-16
#14 cu. = 20 A

Step 2: Calculating OCPD based on conductors
610-53(a); Table 310-16
20 A x 200% = 40 A

Step 3: Finding FLA of Transformer
610-53(a)
FLA = VA / V
FLA = 1,000 VA / 120 V
FLA = 8.3

Step 4: Calculating OCPD based on transformer
610-53(a); 240-6
8.3 A x 200% = 16.6 A

Solution: **The size OCPD using a CB is 15 amps based upon the transformer FLA.**

Note: There are inspectors who may require: 200% x 15 A (for #14), per 240-3(d).

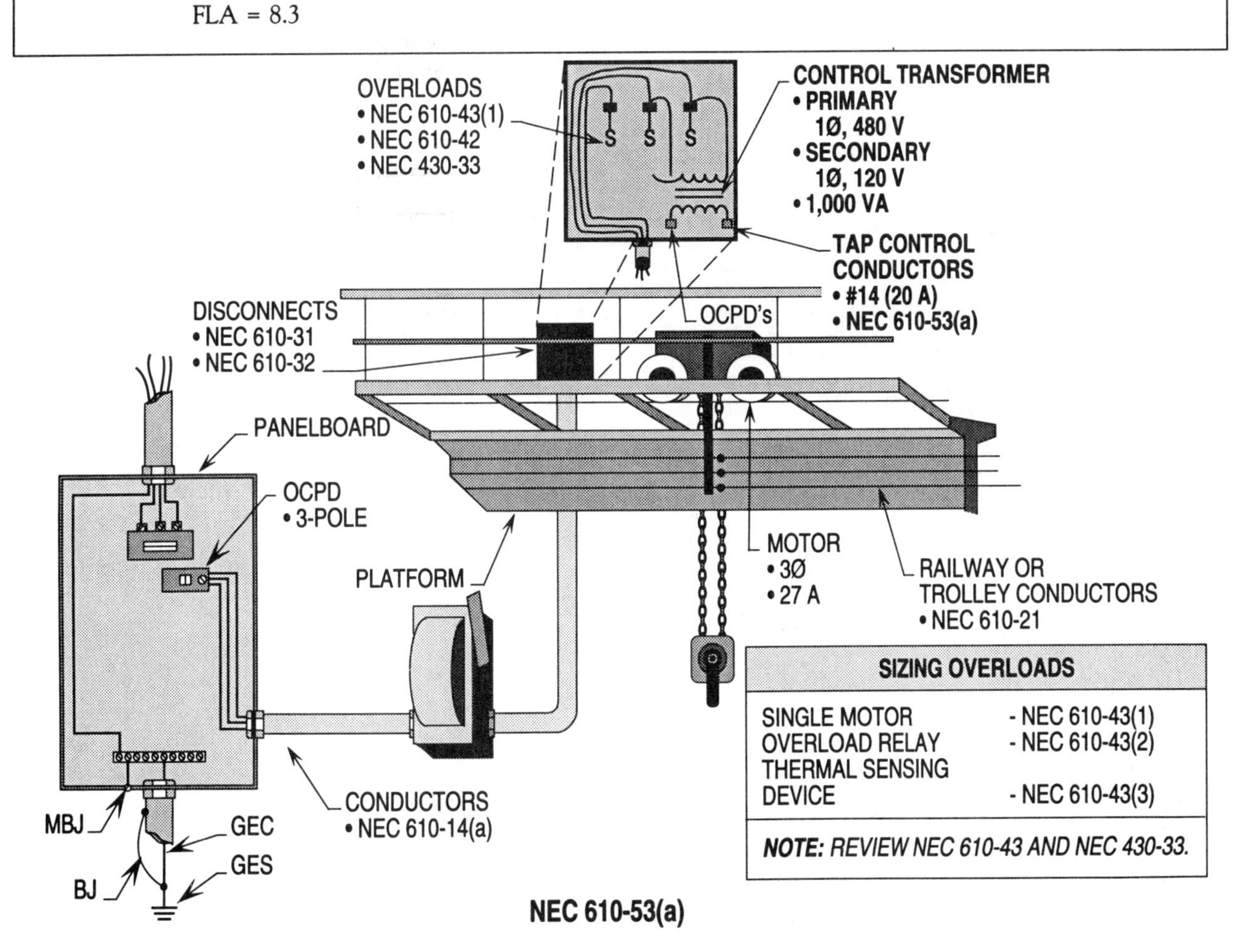

Ill. 21-10

Sizing Conductors for Elevator Loads 620-13(a)

The size of conductors supplying motors that operate elevators, based upon a duty cycle, may be sized with demand factors selected from such classifications.

Example 21-11(a). What size THWN copper conductors are required to supply the intermittent duty motor and loads in Illustration 21-11(a)? (The motor is continuous rated)

Sizing conductors for elevator loads

Step 1: Finding FLA of loads
430-6(a)(1); Table 430-150; 620-22
30 HP = 40 A
Other loads = 12 A

Step 2: Finding percentage
430-22(b); Table 430-22(b)
Continuous rated motor = 140%

Step 3: Calculating conductor
620-13(a); 620-22
40 A x 140% + 12 A = 68 A

Step 4: Selecting conductors
310-10(2); Table 310-16
68 A requires #4 cu.

Solution: The size THWN copper conductors are No. 4.

Note: The FLA on the motor's name-plate is the same as listed in Table 430-150.

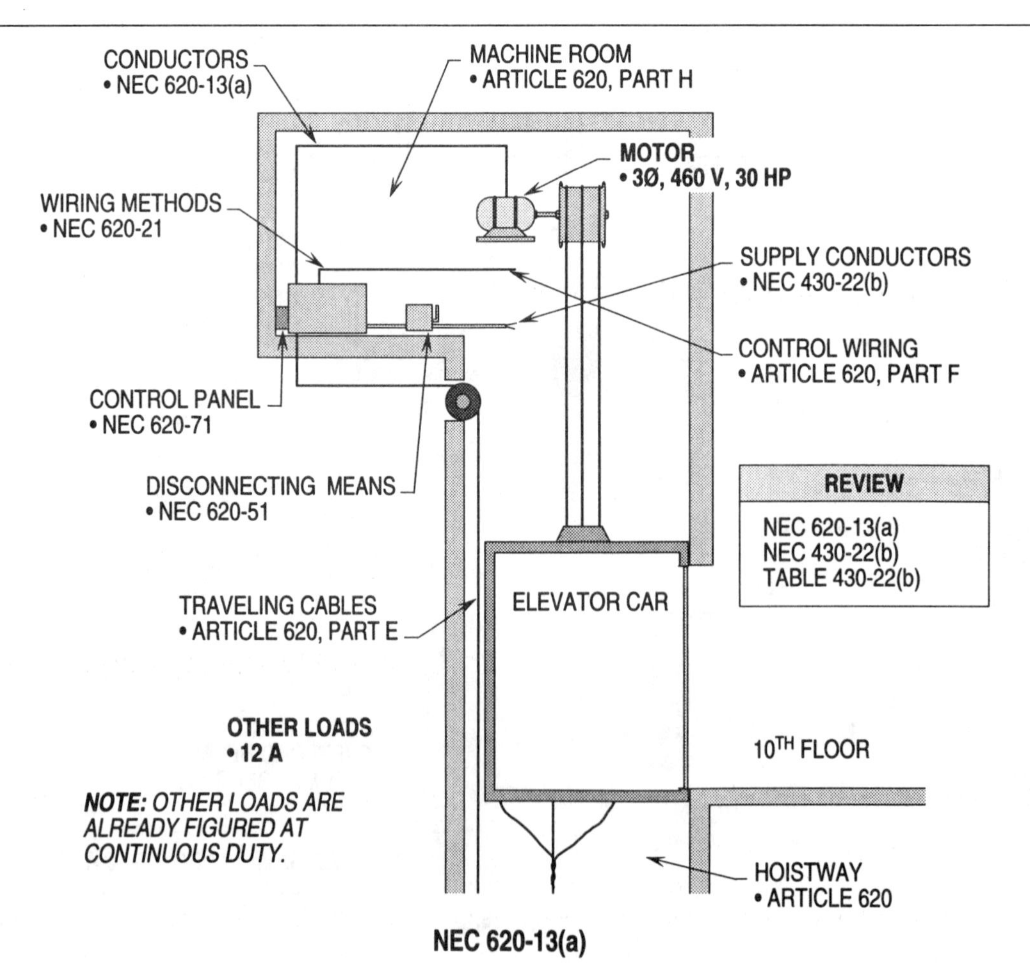

Ill. 21-11(a)

Sizing Conductors for Elevator Loads 620-13(a)

The size of the conductors supplying motors that operate elevators, based upon a duty cycle, may be sized with demand factors selected from such classifications.

Example 21-11(b). What size THWN copper conductors are required to supply the intermittent duty motor in Illustration 21-11(b)? (The motor is 15 minute rated).

Sizing conductors for elevator loads

Step 1: Finding FLA of loads
430-6(a)(1); Table 430-150; 620-22
30 HP = 40 A
Other loads = 12 A

Step 2: Finding percentage
430-22(b); Table 430-22(b)
15 minute rated motor = 85%
(Intermittent)

Step 3: Calculating conductor
620-13(a); 620-22
40 A x 85% + 12 A = 46 A

Step 4: Selecting conductors
310-10(2); Table 310-16
46 A requires #8 cu.

Solution: The size THWN copper conductors are No. 8.

Note: The FLA on the motor's nameplate is the same as listed in Table 430-150.

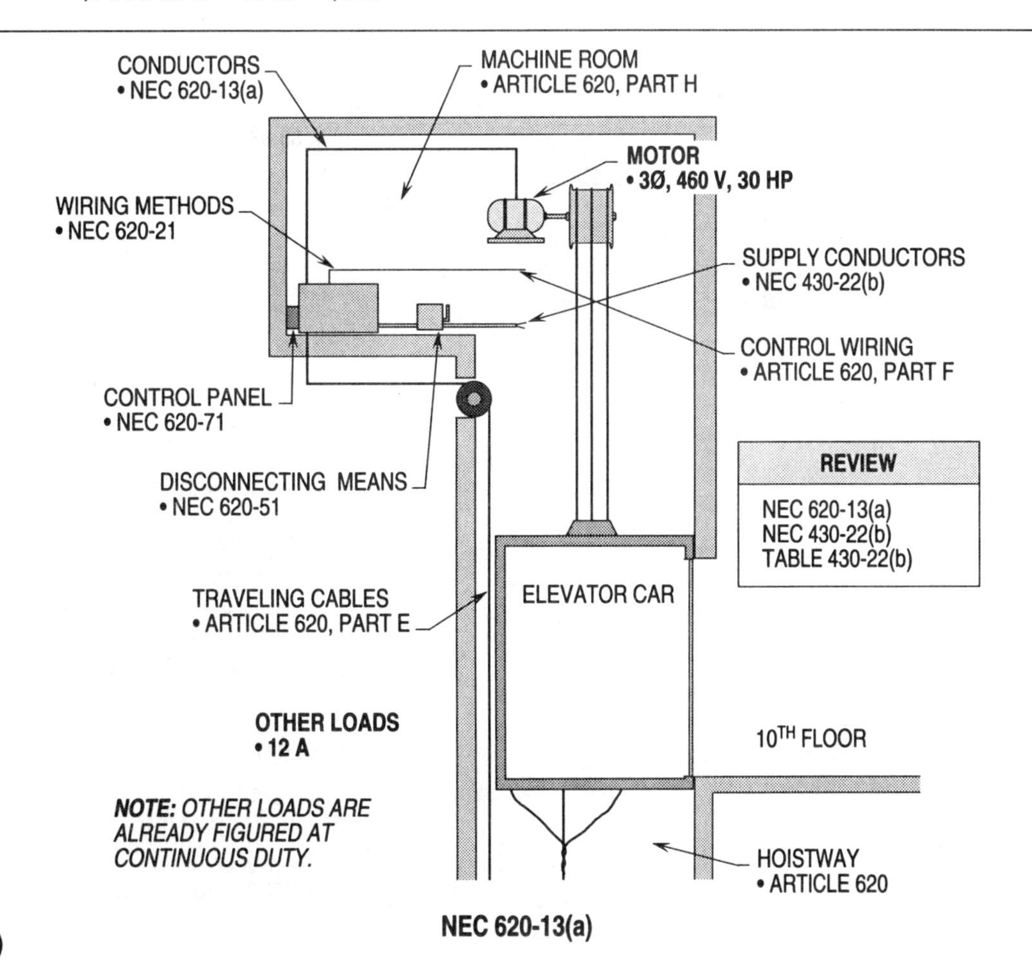

Ill. 21-11(b)

Sizing Conductors for Elevator Loads
620-13(d)

The conductors supplying power to an adjustable speed SCR DC drive shall be sized from the calculation that produces the greater number of amps.

Example 21-12(a). What size THWN copper conductors are required to supply the electrical equipment in the elevator room in Illustration 21-12(a)?

Sizing conductors for elevator loads

Step 1: Calculating sec. A of transformers
620-13(d)
A = 90 x 1,000 / 480 V x 1.732 x 85%
A = 127 A

Step 2: Calculating ratio drive input A to transformer pri.
620-13(d)
A = 210 A x 300 V / 480 V
A = 131

Step 3: Calculating other loads
430-24; 215-2(a)
12 A x 5 x 125% = 75 A

Step 4: Calculating total elevator load
620-13(d)
127 A x 5 + 75 = 710 A

Step 5: Finding demand factor
620-14; Table 620-14
5 elevators = 82%

Step 6: Applying demand factors
620-14; Table 620-14
710 A x 82% = 582 A

Step 7: Paralleling 3 times per phase
310-4; 300-5(i)
582 A / 3 = 194 A

Step 8: Selecting conductors
310-10(2); Table 310-16
194 A requires #3/0 cu.

Solution: The size THWN copper conductors connected in parallel are No. 3/0.

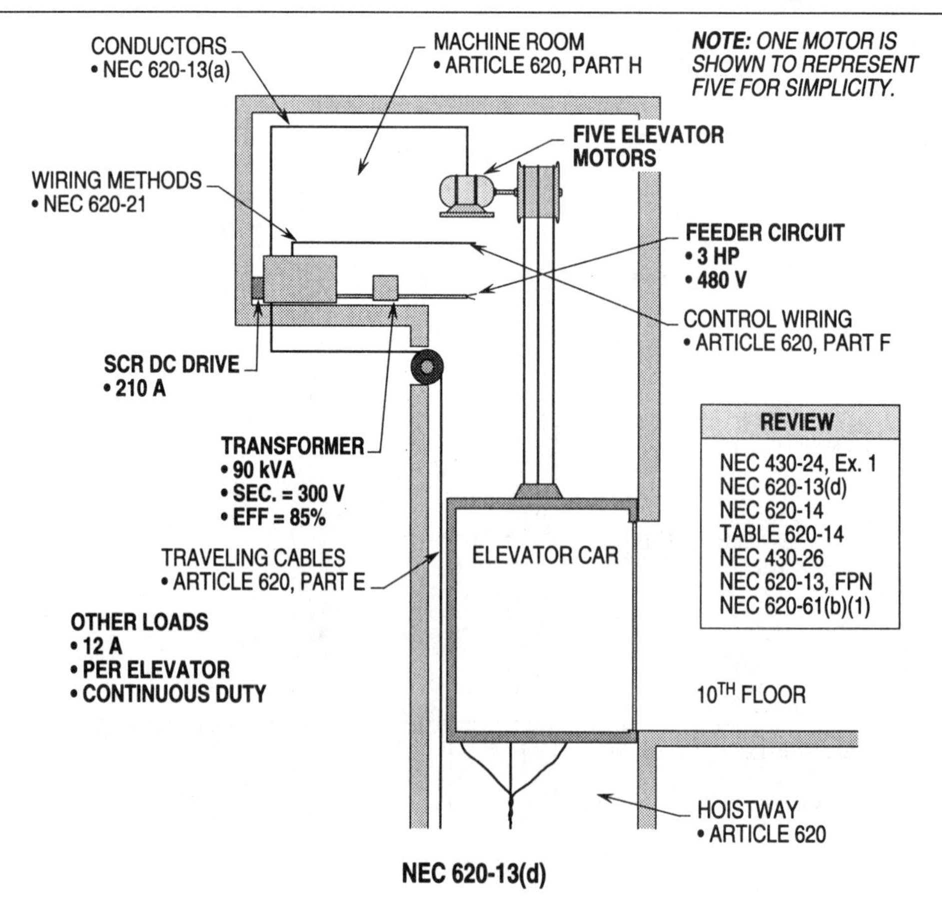

Ill. 21-12(a)

Sizing Conductors for Loads of Elevators 620-14

Conductors supplying more than one motor to operate elevators may be sized using a demand factor (%) which is based upon service classification.

Example 21-12(b). What size THWN copper conductors are required to supply power to the intermittent duty motors and loads in Illustration 21-12(b)? (All elevator motors are continuous rated).

Sizing conductors for loads of elevators

Step 1: Finding FLA of elevators
430-6(a)(1); Table 430-150
30 HP = 40 A
50 HP = 65 A

Step 2: Finding percentage
430-22(b); Table 430-22(b)
Continuous rated motor = 140%

Step 3: Calculating total load
430-22(b); 430-24, Ex. 1;
215-2(a)
65 A x 140% = 91A
40 A x 3 x 140% = 168 A
Motor load = 259 A
Controller load
12 A x 4 x 125% = 60 A

Step 4: Finding demand factor
620-14; Table 620-14
4 elevators = 85%

Step 5: Applying demand factors
620-14; Table 620-14; 430-26
259 A x 85% = 220 A
220 A + 60 A = 280 A

Step 6: Selecting conductors
310-10(2); Table 310-16
280 A requires #300 KCMIL cu.

Solution: The size of the feeder conductors are No. 300 KCMIL THWN copper conductors.

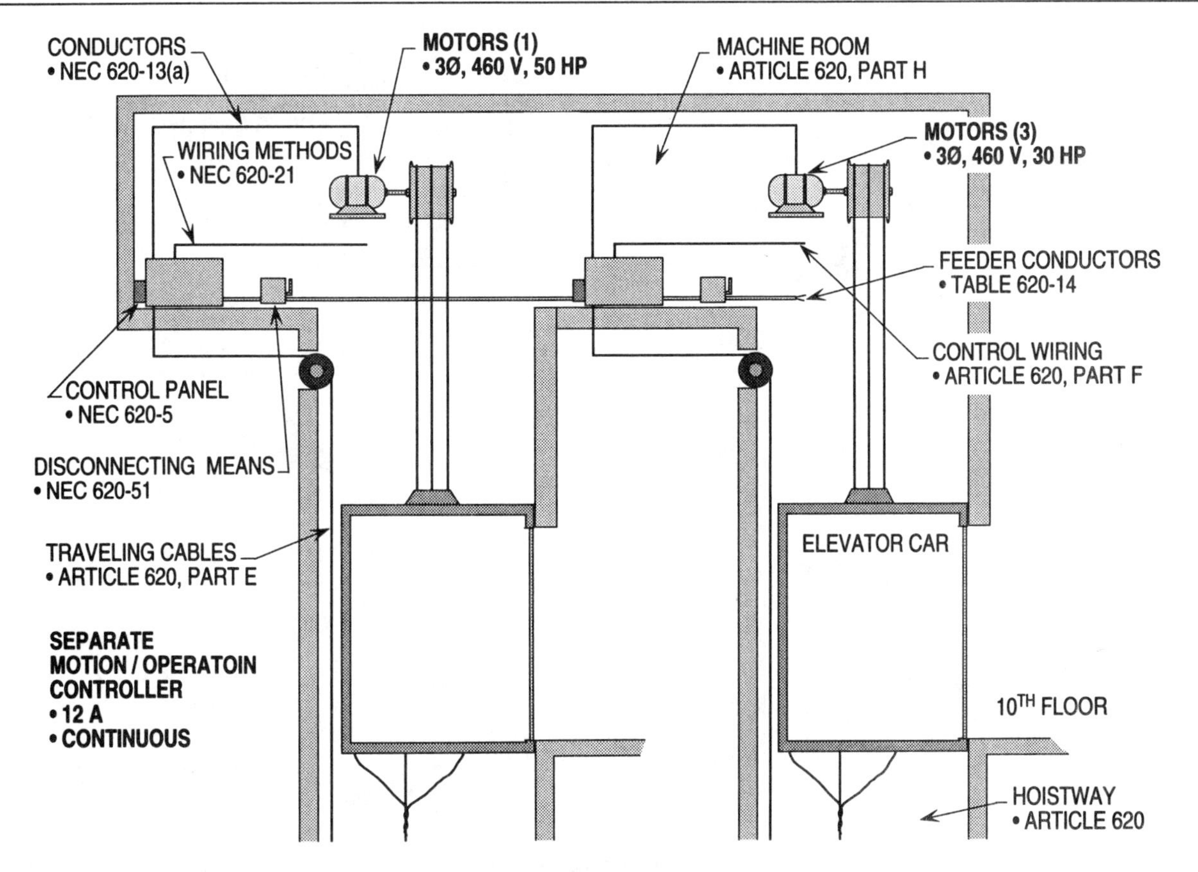

NEC 620-14

Ill. 21-12(b)

Sizing CB for Elevator 620-61(b)(1)

The OCPD between the circuit conductors and elevator motor shall be sized to hold the inrush current of the motor during the starting period.

Example 21-13(a). What size CB is required to allow the elevator motor in Illustration 21-13(a) to start and run?

Sizing CB for elevator

Step 1: Finding FLA of loads
430-6(a)(1); Table 430-150
30 HP = 40 A (same as nameplate)
other loads = 12 A

Step 2: Finding percentage
620-61(b)(1); 430-33; Table 430-152
CB = 250%

Step 3: Calculating CB
620-61(b)(1); 430-33; Table 430-150; 215-3
40 A x 250% = 100 A
9.6 A x 125% = 12 A
100 A + 12 A = 112 A

Step 4: Selecting CB
430-52(c)(1), Ex. 1; 240-6(a)
112 A requires 125 A CB

Solution: **The size OCPD using a CB is 125 amps.**

Note: See NEC 430-52(c)(1), Ex. 1 for selecting next size CB and 430-52(c)(1), Ex. 2 for maximum CB.

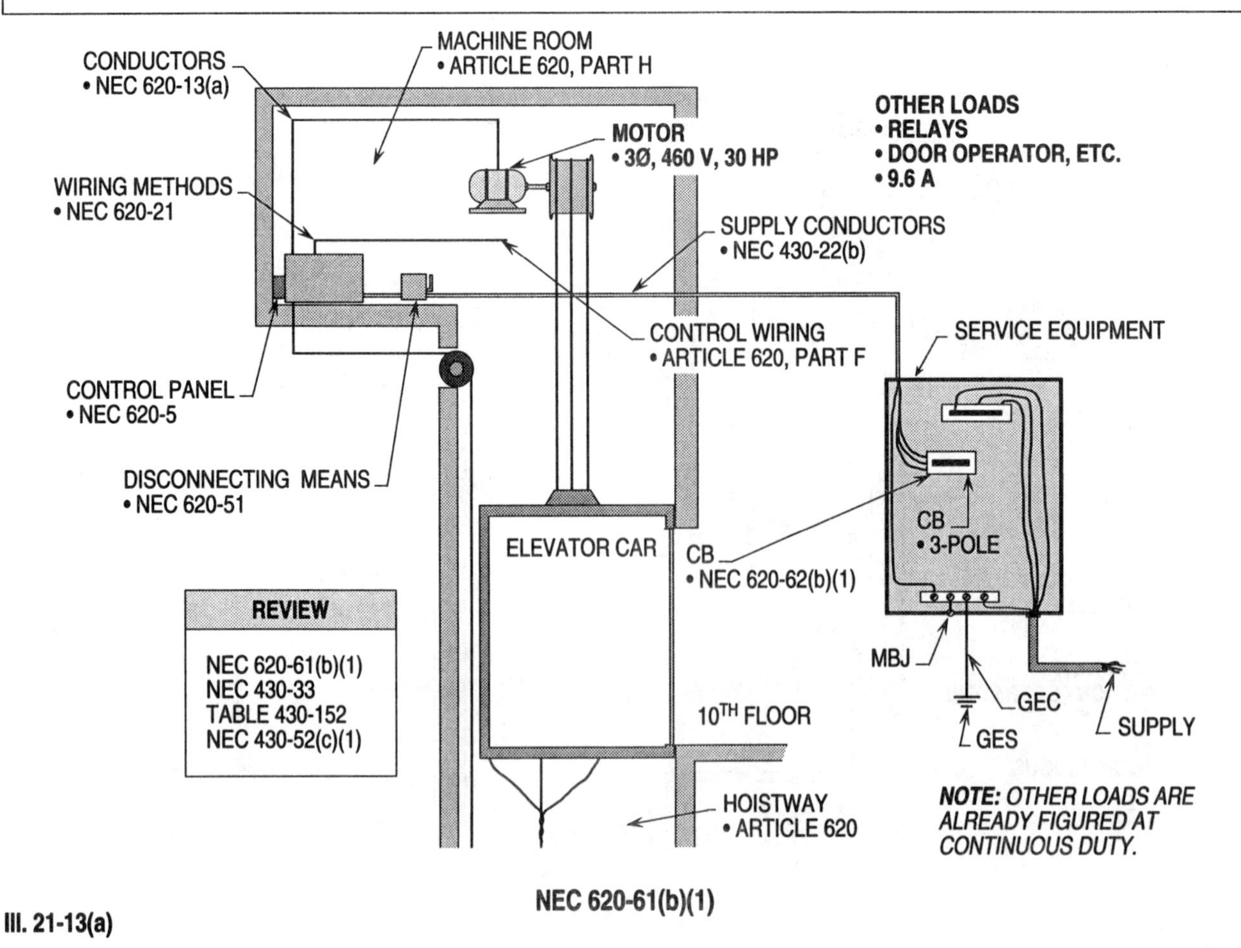

NEC 620-61(b)(1)

Ill. 21-13(a)

Sizing TDF's for Elevators 430-62(b)(1)

Time delay fuses must be sized large enough to permit the motors they supply to start and run each elevator during its operational period.

Example 21-13(b). What size time delay fuses (TDF's) are required to allow the elevator motor in Illustration 21-13(b) to start and run?

Sizing TDF's for elevator

Step 1: Finding FLA for elevators
430-6(a)(1); Table 430-150
30 HP = 40 A
50 HP = 65 A (same as nameplates)

Step 2: Finding percentage
430-62(b)(1); Table 430-152
TDF's = 175%

Step 3: Calculating TDF's
430-62(b)(1); 430-52(c)(1), Ex. 1; 240-6(a)
65 A x 175% = 113.75 A = 125 A
125 A + 40 A = 165 A

Step 4: Selecting TDF's
430-62(b)(1); 240-6(a)
165 A requires 150 A TDF's

Solution: **The size OCPD using time delay fuses is 150 amps.**

Note 1: The 150 amp OCPD will hold 750 amps for ten seconds (150 A x 5 = 750 A) which allows the motors with a LRC of 630 amps (65 A + 40 A x 6 = 630 A) to start and run.

Note 2: In Step 2 the next size OCPD is rounded up to the next size.

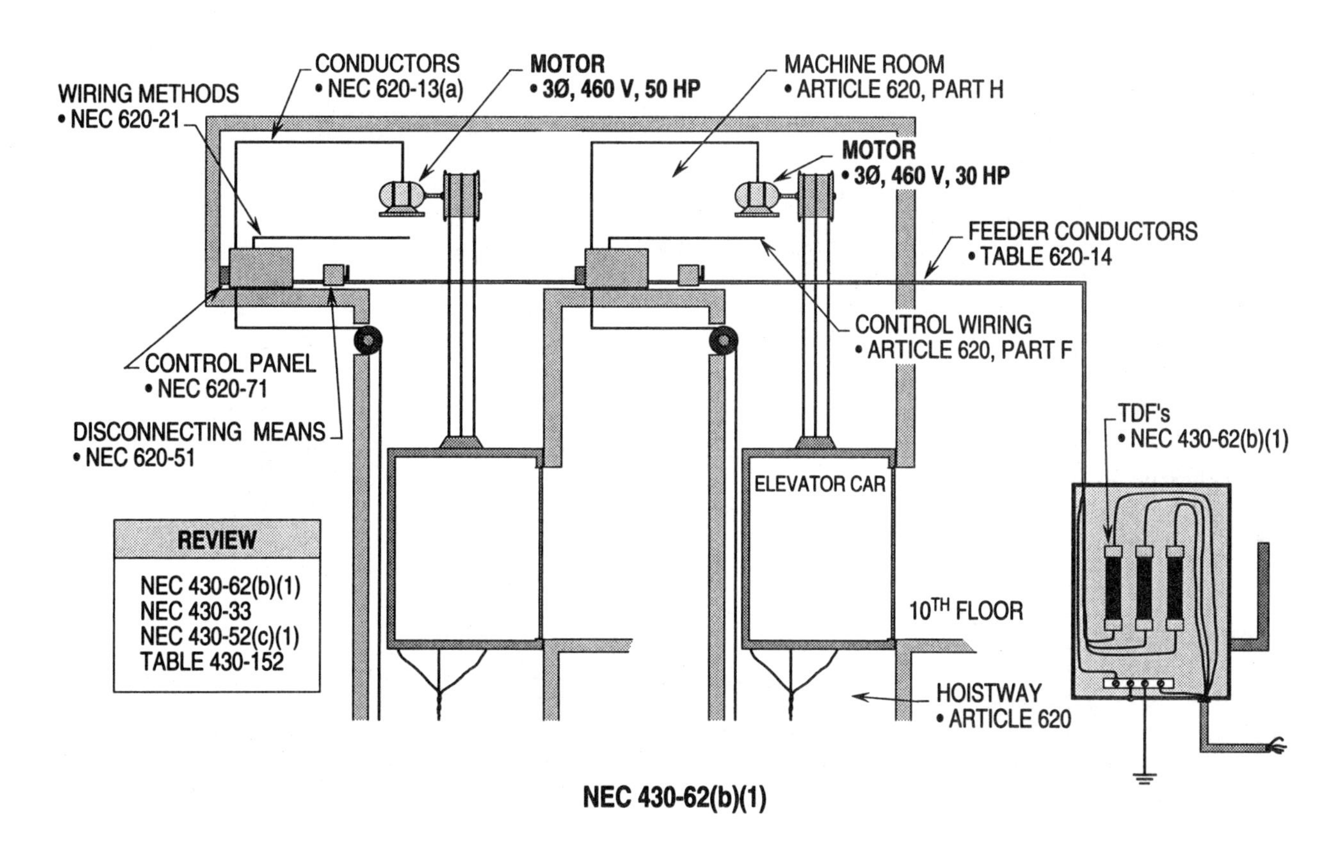

Ill. 21-13(b)

Sizing Disconnects for Elevator Loads 620-51

The disconnect for elevator loads must be sized to hold the size fuses required and to allow the motor to start and run.

Example 21-14. What size disconnecting means is required to disconnect the elevator loads in Illustration 21-14?

Sizing disconnect for elevator loads

Step 1: Finding TDF's for elevator
620-51
TDF's = 80 A

Step 2: Selecting disconnect based upon TDF's
430-57; 240-6(a)
80 A TDF's requires 100 A

Solution: The size disconnecting means based upon the TDF's is 100 amps.

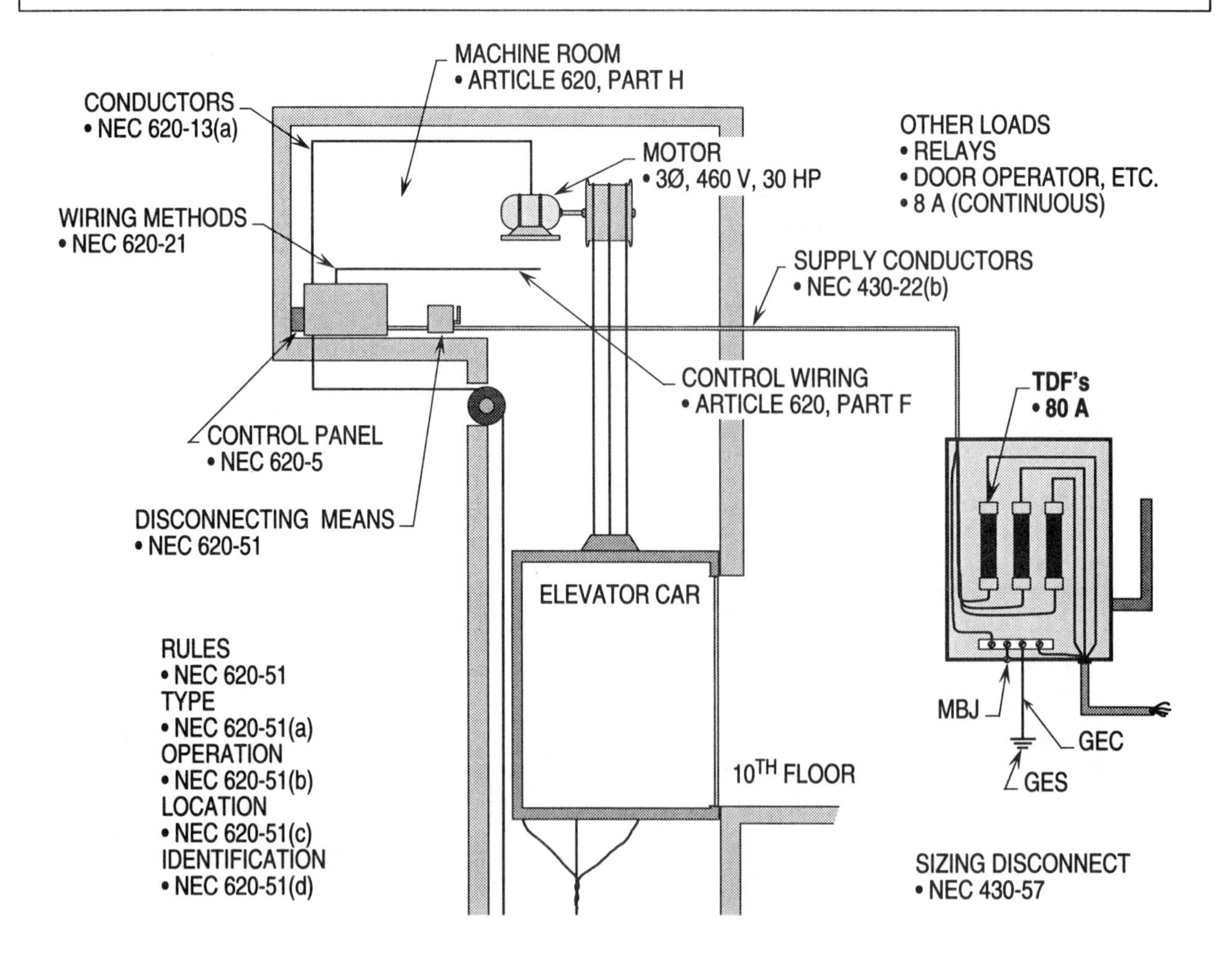

NEC 620-51

Ill. 21-14

22 Electric Welders, Computers, and Nonmedical X-Rays

There are three types of welders used in modern day welding shops. The type used determines how the circuit elements must be designed.

The procedure for calculating the full-load amps for welders is to obtain the duty-cycle factor and select the multiplier. The primary amps of the welder is multiplied by the multiplier to derive full-load amps to size elements.

Computer branch-circuit loads are sized at 125 percent of the total connected load. Feeder-circuit loads are determined by the nameplate's full-load amps or by classifying the load as noncontinuous or continuous operation.

The elements of nonmedical X-ray apparatus is found by multiplying the units based upon the momentary or long-time rating whichever applies due to the installation and use.

Quick Reference

Nonmotor Generator Arc Welders 630-11(a); 630-12(a)

When sizing the branch-circuit conductors for nonmotor generator arc welders, the current-carrying capacity shall not be less than the rated primary current of the welder times a duty cycle factor listed in NEC 630-11(a). The welder's primary full-load current rating listed on the nameplate is selected at not more than 200 percent for the overcurrent protection device to be installed. Branch-circuit conductors must be protected at a rating not exceeding 200 percent of their allowable ampacities per NEC 630-12(b).

Example 22-1. What size THWN cu. conductors and overcurrent protection device are required to supply the nonmotor generator arc welder in Illustration 22-1?

Sizing conductors

Step 1: Finding FLC
630-11(a)
Welder = 68 A

Step 2: Finding multiplier
630-11(a)
50% = .71

Step 3: Calculating A
630-11(a)
68 A x 71% = 48.28 A

Step 4: Selecting conductors
Table 310-16
48.28 A requires #8 cu.

Solution: The size THWN copper conductors are No. 8.

Sizing OCPD

Step 1: Finding FLC
630-12(a); 630-12(b)
Welder = 68 A

Step 2: Finding multiplier
630-12(a); 630-12(b)
Multiplier = 200%

Step 3: Calculating A
630-12(a); 630-12(b)
68 A x 200% = 136 A

Step 4: Selecting OCPD
240-3(g); 240-6(a); 630-12(b)
136 A requires 125 A OCPD

Step 5: Protecting conductors
630-12(b)
- #8 THWN cu. = 50 A
- 50 A x 200% = 100 A
- 100 A requires 100 A OCPD

Solution: The size OCPD required is 100 amps based upon amps of conductors times 200 percent.

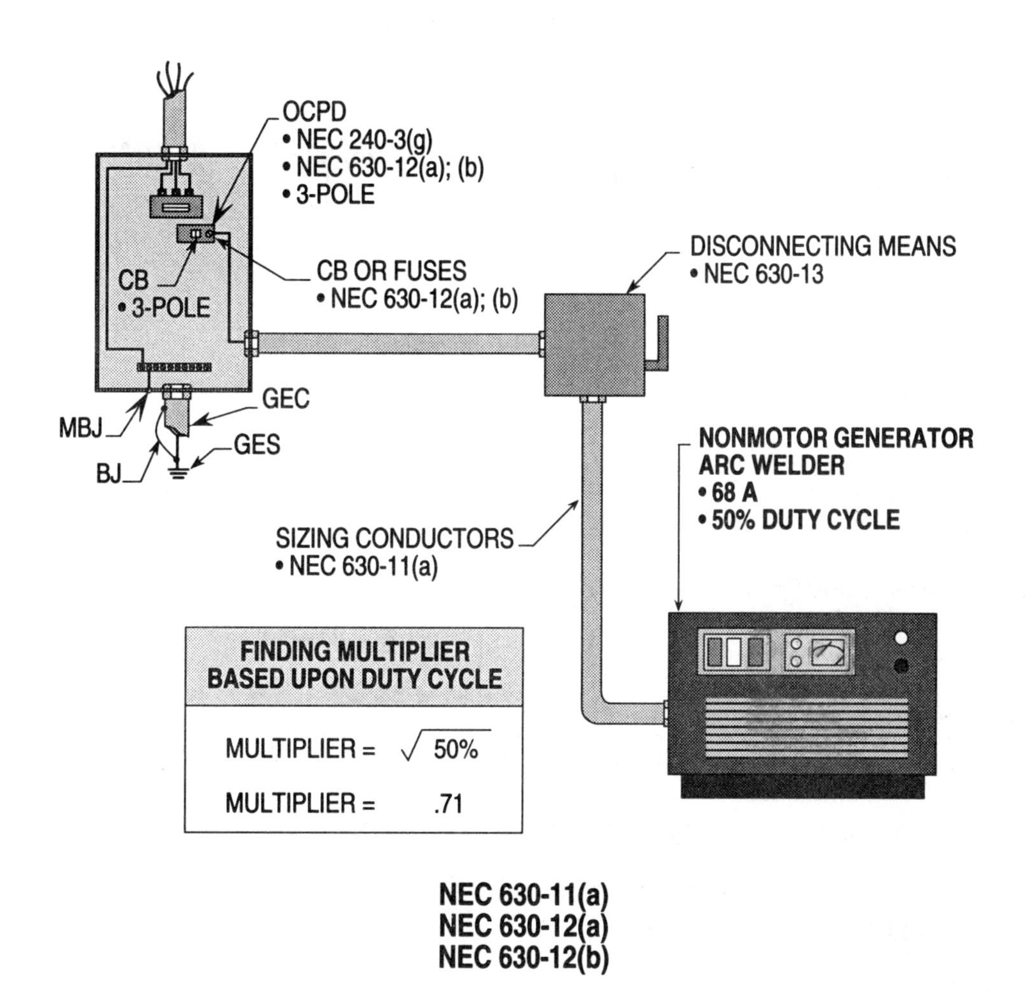
OCPD
• NEC 240-3(g)
• NEC 630-12(a); (b)
• 3-POLE
CB
• 3-POLE
CB OR FUSES
• NEC 630-12(a); (b)
DISCONNECTING MEANS
• NEC 630-13
MBJ
GEC
GES
BJ
NONMOTOR GENERATOR
ARC WELDER
• 68 A
• 50% DUTY CYCLE
SIZING CONDUCTORS
• NEC 630-11(a)
FINDING MULTIPLIER
BASED UPON DUTY CYCLE
MULTIPLIER = √50%
MULTIPLIER = .71
NEC 630-11(a)
NEC 630-12(a)
NEC 630-12(b)

Ill. 22-1

Motor-generator Arc Welders 630-11(a); 630-12(a)

When sizing the branch-circuit conductors for motor-generator arc welders, the current-carrying capacity shall not be less than the rated primary current of the welder times a duty cycle factor listed in Table 630-11(a). The welder's primary full-load current rating listed on the nameplate is selected at not more than 200 percent for the overcurrent protection device to be installed. Branch-circuit conductors must be protected at a rating not exceeding 200 percent of their allowable ampacities per 630-12(b).

Example 22-2. What size THWN cu. conductors and overcurrent protection device are required to supply the motor-generator arc welder in Illustration 22-2?

Sizing conductors

Step 1: Finding FLC
630-11(a)
Welder = 76 A

Step 2: Finding multiplier
630-11(a)
90% = .96

Step 3: Calculating A
630-11(a)
76 A x 96% = 72.96 A

Step 4: Selecting conductors
Table 310-16
72.96 A requires #4 cu.

Solution: The size THWN copper conductors are No. 4.

Sizing OCPD

Step 1: Finding FLC
630-12(a); 630-12(b)
Welder = 76 A

Step 2: Finding multiplier
630-12(a); 630-12(b)
Multiplier = 200%

Step 3: Calculating A
630-12(a); 630-12(b)
76 A x 200% = 152 A

Step 4: Selecting OCPD
240-3(g); 240-6(a); 630-12(a)
152 A requires 150 A OCPD

Solution: The size OCPD required is 150 amps.

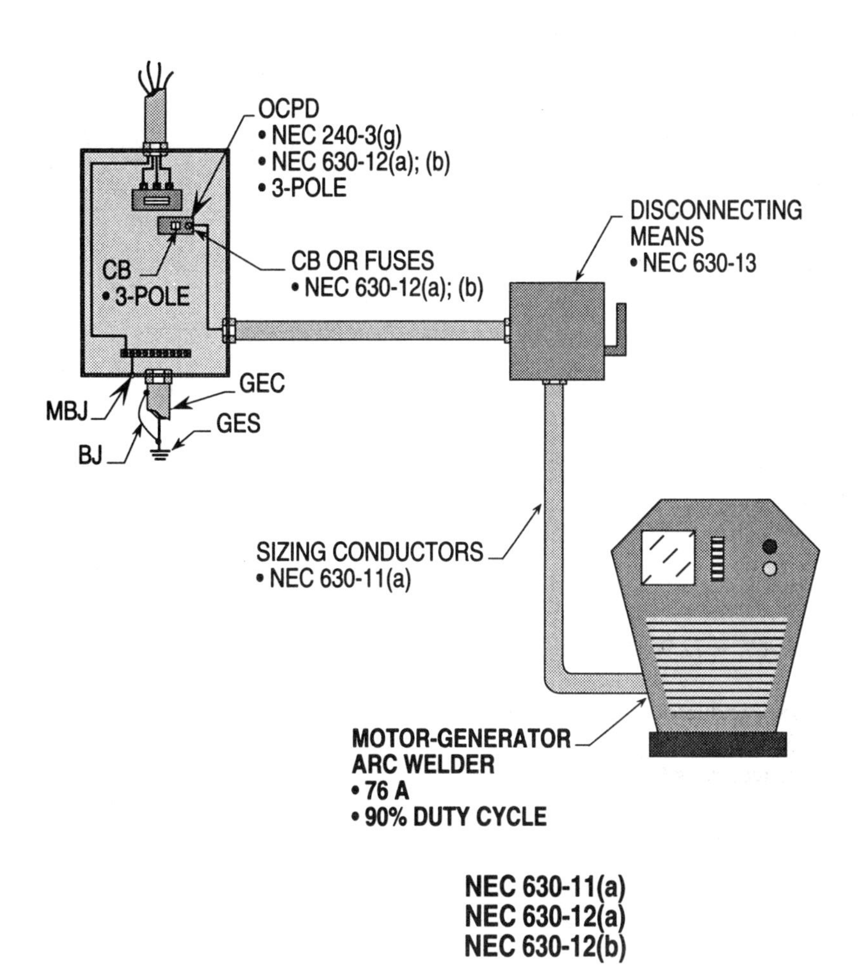
OCPD
• NEC 240-3(g)
• NEC 630-12(a); (b)
• 3-POLE
DISCONNECTING
MEANS
• NEC 630-13
CB OR FUSES
• NEC 630-12(a); (b)
CB
• 3-POLE
GEC
MBJ
GES
BJ
SIZING CONDUCTORS
• NEC 630-11(a)
MOTOR-GENERATOR
ARC WELDER
• 76 A
• 90% DUTY CYCLE
NEC 630-11(a)
NEC 630-12(a)
NEC 630-12(b)

Ill. 22-2

Resistance Welders
630-31(a); 630-32(a)

When sizing the branch-circuit conductors for resistance welders, the current-carrying capacity shall not be less than the rated primary current of the welder times a duty cycle factor listed in NEC 630-31(a). The welder's primary full-load current rating listed on the nameplate is selected at not more than 300 percent for the overcurrent protection device to be installed. Branch-circuit conductors already protected at a rating not exceeding this value are considered protected from overloading per 630-32(b).

Example 22-3(a). What size THWN cu. conductors and OCPD are required to supply the resistance welder in Illustration 22-3(a)?

Sizing conductors

Step 1: Finding FLC
630-31(a)
Welder = 91 A

Step 2: Finding multiplier
630-31(a)
40% = .63

Step 3: Calculating A
91 A x 63% = 57.33 A

Step 4: Selecting conductors
Table 310-16
57.33 A requires #6 cu.

Solution: **The size THWN copper conductors are No. 6.**

Sizing OCPD

Step 1: Finding FLC
630-32(a)
Welder = 91 A

Step 2: Finding multiplier
630-32(a)
Multiplier = 300%

Step 3: Calculating A
630-32(a)
91 A x 300% = 273 A

Step 4: Selecting OCPD
240-3(g); 240-6(a); 630-32(a)
273 A requires 250 A OCPD

Step 5: Protecting conductors
630-32(b)
- #6 THWN cu. = 65 A
- 65 A x 300% = 195 A
- 195 A requires 175 A OCPD

Solution: **The size OCPD required is 175 amp based upon the amps of conductors times 300 percent.**

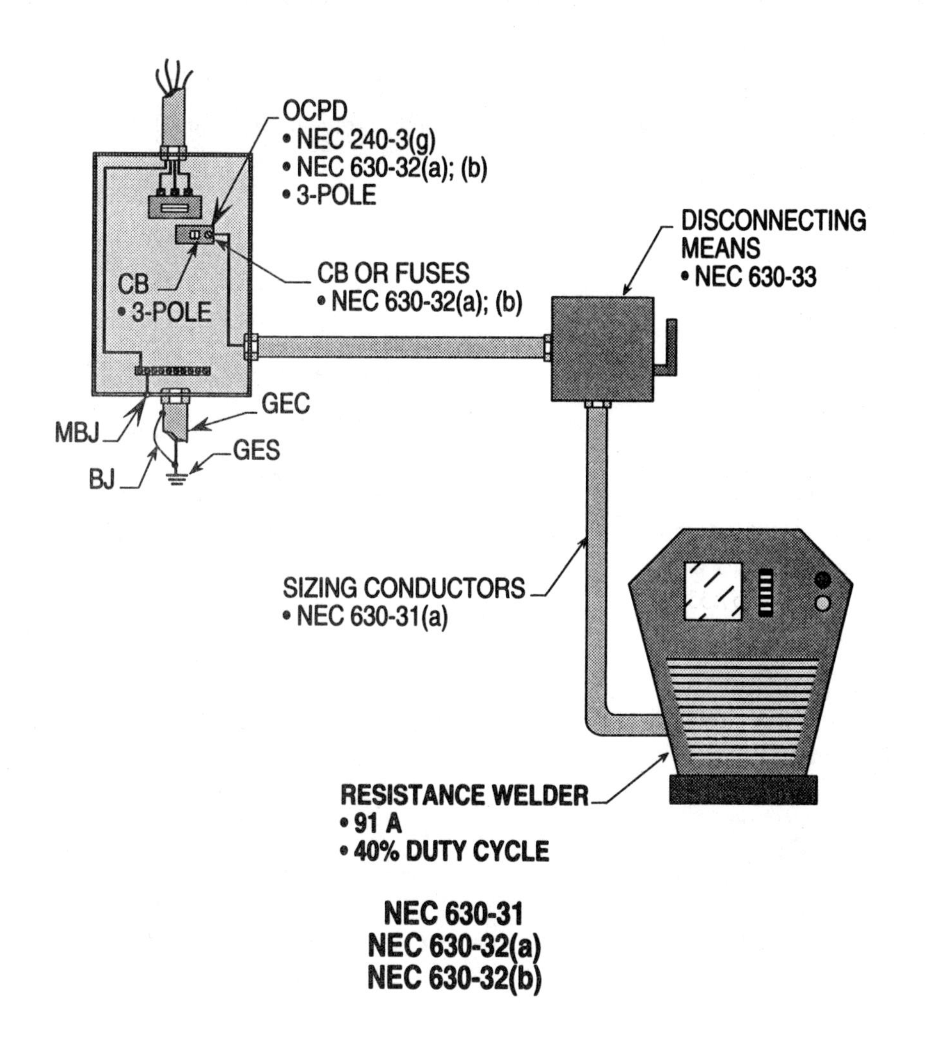

NEC 630-31
NEC 630-32(a)
NEC 630-32(b)

Ill. 22-3(a)

Resistance Welders
630-31(a); 630-32(b)

When sizing the feeder-circuit conductors for a group of welders, the current-carrying capacity shall not be less than the rated primary current of the welders times a duty cycle factor listed in NEC 630-31(a). The welder's primary full-load current rating listed on the nameplate is selected at not more than 100 percent for the largest and 60 percent for all remaining welders of the group. The OCPD shall not exceed 300 percent of the conductor's ampacity or the FLA's of the welders.

Example 22-3(b). What size THWN cu. conductors and OCPD are required to supply the resistance welders in Illustration 22-3(b)?

Sizing conductors

Step 1: Finding FLC
630-31(a)
Largest welder = 91 A
Other welders = 80 A

Step 2: Finding multiplier
630-31(b)
Largest = 71% x 100%
Other = 63% x 60%

Step 3: Calculating A
630-31(b)
91 A x 71% x 100% = 64.61 A
80 A x 63% x 60% = 30.24 A
Total amps = 94.85 A

Step 4: Selecting conductors
Table 310-16
94.85 A requires #3 cu.

Solution: The size THWN copper conductors are No. 3.

Sizing OCPD

Step 1: Finding multiplier
630-32(b)
Multiplier = 300%

Step 2: Calculating A
630-32(b)
100 A x 300% = 300 A

Step 3: Selecting OCPD
240-3(g); 240-6(a); 630-32(b)
300 A requires 300 A OCPD

Solution: The size OCPD required is 300 amp.

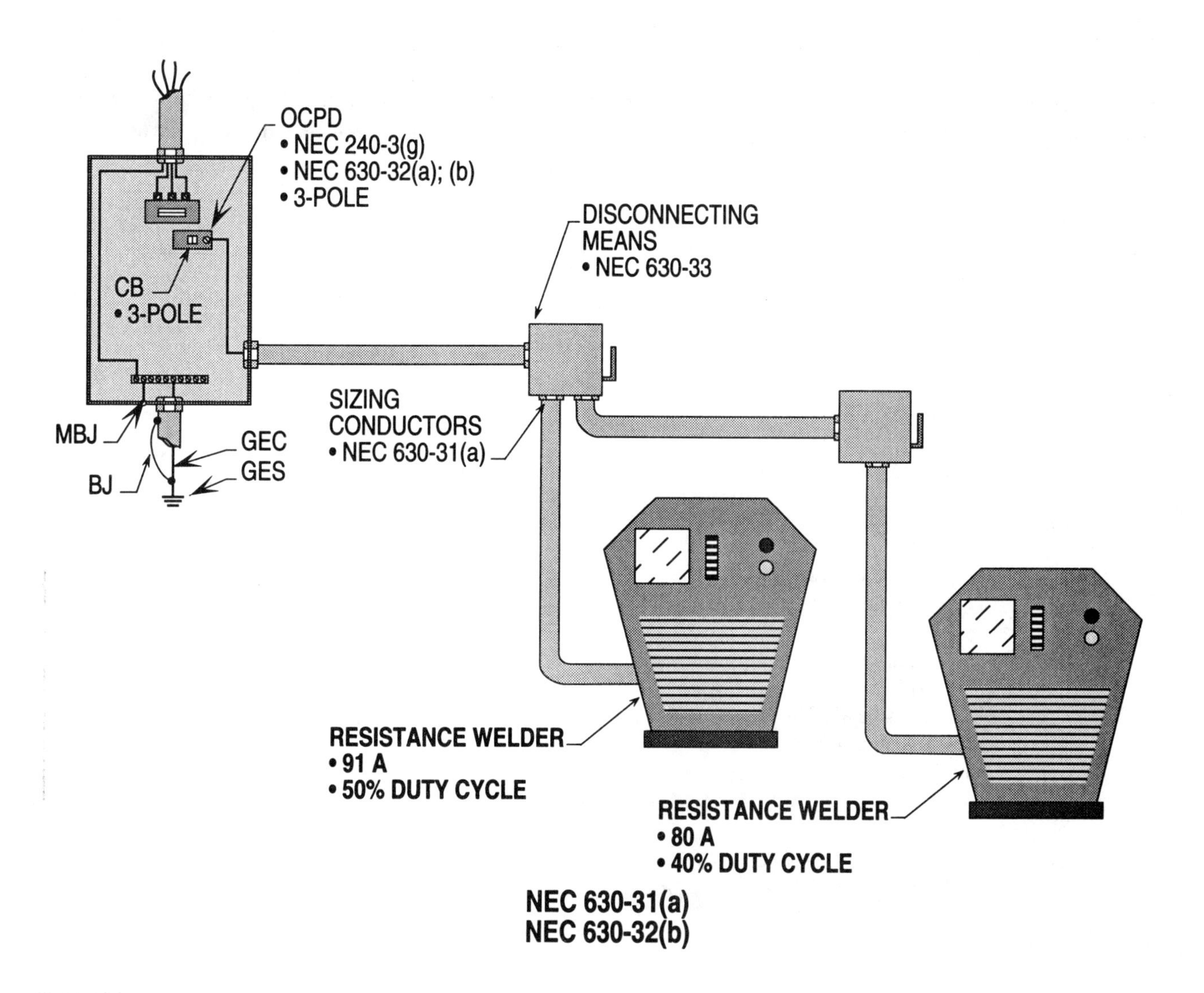
OCPD
• NEC 240-3(g)
• NEC 630-32(a); (b)
• 3-POLE
DISCONNECTING MEANS
• NEC 630-33
CB
• 3-POLE
SIZING CONDUCTORS
• NEC 630-31(a)
MBJ
GEC
GES
BJ
RESISTANCE WELDER
• 91 A
• 50% DUTY CYCLE
RESISTANCE WELDER
• 80 A
• 40% DUTY CYCLE
NEC 630-31(a)
NEC 630-32(b)

Ill. 22-3(b)

Data Processing Equipment
645-5(a)

The OCPD's and conductors supplying power to computer and data processing equipment are sized and based upon continuous and noncontinuous operation. See AHJ for permission to apply demand factors due to operation and use.

Example 22-4. What size OCPD and THWN copper conductors are required for the branch-circuit and feeder-circuit in Illustration 22-4?

Sizing conductors and OCPD for branch-circuit

Step 1: Finding A for branch-circuit (BC)
645-5(a)
BC = 23.5 A

Step 2: Calculating A for BC
645-5(a)
23.5 A x 125% = 29.4 A

Step 3: Selecting conductors and OCPD for BC
310-10(2); 240-3(d); Table 310-16
240-3(b); 240-6(a)
29.4 A requires #10 cu.
29.4 A requires 30 A OCPD

Solution: The size OCPD is 30 amps and the size THWN copper conductors are No. 10.

Sizing conductors and OCPD for feeder-circuit

Step 1: Finding A for feeder-circuit (FC)
FC = 115 A

Step 2: Calculating A for FC
110-3(b); 215-2(a); 215-3;
220-10
115 A x 125% = 143.8 A

Step 3: Selecting conductors and OCPD for FC
310-10(2); Table 310-16;
240-3(b); 240-6(a)
143.8 A requires #1/0 cu.
143.8 A requires 150 A OCPD

Solution: The size OCPD is 150 amps and the THWN copper conductors are No. 1/0.

Note: The AHJ may allow a demand to be applied based upon operation.

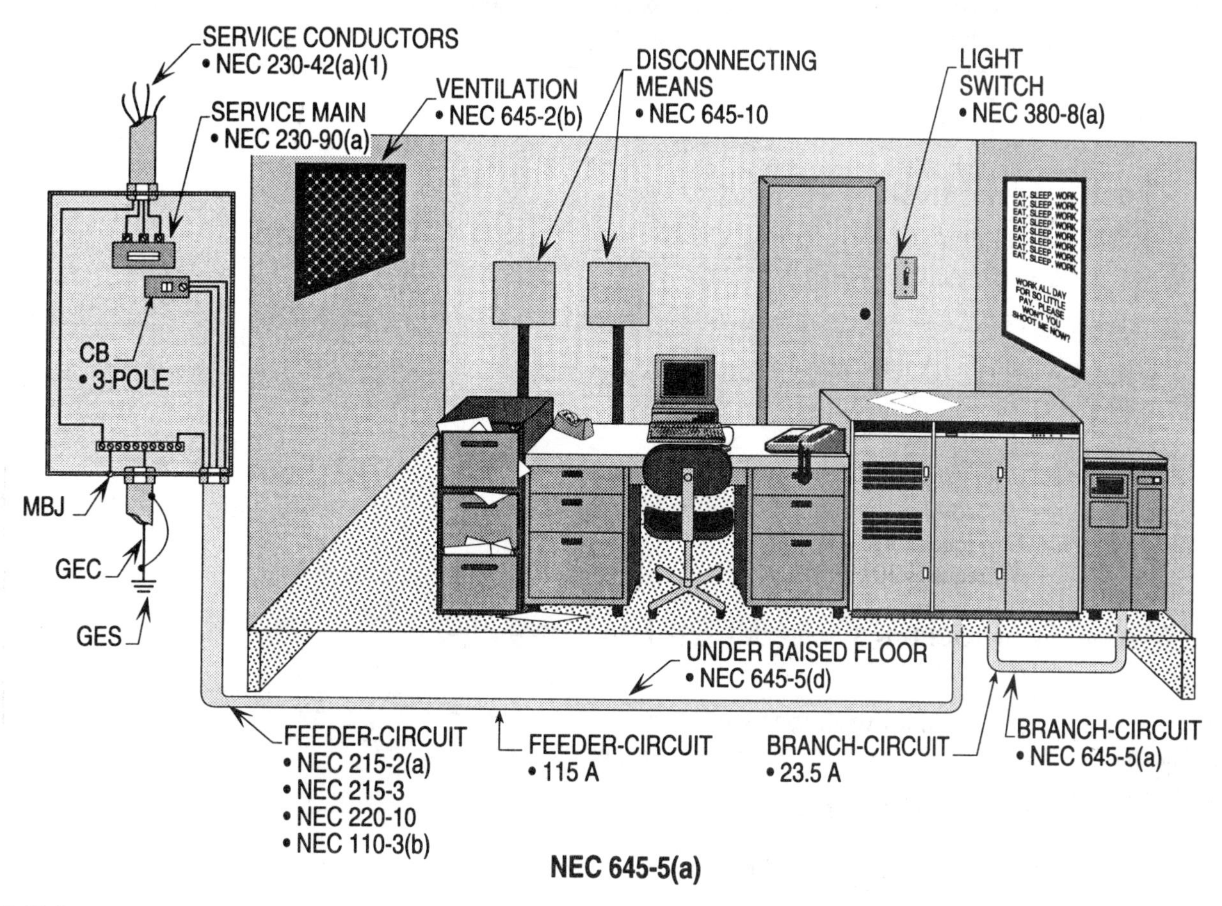

NEC 645-5(a)

Ill. 22-4

X-ray Equipment (Sizing Disconnect) 660-5

The elements of circuits supplying power to X-ray equipment are sized and based upon short or long time operation or manufacturer nameplate rating. For short time rating use 50 percent and for long time rating use 100 percent times the input amps of each unit.

Example 22-5(a). What size disconnecting means is required to disconnect the X-ray equipment in Illustration 22-5(a)? (long time rating)

Sizing disconnect for branch-circuit

Step 1: Finding A for disconnect
660-5
A = MA / 1,000 x Sec. V / Pri. V
A = 25 / 1,000 x 200,000 V / 208 V
A = 24 A

Step 2: Calculating A for disconnect
660-5
24 A x 100% = 24 A

Step 3: Selecting disconnect for BC
660-5; 240-6(a)
24 A requires 25 A CB
24 A requires 30 A switch

Solution: The size disconnecting means for the branch-circuit (BC) is 30 amp CB or 30 amp disconnect switch.

Example 22-5(b). What size disconnecting means is required to disconnect the X-ray equipment in Illustration 22-5(b)? (momentary rating)

Sizing Disconnecting means for branch-circuit

Step 1: Finding A for disconnect
660-5
A = MA / 1,000 x Sec. V / Pri. V
A = 200 / 1,000 x 100,000 V / 208 V
A = 96 A

Step 2: Calculating A for disconnect
660-5
96 A x 50% = 48 A

Step 3: Selecting disconnect for BC
660-5; 240-6(a)
48 A requires 50 A CB
48 A requires 60 A switch

Solution: The size disconnecting means for the branch-circuit (BC) is 50 amps CB or 60 amp disconnect switch.

Note: Momentary rating is sometimes referred to as short time rating.

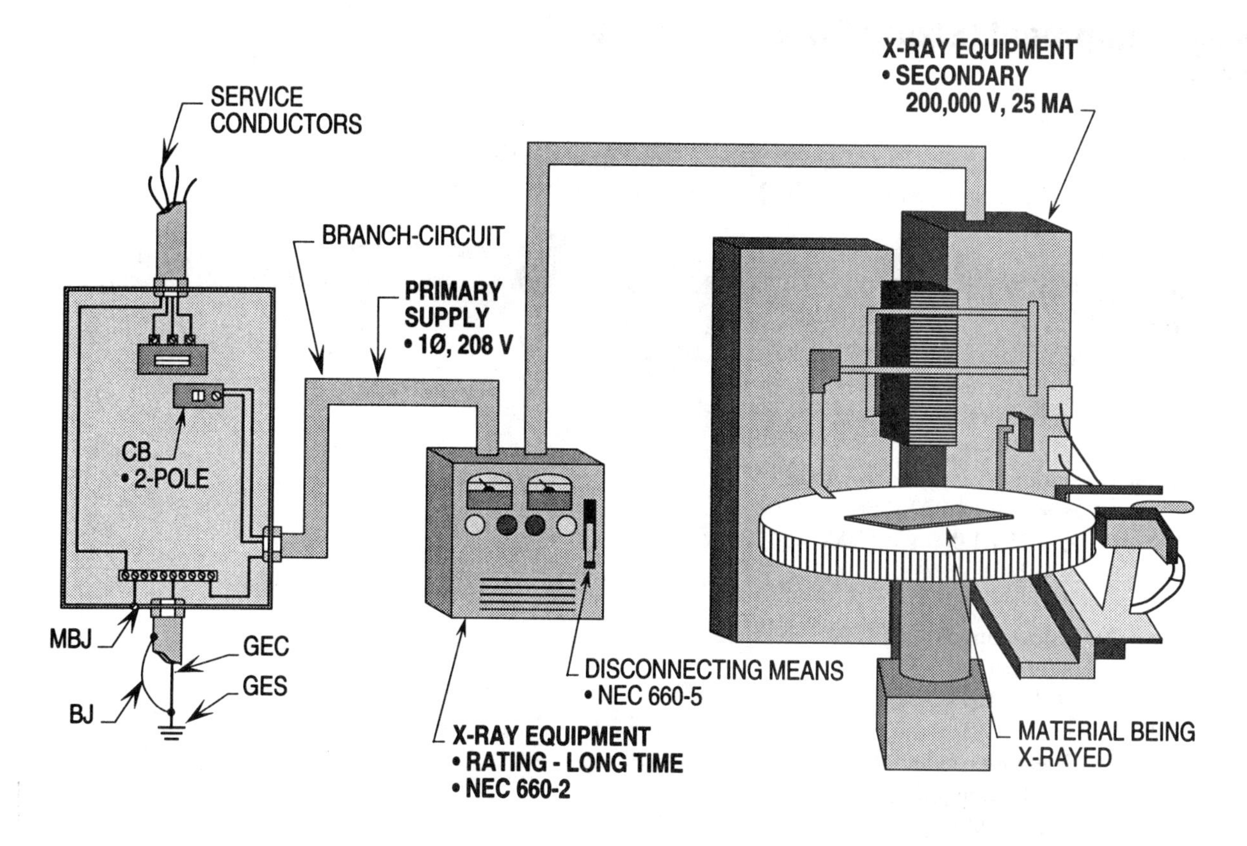

Ill. 22-5(a)

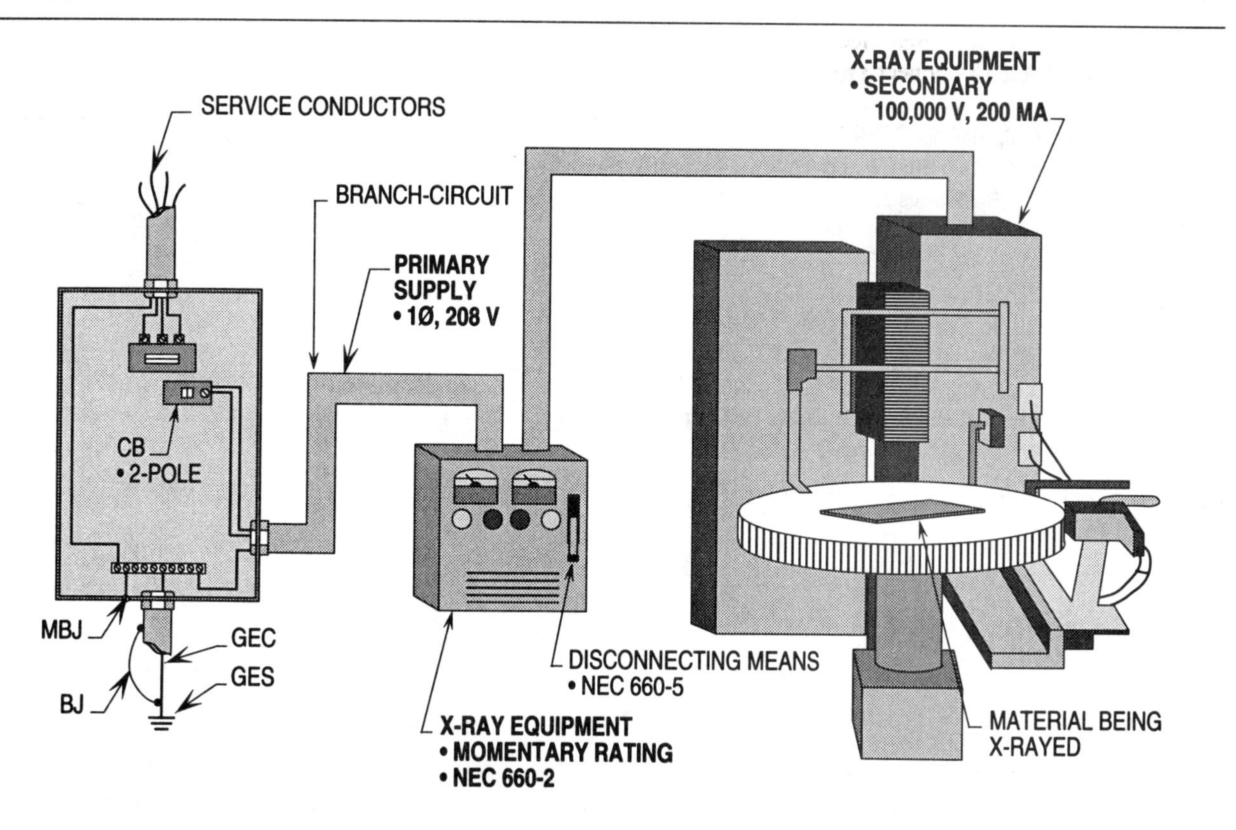

Ill. 22-5(b)

X-ray Equipment (Sizing Conductors And OCPD's) 660-6(a)

The OCPD's and conductors serving power to X-ray equipment are sized and selected at 100 percent of the unit's long time rating. **Note:** The long time rating of the X-ray unit is greater in amps than the momentary rating.

Example 22-6(a). What size OCPD and THWN copper conductors are required for the branch-circuit in Illustration 22-6(a)?

Sizing conductors for branch-circuit

Step 1: Finding A for branch-circuit (BC)
A = MA / 1,000 x Sec. V / Pri. V
A = 25 / 1,000 x 200,000 V / 208 V
A = 24 A

Step 2: Calculating A for BC
660-6(a)
24 A x 100% = 24 A

Step 3: Selecting conductors for BC
310-10(2); Table 310-16; 240-3(d)
24 A requires #10 cu.

Solution: The size THWN copper conductors are No. 10.

Sizing OCPD for branch-circuit

Step 1: Calculating OCPD for BC
660-6(a); Step 2 above
24 A requires 30 A
(Based on conductors)

Step 2: Selecting OCPD for BC
Table 310-16; 240-3(d); 240-6(a)
30 A (#10 cu.) requires 30 A OCPD

Solution: The OCPD for the BC is 30 amps.

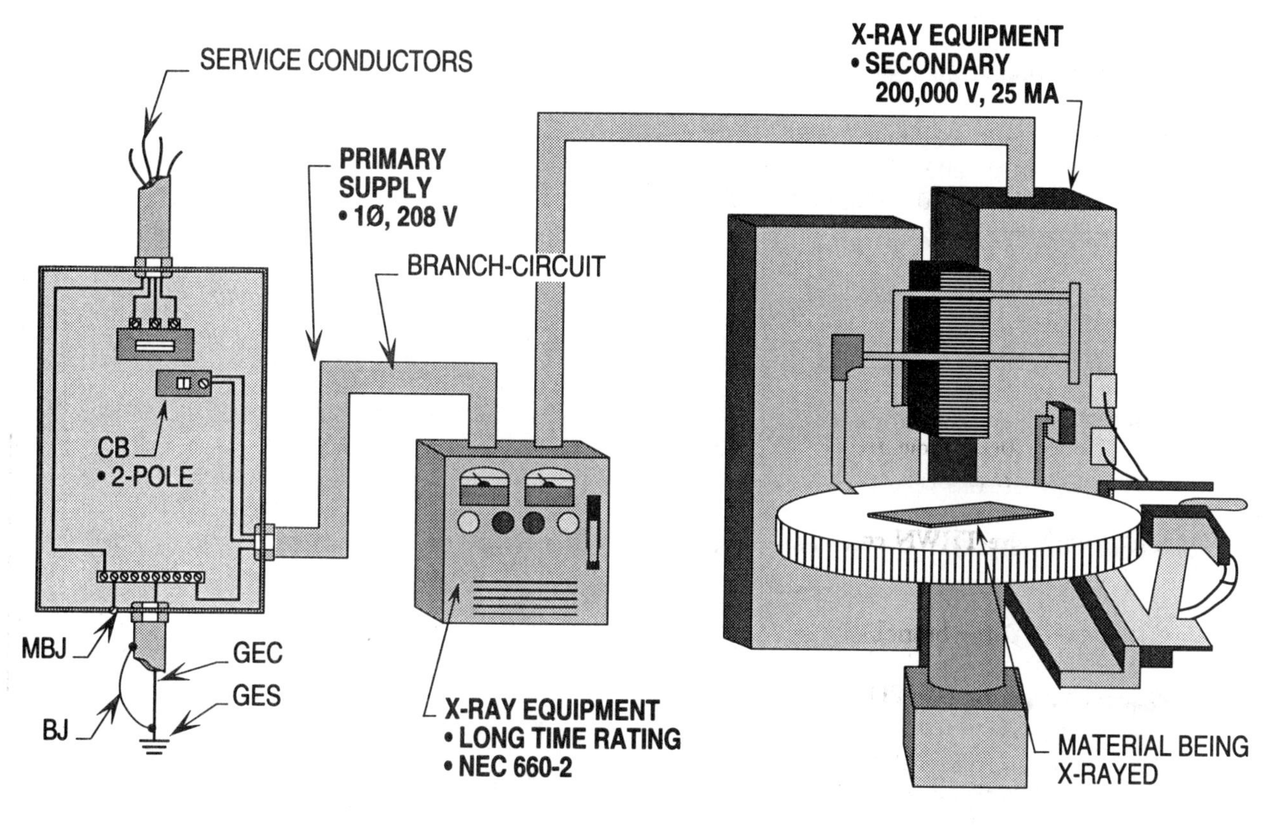
SERVICE CONDUCTORS
PRIMARY SUPPLY
• 1Ø, 208 V
BRANCH-CIRCUIT
X-RAY EQUIPMENT
• SECONDARY
200,000 V, 25 MA
CB
• 2-POLE
MBJ
GEC
GES
BJ
X-RAY EQUIPMENT
• LONG TIME RATING
• NEC 660-2
MATERIAL BEING
X-RAYED
NEC 660-6(a)

Ill. 22-6(a)

X-ray Equipment (Sizing Conductors And OCPD's) 660-6(a)

The OCPD's and conductors serving power to X-ray equipment are sized and selected at 50 percent of the unit's momentary (short time) rating. **Note:** The momentary rating is greater in amps than the long time rating.

Example 22-6(b). What size OCPD and THWN copper conductors are required for the branch-circuit in Illustration 22-6(b)?

Sizing conductors for branch-circuit

Step 1: Finding A for branch-circuit (BC)
A = MA / 1,000 x Sec. V / Pri. V
A = 200 / 1,000 x 100,000 V / 208 V
A = 96 A

Step 2: Calculating A for BC
660-6(a)
96 A x 50% = 48 A

Step 3: Selecting conductors for BC
310-10(2); Table 310-16
48 A requires #8 cu.

Solution: The size THWN copper conductors for the BC is No. 8.

Sizing OCPD for branch-circuit

Step 1: Calculating A for BC
660-6(a)
96 A x 50% = 48 A

Step 2: Selecting OCPD for BC
240-3(b); 240-6(a)
48 A requires 50 A OCPD

Solution: The size OCPD for the BC is 50 amps.

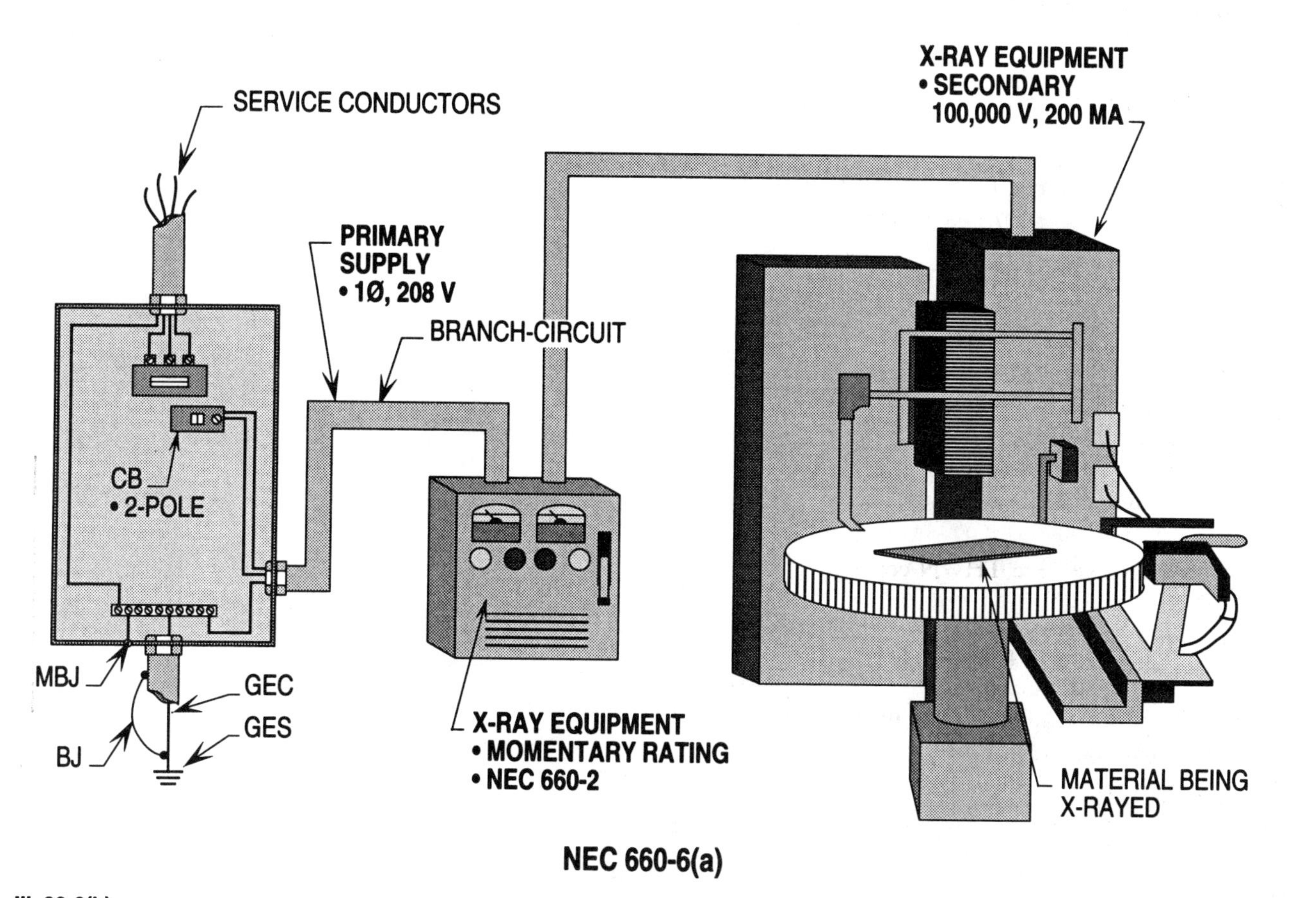
SERVICE CONDUCTORS
PRIMARY SUPPLY
• 1Ø, 208 V
BRANCH-CIRCUIT
X-RAY EQUIPMENT
• SECONDARY
100,000 V, 200 MA
CB
• 2-POLE
MBJ
GEC
GES
BJ
X-RAY EQUIPMENT
• MOMENTARY RATING
• NEC 660-2
MATERIAL BEING X-RAYED
NEC 660-6(a)

Ill. 22-6(b)

X-ray Equipment (Sizing Feeder-circuits) 660-6(b)

The procedure for sizing OCPD's and conductors for feeder-circuits are sized and selected, based upon 50 percent of the two largest units plus 20 percent of the remaining units calculated per NEC 660-6(b). **Note:** All units have momentary ratings.

Example 22-7(a). What size OCPD and THWN copper conductors are required for the feeder-circuit in Illustration 22-7(a)?

Sizing conductors for the feeder-circuit

Step 1: Finding A for feeder-circuit (FC)
660-6(b) Largest units
A = MA / 1,000 x Sec. V / Pri. V
A = 150 / 1,000 x 150,000 V / 208 V
A = 108 A
Other units
A = MA / 1,000 x Sec. V / Pri. V
A = 30 / 1,000 x 125,000 V / 208 V
A = 18 A

Step 2: Calculating A for FC
660-6(b); 660-6(a)
Two largest units
108 A x 2 x 50% x 100% = 108 A
Remaining units
18 A x 3 x 20% = 10.8 A
Total amps = 118.8 A

Step 3: Selecting conductors for FC
310-10(2); Table 310-16
118.8 A requires #1 cu.

Solution: The size THWN copper conductors for the FC is No. 1.

Sizing OCPD for feeder-circuit

Step 1: Selecting OCPD based on Step 2 and 3 above
240-3(b); 240-6(a); Table 310-16
Total A of 118.8 A requires 125 A OCPD
#1 THWN cu. (130 A) allows 150 A OCPD

Solution: The size OCPD for the FC is 125 amps or a 150 amps.

Note: Check with AHJ when installing 150 amp OCPD to protect feeder-circuit conductors and X-ray equipment.

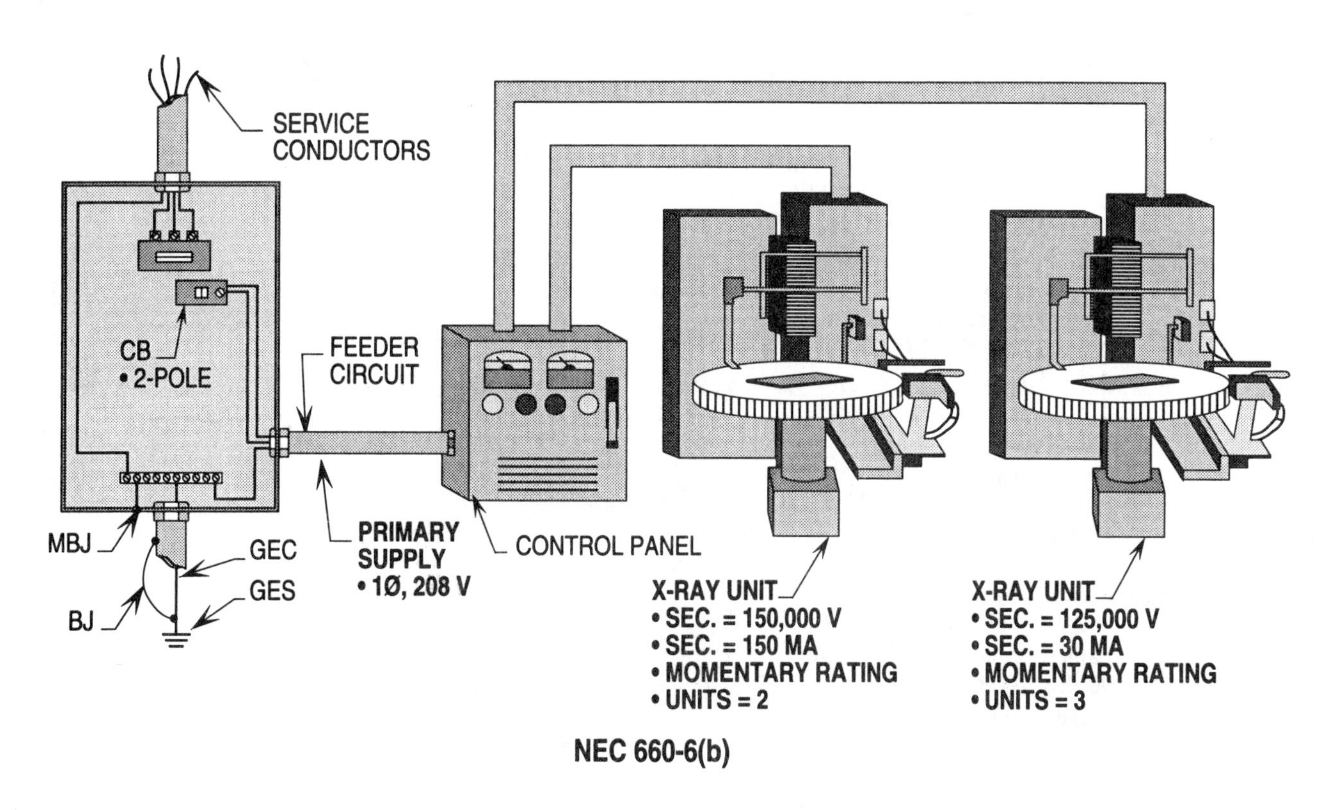
SERVICE CONDUCTORS
CB
• 2-POLE
FEEDER CIRCUIT
MBJ
GEC
GES
BJ
PRIMARY SUPPLY
• 1Ø, 208 V
CONTROL PANEL
X-RAY UNIT
• SEC. = 150,000 V
• SEC. = 150 MA
• MOMENTARY RATING
• UNITS = 2
X-RAY UNIT
• SEC. = 125,000 V
• SEC. = 30 MA
• MOMENTARY RATING
• UNITS = 3
NEC 660-6(b)

Ill. 22-7(a)

X-ray Equipment (Sizing Feeder-circuits) 660-6(b)

The procedure for sizing OCPD's and conductors for feeder-circuits are sized and selected, based upon 100 percent of the two largest units plus 20 percent of the remaining units calculated per NEC 660-6(b). **Note:** All units do not have long time ratings.

Example 22-7(b). What size OCPD and THWN copper conductors are required for the feeder-circuit (FC) in Illustration 22-7(b)?

Sizing conductors for the feeder-circuit

Step 1: Finding A for feeder-circuit (FC)
660-6(b)
Longtime rating
A = MA / 1,000 x Sec. V / Pri. V
A = 125 / 1,000 x 150,000 V / 208 V
A = 90 A

Largest units
Momentary rating of largest units
A = 40 / 1,000 x 175,000 V / 208 V
A = 33.7 A

Other units
Momentary rating of remaining units
A = 35 / 1,000 x 150,000 V / 208 V
A = 25 A

Step 2: Calculating A for FC
660-6(b); 660-6(a)

Largest units (LTR) = 90 A x 100%	= 90 A
Largest units (MR) = 33.7 A x 100%	= 33.7 A
Remaining units (MR) 25 A x 2 x 20%	= 10 A
Total amps	= 133.7 A

Step 3: Selecting conductors for FC
310-10(2); Table 310-16
133.7 A requires #1/0 cu.

Solution: The size THWN copper conductors for the FC is No. 1/0 cu.

Sizing OCPD for the feeder-circuit

Step 1: Selecting OCPD based on
Step 2 and 3 above
240-3(b); 240-6(a); Table 310-16
133.7 A requires 150 A OCPD

Solution: The size OCPD for the FC is 150 amps based upon load or amps for conductors.

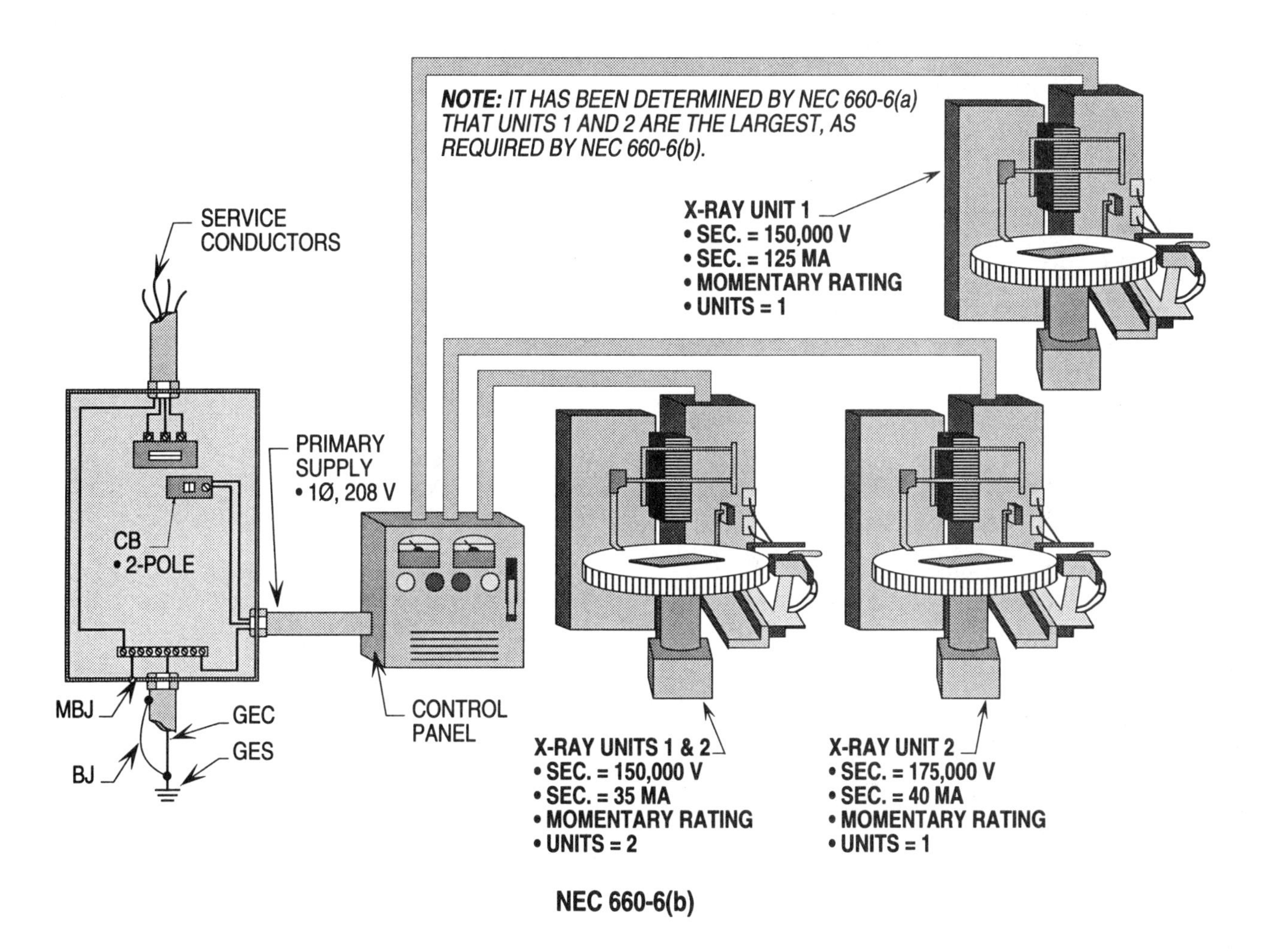

NOTE: IT HAS BEEN DETERMINED BY NEC 660-6(a) THAT UNITS 1 AND 2 ARE THE LARGEST, AS REQUIRED BY NEC 660-6(b).
X-RAY UNIT 1
• SEC. = 150,000 V
• SEC. = 125 MA
• MOMENTARY RATING
• UNITS = 1
SERVICE CONDUCTORS
PRIMARY SUPPLY
• 1Ø, 208 V
CB
• 2-POLE
MBJ
GEC
GES
BJ
CONTROL PANEL
X-RAY UNITS 1 & 2
• SEC. = 150,000 V
• SEC. = 35 MA
• MOMENTARY RATING
• UNITS = 2
X-RAY UNIT 2
• SEC. = 175,000 V
• SEC. = 40 MA
• MOMENTARY RATING
• UNITS = 1
NEC 660-6(b)

Ill. 22-7(b)

23

Electrolytic Cells, Electroplating, and Industrial Machinery

The area around or near electrolytic cells which is considered the workplace for personnel serving such systems has to be insulated and isolated. In other words, it is the workplace that must be insulated at all times instead of the exposed conductors or surfaces of the electrical equipment. However, as a corollary to this concept, to maintain process safety and continuity, the overheating of conductors, overloading of motors, leakage currents, etc. may be required in the cells. Article 668 of the National Electrical Code® (NEC®) must be reviewed and applied to provide such safety and reliability of these electrical systems.

The provisions of Article 669 of the NEC shall be applied to all systems that supply power to electrical components and accessory equipment used for controls in electroplating, anodizing, electropolishing and electrostripping process.

The electrical elements and accessories for industrial machinery shall comply with all the provisions listed in Article 670 of the NEC. Such components shall be calculated at noncontinuous or continuous operation or a duty cycle service is applied.

Quick Reference

Electrolytic Cells (Sizing Conductors) 668-12(b)

The elements of circuits supplying power to electrolytic cells are sized 125 percent times the amps of the cells.

Example 23-1. What size OCPD and THWN copper conductors are required to supply the circuit in Illustration 23-1?

Sizing conductors for the cells

Step 1: Finding A for cells
Load = 75 A

Step 2: Calculating A for conductors
384-16(d); 210-19(a); 20(a); Art. 100
75 A x 125% = 93.8 A

Step 3: Selecting conductors for cell
310-10(2); 668-12(b); Table 310-16
93.8 A requires #3 cu.

Solution: **The size THWN copper conductors to the cells is No. 3.**

Sizing OCPD for the cells

Step 1: Selecting the OCPD for conductors
240-3(b); 240-6(a); Step 2
93.8 A requires 100 A OCPD

Solution: **The size OCPD for the cell conductors is 100 amps under normal working conditions.**

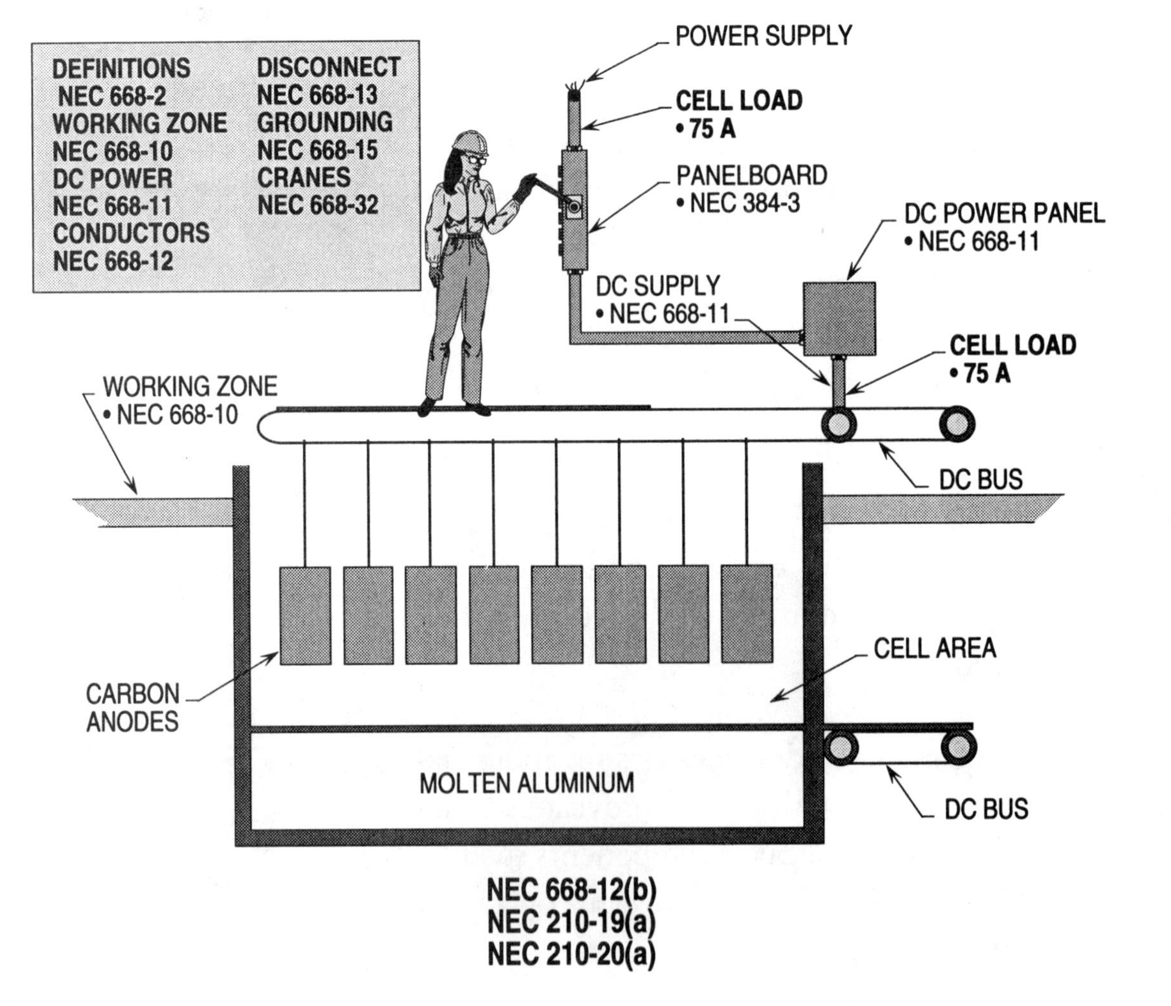

Ill. 23-1

Electroplating Process (Sizing Conductors and OCPD's) 669-5

The elements of circuits supplying power to electroplating process units are sized at 125 percent times the amps of the process units.

Example 23-2. What size OCPD and THWN copper conductors are required to supply the circuit in Illustration 23-2?

Sizing conductors for electroplating process

Step 1: Finding A for electroplating
669-5
Load = 96 A

Step 2: Calculating A for conductors
669-5
96 A x 125% = 120 A

Step 3: Selecting conductor for electroplating
310-10(2); Table 310-16
120 A requires #1 cu.

Solution: The size THWN copper conductors are No. 1.

Sizing OCPD for electroplating process

Step 1: Selecting OCPD for electroplating
669-5; 240-3(b); 240-6(a);
Steps 2 and 3
120 A requires 125 A OCPD
130 A (#1 THWN cu.) allows
150 A OCPD

Solution: The size OCPD for the electroplating process is 125 amps or 150 amps if necessary due to operation.

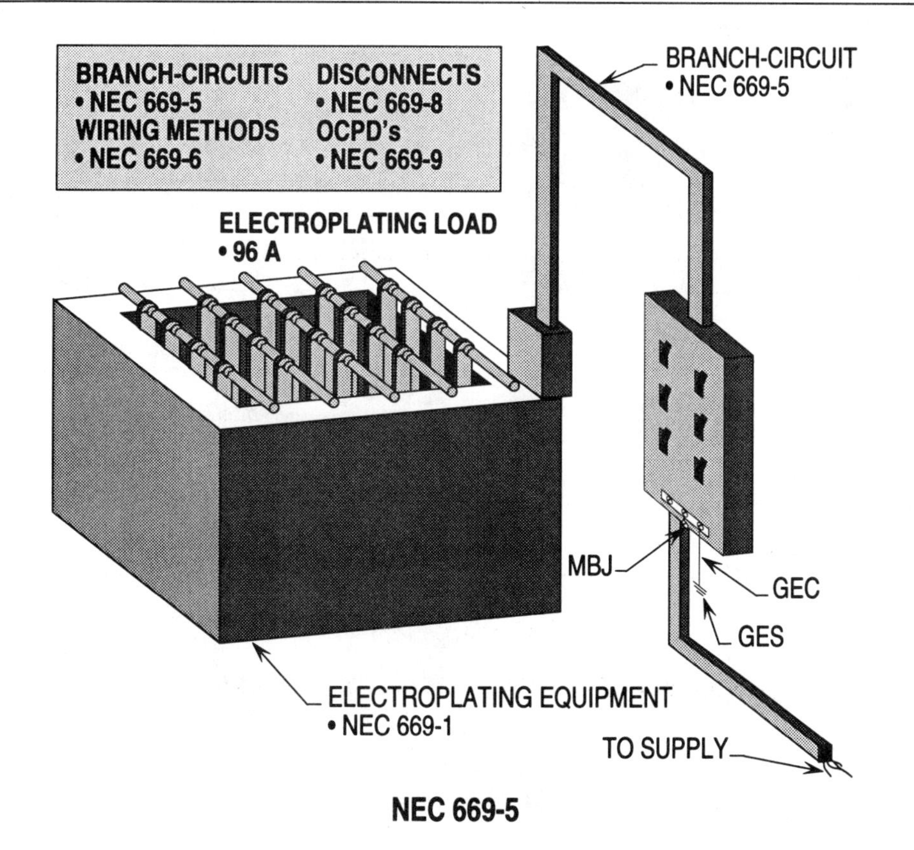

Ill. 23-2

Electropating Process (Sizing Bus Bars)
669-5

The continuous current carried in bare copper bars in electroplating process units must not exceed 1,000 amps per square inch of cross-section of the conductor. Use 700 amps per inch of cross-section of conductor for aluminum.

Example 23-3. What size copper bus bars are required for the electroplating process in Illustration 23-3?

Sizing bus bars for the electroplating process

Step 1: Finding A for electroplating
669-5
Load = 96 A

Step 2: Calculating A for bus bars
669-5; 374-6
1/8" x 1" x 1,000 A = 125 A

Solution: **A bus bar of 125 amps per phase will carry the 96 amp load.**

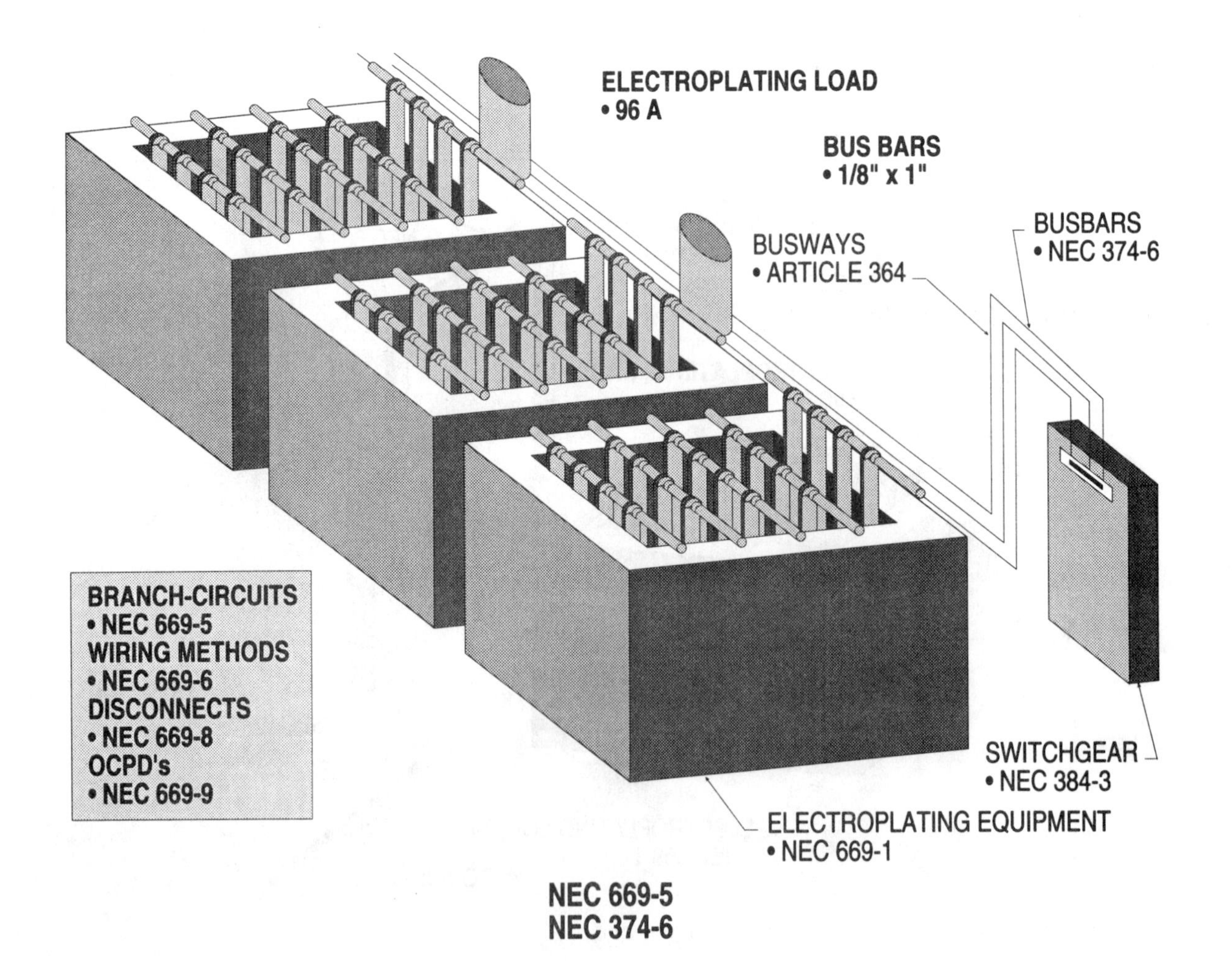

Ill. 23-3

Industrial Machinery (Sizing Conductors and OCPD's) 670-4(a); (b)

The elements of circuits supplying power to a motor used in industrial machinery are sized and selected as any equipment having a motor included in their operation.

Example 23-4. What size OCPD and THWN copper conductors are required to supply the machine in Illustration 23-4?

Sizing conductors for industrial machine

Step 1: Finding A of motor
670-4(a); 430-6(a)(1); Table 430-150
5 HP = 16.7 A

Step 2: Calculating A for conductors
670-4(a); 430-22(a)
16.7 A x 125% = 20.9 A

Step 3: Selecting conductors for motor
310-10(2); 240-3(d)
20.9 A requires #12 cu.

Solution: **The size THWN copper conductors are No. 12.**

Sizing OCPD for motor

Step 1: Calculating A for OCPD using CB
670-4(b); 430-52(c)(1), Ex. 1;
Table 430-152
16.7 A x 250% = 41.8 A

Step 2: Selecting OCPD for motor
240-3(g); 240-6(a); 430-52(c)(1), Ex. 1
41.8 A allows 45 A OCPD

Solution: **The size OCPD using a CB is 45 amps.**

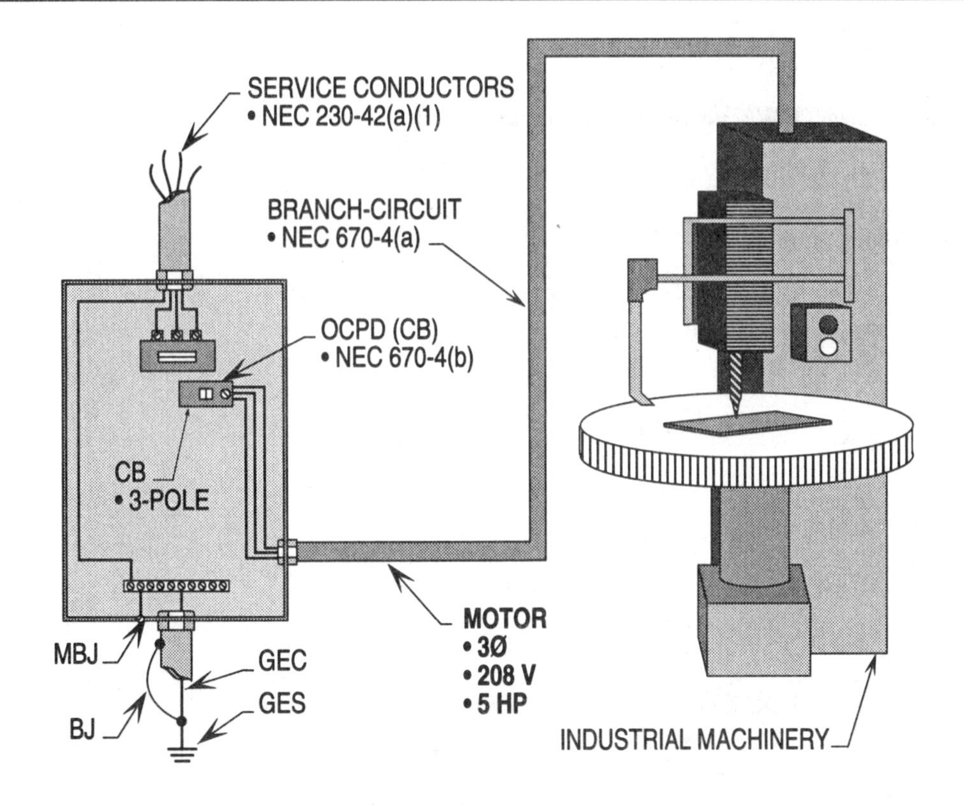

NEC 670-4(a); (b)

Ill. 23-4

Industrial Machinery (Sizing Conductors and OCPD's) 670-4(a); (b)

The elements of circuits supplying power to industrial machinery utilizing motors and heating elements must have the two largest motors multiplied by 125 percent and their total added to the other motors plus 125 percent of the heating element load.

Example 23-5. What size OCPD and THWN copper conductors are required to supply the machine in Illustration 23-5?

Sizing conductors for machine

Step 1: Finding A for machine
670-4(a); 430-6(a)(1); Table 430-150
Heating element = 10 kW x
1,000 / 208 V x 1.732

Heating element	= 27.8 A
Motor 1 - 5 HP	= 16.7 A
Motor 2 - 3 HP	= 10.6 A
Motor 3 - 2 HP	= 7.5 A

Step 2: Calculating A for conductors
670-4(a)

27.8 A x 125%	= 34.75 A
16.7 A x 125%	= 20.88 A
10.6 A x 100%	= 10.6 A
7.5 A x 100%	= 7.5 A
Total A	= 73.73 A

Step 3: Selecting conductor for machine
310-10(2); Table 310-16
73.73 A requires #4 cu.

Solution: The size THWN copper conductors are No. 4.

Sizing OCPD for conductors

Step 1: Finding A for OCPD
670-4(b); 430-6(a)(1); Table 430-150
Steps 1 & 2 above

Heating element load	= 34.75 A
Largest OCPD (Mt. 1)	= 40 A
Motor 2	= 10.6 A
Motor 3	= 7.5 A
Total A	= 92.85 A

Step 2: Selecting OCPD for conductors
670-4(b); 240-3(g); 240-6(a)
92.85 A requires 90 A OCPD
85 A (#4 THWN cu.) requires 90 A OCPD

Solution: The size OCPD for the conductors to the machine is 90 amps.

Note: The 40 amp OCPD for the largest motor was calculated as follows:
- Table 450-152

16.7 A x 250% = 41.75 A
- 430-52(c)(1); 240-6(a)

41.25 A allows a 40 A OCPD

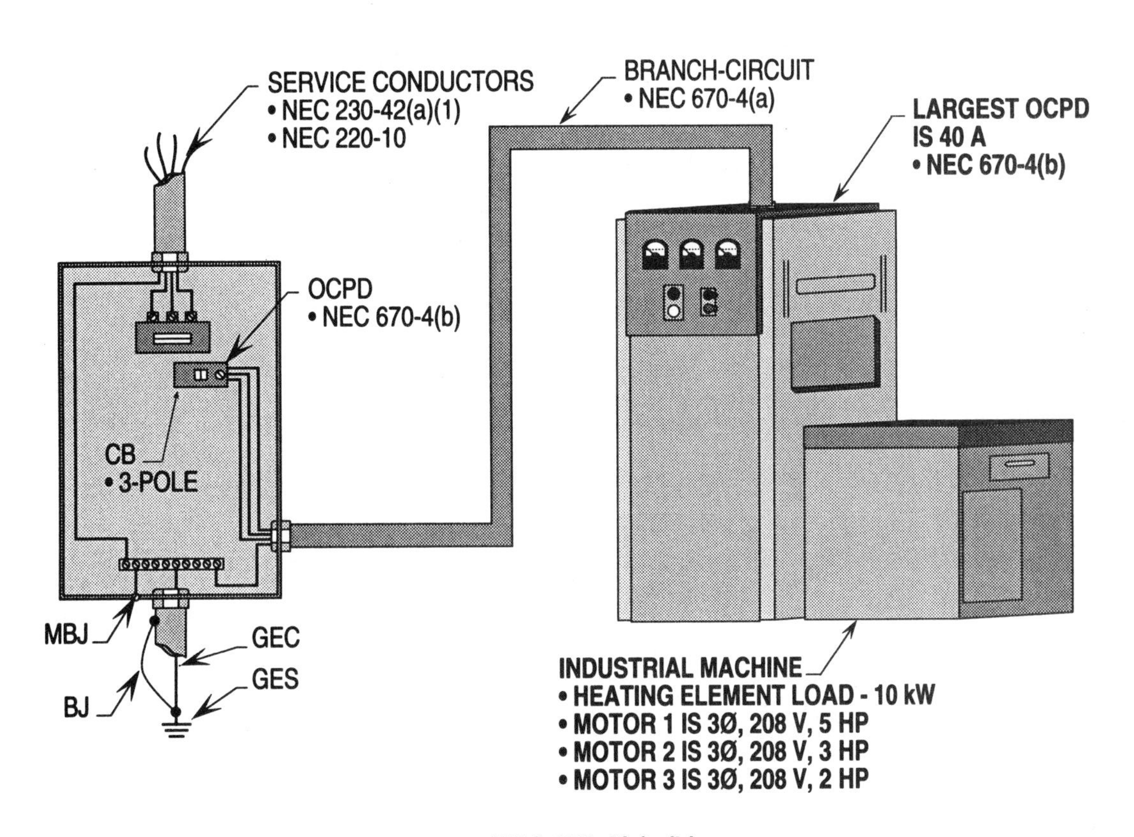

NEC 670-4(a); (b)

Ill. 23-5

24 Irrigation Machines, Swimming Pools, and Solar Photovoltaic Systems

Overcurrent protection devices (OCPD's), conductors, and controllers used to supply power to electrically driven or controlled irrigation machines shall be sized and selected basically as motors in Article 430 of the National Electrical Code® (NEC®). The provisions listed in Article 675 of the NEC are in addition to, or amendatory of, the requirements found in Article 430.

The OCPD's and conductors utilized to serve heating elements and accessories for swimming pool equipment are computed at continuous operation during use to prevent overheating components and tripping open the circuits. When sizing and selecting these elements, the heating elements and other loads must be evaluated and computed together to ensure safe operating electrical systems.

The electrical components that make up solar photovoltaic systems are required to be calculated with extreme caution. All components shall be capable of interrupting the amount of current they may be called to clear during a short-circuit condition. Article 690 of the NEC must be reviewed very carefully when designing such systems.

Quick Reference

Irrigation Machines (Sizing Conductors and OCPD's) 675-7; 675-7(a)

The elements of circuits supplying power to motors operating irrigation machines are sized as motors used on other types of equipment per NEC 430-24 and 430-62(a).

Example 24-1. What size OCPD and THWN copper conductors are required for the circuit in Illustration 24-1?

Sizing conductors for irrigation machine

Step 1: Finding FLA of motors
675-7; 675-7(a) Table 430-150
10 HP = 14 A
7 1/2 HP = 11 A
5 HP = 7.6 A

Step 2: Finding LRA of motors
675-7(b); Table 430-151(B)
10 HP = 81 A x 100% = 81 A
7 1/2 HP = 63.5 A x 100% = 63.5 A
5 HP = 7.6 A x 100% = 7.6 A
Total LRC =152.1 A

Step 3: Calculating A for conductors
675-7; 675-7(a); 430-24
14 A x 125% = 17.5 A
11 A x 100% = 11 A
7.6 A x 100% = 7.6 A
Total A = 36.1 A

Step 4: Selecting conductors for motors
310-10(2); Table 310-16
36.1 A requires #8 cu.

Solution: The size THWN copper conductors are No. 8.

Sizing OCPD for conductors

Step 1: Calculating A for OCPD
675-7; 675-7(a); Table 430-152
14 A x 175% = 20 A
• 430-62(a) + 11 A
(24.5 A = 20 A) + 7.6 A
Total = 38.6 A

Step 2: Selecting OCPD for motors
240-3(g); 240-6(a); 430-62(a)
38.6 A requires 40 A

Solution: The size time delay fuses are 40 amps.

Note 1: The OCPD was selected by rounding up, per NEC 675-22(b).

Note 2: A 40 amp TDF holds 200 amps (40 A x 5 = 200 A) for 10 seconds which is more than enough to start and run all three motors.

Note 3: The AHJ may allow the 24.5 amps to be rounded up to 25 amps per Examples, Appendix D.

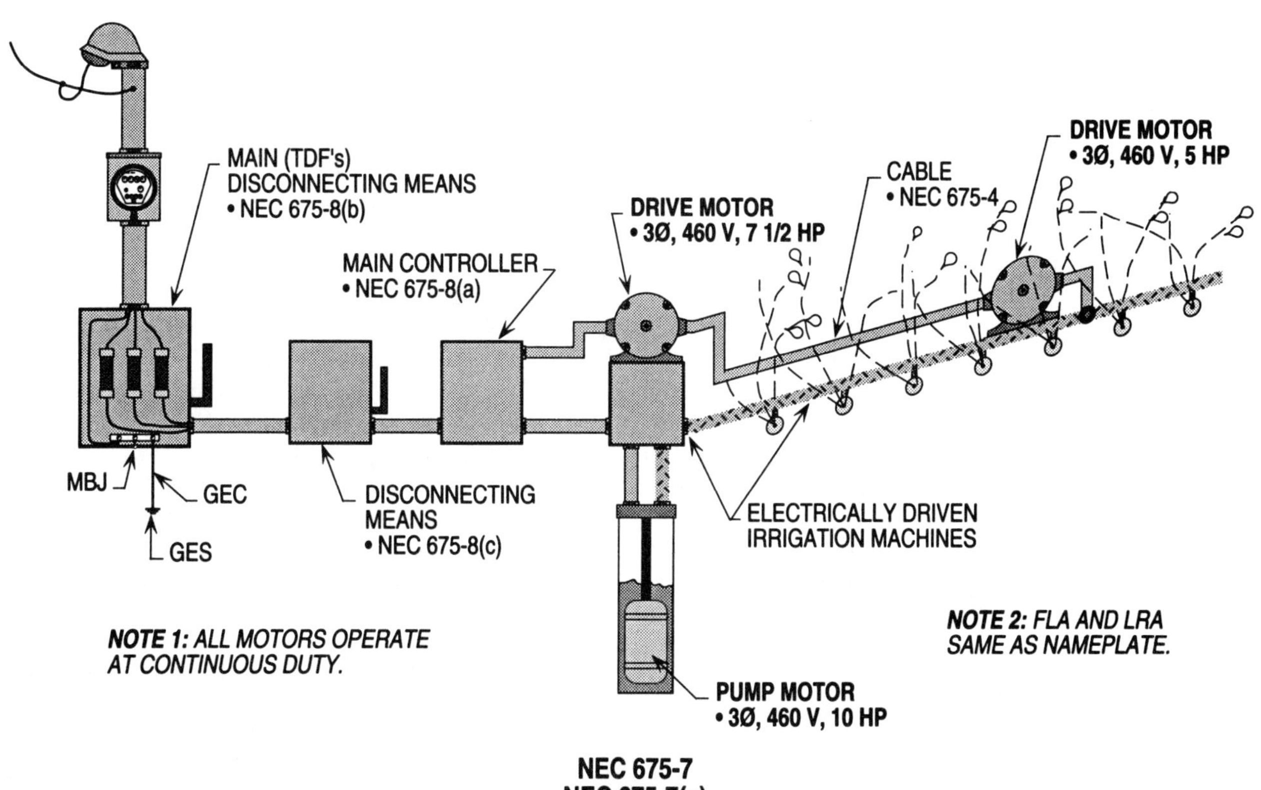

MAIN (TDF's) DISCONNECTING MEANS
• NEC 675-8(b)
MAIN CONTROLLER
• NEC 675-8(a)
DRIVE MOTOR
• 3Ø, 460 V, 7 1/2 HP
CABLE
• NEC 675-4
DRIVE MOTOR
• 3Ø, 460 V, 5 HP
MBJ
GEC
GES
DISCONNECTING MEANS
• NEC 675-8(c)
ELECTRICALLY DRIVEN IRRIGATION MACHINES
NOTE 1: ALL MOTORS OPERATE AT CONTINUOUS DUTY.
NOTE 2: FLA AND LRA SAME AS NAMEPLATE.
PUMP MOTOR
• 3Ø, 460 V, 10 HP
NEC 675-7
NEC 675-7(a)
NEC 430-24

Ill. 24-1

Irrigation Machines (Sizing Conductors and OCPD's) 675-7(a)

Irrigation machines that are equipped with motors that operate at continuous duty and with duty cycles shall have their elements sized per NEC 430-22(b) and Table 430-22(b).

Example 24-2. What size minimum OCPD and THWN copper conductors are required for the circuit in Illustration 24-2?

Sizing conductors for irrigation machine

Step 1: Finding FLA of motors
675-7; 675-7(a); Table 430-150
10 HP = 14 A
7 1/2 HP = 11 A
5 HP = 7.6 A

Step 2: Finding LRA of motors
675-7(b); Table 430-151(B)
10 HP = 81 A x 100% = 81 A
7 1/2 HP = 63.5 A x 100% = 63.5 A
5 HP = 7.6 A x 100% = 7.6 A
Total LRC =152.1 A

Step 3: Calculating A for conductors
675-7(a); Table 430-22(b)
14 A x 125% = 17.5 A
11 A x 85% = 9.35 A
7.6 A x 85% = 6.46 A
Total A = 33.31 A

Step 4: Selecting conductors for motors
310-10(2); Table 310-16
33.31 A requires #10 cu.

Solution: The size THWN copper conductors are No. 10.

Sizing OCPD for conductors

Step 1: Selecting OCPD for conductors
675-7(a); 240-6(a); Step 3
33.31 A requires 35 A OCPD

Solution: The size OCPD using TDF's for the circuit is 35 amps.

Note 1: A 35 amp TDF holds 175 amps (35 A x 5 = 175 A) for 10 seconds which is greater than the LRC of 152.1 amps.

Note 2: Selecting OCPD per NEC 430-52(c)(1), Ex. 1; Table 430-152
17.5 A x 175% = 30.6 = 35.0 A
other motors + 9.35 A
+ 6.46 A
Total amps = 50.81 A
50.81 amps requires 50 amp OCPD.

Note 3: The Ex. 1 to Sec 430-52(c)(1) is applied to select the largest OCPD.

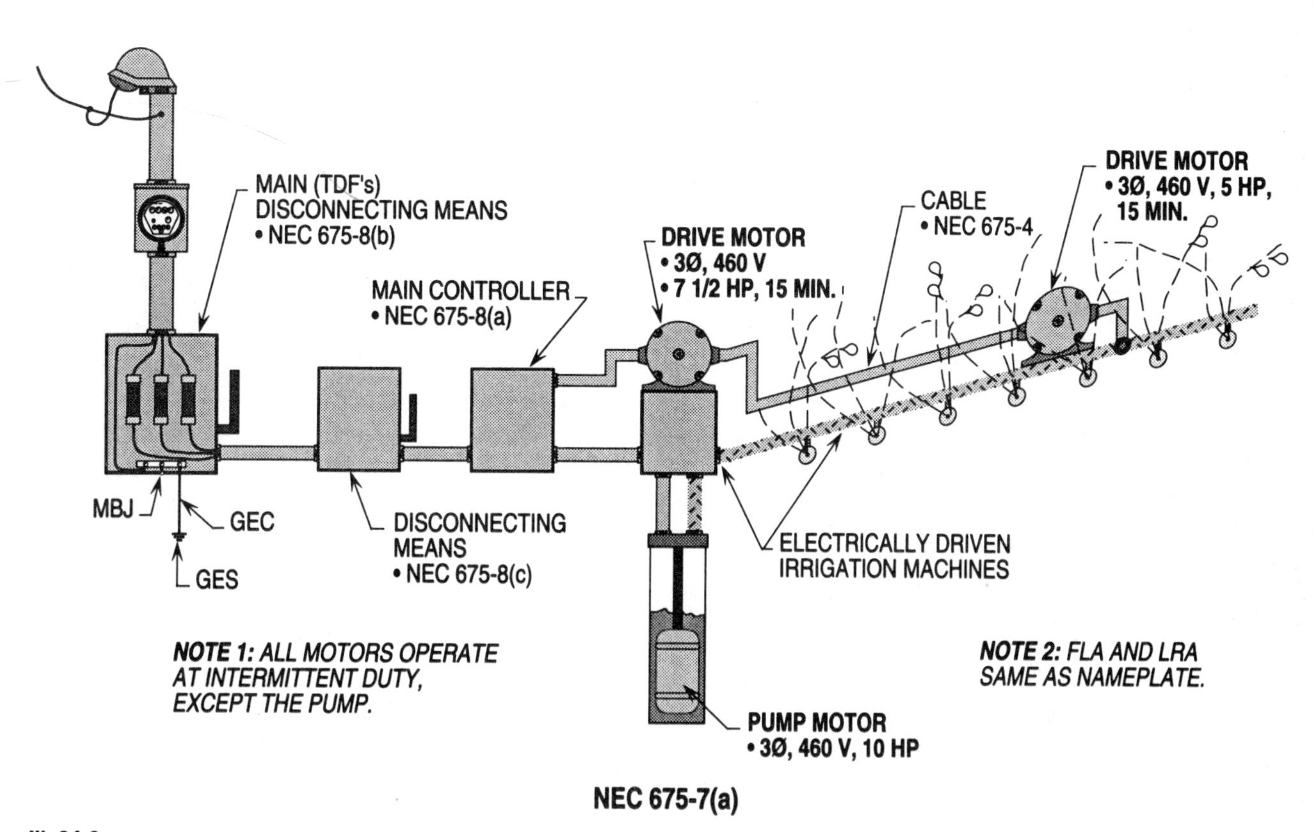

NEC 675-7(a)

Ill. 24-2

Irrigation Machines (Sizing Conductors and OCPD's) 675-22(a)

The OCPD's and conductors supplying power to a center pivot irrigation machine must be sized at 125 percent of the largest motor plus the remaining motors computed at 60 percent of their FLA rating.

Example 24-3. What size OCPD and THWN copper conductors are required for the circuit in Illustration 24-3?

Sizing conductors for irrigation machine

Step 1: Finding FLA of motors
675-22(a); Table 430-150
10 HP = 14 A
7 1/2 HP = 11 A
5 HP = 7.6 A
3 HP = 4.8 A
FLA same as nameplate

Step 2: Finding LRA of motors
675-22(b); Table 430-151
10 HP - 84 A x 2 = 168.0 A
7 1/2 HP - 11 A x 80% = 8.8 A
5 HP - 7.6 A x 80% = 6.08 A
3 HP - 4.8 A x 80% = 3.84 A
Total LRA = 186.72 A

Step 3: Calculating A for conductors
675-22(a)
14 A x 125% = 17.5 A
11 A x 60% = 6.6 A
7.6 A x 60% = 4.56 A
4.8 A x 60% = 2.88 A
Total A = 31.54 A

Step 4: Selecting conductors for motors
310-10(2); Table 310-16
31.54 A requires #10 cu.

Solution: The size THWN copper conductors are No. 10.

Sizing OCPD for irrigation machine

Step 1: Sizing OCPD for conductors
675-22(a); 240-6(a)
31.54 A requires 35 A OCPD

Solution: The size OCPD for the circuit is 35 amps.

Note: A 35 amp TDF holds 175 A (35 A x 5 = 175 A) for 10 seconds which is less than the LRC of 186.72 amps. However, due to operation all motors do not operate together. If the OCPD trips open, apply NEC 430-62(a).

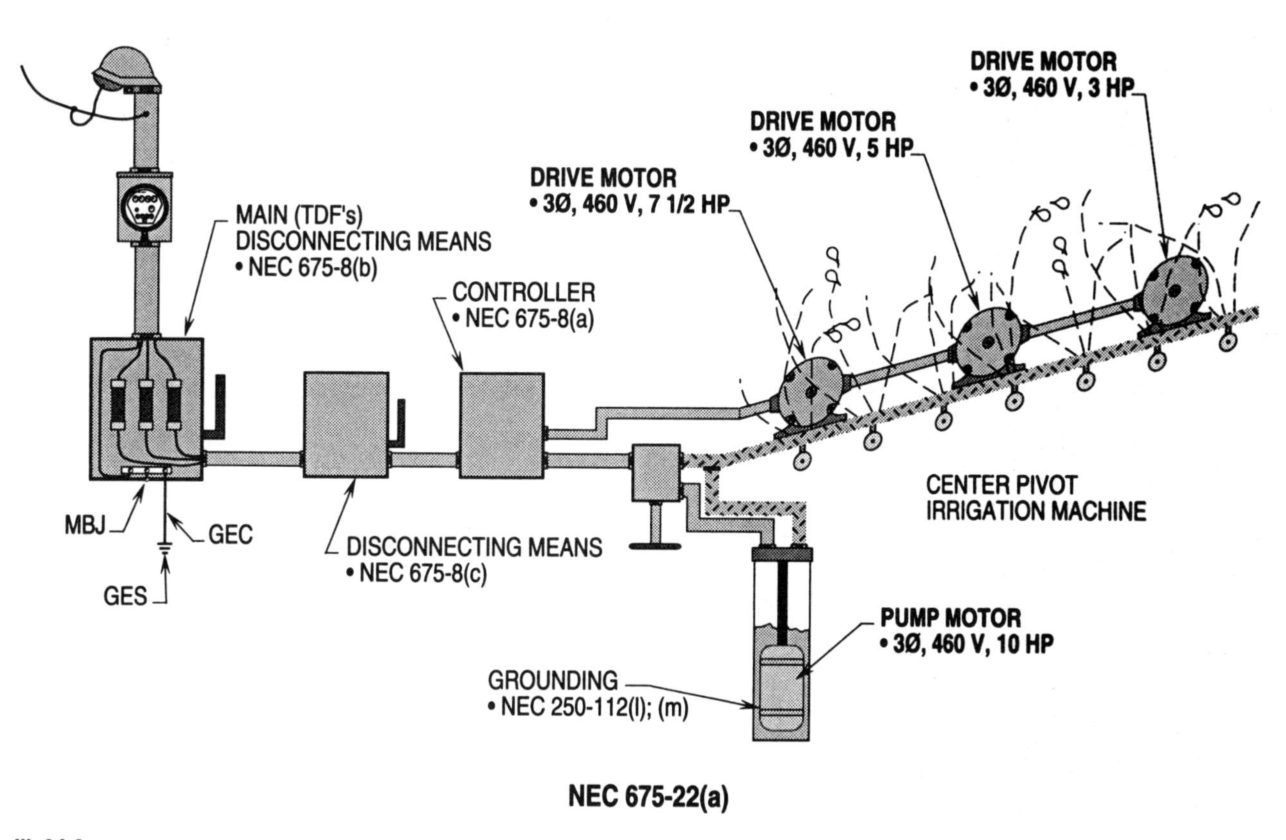

NEC 675-22(a)

Ill. 24-3

Swimming Pools (Sizing Conductors and OCPD's) 680-9

The heating elements used to heat swimming pool water are sized at 125 percent of the FLA's and all circuit components are selected on the amps derived from this calculation. OCPD's and conductors supplying pump motors to circulate water are sized per NEC 430-22(a) and 430-52(a).

Example 24-4. What size OCPD and conductors are required for the circuits in Illustration 24-4?.

Sizing OCPD and conductors for circuit #1

Step 1: Finding A for elements & motor
680-9; Table 430-148
Elements = 10 kW x 1,000 / 240 V
Elements = 41.7

Step 2: Calculating A for conductors
680-9; 430-22(a)
41.7 A x 125% = 52 A

Step 3: Selecting OCPD and conductors
680-9; 240-6(a); 310-10(2); Table 310-16
52 A requires 60 A OCPD
52 A requires #6 cu.

Solution: The size OCPD is 60 amps and the size conductors are No. 6 THWN copper.

Note: A 70 amp OCPD may be used per NEC 240-3(b).
(#6 THWN cu. = 65 A) Check with AHJ.

Sizing OCPD and conductors for circuit #2

Step 1: Finding A for motor
430-6(a)(1); Table 430-148
1 HP = 8 A

Step 2: Calculating OCPD for motor
430-6(a)(1); 430-52(c)(1);
Table 430-152
8 A x 250% = 20 A

Step 3: Calculating conductors for motor
430-6(a)(1); 430-22(a)
8 A x 125% = 10 A

Step 4: Selecting conductors for motor
310-10(2); Table 310-16
10 A requires #14 cu.

Solution: The size OCPD using a CB is 20 amps and the size THWN copper conductors are No. 14.

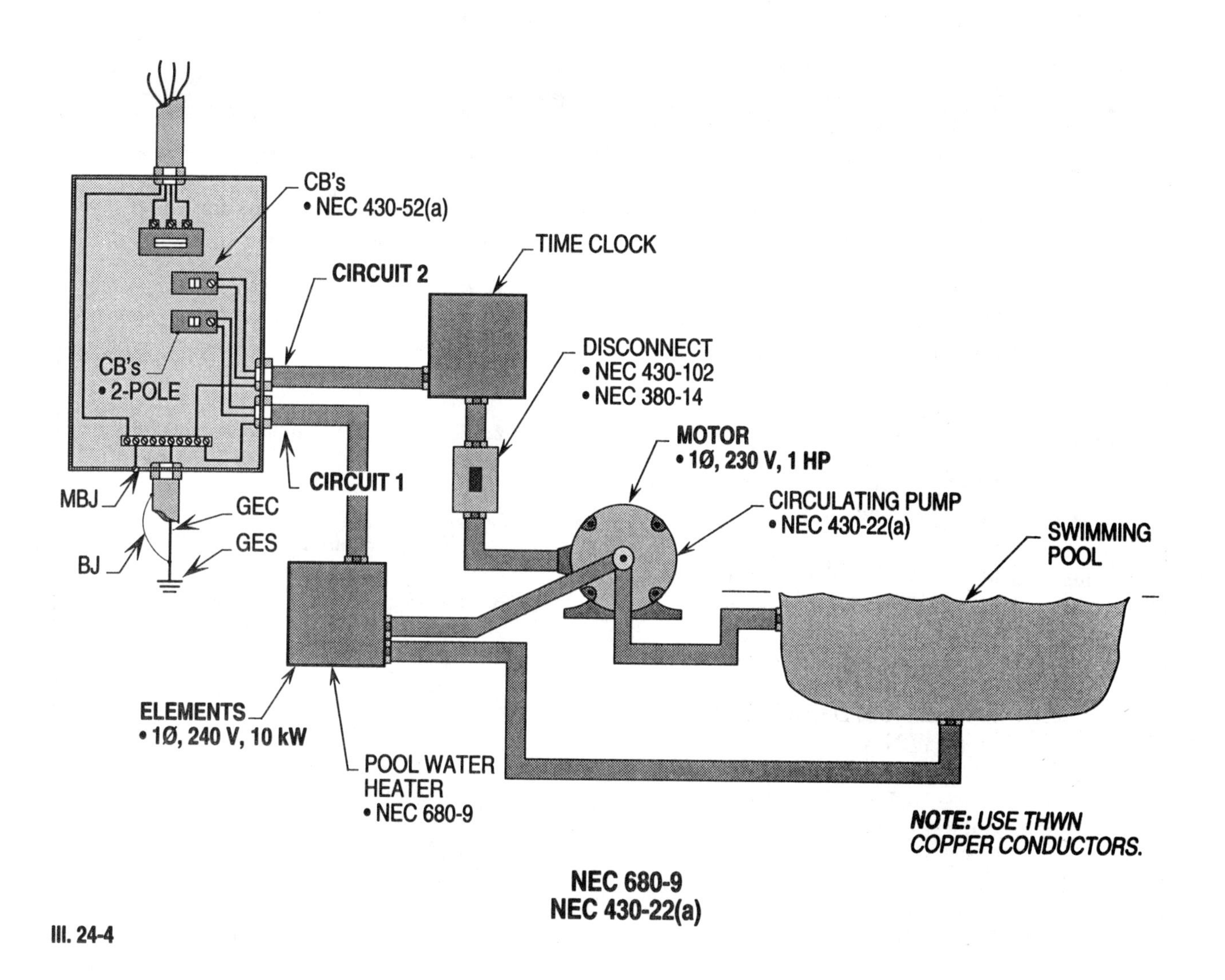
CB's
• NEC 430-52(a)
TIME CLOCK
CIRCUIT 2
DISCONNECT
• NEC 430-102
• NEC 380-14
CB's
• 2-POLE
MOTOR
• 1Ø, 230 V, 1 HP
CIRCUIT 1
CIRCULATING PUMP
• NEC 430-22(a)
MBJ
GEC
SWIMMING
POOL
GES
BJ
ELEMENTS
• 1Ø, 240 V, 10 kW
POOL WATER
HEATER
• NEC 680-9
NOTE: USE THWN
COPPER CONDUCTORS.
NEC 680-9
NEC 430-22(a)

Ill. 24-4

Solar Photovoltaic Systems 690-8(b)

The elements of circuits supplying and protecting solar photovoltaic systems must be sized to carry the normal and short circuit currents that the system is capable of developing during normal or abnormal conditions.

Example 24-5. What size OCPD and THWN copper conductors are required for the circuits in Illustration 24-5?

Sizing OCPD and conductors for photovoltaic source circuits

Step 1: Finding A for OCPD and conductors
690-8(a); (b)
Photovoltaic source circuits = 15 A x 3 = 45 A SCC

Step 2: Calculating A for OCPD and conductors
690-8(a)(1); (b)
45 A x 125% = 56.25 A SCC

Step 3: Selecting OCPD and conductors
310-10(2); Table 310-16; 240-3(b); 240-6(a)
56.25 A requires #6 cu. per run
56.25 A requires 60 A OCPD
65 A (#6 THWN cu.) allows
70 A OCPD

Solution: The size OCPD is 60 or 70 amps and the size conductors are No. 6 THWN copper.

Sizing OCPD and conductors for photovoltaic output circuit

Step 1: Calculating A for OCPD and conductors
690-8(b)
135 A x 125% = 168.75 A

Step 2: Selecting OCPD and conductors
310-10(2); Table 310-16; 240-3(b); 240-6(a)
168.75 A requires #2/0 cu.
168.75 A requires 175 A OCPD

Solution: The size OCPD is 175 amps and the size THWN copper conductors are No. 2/0.

Sizing OCPD and conductors for power conditioning unit output circuit

Step 1: Finding A for OCPD and conductors
690-8(a)(1); (b)
15 A for each run

Step 2: Calculating A for OCPD and conductors
690-8(a)(1); (b)
15 A x 125% = 18.75 A

Step 3: Selecting A for OCPD and conductors
310-10(2); 240-3(d); 240-3(b); 240-6(a)
18.75 A requires #12 cu.
18.75 A requires a 20 A OCPD
25 A (#12 THWN cu.) allows 20 A

Solution: The size OCPD is 20 amps and the size conductors are No. 12 THWN copper.

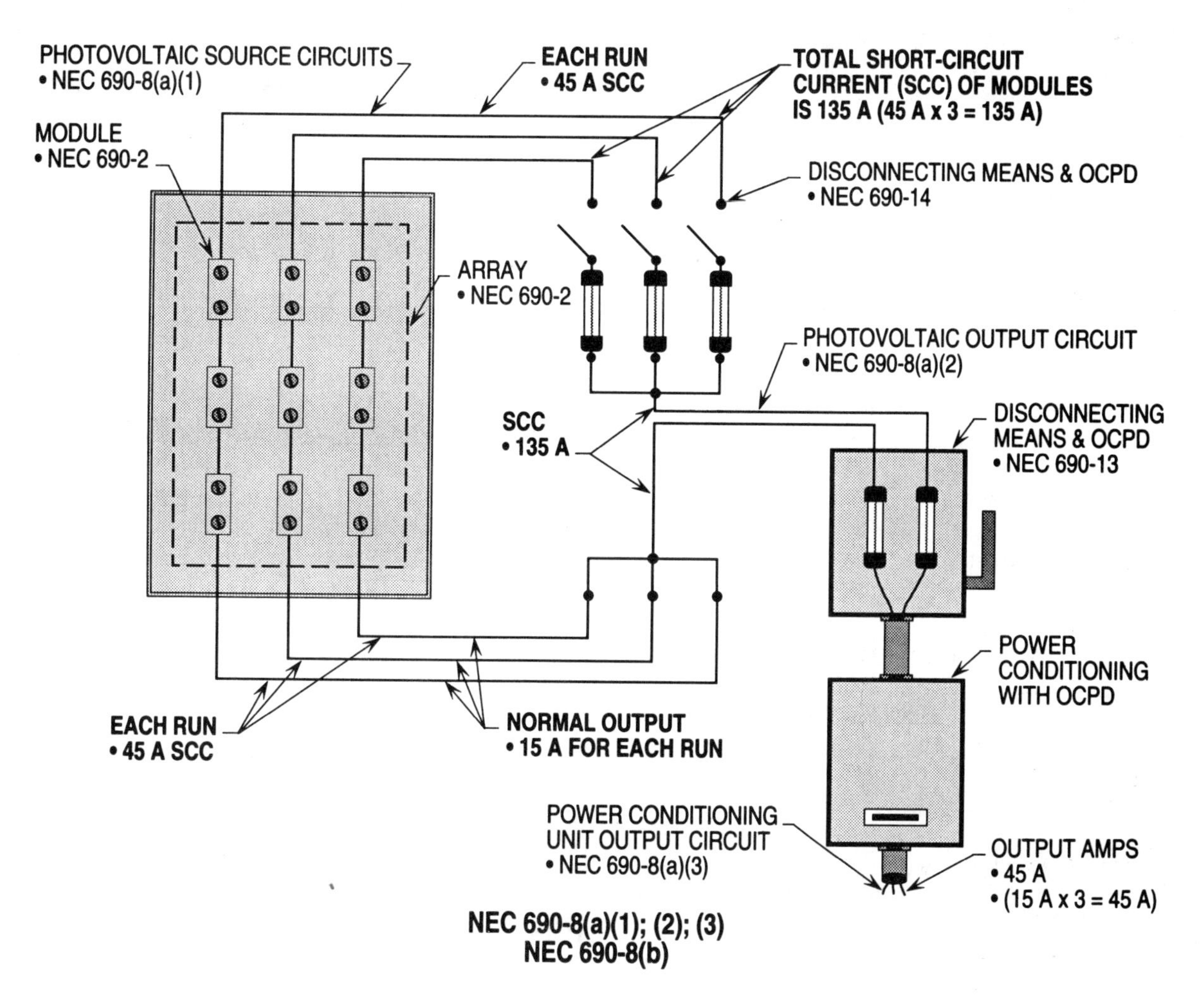
PHOTOVOLTAIC SOURCE CIRCUITS
• NEC 690-8(a)(1)
EACH RUN
• 45 A SCC
TOTAL SHORT-CIRCUIT CURRENT (SCC) OF MODULES IS 135 A (45 A x 3 = 135 A)
MODULE
• NEC 690-2
DISCONNECTING MEANS & OCPD
• NEC 690-14
ARRAY
• NEC 690-2
PHOTOVOLTAIC OUTPUT CIRCUIT
• NEC 690-8(a)(2)
SCC
• 135 A
DISCONNECTING MEANS & OCPD
• NEC 690-13
POWER CONDITIONING WITH OCPD
EACH RUN
• 45 A SCC
NORMAL OUTPUT
• 15 A FOR EACH RUN
POWER CONDITIONING UNIT OUTPUT CIRCUIT
• NEC 690-8(a)(3)
OUTPUT AMPS
• 45 A
• (15 A x 3 = 45 A)
NEC 690-8(a)(1); (2); (3)
NEC 690-8(b)

Ill. 24-5

Fire Pump Installations
695-5(b)

Transformers dedicated to supply power to fire pumps are permitted to have OCPD provided in their primary side. The OCPD in the primary must be capable of carrying the locked rotor current of the fire pump plus accessory loads indefinitely.

Example 24-6. What is the maximum size OCPD permitted in the primary side of the fire pump in Illustration 24-6?

Sizing primary OCPD

Step 1: Finding LRA of fire pump
695-5(b); Table 430-151(B)
40 HP requires 290 A LRC

Step 2: Calculating primary to secondary ratio
695-5(b)
Ratio = primary V / secondary V
Ratio = 4,160 V / 480 V
Ratio = 8.67 A

Step 3: Calculating total A
695-5(b)

LRC	=	290 A
Acc. equip.	=	30 A
Total	=	320 A

Step 4: Calculating OCPD
OCPD = Total A / Ratio
OCPD = 320 A / 8.67
OCPD = 36.9 A

Step 5: Selecting OCPD
695-5(b); 695-3(c)(2)
36.9 requires 40 A OCPD

Solution: The size OCPD in the primary side is 40 amps.

Note: The OCPD must be sized above the LRC of the fire pump and accessory loads.

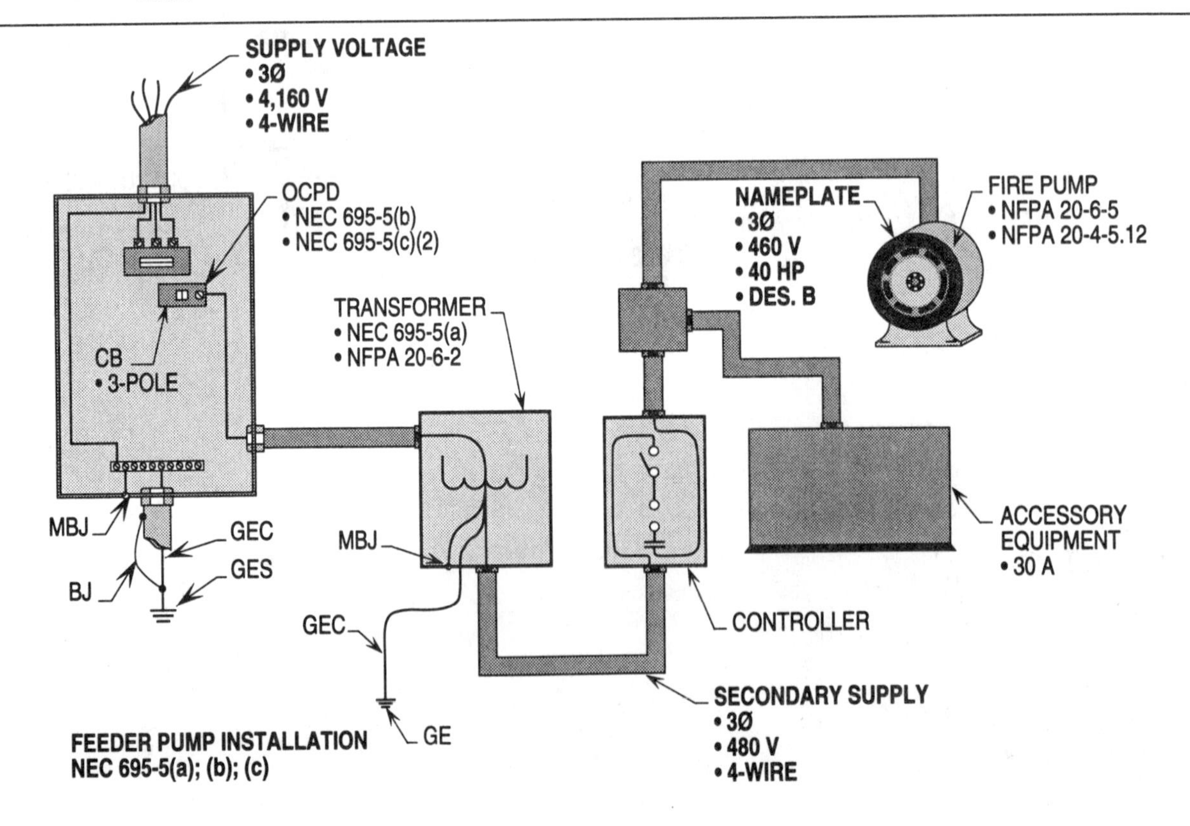

NEC 695-5(b)

Ill. 24-6

25

Control, Signaling, Power-limited Circuits, Fire-Protection Signaling System

Control, signaling, and power-limited circuits are divided into three types of circuits per Article 725 of the National Electrical Code (NEC®) and they are as follows:

(1) Class 1
(2) Class 2
(3) Class 3

Each classification is determined and used based upon their output rated in voltage and volt-amps. Overcurrent protection devices (OCPD's) and conductors supplying such circuits are sized and selected according to this classification of service.

Fire protective signaling systems are regulated by the requirements listed in Article 760 of the NEC. The rules require certain types of conductors and cables to be sized and used which are determined by the locations where they are run and installed.

Communication circuits used for telephones, and outside wiring for fire alarms, burglar alarms, and similar central station systems are sized and selected by the rules and regulations listed in Article 800 of the NEC. OCPD's and conductors or cables are sized according to very low outputs derived from such systems.

Quick Reference

Control Circuits
725-23

Remote control circuits of this type are derived from a separate power source other than what supplies the controller and motor. Such control circuits are required to be protected where they are connected to their power source.

Example 25-1. What size OCPD is required at the service equipment in Illustration 25-1 to protect the control circuit conductors?

Sizing OCPD for control circuit

Step 1: Finding A of control circuit
310-10(2); Table 310-16
#14 THWN cu. = 20 A

Step 2: Selecting OCPD for control conductors
240-3; 240-3(d); 240-6(a)
#14 cu. requires 15 amp OCPD

Solution: The size OCPD at the service equipment is 15 amps.

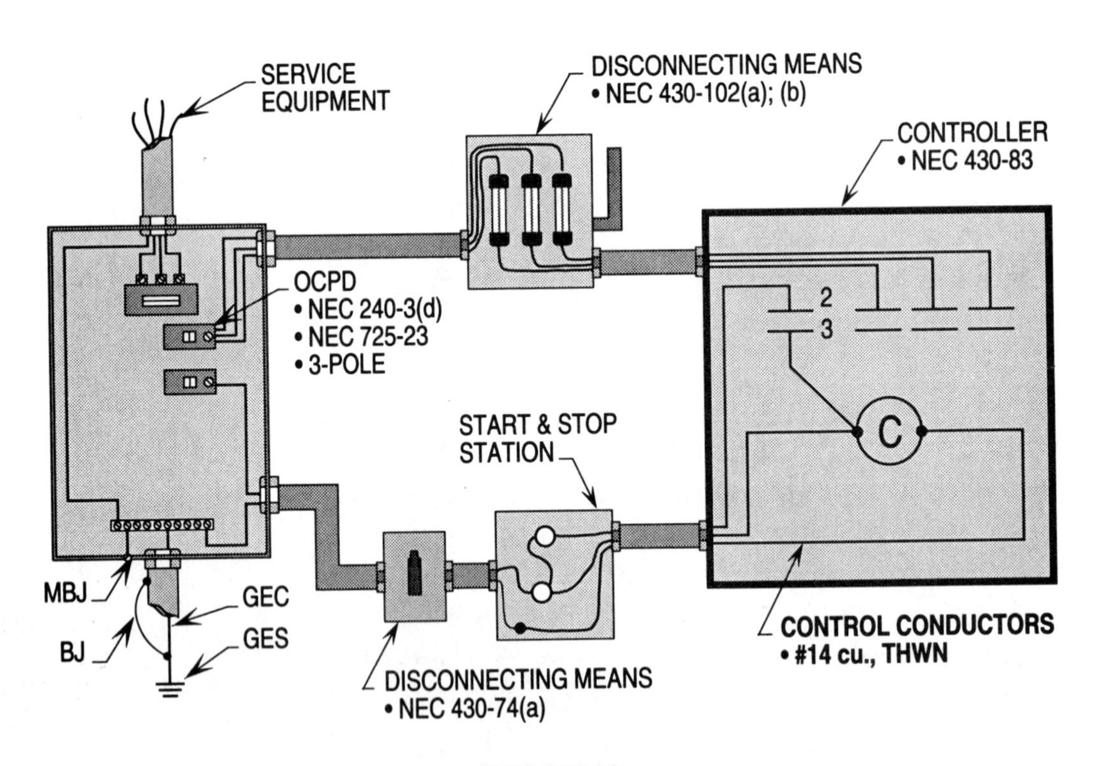

NEC 725-23

Ill. 25-1

Control Circuits
725-24, Ex. 2

The primary OCPD may be used to provide the protection for the conductors tapped from the secondary side of a control transformer, if its rating does not exceed the secondary to primary voltage ratio of the transformer.

Example 25-2. What size OCPD is required at the service equipment in Illustration 25-2 that can be used to protect the primary and secondary side of the control circuit?

Sizing OCPD for the control circuit

Step 1: Finding ratio of transformer
Pri. and Sec.
725-24, Ex. 2
Ratio = 240 V Sec. / 480 V Pri. x 35 A
Ratio = 17.5 A

Step 2: Selecting OCPD for control circuit
725-24, Ex. 2; 240-6(a)
17.5 A requires 15 A OCPD

Step 3: Checking ratio
725-24, Ex. 2; 240-6(a)
Ratio = 480 V Pri. / 240 V
Sec. = 2 to 1
OCPD = 35 A (#10) / 2 (Ratio)
OCPD = 17.5 A

Solution: **The size OCPD for the control circuit is 15 amps which complies with the NEC.**

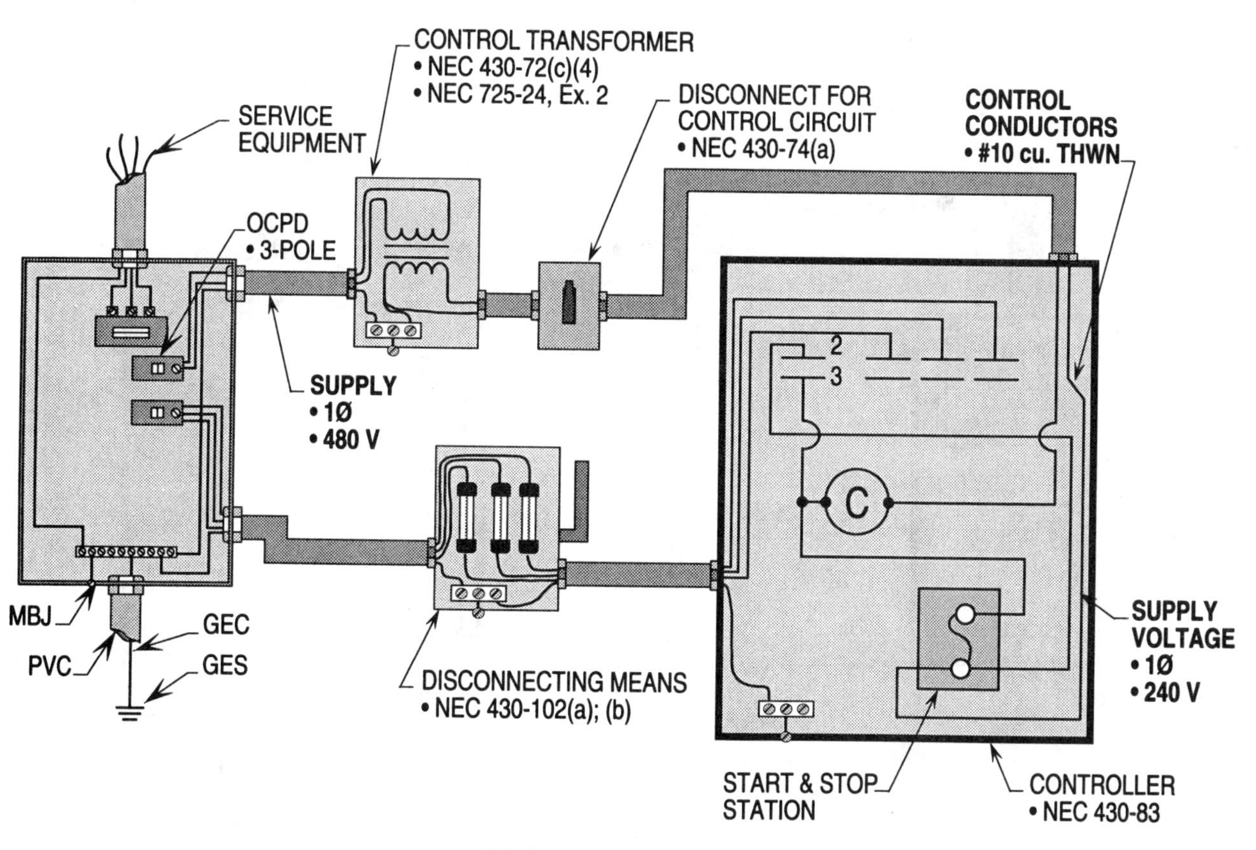

NEC 725-24, Ex. 2

Ill. 25-2

Control Circuits
725-24, Ex. 4

Class 1 remote control circuits controlling a light and power contactor are considered protected by the branch-circuit OCPD, if its rating does not exceed 300 percent of the ampacity of the control circuit conductors.

Example 25-3. What size OCPD is required at the service equipment in Illustration 25-3?

Sizing OCPD for the branch-circuit (BC) and control circuit

Step 1: Finding A of control conductors
725-23, Ex. 4; Table 310-16
#14 THWN cu. = 20 A

Step 2: Calculating size OCPD
725-23, Ex.; 725-24, Ex. 4
20 A x 300% = 60 A

Step 3: Selecting size OCPD
240-3; 240-6(a)
60 A allows 60 A OCPD

Solution: The size OCPD for the branch-circuit and control conductors is 60 amps.

Note: The AHJ may require a 45 amp OCPD (15 A OCPD x 300% = 45 A) to be used to protect control conductors per Table 430-72(b).

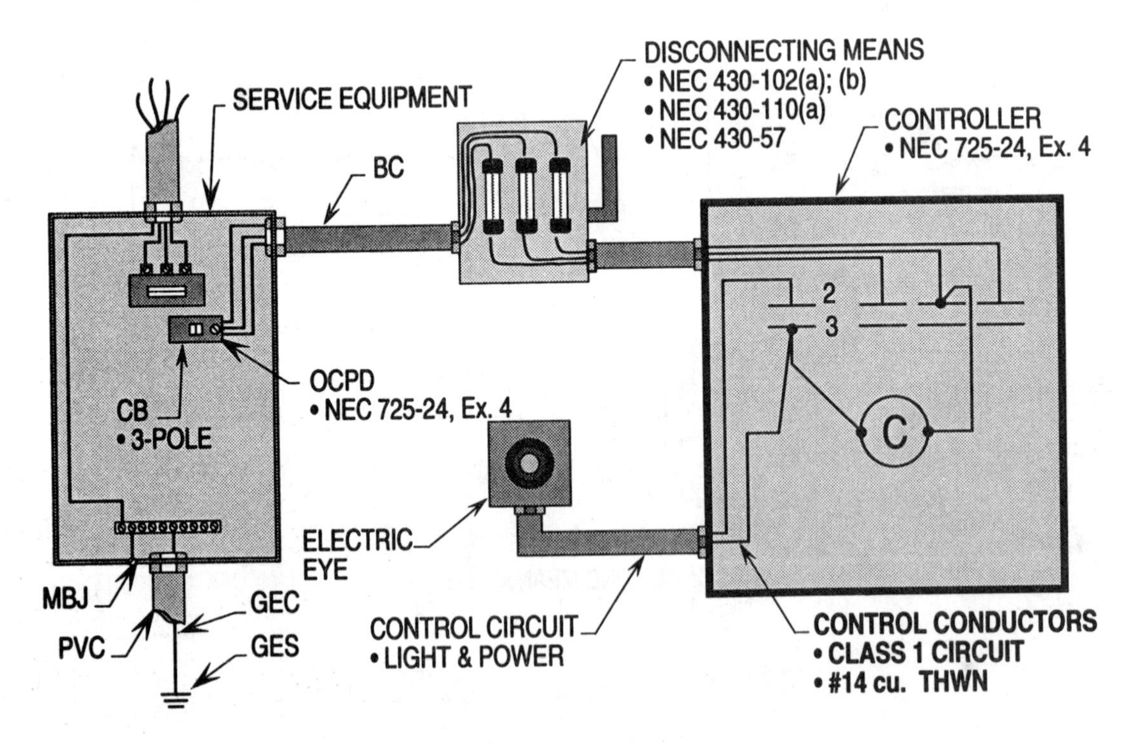

NEC 725-24, Ex. 4

Ill. 25-3

Fire-signaling Circuits 760-23

Nonpower-limited fire signaling circuit conductors are required to be protected at their ampacity rating to ensure reliability when they are energized to sound an alarm.

Example 25-4. What size OCPD is required for the size copper conductor utilized for the fire-signaling circuits in Illustration 25-4?

Sizing OCPD for fire-signaling circuits

Step 1: Finding conductor A
760-23; Table 310-16
#14 cu. = 20 A
#16 cu. = 10 A

Step 2: Selecting OCPD for signaling circuits
240-3(d); 760-23
20 A (#14 cu.) requires 15 A OCPD
10 A (#16 cu.) requires 10 A OCPD

Solution: **The size OCPD for the No. 14 THWN copper conductors is 15 amps and 10 amps for the No. 16 THWN copper conductors.**

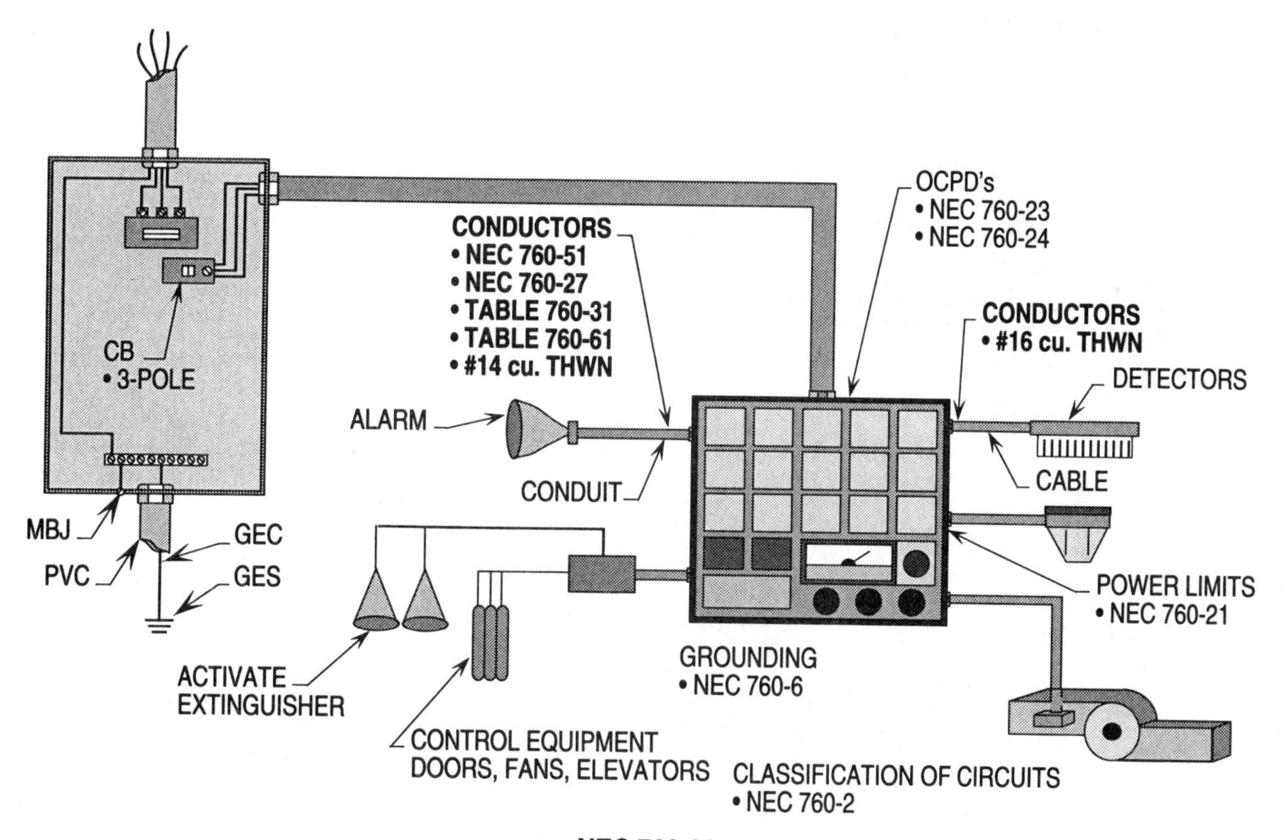

Ill. 25-4

Appendix

Tables and Appendixes of this appendix are included to provide information which is useful to engineers and electricians performing various calculations using the requirements of the National Electrical Code. For example, if the circular mills (CM) rating of a No. 12 copper conductor is needed to calculate a value pertaining to conductor sizing, Table C to this handy Appendix shows a No. 12 having a CM rating of 6,530. On the other hand, if the DC resistance of an uncoated No. 12 stranded copper conductor is needed for "R" in the voltage drop formula.

($VD = \frac{2 \times R \times L \times I}{1,000}$) to compute the VD of a circuit, then 1.98 ohm/KFT is selected from Table C. Note that the area in square inches of an insulated No. 12 copper conductor is 0.006 from sixth column of Table C. Appendix E is used to convert a certain value to another value by applying a particular multiplier.

For example, meters can be converted by multiplying the value in meters by the conversion multiplier of 39.37 and inches can be converted by multiplying by .0254 per Appendix E. Appendix F can be used to convert temperatures in degrees centigrade (°C) to degrees in fahrenheit (°F) without applying one of the formulas below the Table. For example, 30°C is converted to 86°F per Appendix F. Notice that handy calculation tips are listed in the Appendix I for easy reference.

These calculation tips are designed to alert personnel using the NEC to be aware of the different interpreta-

Quick Reference

tions of certain sections of the code. It is these different interpretations by the authority having jurisdiction that hinders correct computations from being calculated by the NEC. Such tips should be reviewed very carefully before performing the various calculations using the rules of the NEC.

It is the suggestion of the author that each Appendix be reviewed, so that the information, if needed, can be utilized when calculations are made using the requirements of the NEC.

Table A

Size AWG/ KCMIL	Ohm's to Neutral per 1,000 feet					
	AC Resistance for uncoated Copper Wires			AC Resistance for Aluminum Wires		
	PVC Conduit	Alu. Conduit	Steel Conduit	PVC Conduit	Alu. Conduit	Steel Conduit
14	3.1	3.1	3.1	—	—	—
12	2.0	2.0	2.0	3.2	3.2	3.2
10	1.2	1.2	1.2	2.0	2.0	2.0
8	0.78	0.78	0.78	1.3	1.3	1.3
6	0.49	0.49	0.49	0.81	0.81	0.81
4	0.31	0.31	0.31	0.51	0.51	0.51
3	0.25	0.25	0.25	0.40	0.41	0.40
2	0.19	0.20	0.20	0.32	0.32	0.32
1	0.15	0.16	0.16	0.25	0.26	0.25
1/0	0.12	0.13	0.12	0.20	0.21	0.20
2/0	0.10	0.10	0.10	0.16	0.16	0.16
3/0	0.077	0.082	0.079	0.13	0.13	0.13
4/0	0.062	0.067	0.063	0.10	0.11	0.10
250	0.052	0.057	0.054	0.085	0.090	0.086
300	0.044	0.049	0.045	0.071	0.076	0.072
350	0.038	0.043	0.039	0.061	0.066	0.063
400	0.033	0.038	0.035	0.054	0.059	0.055
500	0.027	0.032	0.029	0.043	0.048	0.045
600	0.023	0.028	0.025	0.036	0.041	0.038
750	0.019	0.024	0.021	0.029	0.034	0.031
1000	0.015	0.019	0.018	0.023	0.027	0.025

Note: See Table 9 to Ch. 9 of the NEC.

Table B

(Three Single Conductors) "C" Value for Conductors and Busway						
Copper						
AWG	**Three Single Conductors**					
or	**Conduit**					
KCMIL	**Steel**			**Nonmagnetic**		
	600V	**5KV**	**15KV**	**600V**	**5KV**	**15KV**
14	389	389	389	389	389	389
12	617	617	617	617	617	617
10	981	981	981	981	981	981
8	1557	1551	1557	1558	1555	1558
6	2425	2406	2389	2430	2417	240
4	3806	3750	3695	3825	3789	375
3	4760	4760	4760	4802	4802	4802
2	5906	5736	5574	6044	5826	5809
1	7292	7029	6758	7493	7306	7108
1/0	8924	8543	7973	9317	9033	8590
2/0	10755	10061	9389	11423	10877	10318
3/0	12843	11804	11021	13923	13048	12360
4/0	15082	13605	12542	16673	15351	14347
250	16483	14924	13643	18593	17120	15865
300	18176	16292	14768	20867	18975	17408
350	19703	17385	15678	22736	20526	18672
400	20565	18235	16365	24296	21786	19731
500	22185	19172	17492	26706	23277	21329
600	22965	20567	17962	28033	25203	22097
750	24136	21386	18888	28303	25430	22690
1000	25278	22539	19923	31490	28083	24887
Aluminum						
14	236	236	236	236	236	236
12	375	375	375	375	375	375
10	598	598	598	598	598	598
8	951	950	951	951	951	951
6	1480	1476	1472	1481	1478	1476
4	2345	2332	2319	2350	2341	2333
3	2948	2948	2948	2958	2958	2958
2	3713	3669	3626	3729	3701	3672
1	4645	4574	4497	4678	4631	4580
1/0	5777	5669	5493	5838	5766	5645
2/0	7186	6968	6733	7301	7152	6986
3/0	8826	8466	8163	9110	8851	8627
4/0	10740	10167	9700	11174	10749	10386
250	12122	11460	10848	12862	12343	11847
300	13909	13009	12192	14922	14182	13491
350	15484	14280	13288	16812	15857	1495
400	16670	15355	14188	18505	17321	16233
500	18755	16827	15657	21390	19503	18314
600	20093	18427	16484	23451	21718	19635
750	21766	19685	17686	23491	21769	19976
1000	23477	21235	19005	28778	26108	23482

Table B

(Three Conductor Cable) "C" Value for Conductors and Busway						
Copper						
AWG	Three Single Conductors					
or	Conduit					
KCMIL	Steel			Nonmagnetic		
	600V	5 kV	15 kV	600V	5 kV	15 kV
14	389	389	389	389	389	389
12	617	617	617	617	617	617
10	981	981	981	981	981	981
8	1559	1557	1559	1559	1558	1559
6	2431	2424	2414	2433	2428	2420
4	3830	3811	3778	3837	3823	3798
3	4760	4790	4760	4802	4802	4802
2	5989	5929	5827	6087	6022	5957
1	7454	7364	7188	7579	7507	7364
1/0	9209	9086	8707	9472	9372	9052
2/0	11244	11045	10500	11703	11528	11052
3/0	13656	13333	12613	14410	14118	13461
4/0	16391	15890	14813	17482	17019	16012
250	18310	17850	16465	19799	19352	18001
300	20617	20051	18318	22524	21938	20163
350	22646	21914	19821	24904	24126	21982
400	24253	23371	21042	26915	26044	23517
500	26980	25449	23125	30028	28712	25916
600	28752	27974	24896	32236	31258	27766
750	31050	30024	26932	32404	31338	28303
1000	33864	32688	29320	37197	35748	31959
Aluminum						
14	236	236	236	236	236	236
12	375	375	375	375	375	375
10	598	598	598	598	598	598
8	951	951	951	951	951	951
6	1481	1480	1478	1482	1481	1479
4	2351	2347	2339	2353	2349	2344
3	2948	2956	2948	2958	2958	2958
2	3733	3719	3693	3739	3724	3709
1	4686	4663	4617	4699	4681	4646
1/0	5852	5820	5717	5875	5851	5771
2/0	7327	7271	7109	7372	7328	7201
3/0	9077	8980	8750	9242	9164	8977
4/0	11184	11021	10642	11408	11277	10968
250	12796	12636	12115	13236	13105	12661
300	14916	14698	13973	15494	15299	14658
350	15413	16490	15540	17635	17351	16500
400	18461	18063	16921	19587	19243	18154
500	21394	20606	19314	22987	22381	20978
600	23633	23195	21348	25750	25243	23294
750	26431	25789	23750	25682	25141	23491
1000	29864	29049	26608	32938	31919	29135

Note: These values are equal to one over the impedence per foot for impedence found in IEEE. std. 241-1990,IEEE Recommended Practice for Commercial Building Power System.

Table C

Size AWG/ KCMIL	Area Cir. Mills	Conductors				DC Resistance at 75°C (167°F)		
		Standing		Overall		Copper		Aluminum
		Quantity	Diam. In.	Diam. In.	Area Sq. In.^{2}	Uncoated ohm/kFT	Coated ohm/kFT	ohm/kFT
18	1620	1	—	0.040	0.001	7.77	8.08	12.8
18	1620	7	0.015	0.046	0.002	7.95	8.45	13.1
16	2580	1	—	0.051	0.002	4.89	5.08	8.05
16	2580	7	0.019	0.058	0.003	4.99	5.29	8.21
14	4110	1	—	0.064	0.003	3.07	3.19	5.06
14	4110	7	0.024	0.073	0.004	3.14	3.26	5.17
12	6530	1	—	0.081	0.005	1.93	2.01	3.18
12	6530	7	0.030	0.092	0.006	1.98	2.05	3.25
10	10380	1	—	0.102	0.008	1.21	1.26	2.00
10	10380	7	0.038	0.116	0.011	1.24	1.29	2.04
8	16510	1	—	0.128	0.013	0.764	0.786	1.26
8	16510	7	0.049	0.146	0.017	0.778	0.809	1.28
6	26240	7	0.061	0.184	0.027	0.491	0.510	0.808
4	41740	7	0.077	0.232	0.042	0.308	0.321	0.508
3	52620	7	0.087	0.260	0.053	0.245	0.254	0.403
2	66360	7	0.097	0.292	0.067	0.194	0.201	0.319
1	83690	19	0.066	0.332	0.087	0.154	0.160	0.253
1/0	105600	19	0.074	0.373	0.109	0.122	0.127	0.201
2/0	133100	19	0.084	0.419	0.138	0.0967	0.101	0.159
3/0	167800	19	0.094	0.470	0.173	0.0766	0.0797	0.126
4/0	211600	19	0.106	0.528	0.219	0.0608	0.0626	0.100
250	—	37	0.082	0.575	0.260	0.0515	0.0535	0.0847
300	—	37	0.090	0.630	0.312	0.0429	0.0446	0.0707
350	—	37	0.097	0.681	0.364	0.0367	0.0382	0.0605
400	—	37	0.104	0.728	0.416	0.0321	0.0331	0.0529
500	—	37	0.116	0.813	0.519	0.0258	0.0265	0.0424
600	—	61	0.099	0.893	0.626	0.0214	0.0223	0.0353
700	—	61	0.107	0.964	0.730	0.0184	0.0189	0.0303
750	—	61	0.111	0.998	0.782	0.0171	0.0176	0.0282
800	—	61	0.114	1.03	0.834	0.0161	0.0166	0.0265
900	—	61	0.122	1.09	0.940	0.0143	0.0147	0.0235
1000	—	61	0.128	1.15	1.04	0.0129	0.0132	0.0212
1250	—	91	0.117	1.29	1.30	0.0103	0.0106	0.0169
1500	—	91	0.128	1.41	1.57	0.00858	0.00883	0.0141
1750	—	127	0.117	1.52	1.83	0.00735	0.00756	0.0121
2000	—	127	0.126	1.63	2.09	0.00643	0.00662	0.0106

Note: See Table 8 to Ch. 9 of the NEC.

Appendix D
Weights and Measurements

Appendix D contains information pertaining to " U.S. Weights and Measures." All major weights and measures are illustrated for fact and easy use. For Example, according to the Linear Measure in the Table, 12 inches is equal to 1 foot and 3.048 decimeters. For applying conversion Tables and using multipliers, see Appendix E.

Linear Measurements

			1	Inch	=	2.540	Centimeters
12	Inches	=	1	Foot	=	3.048	Decimeters
3	Feet	=	1	Yard	=	9.144	Decimeters
5.5	Yards	=	1	Rod, Pole, or Perch	=	5.029	Meters
40	Rods	=	1	Furlong	=	2.018	Hectometers
8	Furlongs	=	1	Mile	=	1.609	Kilometers

Mile Measurement

1	Statute Mile	=	5,280	Feet
1	Scots Mile	=	5,952	Feet
1	Irish Mile	=	6,720	Feet
1	Russian Verst	=	3,504	Feet
1	Italian Mile	=	4,401	Feet
1	Spanish Mile	=	15,084	Feet

Other Linear Measurements

1 Hand	=	4	Inches
1 Span	=	9	Inches
1 Chain	=	22	Yards
1 Knot	=	1	Nautical Mile
	=	6,080	Feet

1 Link	=	7.92	Inches
1 Fatom	=	6	Feet
1 Furlong	=	10	Chains
1 Cable	=	608	Feet

Square Measurements

144	Square Inches	=	1	Square Foot
9	Square Feet	=	1	Square Yard
30 1/4	Square Yards	=	1	Square Rod
		=	1	Square Pole
		=	1	Square Perch
40	Rods	=	1	Rood
4	Roods	=	1	Acre
640	Acres	=	1	Square Mile
1	Square Mile	=	1	Section
36	Sections	=	1	Township

Cubic Or Solid Measure

1 cu. Foot	=	1,728	cu. Inches
1 cu. Yard	=	27	cu. Feet
1 cu. Foot	=	7.48	Gallons
1 Gallons (water)	=	8.34	Lbs.
1 Gallon (U.S.)	=	231	cu. Inches of water
1 Gallon (Imperial)	=	277 1/4	cu. Inches of water

Liquid Measure

1	Pint	=	4	Gills
1	Quart	=	2	Pints
1	Gallon	=	4	Quarts
1	Firkin	=	9	Gallons (ale or beer)
1	Barrel	=	42	Gallons (petroleum or crude oil)

Dry Measure

1	Quart	=	2	Pints
1	Peck	=	8	Quarts
1	Bushel	=	4	Pecks

Weight Measurement (Mass.)

• **AVOIRDUPOIS WEIGHT:**

1	Ounce	=	16	Drams
1	Pound	=	16	Ounces
1	Hundredweight	=	100	Pounds
1	Ton	=	2,000	Pounds

• **TROY WEIGHT:**

1	Carat	=	3.17	Grains
1	Pennyweight	=	20	Grains
1	Ounce	=	20	Pennyweights
1	Pound	=	12	Ounces
1	Long Hundred-Weight	=	112	Pounds
1	Long Ton	=	20	Long Hundredweights
		=	2,240	Pounds

• **APOTHECARIES WEIGHT:**

1	Scruple	=	20	Grains	=	1.296	Grams
1	Dram	=	3	Scruple	=	3.888	Grams
1	Ounce	=	8	Drams	=	31.1035	Grams
1	Pound	=	12	Ounces	=	373.2420	Grams

• **KITCHEN WEIGHTS AND MEASUREMENTS:**

1	U.S. Pint	=	16	Fl. Ounces
1	Standard Cup	=	8	Fl. Ounces
1	Tablespoon	=	0.5	Fl. Ounces (15 cu. CMS.)
1	Teaspoon	=	0.16	Fl. Ounces (5 cu. CMS.)

Metric System

• CUBIC MEASURE:
THE UNIT IS THE "METER" = 39.37 INS.:

1	cu.	Centimeter	=	1,000 cu.	Millimeters	=	0.06125	cu. In.
1	cu.	Decimeter	=	1,000 cu.	Centimeters	=	61.1250	cu. Ins.
1	cu.	Meter	=	1,000 cu.	Decimeters	=	35.3156	cu. Ft.
			=	1 Stere		=	1.30797	cu. Yds.

1	cu.	Centimeter (water)				=	1 Gram
1,000	cu.	Centimeters (water)	=	1	Liter	=	1 Kilogram
1	cu.	Meter (1,000 liters)				=	1 Metric Ton

• MEASURES OF WEIGHT:
THE UNIT IS THE GRAM = 0.035274 OUNCES

1	Milligram				=	0.015432	Grains
1	Centigram	=	10	Milligrams	=	0.015432	Grains
1	Decigram	=	10	Centigrams	=	0.15432	Grains
1	Gram	=	10	Decigrams	=	15.4323	Grains
1	Dekagram	=	10	Grams	=	5.6438	Drams
1	Hectogram	=	10	Dekagrams	=	3.5274	Ounces
1	Kilogram	=	10	Hectograms	=	2.2046223	Pounds
1	Myriagram	=	10	Kilograms	=	22.046223	Pounds
1	Quintal	=	10	Myriagrams	=	1.986412	Cwt.
1	Metric Ton	=	10	Quintal	=	2,204.622	Pounds

1	Gram	=	0.56438	Drams
1	Dram	=	1.77186	Grams
		=	27.3438	Grains
1	Metric Ton	=	2,204.6223	Pounds

• MEASURE OF CAPACITY:
THE UNIT IS THE "LITER" = 1.0567 LIQUID QUARTS:

1	Centiliter	=	10	Millimeters	=	0.338	Fluid Ounces
1	Deciliter	=	10	Centiliter	=	3.38	Fluid Ounces
1	Liter	=	10	Deciliter	=	33.8	Fluid Ounces
1	Dekaliter	=	10	Liters	=	0.284	Bushel
1	Hectoliter	=	10	Dekaliters	=	2.84	Bushels
1	Kiloliter	=	10	Hectoliters	=	264.2	Gallons

Author's Note: $\frac{\text{Kilometers}}{8} \times 5 = \text{Miles}$ $\quad \frac{\text{Miles}}{5} \times 8 = \text{Kilometers}$

Appendix E
Applying Conversion Factors

Conversion Tables can be used to convert a certain value to another value by applying multipliers. For Example, circuit mils can be converted to square mils by multiplying by .7854 and be transformed back by multiplying by 1.272. See circuit mils in Column 1 of To Convert.

Applying Conversion Factors			
For Conversion Of	**Into**	**Multiply By**	**To Convert Back, Multiply By**
Circular Mils	Square Mils	.7854	1.272
Circular Mils	Square Inches	7.854×10^{-7}	
Centigrade	Fahrenheit	°C x 9/5 plus 32	°F - 32 x 5/9
Cubic Inches	Cubic Centimeters	16.39	6.102×10^{-2}
Cubic Inches	Cubic Feet	5.787×10^{-4}	1728
Cubic Inches	Cubic Meters	1.639×10^{-5}	61,023
Cubic Inches	Cubic Yards	2.143×10^{-5}	46,656
Farads	Microfarads	10^{-6}	
Horsepower	Btu Per Minute	42.40	.02357
Horsepower	Foot-Pounds Per Minute	33,000	3.030×10^{-5}
Inches	Centimeters	2.540	.3937
Inches	Feet	8.333×10^{-2}	12
Inches	Miles	1.578×10^{-5}	6.336×10^{-4}
Inches	Mils	10^{-3}	10^{-3}
Inches	Yards	2.778×10^{-2}	36
Joules	Btu	.0009488	1054.5
Joules	Foot-Pounds	.7375	1.356
Kilometers	Miles	.62137	1.6094
Kilowatts	Watts	10^{-3}	
Kilowatt Hours	Btu	3413	2.930×10^{-4}
Kilowatt Hours	Foot-Pounds	2.656×10^{-6}	3.766×10^{-7}
Kilowatt Hours	Horsepower Hours	1.341	.7455
Kilowatt Hours	Joules	3.6×10^{-6}	2.778×10^{-7}
Lumens Per Sq. Ft.	Foot-Candles	1	
Megohms	Ohms	10^{-6}	
Meters	Centimeters	100	.01
Meters	Feet	3.2808	.3048
Meters	Inches	39.37	.0254
Meters	Kilometers	10^{-3}	10^{-3}
Meters	Miles	6.214×10^{-4}	1.609×10^{-3}
Meters	Yards	1.094	.9144
Microhms	Megohms	10^{-12}	
Microhms	Ohms	10^{-6}	
Miles (nautical)	Feet	6080.2	1.667×10^{-4}
Miles (nautical)	Miles (statute)	1.1516	.86836
Miles (statute)	Feet	5280	1.894×10^{-4}
Miles (statute)	Yards	1760	5.682×10^{-4}
Ohms	Megohms	10^{-6}	
Ohms	Microhms	10^{-6}	
Square Inches	Circular Mils	1.273×10^{-6}	7.854×10^{-7}
Square Inches	Square Feet	6.944×10^{-3}	144
Square Inches	Square Mils	10^{-6}	10^{-6}
Square Inches	Square Yards	7.716×10^{-4}	1296
Watts	Btu Per Minute	.05688	17.58
Watts	Horsepower	1.341×10^{-3}	
Watts	Horsepower	1.341×10^{-3}	745.7
Watts	Kilowatts	10^{-3}	
Watt Hours	Horsepower Hours	1.341×10^{-3}	
Watt Hours	Kilowatt Hours	10^{-3}	

Appendix F
Applying °C and °F Table

The centigrade (°C) and fahrenheit (°F) Table can be used to convert degrees in centigrade to fahrenheit and degrees in fahrenheit to centigrade without having to apply one of the formulas (1 or 2) below the Table. For example, to convert 30°C to fahrenheit use Columns 3 and 4 in the Table and the degree is 86°F.

Centigrade And Fahrenheit Thermometer Scales							
Deg - °C	Deg - °F	Deg -°C	Deg -°F	Deg - °C	Deg - °F	Deg -°C	Deg - °F
0	32						
1	33.8	26	78.8	51	123.8	76	168.8
2	35.6	27	80.6	52	125.6	77	170.6
3	37.4	28	82.4	53	127.4	78	172.4
4	39.2	29	84.2	54	129.2	79	174.2
5	41	30	86	55	131	80	176
6	42.8	31	87.8	56	132.8	81	177.8
7	44.6	32	89.6	57	134.6	82	179.6
8	46.4	33	91.4	58	136.4	83	181.4
9	48.2	34	93.2	59	138.2	84	183.2
10	50	35	95	60	140	85	185
11	51.8	36	96.8	61	141.8	86	186.8
12	53.6	37	98.6	62	143.6	87	188.6
13	55.4	38	100.4	63	145.4	88	190.4
14	57.2	39	102.2	64	147.2	89	192.2
15	59	40	104	65	149	90	194
16	60.8	41	105.8	66	151.8	91	195.8
17	62.6	42	107.6	67	153.6	92	197.6
18	64.4	43	109.4	68	155.4	93	199.4
19	66.2	44	111.2	69	157.2	94	201.2
20	68	45	113	70	159	95	203
21	69.8	46	114.8	71	160.8	96	204.8
22	71.6	47	116.6	72	162.6	97	206.6
23	73.4	48	118.4	73	164.4	98	208.4
24	75.2	49	120.2	74	166.2	99	210.2
25	77	50	122	75	168	100	212

1. Temp. °C = 5/9 x (Temp. °F - 32)

2. Temp. °F = (9/5 x Temp. °C) + 32

3. Ambient temperature is the temperature of the surrounding cooling medium.

4. Rated temperature rise is the permissible rise in temperature above ambient when operating under load.

Appendix G
Characteristics Of Metals

The Table below can be used to obtain the various characteristics of metals. For example, according to Column 5 of the Table, rolled copper will melt at about 1,981 degrees fahrenheit.

Characteristics Of Metals

Metal	Symbol	Spec. Grav.	Melting Point °C	Melting Point °F	Elec. Cond. % Copper	Lbs. cu."
Aluminum	AL.	2.710	660	1220	64.9	.0978
Antimony	SB	6.620	630	1167	4.42	.2390
Arsenic	AS	5.730	—	—	4.9	.2070
Beryllium	BE	1.830	1280	2336	9.32	.0660
Bismuth	BI	9.800	271	520	1.50	.3540
Brass (70-30)		8.510	900	1652	28.0	.3070
Bronze (5% SN)		8.870	1000	1382	18.0	.3200
Cadmium	CD	8.650	321	610	22.7	.3120
Calcium	CA	1.550	850	1562	50.1	.0560
Cobalt	CO	8.900	1495	2723	17.8	.3210
Copper	CU					
Rolled		8.890	1083	1981	100.00	.3210
Tubing		8.950	—	—	100.00	.3230
Gold	AU	19.30	1063	1945	71.2	.6970
Graphite		2.25	3500	6332	10^{-3}	.0812
Indium	IN	7.30	156	311	20.6	.2640
Iridium	IR	22.40	2450	4442	32.5	.8090
Iron	FE	7.20	1200 to 1400	2192 to 2552	17.6	.2600
Malleable		7.20	1500 to 1600	2732 to 2912	10	.2600
Wrought		7.70	1500 to 1600	2732 to 2912	10	.2780
Lead	PB	11.40	327	621	8.35	.4120
Magnesium	MG	1.74	651	1204	38.7	.0628
Manganese	MN	7.20	1245	2273	0.9	.2600
Mercury	HG	13.65	-38.9	-37.7	1.80	.4930
Molybdenum	MO	10.20	2620	4748	36.1	.3680
Monel (63-37)		8.87	1300	2372	3.0	.3200
Nickel	NI	8.90	1452	2646	25.0	.3210
Phosphorus	P	1.82	44.1	111.4	10^{-17}	.0657
Platinum	PT	21.46	1773	3221	17.5	.7750
Potassium	K	0.860	62.3	144.1	28	.0310
Selnium	SE	4.81	220	428	14.4	.1740
Silicon	SI	2.40	1420	2588	10^{-5}	.0866
Silver	AG	10.50	960	1760	106	.3790
Steel (Carbon)		7.84	1330 to 1380	2436 to 2516	10	.2830
Stainless						
(18-8)		7.92	1500	2732	2.5	.2860
(13-CR)		7.78	1520	2768	3.5	.2810
(18-CR)		7.73	1500	2732	3.0	.2790

Characteristics Of Metals cont.						
Metal	Symbol	Spec. Grav.	Melting Point		Elec. Cond. % Copper	Lbs. cu."
			°C	°F		
Tantalum	TA	16.6	2900	5414	13.9	.5990
Tellurium	TE	6.2	450	846	10^{-5}	.2240
Thorium	TH	11.70	1845	3353	9.10	.422
Tin	SN	7.30	232	449	15.00	.264
Titanium	TI	4.50	1800	3272	2.10	.162
Tungsten	W	19.30	3410	—	31.50	.697
Uranium	U	18.70	1130	2066	2.80	.675
Vanadium	V	5.96	1710	3110	6.63	.215
Zinc	ZN	7.14	419	786	29.10	.258
Zirconium	AZ	6.40	1700	3092	4.20	.231

Appendix H
Specific Resistance (K) Factors

The Table below can be used to obtain the various specific resistances of metals which are called (K) factors to be applied in a certain formula. For example, 10.8 is the (K) factor used for copper per Column 2 of the Table. Note that 12 is used for copper in the formulas of this book.

Specific Resistance (K)			
The Specific Resistance (K) of a material is the resistance offered by a wire of this material which is one foot long with a diameter of one Mil.			
Material	"K"	Material	"K"
Brass	43.0	Aluminum	17.0
Constantan	295	Monel	253
Copper	10.8	Nichrome	600
German Silver 18%	200	Nickel	947
Gold	14.7	Tantalum	93.3
Iron (Pure)	60.0	Tin	69.0
Magnesium	276	Tungsten	34.0
Manganin	265	Silver	9.7

Author's Note:

1. The resistance of a wire is directly proportional to the specific resistance of the material.
2. "K" = Specific Resistance
3. In the formulas used in this book, 12 is used as the (K) factor for copper and 18 is used for aluminum.

Appendix I
CALCULATIONS TIPS

These calculation tips were designed to alert personnel using the NEC to be aware of the different interpretations of certain sections of the code. It is these different interpretations by the authority having jurisdiction that hinders correct computations from being calculated by the NEC.

The following tips should be studied carefully and the authority having jurisdiction should be consulted as to the interpretation of each particular rule and procedure to be used in calculating volt-amps or ampacities by the NEC.

Calculation Tip 1. The AHJ may require the general lighting calculated at 3 VA per sq. ft. to be compared to the total VA rating of the lighting fixtures and the larger load selected. This procedure is applied where the listed occupancy and the number of fixtures with VA ratings are available.

Calculation Tip 2. The AHJ may permit all cooking equipment rated over 1 3/4 kW and up to 8 3/4 kW to be grouped and the percentages in Columns B and C used to determine the demand load.

Calculation Tip 3. The AHJ may interpret the NEC to allow Notes 3 to Table 220-19 to be applied to cooking units grouped in Columns B and C. After demand factors are applied, the smaller load is used in the calculation.

Calculation Tip 4. The AHJ may permit the demand load for service-entrance conductors and equipment to be selected from Note 4 to Table 220-19 is utilized.

Calculation Tip 5. Four or less dryers must be calculated at 100 percent of 5,000 VA or the nameplate value, whichever is greater.

Calculation Tip 6. The AHJ may rule that the A/C load is eligible as the largest motor load, even when the A/C is dropped due to the heating unit being the largest load per NEC 220-21.

Calculation Tip 7. The AHJ may require the A/C load to be calculated at 125 percent compared to 100 percent of the heating load, and the smaller load dropped per NEC 220-21. Note that 100 percent for both is used by most of the AHJ's.

Calculation Tip 8. NEC 220-21 allows the VA of more than one noncoincident load to be dropped due to a load having a greater VA rating.

Calculation Tip 9. The AHJ may require the blower motor to be calculated as part of the heating load and added to the service load at 100 percent per NEC 220-15. However, other AHJ's may require the blower motor to be considered eligible for the largest motor per NEC 220-14.

Calculation Tip 10. The AHJ may require a sign less than 1,200 VA to be increased to a minimum of 1,200 VA.

Calculation Tip 11. Derating the current-carrying ability of the (OCPD's) by 80 percent is the same as multiplying the load times 125 percent. The 80 percent is derived by dividing 1 by 125 percent (1/125 = 80%). **Note:** If the OCPD is derated by 80% and the load is added to this value at 125 percent, the total rating for OCPD is calculated at 150 percent and not 125 percent.

Calculation Tip 12. The VA rating of a general-purpose receptacle load can be added to the VA rating of a multioutlet receptacle load and the demand factors of Table 220-13 applied per NEC 220-13.

The following is an example on how to use Calculation Tips.

For Example: If a 25 kW heating unit and a 6 kVA air conditioning unit are compared per NEC 220-21, the A/C unit is dropped. If the AHJ requires the A/C unit to be considered as the largest motor, 6 kVA must be compared with other motor related loads and should be selected as the largest motor, 25 percent of the 6 kVA must be added to the total calculated load.

Stallcup's Electrical Calculations Simplified

Topic Index

Symbols

A

B

C

D

E

F

G

H

I

K

L

M

N

O

P

R

Additional Help Index

CALCULATING BRANCH-CIRCUITS

CALCULATING FEEDER-CIRCUITS

GROUNDING AND BONDING

CALCULATING CONDUCTORS

MOTORS, MOTOR CIRCUITS AND CONTROLLERS

GENERATORS AND TRANSFORMERS

Section Number Index

Section Number	Page #

P. 23-4 BUSS BARS